Handbook of Zirconia-Based Materials

Recent decades have witnessed significant progress in zirconia-based materials research. This comprehensive handbook explores the latest advancements and applications of these materials, providing insights into cutting-edge manufacturing and materials science research while covering diverse industrial uses, properties, and synthesis methods. This handbook is structured into five key sections: I. Fundamentals and Materials Science, II. Industrial and Engineering Applications, III. Energy and Catalytic Applications, IV. Medical and Dental Applications, and V. Emerging Technologies.

Handbook of Zirconia-Based Materials: Manufacturing, Electrochemical, and Biomedical Engineering examines the notable characteristics and wide-ranging applications of zirconia-based materials. Zirconia's distinctive properties like optical transparency, thermal stability, mechanical strength, and electrical conductivity have attracted engineers and researchers across numerous fields, including energy production, biomedical engineering, and environmental remediation. These materials find significant applications in environmental protection, sustainable energy advancement, dental implants, solid oxide fuel cells, and catalytic converters.

This handbook thoroughly investigates zirconia's crystal structure, phase transitions, and surface chemistry while exploring synthesis methodologies for nanostructured materials, from conventional to cutting-edge approaches. It also highlights the practical industries utilizing zirconia to demonstrate its importance. This resource provides accessible insights suitable for readers with varying levels of expertise, including industrial researchers, students, professionals, and scholars in manufacturing, materials science, chemistry, and physics.

Innovations in Smart Manufacturing for Long-Term Development and Growth

Series Editors: Atul Babbar, Gursel Alici, Yu Dong, and Ankit Sharma

Fabrication Techniques and Machining Methods of Advanced Composite Materials
Vikas Dhawan, Atul Babbar, Inderdeep Singh, and Jonathan M. Weaver

Advances in Pre- and Post-Additive Manufacturing Processes: Innovations and Applications
Naveen Mani Tripathi and Ankit Sharma

Handbook of 3D Printing in Pharmaceutics: Innovations and Applications
Prakash Katakam, Ranvijay Kumar, Nishant Ranjan, and Atul Babbar

Intelligent Machinery Fault Diagnostics and Prognostics: The Future of Smart Manufacturing
Deepam Goyal, Ankit Sharma, and Mohamad Abou Houran

Nanomaterials for Smart Manufacturing: Fundamental Principles, Theory, Case Studies, and Advanced Applications
Dinesh Deshwal, Anil Kumar Narwal, Vivek Srivastava, Atul Babbar, and Jonathan M. Weaver

Sensor Technologies and Wireless Communications in Industry 5.0: A Solution for Smart Manufacturing
Edited by Ramkumar K.R., Bhupendra Singh, Lei Pan, and Ankit Sharma

Handbook of Smart Manufacturing Technologies: Industrial Engineering in the Digital Age
Edited by Mahendra U. Gaikwad, Ankit Sharma, Muthuramalingam Thangaraj, and Sachin Salunkhe

Handbook of Bio-Based Materials for Smart Manufacturing
Edited by Devendra Singh, Naveen Mani Tripathi, Dharminder Singh, and Ankit Sharma

Smart Manufacturing with AIoT: Signal Processing, Computer Vision, and Data Analytics
Edited by Keyurkumar Patel, Ankit D. Oza, Ankit Sharma, Archana Mantri, and Hreetabh Kishore

Handbook of Zirconia-Based Materials: Manufacturing, Electrochemical, and Biomedical Engineering
Edited by Jashanpreet Singh, Hitesh Vasudev, Ankit Sharma, and Sunaina

For more information about this series, please visit: www.routledge.com/Innovations-in-Smart-Manufacturing-for-Long-Term-Development-and-Growth/book-series/CRCISMLDG

Handbook of Zirconia-Based Materials

Manufacturing, Electrochemical, and Biomedical Engineering

Edited by
Jashanpreet Singh, Hitesh Vasudev, Ankit Sharma, and Sunaina

CRC Press
Taylor & Francis Group
Boca Raton London New York

CRC Press is an imprint of the
Taylor & Francis Group, an **informa** business

Designed cover image: Shutterstock

First edition published 2026
by CRC Press
2385 NW Executive Center Drive, Suite 320, Boca Raton FL 33431

and by CRC Press
4 Park Square, Milton Park, Abingdon, Oxon, OX14 4RN

CRC Press is an imprint of Taylor & Francis Group, LLC

ISBN: 978-1-032-85587-5 (hbk)
ISBN: 978-1-032-85662-9 (pbk)
ISBN: 978-1-003-51930-0 (ebk)

DOI: 10.1201/9781003519300

Typeset in Times
by codeMantra

This book is dedicated to our families, friends, teachers, and research supervisors.

Contents

SECTION III Energy and Catalytic Applications

SECTION IV Medical and Dental Applications

Preface

The *Handbook of Zirconia-Based Materials: Manufacturing, Electrochemical, and Biomedical Engineering* provides an exploration of the notable characteristics, synthesis techniques, and wide-ranging uses of zirconia-based materials. The distinctive qualities of optical transparency, thermal stability, mechanical strength, and electrical conductivity exhibited by zirconia have garnered the interest of engineers, researchers, and innovators in numerous fields. Zirconia has emerged as a prominent contender in fields such as energy production, biomedical engineering, and environmental remediation, as the demand for sophisticated materials continues to increase. It provides solutions to a vast array of challenges. Zirconia-based materials have significant applications in various sectors, including energy generation, biomedical engineering, environmental remediation, and advanced manufacturing. The text of this handbook is structured into five thematic sections, namely, *fundamentals and material science, industrial and engineering applications, energy and catalytic applications, medical and dental applications,* and *emerging technologies.* Additionally, this book illustrates many practical and revolutionary industries in which zirconia is utilized to emphasize its significance. This book offers an all-encompassing resource that accommodates the needs of scholars, practitioners, and students interested in the science and technology of zirconia-based materials. The consideration is that it seeks to support interdisciplinary knowledge and encourage innovation in an academic and industrial environment. It aims to facilitate interdisciplinary understanding and foster innovation in both academic and industrial environments.

About the Editors

Dr. Jashanpreet Singh is an assistant professor at the Department of Mechanical Engineering, University Center for Research and Development, Chandigarh University, Mohali. He earned his PhD in Slurry Erosion and Thermal Spray Coatings from Thapar Institute of Engineering and Technology in 2019. With over 3 years of teaching, 3 years of research, and 2 years of industry experience, his research focuses on tribology, thermal spray coatings, CFD simulations, biomedical engineering materials and coatings, additive manufacturing, and materials characterization. He has published over 100 research papers, including more than 50 SCIE/ESCI-indexed articles and over 92 in SCOPUS. He has authored a book and contributed 25 chapters to different books. He is also a series editor of Advances in Surface Engineering and Tribology from CRC Press/Taylor & Francis. He is listed in Stanford University's Top 2% Scientists (2022–2024), and he received the "Outstanding Reviewer Award" from IOP in 2021. He is a life member of AICTSD, IAENG, and NITTSD.

Prof. (Dr.) Hitesh Vasudev is a professor in the School of Mechanical Engineering at Lovely Professional University, Phagwara, Punjab. He is also the head of the Research and Development Cell. He earned his PhD in 2018 from Guru Nanak Dev Engineering College, Ludhiana. His research interests include surface engineering and thermal spraying techniques, such as HVOF, flame spray, cold and plasma spray, and microwave cladding. He focuses on developing nanostructured, multimodal, and high-entropy alloy coatings for high-temperature applications. He has over 190 indexed publications, along with 10 books and 50 book chapters. He is also a series editor of Advances in Surface Engineering and Tribology from CRC Press/Taylor & Francis. A dedicated reviewer, he has received the "Top Cited Paper Award India 2022" by IOP Publishing and LPU's Research Excellence Awards (2019–2023). He holds a patent and has received research grants from LPU and SERB for biomedical coatings and conference organization.

Dr. Ankit Sharma holds the position of Associate Director (Research) of the Center for Research, Impact and Outcome Department, Chitkara University, Punjab, India. He received his doctorate in Mechanical Engineering from the Thapar Institute of Engineering and Technology, India. He has more than 10 years of experience in academics, research, consulting, training, and industry. He has authored numerous national and international publications and has filed/published 30+ international patents. He is also a series editor of Innovations in Smart Manufacturing for Long-Term Development and Growth from CRC Press/Taylor & Francis. He is also the managing and special issue editor of several international journals and has delivered several invited seminars and keynote talks on international (USA, China, India) platforms. Along with that, he is also a reviewer board member of several international journals.

Dr. Sunaina is an assistant professor in the Department of Chemistry at Guru Nanak Dev University, Amritsar. She previously worked as a postdoctoral associate at the University of Calgary, focusing on nanobubbles for wastewater treatment. She earned her PhD in 2021 through a joint program between Panjab University and the Institute of Nano Science and Technology, India. Her research spans nanomaterials, including metal oxide–chalcogenide heterostructures for photocatalysis and photodetection, metal ferrites from slime waste for water treatment, and advanced field emission materials. She has published over 31 research articles and 10 book chapters. She has received multiple best paper awards and has hands-on experience in XRD, SAXS, and UV-Vis techniques. She was awarded synchrotron beam time at Photon Factory, Japan, highlighting her global research recognition. Her work reflects strong contributions to environmental sustainability and materials innovation.

Contributors

Mohamed Aboughaly
Faculty of Energy Systems and Nuclear Science
University of Ontario Institute of Technology
Oshawa, Canada

Shinar Athwal
Department of Biotechnology
Thapar Institute of Engineering and Technology
Patiala, India

Nacer Badi
Department of Physics
Faculty of Science
University of Tabuk
Tabuk, Kingdom of Saudi Arabia

Sanganna Burli
Department of Pharmacology
Ashokrao Mane College of Pharmacy
Kolhapur, India

Yanhai Cheng
School of Mechanical and Electrical Engineering
China University of Mining and Tcchnology
Xuzhou, China

Vikas R. Dhole
Department of Pharmaceutical Pharmacology
Ashokrao Mane College of Pharmacy
Kolhapur, India

Kashish Garg
University Institute of Biotechnology
Chandigarh University
Mohali, India

Abdul Wahab Hashmi
School of Mechanical Engineering
Shandong University of Technology
Zibo, China

Pulkit Jain
Department of Electronics and Communication Engineering
Chandigarh University
Mohali, India

Jeyaprakash Natarajan
School of Mechanical and Electrical Engineering
China University of Mining and Technology
Xuzhou, China

Hardeep Kaur
University Institute of Biotechnology
Chandigarh University
Mohali, India

Perumal Kumar
Department of Chemical and Energy Engineering
Faculty of Engineering and Science
Curtin University
Sarawak, Malaysia

Prasann Kumar
Department of Agronomy
School of Agriculture
Lovely Professional University
Phagwara, India

Shivam Kumar
WA School of Mines: Minerals, Energy and Chemical Engineering
Curtin University
Perth, Australia

Sawanta S. Mali
School of Chemical Engineering
Chonnam National University
Gwangju, South Korea

Amrinder Mehta
Research and Development Cell
School of Mechanical Engineering
Lovely Professional University
Phagwara, India

Moniya
Department of Aerospace Engineering
Chandigarh University
Mohali, India

Nigar K. Mujawar
Department of Pharmaceutics
Women's College of Pharmacy
Kolhapur, India

Sule Emmanuel Oshioriameh
Industrial Engineering and Management
American University of Beirut
Beirut, Lebanon

Mastu Patel
Department of Agronomy
School of Agriculture
Lovely Professional University
Phagwara, India

Amit Raj
Department of Agronomy
School of Agriculture
Lovely Professional University
Phagwara, India

Supriya Rana
Hydron Energy Inc., R&D
North Vancouver, Canada

Raghavendra Sagar
Department of Physics
Mangalore Institute of Technology & Engineering
Mangalore, India

Vikas Sharma
University Centre for Research and Development
Chandigarh University
Mohali, India

Milind Shrivastava
Department of Biotechnology
Thapar Institute of Engineering and Technology
Patiala, India

Akshita Singh
Department of Biotechnology
Thapar Institute of Engineering and Technology
Patiala, India

Bhupinder Singh
Department of Chemical and Environmental Engineering
University of Nottingham
Nottingham, United Kingdom

Gurmeet Singh
Department of Mechanical Engineering
Chandigarh University
Mohali, India

Manpreet Singh
Department of Mechanical Engineering
College of Engineering and Information Technology
University of Maryland
College Park, MD, United States

Rajwinder Singh
Department of Materials Science and Engineering
Uppsala University
Uppsala, Sweden

Ugyen Tshering
Agriculture Research and Development Centre
Mongar, Bhutan

Praveen Kumar Verma
Research and Development Cell
School of Mechanical Engineering
Lovely Professional University
Phagwara, India

Krishna K. Yadav
Department of Chemical Engineering
Sami Shamoon College of Engineering
Beer-Sheva, Israel

Che Hua Yang
School of Mechanical and Electrical Engineering
National Taipei University of Technology
Taipei, Taiwan

Jibrin Muhammad Yelwa
Department of Scientific and Industrial Research
National Research Institute for Chemical Technology
Zaria, Nigeria

Rassim Younes
Faculty of Technology
University of Bejaia
Bejaia, Algeria

Noushi Zaidi
Department of Basic Science
College of Science
King Faisal University
Al-Ahsa, Saudi Arabia

Section I

Fundamentals and Materials Science

1 Introduction to Zirconia and Zirconia-Based Materials

Jashanpreet Singh, Moniya, and Vikas Sharma

1.1 INTRODUCTION TO ZIRCONIA

Zirconia, also known as zirconium dioxide (ZrO_2), is a high-performance ceramic material with remarkable hardness, fracture toughness, wear resistance, and biocompatibility. Due to its significant mechanical properties, zirconia is often dubbed "ceramic steel", and zirconia as a transition metal oxide is used throughout advanced ceramics, solid oxide fuel cells (SOFCs), and semiconductor technologies [1]. While baddeleyite contains natural zirconia, most commercial zirconia is obtained from the mineral zircon ($ZrSiO_4$) using high-temperature processing methods. One of the salient properties of zirconia is its ability to exist in monoclinic, tetragonal, and cubic phases. Its mechanical properties are significantly improved when stabilized with oxides like yttria (Y_2O_3), which guards against phase transformations that would otherwise lead to cracking and structural failure. The origin of zirconia's development can be traced back to 1789, when Martin Heinrich Klaproth, a German chemist, discovered zirconium oxide from the mineral zircon [2]. However, the unique combination of mechanical strength, chemical stability, thermal resistance, and industrial interest in zirconia ceramics only emerged in the 20th century [3]. Initially, its applications were primarily focused on refractories and pigments, but this research advanced significantly in the 1970s with the finding of transformation toughening, a process that added immense value to the material's resistance to crack propagation. This advancement in research propelled the use of zirconia in structural ceramics, biomedical implants, and thermal barrier coatings (TBCs), reinforcing zirconia's position as one of the most pivotal ceramics in contemporary materials science. The most critical source for zirconium, the elemental basis of zirconia, is the mineral zircon ($ZrSiO_4$), which is extensively found in heavy mineral sand, placer deposits, and igneous rocks. Australia and South Africa are the main producers of zircon worldwide, whereas baddeleyite (ZrO_2) is naturally occurring, but less so, and is mainly mined from Russia's Kola Peninsula [4]. The commercial use of zirconia involves high-temperature steps such as pyrogenic decomposition of zircon sand, followed by concentrating crude zirconia using white zircon silicate, removing silica and chlorine from pure zircon. These minerals are particularly important for meeting the increasing demand for ultrapure zirconia needed in constructing advanced technological systems. Materials scientists continue to have considerable interest in

DOI: 10.1201/9781003519300-2

zirconia because no other single ceramic material offers such diverse properties [5]. The exceptional fracture toughness of zirconia ceramic improves its durability while simultaneously preventing crack propagation due to its stress-induced phase transformations. Zirconia's cutting-tool-grade wear resistance and extraordinary hardness render it usable even in mechanical components [6]. Its thermal stability further allows it to withstand high temperatures, making it extremely useful as a TBC for systems that operate at high temperatures [7]. Moreover, zirconia is electronically neutral, which, along with its biocompatibility, makes it suitable in the field of medicine, especially in orthopaedic implants and dental crowns [8]. Another noteworthy property is zirconia's ionic conductivity, making it useful in SOFCs and oxygen sensors [9,10]. Due to the aforementioned properties, zirconia has gained active, ongoing, multidisciplinary research focused on enhancing its performance in extreme environments, improving its mechanical reliability, and developing new advanced nanostructured composites for aerospace, energy, biomedicine, and electronics industries [11,12].

1.2 TEMPERATURE-INDUCED PHASE TRANSFORMATION AND THEIR IMPLICATIONS

Reversible structural phase transformations in zirconium and zirconia-based materials are transformative to their mechanical properties, enduring stability under high-temperature and high-stress environments [14]. As shown in Figure 1.1, transformations controlled by external control parameters such as temperature, pressure, and stress are highly critical for the structural integrity and employability of these materials in modern high-tech industries such as aerospace, nuclear energy, and advanced ceramic engineering.

1.2.1 Phase Transformations in Zirconium

Zirconium has several phase transformations that can be attributable to temperature increases, depending on the purity and pressure of the substance [15]. At room temperature and pressure, zirconium is present in the α phase, which is characterized by hexagonal close packing (hcp). This twin structure remains stable until

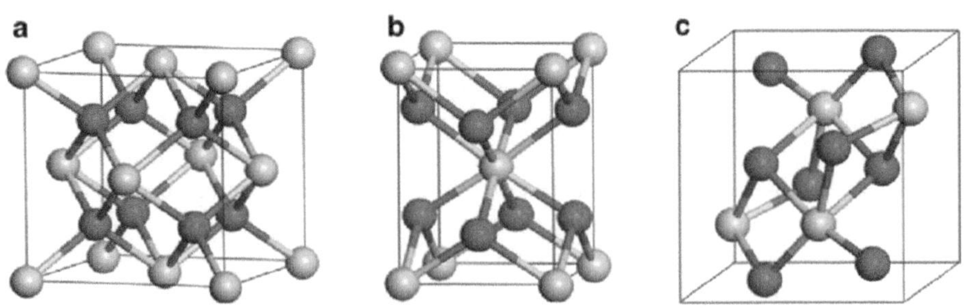

FIGURE 1.1 Crystal structures of ZrO_2: (a) cubic, (b) tetragonal, and (c) monoclinic. Darker and light grey spheres correspond to oxygen and zirconium atoms, respectively [13].

approximately 1135 K. Beyond this temperature, phase transformation shifts to the β phase, known as body-centred cubic (bcc). Under extensive pressure, zirconium will shift to the ω phase, which is hexagonal but not close packed, at roughly 2.10 GPa with a temperature of 300 K and 6.35 GPa and a predicted triple point at 910 K [16]. The most important phase transformation is the shift from the ω to the α phase due to increasing hardness and changes in mechanical properties. This transition can happen directly ($\alpha \rightarrow \omega$) or from an intermediate β phase ($\alpha \rightarrow \beta \rightarrow \omega$). At room temperature, the ω phase is regarded as metastable and will return to the α phase when heated over 470 K [17]. Transformations that occur under shock loading that exceeds 7 GPa can shift up to 80% of the substance to the ω phase. These alarming indicators could be highly beneficial in applications that need impact resistance and mechanical strengthening [18,19]. The result of this shift is startling, considering the changes it can induce in mechanical properties such as stress and hardness. The development of the ω phase improves strength and wear resistance, which is useful for nuclear and aerospace applications. On the other hand, such a transformation also introduces residual stresses, which can pose a threat to material stability over time [20]. High-pressure sintering at 1273 and 1473 K has been shown to reduce residual stresses, thus stabilizing the α phase in relation to the ω phase and enhancing the performance of zirconium-based materials [14,21].

1.2.2 Phase Transformation in Zirconia-Based Materials

Pure zirconia undergoes the following reversible phase transformations upon heating and cooling:

$$m-\mathrm{ZrO_2} \underset{\text{Cooling}}{\overset{\text{Heating at } 1170°C}{\leftrightarrow}} t-\mathrm{ZrO_2} \underset{\text{Cooling}}{\overset{\text{Heating at } 2370°C}{\leftrightarrow}} c-\mathrm{ZrO_2} \tag{1.1}$$

Zirconia-based materials undergo remarkable temperature-induced phase transformations, which affect their mechanical properties and thermal stability. Both temperature and external stress factors result in numerous crystallographic phase changes in zirconia. The changes that have the highest impact on strength and toughness are stress-induced and include the tetragonal-to-monoclinic ($t \rightarrow m$) phase transition, which is the most critical transformation. This transformation can be studied using X-ray diffraction (XRD) and Raman spectroscopy. A high-temperature cubic phase in combination with the tetragonal phase and its transformation toughening mechanism greatly improves the mechanical performance of zirconia. In yttria-stabilized zirconia (YSZ), the metastable tetragonal (T′) phase decomposes at high temperatures into thermodynamically stable tetragonal (T) and cubic (C) phases. During thermal cycling, the T phase can undergo further transformation to the monoclinic (m) phase, which results in 5%–7% volume expansion. This transformation can lead to degradation in mechanical properties, which makes phase stability a crucial requirement for TBCs. Research on nanoindentation techniques has shown that the cubic phase, compared to the tetragonal phase, has higher hardness and Young's modulus, which affects the durability of TBCs. In addition, the existence of the metastable T′ phase results in ferroelastic toughening, a mechanism that reduces crack propagation by *c*-axis reorientation in the crystal lattice, absorbing elastic energy and enhancing resistance to fracturing. Furthermore, the $\alpha \rightarrow \omega$ transformation of zirconium under high pressure is affected by the presence of interstitial oxygen and the density of lattice defects. The ω phase is formed by a pressure-induced phase transition, which increases hardness and resistance to deformation. These phases are unstable depending on the adopted heating method, as materials that undergo high-pressure high-temperature (HPHT) sintering are transformed differently than those subjected to shock-wave compression.

1.2.3 Implications for Advanced Applications

Phase stabilization and transformations are shown in Figure 1.2. Phase stabilization and transformations of zirconium and zirconia materials induced by temperature have a significant impact on their thermal and mechanical properties as well as their structural applications. The capability of these materials to tolerate high thermal and mechanical stress makes them suitable for use in aerospace systems, nuclear energy, TBCs, and dental ceramics. Zirconia's phase transformation from the tetragonal to monoclinic form is used for transformation toughening in dental implants and high-performance ceramics to improve fracture toughness. On the other hand, the α to ω transformation phase of zirconium further supports its use in the cladding of nuclear reactors and in aerospace components, where susceptibility to thermal shock and high impact is critical. In summary, in order to optimize the properties of zirconium and zirconia-based materials, it is critical to understand the phase stability, transformation mechanisms, and stress factors of the components, especially for dynamic applications. Ongoing investigations into nanoindentation methods, phase mapping, and tempered doping will further strengthen the performance and reliability of these materials under harsh conditions.

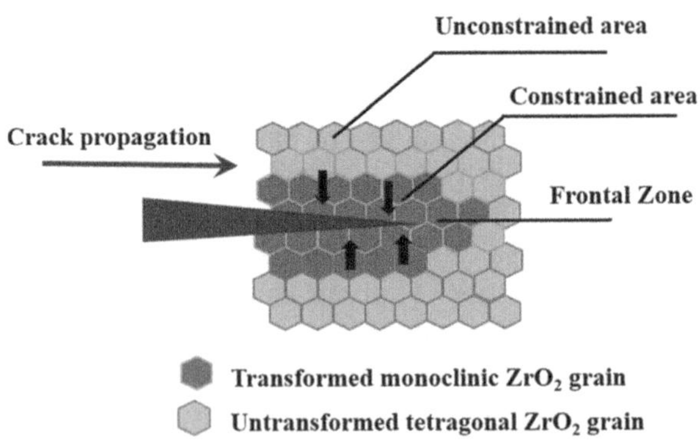

FIGURE 1.2 Transformation toughening mechanism in zirconia [25].

1.3 TRANSFORMATION TOUGHENING MECHANISM

Among various processes associated with zirconia, its transformation toughening mechanism is arguably the most important factor of interest due to its ability to improve resistance to crack propagation [22]. This type of toughening is attributable to the so-called tetragonal phase's capability to transform to the monoclinic phase due to stress relief when mechanical stress is applied, especially at crack tips [23]. There is also a claimed increase in the volume of about 3%–5% during this process, which helps generate compressive stresses around the crack front [17]. These forces help resist the tensile forces responsible for crack formation, thus stopping or slowing down fracture advancement. This mechanism accounts for the exceptional combination of properties such as hardness and fracture toughness, which are usually not present together in ceramics. Because of this, transformation toughening is presumed to be the major reason why zirconia is considered a structural ceramic material capable of high-stress and high-wear applications such as cutting tools, dental implants, and components for aerospace technology [24].

1.4 PHASE DIAGRAM AND STABILIZATION OF ZIRCONIA

Zirconia undergoes polymorphic changes within its three crystal forms: monoclinic, tetragonal, and cubic, which depend on temperature. Although in zirconia, polymorphic behaviour occurs in a controlled manner, transformations pose significant problems for practical usage because of the volume changes associated with cooling.

This is especially true for the transformation from the tetragonal to monoclinic phase. It has been observed that volume expansion during this transformation is around 3%–5% [26]. The abrupt expansion leads to the formation of internal stresses and cracks, making pure zirconia mechanically unstable under normal operating conditions during cyclic thermal operations. Structural stability over a wide range of temperatures is ensured by the addition of suitable dopant oxides to zirconia or, in other words, by stabilizing it.

1.4.1 Stabilization through Dopants

In order to retain high-temperature tetragonal or cubic phases at room temperature, small amounts of divalent or trivalent metal oxides are used as stabilizers to avoid harmful phase changes. These dopants induce oxygen vacancies within the bone structure, which alters the stability of the phases and helps retain the sired phases over a broader temperature range. The choice of the stabilizer and its concentration determine which phase of zirconia dominates at room temperature and influence the final material properties [27]. Yttria (Y_2O_3), magnesia (MgO), and calcia (CaO) are the most commonly used stabilizers for zirconia [16].

1.4.2 Types of Stabilized Zirconia

YSZ typically contains 3–8 mol% Y_2O_3. It stabilizes the tetragonal or cubic phase, providing excellent thermal stability, ionic conductivity, and mechanical toughness. YSZ is commonly used in SOFCs, TBCs, and biomedical implants, whereas magnesia-stabilized zirconia (MSZ) is stabilized with 8–10 mol% MgO. MSZ generally maintains a cubic phase with good thermal shock resistance and moderate toughness [28]. It is used in refractory materials and high-temperature ceramics. Calcia-stabilized zirconia (CSZ) is doped with 8–12 mol% CaO. CSZ also stabilizes the cubic phase and provides good performance in high-temperature environments, such as furnaces and insulating coatings [29].

1.4.3 Phase Diagram Explanation

ZrO_2–dopant binary phase diagrams illustrate how different concentrations of stabilizers affect the phase stability of zirconia across various temperatures, which are shown in Figure 1.3. For example, in the ZrO_2–Y_2O_3 phase diagram, adding

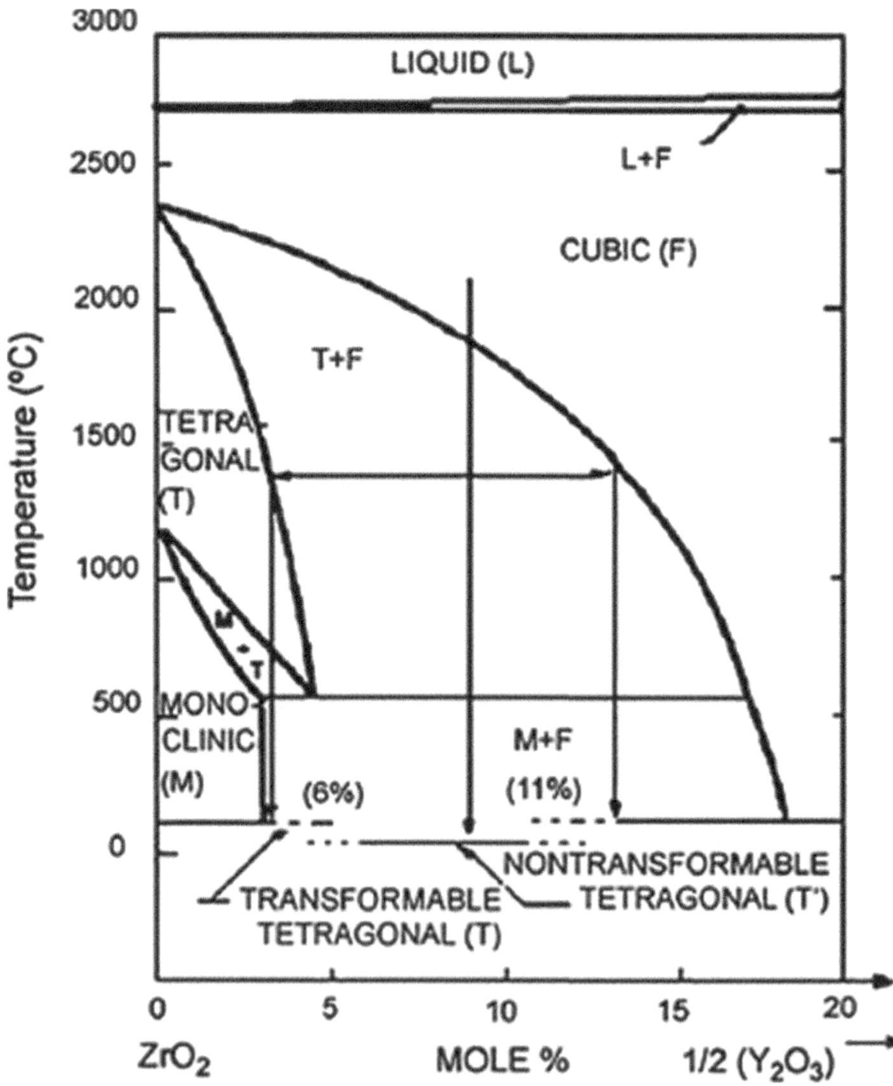

FIGURE 1.3 ZrO_2–Y_2O_3 phase diagram [28].

increasing amounts of yttria shifts the stability regions of the tetragonal and cubic phases down to lower temperatures. This allows high-temperature phases to remain stable even at room temperature, depending on the concentration of yttria.

For instance, 3 mol% Y_2O_3 stabilizes mostly tetragonal zirconia at room temperature (used in Y-TZP), and 8 mol% Y_2O_3 stabilizes the cubic phase and enhances its ionic conductivity (ideal for SOFCs). These diagrams are crucial in ceramic processing as they guide the choice of stabilizer content required to achieve desired properties such as fracture toughness, thermal stability, and conductivity.

1.4.4 Zirconia-Based Ceramics: Classification, Properties, and Applications

YSZ [30] is the most widely used zirconia ceramic due to its exceptional toughness, thermal insulation, and ionic conductivity, making it ideal for applications such as TBCs in jet engines, SOFCs, and dental implants [31], as shown in Table 1.1. MSZ is

TABLE 1.1
Types of Zirconia-Based Ceramics and Applications

Type of Zirconia Ceramic	Composition	Key Properties	Applications
Yttria-stabilized zirconia (YSZ)	ZrO_2 + 3–8 mol% Y_2O_3	High fracture toughness, thermal insulation, ionic conductivity	Thermal barrier coatings, solid oxide fuel cells (SOFCs), dental implants
Magnesia-stabilized zirconia (MSZ)	ZrO_2 + MgO	High strength, wear resistance, moderate toughness	Cutting tools, grinding media, wear-resistant components
Calcia-stabilized zirconia (CSZ)	ZrO_2 + CaO	Low thermal conductivity, moderate strength	Refractory materials, thermal insulation
Ceria-stabilized zirconia (CeSZ)	ZrO_2 + CeO_2	High fracture toughness, excellent wear resistance	Biomedical implants, oxygen sensors, engine components
Alumina-toughened zirconia (ATZ)	ZrO_2 + Al_2O_3	High hardness, improved fracture toughness	Biomedical implants, cutting tools, wear-resistant coatings
Zirconia-toughened alumina (ZTA)	Al_2O_3 + ZrO_2	High strength, improved toughness compared to pure alumina	Biomedical implants, cutting tools, industrial wear parts
Partially stabilized zirconia (PSZ)	ZrO_2 + 2–3 mol% Y_2O_3 or MgO	High strength, transformation toughening	Structural ceramics, cutting tools, grinding media
Tetragonal zirconia polycrystal (TZP)	ZrO_2 + 2–3 mol% Y_2O_3	High fracture toughness, excellent wear resistance	Dental implants, scissors, knives, biomedical devices

valued for its high strength and wear resistance, making it suitable for cutting tools and grinding media in the industrial sector [32]. Ceria-stabilized zirconia (CeSZ) offers high fracture toughness, making it a preferred choice for biomedical implants and oxygen sensors due to its resistance to cracking and wear. Alumina-toughened zirconia (ATZ) and zirconia-toughened alumina (ZTA) combine the hardness of alumina with the toughness of zirconia, making them ideal for biomedical implants and wear-resistant components [33]. Tetragonal zirconia polycrystal (TZP) is known for its high fracture toughness and wear resistance, widely used in dental implants and cutting tools where mechanical strength and durability are critical.

1.5 SYNTHESIS AND PROCESSING METHODS

Zirconia can be synthesized using various techniques, each with distinct advantages in controlling particle size, morphology, and phase composition [34–36].

1.5.1 The Sol-Gel Processing Technique

The sol-gel approach is commonly used in the synthesis of inorganic nanomaterials. In recent years, sol-gel processing has become highly popular in the fabrication of inorganic materials due to its ease of operation and several other important benefits [37]. Materials scientists and designers are attracted to the major advantages of sol-gel processing, which are (i) ambient temperature of sol preparation and gel processing, (ii) high purity of precursors, (iii) homogeneity of the product, (iv) low temperature of sintering, (v) ease in the fabrication of multicomponent materials, and (vi) good control on shape, size, and size distribution of powder particles. These advantages would be helpful in the synthesis of materials for optical and electronic uses. A sol is a type of dispersion where ultrafine solid particles are bundled up in a liquid, ranging in size from a nanometre to a micrometre. The sol-gel process is a common method for the synthesis of zirconia nanoparticles due to high purity and uniformity, especially using alkoxide-based sols, where the hydrolysis of zirconium alkoxides and polycondensation in a solvent, like ethanol or water, occur in ZrO_2 [38]. Based on their formation, sols can be classified as colloidal (particulate) sols, which have individual distinct nanoparticles, or polymeric sols, where structures resembling polymers are formed and developed within the liquid phase. Gradually, the sol changes into a gel, which is a semisolid structure forming a network with miniature nanopores holding residual liquid. In colloidal sols, gelation is attributable to the dispersion medium evaporation, leading to weak intermolecular forces such as hydrogen bonding that work to pull together nanoparticles. In contrast, structural gel formation is brought about in polymeric sols due to continuous polycondensation and subsequent growth of interconnected clusters. The sol-gel method is precise in controlling the geometrical features and morphology of particles, as shown in Figure 1.4. In order to enhance the crystallization of zirconia while the organic material is being removed, the sol undergoes conversion to a three-dimensional gel macromolecular porous network, followed by directed heating, and then, the material is calcinated at a temperature of 400°C to 800°C. This makes it possible to fabricate nanoparticles with particular properties. For instance, some dopants, like yttria, can be added in the form of nitrogen or acetate

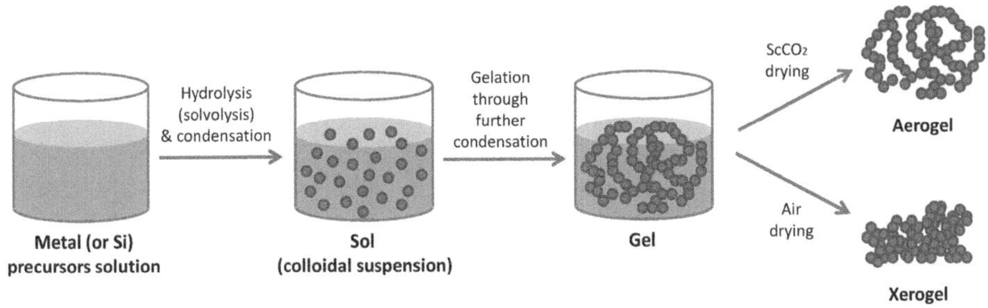

FIGURE 1.4 General scheme of the sol-gel synthesis method [41].

solutions to stabilize some forms of zirconia crystal and enhance functionality [39]. The nanostructured zirconia particles obtained via the sol-gel technology are of the highest purity with controlled dimensions and form and thus are suitable for catalysis, coatings, biomedical implants, and advanced materials [40].

1.5.2 Hydrothermal Synthesis

Hydrothermal treatment is a sophisticated fabrication method that is widely used in the manufacture of nanomaterials and composites with desired characteristics for use in many branches of electronics, optoelectronics, catalysis, and biomedicine [42]. The term "hydrothermal", first introduced by Sir Roderick Murchison for geological purposes, pertains to the formation of rocks and minerals through the action of water being brought to high temperatures and pressure. Hydrothermal synthesis occurs through the crystallization of the existing compounds from a water solution at higher than medium temperatures and then applying pressure in order to achieve unparalleled purity, homogeneity, and structural quality.

It is advantageous in achieving fine size, low-temperature synthesis, accurate composition, and high-density materials in short periods of time [43]. In a solution-based approach, hydrothermal synthesis utilizes the solvability of inorganic solids in water under extreme conditions for direct precursor-to-product conversion without requiring a defining agent. The phase diagram of water is important to this, as the boundaries of liquid and gas above certain temperatures and below certain pressures dictate the synthesis. Hydrothermal processes are less harmful to the environment and offer the unique benefit of producing nanocrystalline materials, making them favoured approaches for many inorganic compounds. Similarly, the precipitation method for zirconia synthesis is simple and cost-effective. It consists of dissolving a zirconium salt in water and then adding a precipitating reagent, like ammonia or sodium hydroxide, causing Zr $(OH)_4$ to precipitate out of solution, which is shown in Figure 1.5.

1.5.3 Precipitation Method

The precipitate is then aged to improve crystallinity, washed to remove any impurities, and then heated in a furnace at 400°C–800°C for the formation of zirconia. The precipitation technique is quite versatile as it allows for the preparation of zirconia

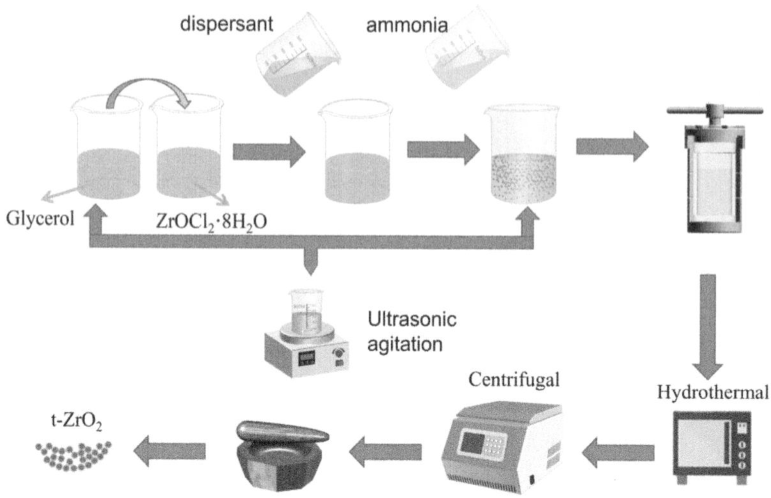

FIGURE 1.5 Flowchart of the preparation of tetragonal phase nano-zirconia powder using the hydrothermal synthesis technique [43].

used in catalysts, ceramics, and biomedical devices, while being controlled in terms of the shape and size of the produced particles. Choosing an appropriate method for the synthesis of zirconia materials is imperative, pertaining to a specific purpose or need, as different methods will yield differing properties.

1.6 CHARACTERISTICS OF ZIRCONIA

1.6.1 Physical Attributes

Ageing phenomenon, or low-temperature deterioration, is one of the most detrimental physical characteristics of zirconia. At temperatures more than 200°C in the presence of water vapours, mechanical characteristics deteriorate as a result of the increasing spontaneous changes in the metastable tetragonal phase into a monoclinic one [44]. The material's density, hardness, and strength are all drastically reduced under these circumstances. So, decreasing the grain size of zirconia and/or increasing the concentration of stabilizing oxides may slow down the transformation rate and, by extension, prevent the crystalline structure from degrading. In addition, zirconia is believed to have a poor thermal conductivity, a boiling point that is almost twice as high as its melting temperature (2700°C), and an exceptionally high melting point overall [45].

1.6.2 Chemical Characteristics

Although it is chemically inert, zirconia, an amphoteric metal oxide, may show anion- and cation-exchange capabilities in solutions of varying pH and buffer types [46]. Zirconia has potential uses as a catalyst or catalytic support due to its acidic and basic chemical characteristics. Functional groups, such as phosphate esters, are added to zirconia to make it usable in many contexts because of its chemical inertness and non-polar surface [47].

1.6.3 Characteristics of Machinery

As summarized in Table 1.2, with a Young's modulus of 210 GPa (as opposed to 193 GPa), a traction resistance between 900 and 1200 MPa, and a compression resistance of up to 2000 MPa, zirconia's mechanical characteristics are quite similar to those of stainless steels [48]. Nevertheless, its mechanical characteristics may differ based on the crystalline structure's phase, grain size, and the kind and amount of additives used to stabilize each phase [49]. Therefore, zirconia-based ceramics stabilized with magnesia and calcium oxides have inferior characteristics to YSZ-based ceramics, with the metastable tetragonal phase being the most desirable and hardest form [50].

1.6.4 Shape Memory Effect (SME)

Most notably, the T→M martensitic transition that occurs with SME is of interest among the three polymorphic phases of ZrO_2 (monoclinic, tetragonal, and cubic, as shown in Figure 1.1).

TABLE 1.2
Thermal, Electrical, and Mechanical Properties of Zirconia [4]

Properties	Amount	Units
Density	6.05	g/cm^3
Hardness	1200	HV
Bend strength	900–1200	MPa
Compressive strength	2000	MPa
Fracture toughness	7–10	MPam½
Young's modulus	210	GPa
Thermal expansion coefficient	11×10^{-6} 1/K	W/(m K^{-1})
Thermal conductivity	1.7–2.7	10^{-6} K^{-1}
Thermal expansion	2.3–12.2	(J/kg·K)
Specific heat capacity	420–540	K
Melting point	2823–2973	kJkg 1
Latent heat of fusion	700–820	W/(m K)
Breakdown potential	4–6	M/V
Dielectric constant	10–23	–
Resistivity	3.16 10^{13}–3.16 10^{18}	$10^{-8\text{-ohm}}$ m
Atomic volume (average)	0.02–0.021	m^3/kmol
Energy content	200–300	MJ/kg
Bulk modulus	72.3–212	GPa
Compressive strength	1200–5200	MPa
Ductility	0.00066–0.0035	–
Elastic limit	115–711	MPa
Endurance limit	107–640	MPa

One example of SME is the capacity to change to a previously determined shape by applying heat [51]. Martensitic changes, however, have the potential to break zirconia-based materials in an undesired way. Alternatively, zirconia may have ceria, magnesia, or yttria added to it as additives to make it stronger, or small-scale ZrO_2 can be used to relieve tension during martensitic transformation by reducing mismatch stress at grain borders and increasing surface area [52]. Applications that can use the form memory effect of zirconia-based materials include energy damping, artificial muscles, shape memory ceramics made of bulk polycrystalline zirconia, and micro-electro-mechanical systems (MEMSs) [53].

1.6.5 Properties Related to Biology

Zirconia is physiologically inert, which means that it is biocompatible. This is because zirconia does not exhibit cytotoxic effects, and chances of unfavourable interactions with tissues are zero [54]. When a material is biocompatible, it can be used by various living systems without causing any negative reactions, such as inflammatory, irritative, allergic, toxic, mutagenic, or carcinogenic ones (bio-inertia), and conversely, it can maximize therapeutic performance by eliciting the most beneficial cellular or tissue response possible in a given context (bioactivity) [55]. So, biocompatibility testing is the first step for any substance that will be introduced into the human body. *In vitro* tests were conducted 30 years ago to determine the biocompatibility of zirconia in various physical forms with supporting connective tissue cells such as fibroblasts and fibrocytes [56,57]. Similar studies comparing zirconia with titanium were conducted. It is unclear from the available data whether the various zirconia powders tested were cytotoxic to these cell lines; it seems that only the wear products of zirconia might be poisonous [58]. Finally, numerous *in vivo* biocompatibility studies on different types of zirconia showed that zirconia is superior to other materials when applied to living tissues, including bone and muscle [59,60].

1.7 MICROSTRUCTURE OF ZIRCONIA

The microstructure of zirconia is shown in Figure 1.6. Microstructure is a crucial factor that directly influences its mechanical properties, thermal stability, and overall performance in advanced applications [61]. One of the most significant aspects of zirconia's microstructure is grain size as it plays a vital role in stabilizing the tetragonal phase at room temperature. Finer grains help maintain the tetragonal phase, which is essential for ensuring toughness through the transformation toughening mechanism. However, when the grain size increases beyond a critical size (typically around 0.3 to 0.5 μm), the tetragonal phase becomes unstable and tends to transform into the monoclinic phase, even without external stress, leading to volume expansion and microcracking. Dopants such as yttria, magnesia, and calcia not only stabilize specific zirconia phases but also control grain growth during sintering by inhibiting excessive coarsening of grains. This control ensures that the material retains an optimized microstructure, balancing strength, toughness, and thermal resistance. Furthermore, maintaining microstructural stability at high temperatures is vital for long-term performance, especially in applications such as TBCs and SOFCs,

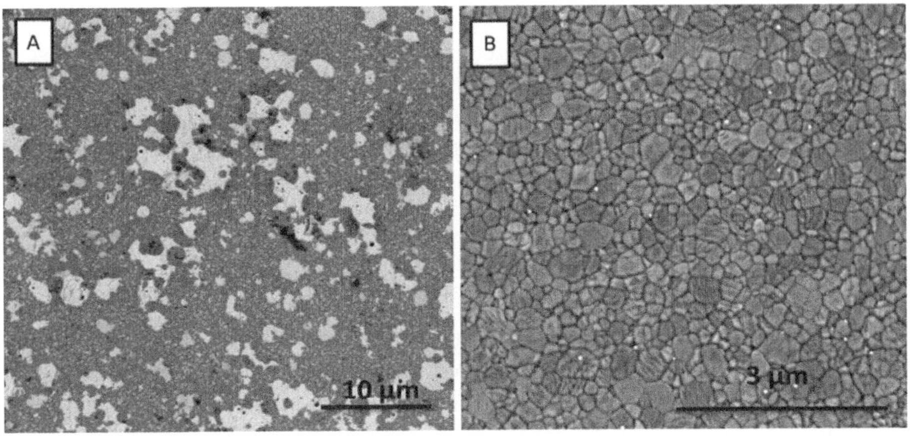

FIGURE 1.6 SEM micrographs of a polished surface of the zirconia–zirconium composite (a) (ceramic matrix: dark; zirconium particles: bright) and thermally etched monolithic zirconia (b) [63].

TABLE 1.3
Purity and Particle Sizes of Zirconium Powders [26,64]

Powder Determination	Zr 1	Zr 2
Supplier/supply procedure	BIMOTECH, Poland, in an aqueous suspension	KAMB, Poland, in an aqueous suspension
Purity (%)	99.9	98.86
Particle size by the producer (µm)	60	<451
Measured particle size median (µm)	12	6
Oxygen content (%)	1.6 ± 0.08	0.75 ± 0.01
Nitrogen content (ppm)	121.0 ± 8.8	446.5 ± 18.3
Hydrogen content (%)	0.155 ± 0.02	0.381 ± 0.002

where grain growth and phase destabilization can degrade material properties [62]. Therefore, careful design of the zirconia microstructure through controlled doping and processing methods is essential to ensure its reliability under demanding conditions [47] (Table 1.3).

1.8 FUTURE TRENDS IN ZIRCONIA RESEARCH

Research on zirconia is concentrated on further enhancing its properties and expanding its applications. Nanostructured zirconia is a prominent zone of research because it is presumed that reducing the grain size to its nanoscale will improve mechanical properties, phase stability, and surface area and make it more suitable for catalysis, sensors, and biomedical devices [65]. Another significant direction of research is the development of zirconia composites, which includes alumina or silicon carbide into zirconia to make hybrid systems with improved toughness, wear resistance, and thermal conductivity [66,67]. Further, the development of long-term stability and toughness of zirconia during exposure to stress such as high temperatures, cyclic mechanical loading, and corrosive environments is another major area of focus. Improvements in dopant chemistry, grain boundary modification, and surface alteration techniques are the efforts being pursued to remove these restrictions and ensure zirconia's position as a leader in high-performance ceramics for advanced technologies [48,68].

1.8.1 Future Trends in AI-Driven Zirconia-Based Dentistry

The future of prosthetic dentistry is projected to shift with the adoption of artificial intelligence (AI) and machine learning (ML) technologies in zirconia-based dental practices. AI-driven models will be further developed to augment treatment planning, shifting towards even deeper levels of personalization and greater reliance on existing treatment data. Moreover, computer-aided design (CAD) and computer-aided manufacturing (CAM) technologies will experience further advancements due to AI, which will enhance the precision, efficiency, and automation of the design and fabrication of zirconia restorations.

In the next few years, AI-driven wear prediction models will be further enhanced, allowing more accurate prediction of wear and failure risks associated with zirconia restorations. AI-assisted colour matching solutions will also ensure more natural-looking results, improving the harmony between prosthetic zirconia and natural teeth [69]. In addition, the use of robotics-assisted techniques in dentistry has the potential to fully automate restorative procedures, minimizing human error and improving the quality of care. The integration of AI in diagnosis, material selection, and dynamic procedural treatment optimization will ensure further development of zirconia-based restorations in terms of reliability, aesthetic appeal, and customization tailored to individual patients.

1.9 CONCLUSION

Zirconia is one of the most common and advanced ceramics because of its unparalleled structural, mechanical, thermal, and biological features. Its stress-induced phase transformation capabilities coupled with dopant stabilization make zirconia fracture-resistant and thermally stable while also providing high wear resistance. Zirconia's unmatched qualities have led to its widespread adoption in biomedical implants, TBCs, cutting tools, and energy systems. As researchers develop nanostructured and composite materials, zirconia will secure its title as a high-performance ceramic, guaranteeing improvements in materials science, engineering, and innovations.

REFERENCES

1. C. Piconi and G. Maccauro, "Zirconia as a Ceramic Biomaterial," *Biomaterials*, vol. 20, no. 1, pp. 1–25, 1999.
2. B. Arnold, *Zircon, Zirconium, Zirconia – Similar Names, Different Materials.* Springer, Berlin, Heidelberg, 2022. doi: 10.1007/978-3-662-64269-6.

3. P. Pittayachawan, "Comparative study of physical properties of zirconia based dental ceramics," 2014. [Online]. Available: https://www.researchgate.net/publication/32897236
4. G. Albarghouti, S. Mahmoud, and F. Darwish, "Processing of Zirconia," [Online]. Available: www.intechopen.com
5. Z. Liang et al., "Zirconium-Aluminum-Oxide Dielectric Layer with High Dielectric and Relatively Low Leakage Prepared by Spin-Coating and the Application in Thin-Film Transistor," *Coatings*, vol. 10, no. 3, Mar. 2020, doi: 10.3390/coatings10030282.
6. S. G. Giniyatova, A. L. Kozlovskiy, R. I. Shakirzyanov, N. O. Volodina, D. I. Shlimas, and D. B. Borgekov, "Structural, Dielectric, and Mechanical Properties of High-Content Cubic Zirconia Ceramics Obtained via Solid-State Synthesis," *Applied Sciences*, vol. 13, no. 19, Oct. 2023, doi: 10.3390/app131910989.
7. K. Jitwirachot, P. Rungsiyakull, J. A. Holloway, and W. Jia-Mahasap, "Wear Behavior of Different Generations of Zirconia: Present Literature," 2022. doi: 10.1155/2022/9341616.
8. A. Sidorkin, L. Nesterenko, Y. Gagou, P. Saint-Gregoire, E. Vorotnikov, and N. Popravko, "Dielectric Properties and Switching Processes of Barium Titanate-Barium Zirconate Ferroelectric Superlattices," *Materials*, vol. 11, no. 8, Aug. 2018, doi: 10.3390/ma11081436.
9. F. Murillo-Gómez et al., "Mechanical, Adhesive and Surface Properties of a Zirconia-Reinforced Lithium Silicate CAD/CAM Ceramic Exposed to Different Etching Protocols," *Materials*, vol. 17, no. 20, Oct. 2024, doi: 10.3390/ma17205039.
10. J. Singh, S. Kumar, and S. K. Mohapatra, "Tribological Performance of Yttrium (III) and Zirconium (IV) Ceramics Reinforced WC–10Co4Cr Cermet Powder HVOF Thermally Sprayed on X2CrNiMo-17-12-2 Steel," *Ceramics International*, vol. 45, no. 17, pp. 23126–23142, Dec. 2019, doi: 10.1016/j.ceramint.2019.08.007.
11. J. Xie et al., "Research Progress of High Dielectric Constant Zirconia-Based Materials for Gate Dielectric Application," Jul. 01, 2020. doi: 10.3390/coatings10070698.
12. R. D. Radu and D. Drăgănescu, "Present and Future of ZrO_2 Nanostructure as Reservoir for Drug Loading and Release," Jul. 01, 2023. doi: 10.3390/coatings13071273.
13. Y. Han and J. Zhu, "Surface Science Studies on the Zirconia-Based Model Catalysts," *Topics in Catalysis*, vol. 56, no. 15–17, pp. 1525–1541, 2013, doi: 10.1007/s11244-013-0156-5.
14. F. Namavar et al., "Thermal Stability of Nanostructurally Stabilized Zirconium Oxide," *Nanotechnology*, vol. 18, no. 41, Oct. 2007, doi: 10.1088/0957-4484/18/41/415702.
15. H. H. Park, C. H. Kim, T. G. Lee, J. H. Park, M. S. Park, and J. S. Lee, "Effects of Zirconium-Based Crosslinkers with Different Zirconium Contents on Pigment Coating in Paper," *Applied Sciences*, vol. 14, no. 20, Oct. 2024, doi: 10.3390/app14209187.
16. N. Song et al., "Evaluation of Phase Transformation and Mechanical Properties of Metastable Yttria-Stabilized Zirconia by Nanoindentation," *Materials*, vol. 12, no. 10, 2019, doi: 10.3390/MA12101677.
17. P. A. Seçkiner, F. Gönüldaş, B. Akat, A. Buyuksungur, and K. Orhan, "Investigation of Phase Transformation and Fracture Pattern as a Result of Long-Term Chewing Simulation and Static Loading of Reduced-Diameter Zirconia Implants," *Materials*, vol. 17, no. 19, Oct. 2024, doi: 10.3390/ma17194719.
18. M. P. de A. Mafra, N. Silva Júnior, C. dos Santos, J. L. de A. Ferreira, J. A. Araújo, and C. R. M. da Silva, "Evaluation of the Mechanical Properties and Fatigue Resistance of the $ZrO_2CeYAl_2O_3$ Composite," *Ceramics*, vol. 7, no. 4, pp. 1600–1615, Dec. 2024, doi: 10.3390/ceramics7040103.
19. P. M. Shah, J. W. H. Burnett, D. J. Morgan, T. E. Davies, and S. H. Taylor, "Ceria–Zirconia Mixed Metal Oxides Prepared via Mechanochemical Grinding of Carbonates for the Total Oxidation of Propane and Naphthalene," *Catalysts*, vol. 9, no. 5, May 2019, doi: 10.3390/catal9050475.

20. A. L. Kozlovskiy, M. Alin, and D. B. Borgekov, "Study of Polymorphic Transformation Processes and Their Influence in Polycrystalline ZrO_2 Ceramics upon Irradiation with Heavy Ions," *Ceramics*, vol. 6, no. 1, pp. 686–706, Mar. 2023, doi: 10.3390/ceramics6010042.
21. T. N. Rao, I. Hussain, J. E. Lee, A. Kumar, and B. H. Koo, "Enhanced Thermal Properties of Zirconia Nanoparticles and Chitosan-Based Intumescent Flame Retardant Coatings," *Applied Sciences*, vol. 9, no. 17, Sep. 2019, doi: 10.3390/app9173464.
22. I. Žmak, D. Ćorić, V. Mandić, and L. Ćurković, "Hardness and Indentation Fracture Toughness of Slip Cast Alumina and Alumina-Zirconia Ceramics," *Materials*, vol. 13, no. 1, Jan. 2020, doi: 10.3390/ma13010122.
23. U. Betke, M. Scheunemann, and M. Scheffler, "Refitting of Zirconia Toughening into Open-Cellular Alumina Foams by Infiltration with Zirconyl Nitrate," *Materials*, vol. 12, no. 12, Jun. 2019, doi: 10.3390/ma12121886.
24. X. Fu, F. Zhang, W. Zhu, and Z. Pi, "Mechanical Properties and Toughening Mechanisms of Promising Zr-Y-Ta-O Composite Ceramics," *Coatings*, vol. 13, no. 5, May 2023, doi: 10.3390/coatings13050855.
25. P. Palmero, "Structural Ceramic Nanocomposites: A Review of Properties and Powders' Synthesis Methods," Apr. 28, 2015. doi: 10.3390/nano5020656.
26. L. Jaworska, J. Cyboron, S. Cygan, A. Zwolinski, B. Onderka, and T. Skrzekut, "Zirconium Phase Transformation under Static High Pressure and ω-Zr Phase Stability at High Temperatures," *Materials*, vol. 12, no. 14, Jul. 2019, doi: 10.3390/ma12142244.
27. F. Rahmawati, I. Permadani, D. G. Syarif, and S. Soepriyanto, "Electrical Properties of Various Composition of Yttrium Doped-Zirconia Prepared from Local Zircon Sand," *International Journal of Technology*, vol. 8, no. 5, pp. 939–946, Oct. 2017, doi: 10.14716/ijtech.v8i5.876.
28. J. Smialek, "Phase Stability in Plasma-Sprayed Partially Stabilized Zirconia Yttria," 1981. [Online]. Available: https://www.researchgate.net/publication/23589949
29. S. Tailor, M. Singh, and A. V. Doub, "Synthesis and Characterization of Yttria-Stabilized Zirconia (YSZ) Nano-Clusters for Thermal Barrier Coatings (TBCs) Applications," *Journal of Cluster Science*, vol. 27, no. 4, pp. 1097–1107, Jul. 2016, doi: 10.1007/s10876-016-1014-y.
30. J. Wu, J. Xie, M. He, J. Zhang, and S. Li, "Yttria-Stabilized Zirconia Composite Coating as Barrier to Reduce Hydrogen Permeation into Steel," *Materials*, vol. 17, no. 12, Jun. 2024, doi: 10.3390/ma17123017.
31. H. W. Park et al., "Synthesis of Yttria-Stabilized Zirconia Nanospheres from Zirconium-Based Metal–Organic Frameworks and the Dielectric Properties," *Nanomaterials*, vol. 13, no. 1, Jan. 2023, doi: 10.3390/nano13010028.
32. R. B. Wijayanti, I. Rosmayanti, K. Wahyudi, E. Maryani, H. Hernawan, and R. Septawendar, "Preparation of Magnesia Partially Stabilized Zirconia Nanomaterials from Zirconium Hydroxide and Magnesium Carbonate Precursors Using Peg as a Template," *Crystals*, vol. 11, no. 6, Jun. 2021, doi: 10.3390/cryst11060635.
33. M. Grabowy et al., "Alumina-Toughened-Zirconia With Low Wear Rate in Ball-on-Flat Tribological Tests at Temperatures to 500 °C," *Materials*, vol. 14, no. 24, Dec. 2021, doi: 10.3390/ma14247646.
34. M. L. Matias et al., "A Comparison between Solution-Based Synthesis Methods of ZrO_2 Nanomaterials for Energy Storage Applications," *Energies*, vol. 15, no. 17, Sep. 2022, doi: 10.3390/en15176452.
35. S. Yilmaz, S. Cobaner, E. Yalaz, and B. A. Horri, "Synthesis and Characterization of Gadolinium-Doped Zirconia as a Potential Electrolyte for Solid Oxide Fuel Cells," *Energies*, vol. 15, no. 8, Apr. 2022, doi: 10.3390/en15082826.
36. H. Yan, J. Di, J. Li, Z. Liu, J. Liu, and X. Ding, "Synthesis of Zirconia Micro-Nanoflakes with Highly Exposed (001) Facets and Their Crystal Growth," *Crystals*, vol. 11, no. 8, Aug. 2021, doi: 10.3390/cryst11080871.

37. C. J. Brinker, and G. W. Scherer, *Sol-Gel Science: The Physics and Chemistry of Sol-Gel Processing*. Academic Press, 1990.
38. C. R. Babu, N. R. M. Reddy, and K. Reddy, "Synthesis and Characterization of High Dielectric Nano Zirconium Oxide," *Ceramics International*, vol. 41, no. 9, pp. 10675–10679, Nov. 2015, doi: 10.1016/j.ceramint.2015.04.168.
39. M. A. Alam, U. A. Samad, A. Anis, E. S. M. Sherif, H. S. Abdo, and S. M. Al-Zahrani, "The Effect of Zirconia Nanoparticles on Thermal, Mechanical, and Corrosion Behavior of Nanocomposite Epoxy Coatings on Steel Substrates," *Materials*, vol. 16, no. 13, Jul. 2023, doi: 10.3390/ma16134813.
40. A. M. Barbasz and B. Dyba, "Direct Interaction of Zirconia Nanoparticles with Human Immune Cells," *Biophysica*, vol. 4, no. 1, pp. 83–91, Mar. 2024, doi: 10.3390/biophysica4010006.
41. Y. Tao and P. P. Pescarmona, "Nanostructured Oxides Synthesised via $scCO_2$-Assisted Sol-Gel Methods and Their Application in Catalysis," May 01, 2018. doi: 10.3390/catal8050212.
42. L. Liu et al., "From Modification to Mechanism: Supercritical Hydrothermal Synthesis of Nano-Zirconia," Feb. 15, 2022. doi: 10.1016/j.ceramint.2021.11.028.
43. S. Liu et al., "Tetragonal Nanosized Zirconia: Hydrothermal Synthesis and Its Performance as a Promising Ceramic Reinforcement," *Inorganics*, vol. 11, no. 5, May 2023, doi: 10.3390/inorganics11050217.
44. N. Kumari et al., "Zirconia-Based Nanomaterials: Recent Developments in Synthesis and Applications," Aug. 23, 2022. doi: 10.1039/d2na00367h.
45. A. Scarano, M. Petrini, F. Mastrangelo, S. Noumbissi, and F. Lorusso, "The Effects of Liquid Disinfection and Heat Sterilization Processes on Implant Drill Roughness: Energy Dispersion X-ray Microanalysis and Infrared Thermography," *Journal of Clinical Medicine*, vol. 9, no. 4, Apr. 2020, doi: 10.3390/jcm9041019.
46. M. Diallo and D. Dewez, "Chemical Characteristics of Zirconium Chloride and Zirconium Oxide Nanoparticles Driving Toxicity on Lemna Minor," *Environments*, vol. 11, no. 10, Oct. 2024, doi: 10.3390/environments11100222.
47. E. Matei et al., "Environmental Sustainability Based on Zirconium Dioxide Utilization in Non-Conventional Energy Applications," Dec. 01, 2024. doi: 10.3390/environments11120265.
48. P. Sokola et al., "Comprehensive Study of Stereolithography and Digital Light Processing Printing of Zirconia Photosensitive Suspensions," Dec. 01, 2024. doi: 10.3390/ceramics7040104.
49. S. Lamnini, D. Pugliese, and F. Baino, "Zirconia-Based Ceramics Reinforced by Carbon Nanotubes: A Review with Emphasis on Mechanical Properties," Sep. 01, 2023. doi: 10.3390/ceramics6030105.
50. C. Y. Lee, S. Lee, J. H. Ha, J. Lee, I. H. Song, and K. S. Moon, "Effect of the Zirconia Particle Size on the Compressive Strength of Reticulated Porous Zirconia-Toughened Alumina," *Applied Sciences*, vol. 12, no. 5, Mar. 2022, doi: 10.3390/app12052316.
51. A. M. A. Moncayo, L. Peñate, M. Arregui, L. Giner-Tarrida, and R. Cedeño, "State of the Art of Different Zirconia Materials and Their Indications According to Evidence-Based Clinical Performance: A Narrative Review," Jan. 01, 2023. doi: 10.3390/dj11010018.
52. C. Palade, A. Slav, I. Stavarache, V. A. Maraloiu, C. Negrila, and M. L. Ciurea, "Memory Properties of Zr-Doped ZrO_2 MOS-like Capacitor," *Coatings*, vol. 12, no. 9, Sep. 2022, doi: 10.3390/coatings12091369.
53. R. A. N. Alkhalifa, A. E. A. E. Albadri, R. Ali, A. H. Alluhayb, A. M. Younis, and S. M. Saleh, "Sustainable Synthesis of Zirconium Dioxide (ZrO_2) Nanoparticles Utilizing *Asphodelus fistulosus* Extract for Congo Red Degradation," *Catalysts*, vol. 15, no. 2, p. 123, Jan. 2025, doi: 10.3390/catal15020123.

54. A. Berberi, D. Maroun, W. Kanj, E. Z. Amine, and A. Philippe, "Micromovement Evaluation of Original and Compatible Abutments at the Implant-Abutment Interface," *Journal of Contemporary Dental Practice*, vol. 17, no. 11, pp. 907–913, 2016, doi: 10.5005/jp-journals-10024-1952.
55. E. Borie, E. Rosas, R. F. de Souza, and F. J. Dias, "Zirconia Implants: A Brief Review and Surface Analysis of a Lost Implant," *Coatings*, Aug. 01, 2024. doi: 10.3390/coatings14080995.
56. B. F. Fernandes et al., "Cell Biological and Antibacterial Evaluation of a New Approach to Zirconia Implant Surfaces Modified with MTA," *Biomimetics*, vol. 9, no. 3, Mar. 2024, doi: 10.3390/biomimetics9030155.
57. C. Larsson and P. Vult Von Steyern, "Implant-Supported Full-Arch Zirconia-Based Mandibular Fixed Dental Prostheses. Eight-Year Results from a Clinical Pilot Study," *Acta Odontologica Scandinavica*, vol. 71, no. 5, pp. 1118–1122, Sep. 2013, doi: 10.3109/00016357.2012.749518.
58. C. I. Resende-Gonçalves et al., "Porous Zirconia Blocks for Bone Repair: An Integrative Review on Biological and Mechanical Outcomes," *Ceramics*, Mar. 01, 2022. doi: 10.3390/ceramics5010014.
59. D. Lolos, S. G. Mihali, S. Dinu, M. Mitariu, A. Tudor, and R. Oancea, "Retrospective Long-Term Survival Rate and Clinical Performance of Zirconium Oxide Restorations over the Past 5 Years: A Comparative Study Between Single Crowns and Fixed Dental Prostheses," *Medicina*, vol. 61, no. 2, p. 210, Jan. 2025, doi: 10.3390/medicina61020210.
60. M. F. R. P. Alves, S. Ribeiroa, P. A. Suzuki, K. Strecker, and C. dos Santosc, "Effect of Fe_2O_3 Addition and Sintering Temperature on Mechanical Properties and Translucence of Zirconia Dental Ceramics With Different Y_2O_3 Content," *Materials Research*, vol. 24, no. 2, Mar. 2021, doi: 10.1590/1980-5373-MR-2020-0402.
61. S. S. Alsharari, M. A. Alenezi, M. S. Al Tami, and M. Soliman, "Recent Advances in the Biosynthesis of Zirconium Oxide Nanoparticles and their Biological Applications," *Baghdad Science Journal*, vol. 20, no. 1, pp. 41–57, 2023, doi: 10.21123/bsj.2022.7055.
62. K. Glowka, M. Zubko, P. Świec, K. Prusik, M. Szklarska, and D. Stróż, "Influence of Zirconium on the Microstructure, Selected Mechanical Properties, and Corrosion Resistance of $Ti_{20}Ta_{20}Nb_{20}(HfMo)_{20-x}Zr_x$ High-Entropy Alloys," *Materials*, vol. 17, no. 11, Jun. 2024, doi: 10.3390/ma17112730.
63. M. Díaz, A. Smirnov, C. F. Gutiérrez-González, D. Estrada, and J. F. Bartolomé, "Microstructure and Mechanical Properties of Zirconia (3Y-TZP)/Zr Composites Prepared by Wet Processing and Subsequent Spark Plasma Sintering," *Ceramics*, vol. 3, no. 1, pp. 53–64, Mar. 2020, doi: 10.3390/ceramics3010007.
64. J. Singh, S. Kumar, and S. K. Mohapatra, "An Erosion and Corrosion Study on Thermally Sprayed WC-Co-Cr Powder Synergized with $Mo_2C/Y_2O_3/ZrO_2$ Feedstock Powders," *Wear*, vol. 438–439, Nov. 2019, doi: 10.1016/j.wear.2019.01.082.
65. G. Magnani, P. Fabbri, E. Leoni, E. Salernitano, and F. Mazzanti, "New Perspectives on Zirconia Composites as Biomaterials," *Journal of Composites Science*, vol. 5, no. 9, Sep. 2021, doi: 10.3390/JCS5090244.
66. Q. Dong and J. Tan, "Advances in Zr-Based Alloys," Apr. 01, 2024. doi: 10.3390/cryst14040351.
67. P.-J. Koo, J.-H. Lee, S.-R. Ha, D.-G. Seo, J.-S. Ahn, and Y.-S. Choi, "Changes in the Properties of Different Zones in Multilayered Translucent Zirconia Used in Monolithic Restorations During Aging Process," *Journal of Functional Biomaterials*, vol. 16, no. 3, p. 96, Mar. 2025, doi: 10.3390/jfb16030096.
68. L. P. Webber, H. L. Chan, and H. L. Wang, "Will Zirconia Implants Replace Titanium Implants?" *Applied Sciences*, vol. 11, no. 15, Aug. 2021, doi: 10.3390/app11156776.
69. J. Singh, S. Singh, and A. Verma, "Artificial Intelligence in Use of ZrO_2 Material in Biomedical Science," *International Association of Physical Chemists*, Feb. 19, 2023. doi: 10.5599/jese.1498.

2 Zirconia Nanoparticles as the Strengthening Phase

Vikas Sharma, Jashanpreet Singh, and Shivam Kumar

2.1 INTRODUCTION

Zirconia (ZrO_2) has three polymorphic phases: c-ZrO_2, t-ZrO_2, and m-ZrO_2 (Saenko et al., 2018). The tetragonal phase is one of the important phases due to its high strength and toughness, properties that could be harnessed in composites. The tetragonal to monoclinic phase conversion of zirconia can be achieved by a 3%–5% volume increase, and it is used to improve the mechanical properties of materials through stress-induced phase transformation toughening. Zirconia nanoparticles (NPs) can emerge as an effective strengthening phase due to their small size and high-temperature phases (Chevalier et al., 2009).

The large surface area of zirconia NPs significantly affects their catalytic activity and optical and electronic properties. Oxide NPs show potential in various industrial applications and innovative projects. Nowadays, ZrO_2 NPs are extensively used as catalysts and precursors in ceramics. One of the most widespread techniques to obtain stable tetragonal-phase zirconium dioxide NPs is hydrothermal synthesis with the addition of metal salts (Kumari et al., 2022). On the other hand, hydrothermal synthesis needs high process temperatures (160°C–200°C) in tandem with sophisticated heat-resistant autoclaves that can withstand high pressure. This chapter explores the variations in the phase composition, optical properties, size, and shape of ZrO_2 NPs prepared using hydrothermal synthesis without calcination as a function of various precursors (such as $ZrOCl_2{\cdot}8H_2O$) and synthesis temperatures (110°C–160°C). In addition, the influence of temperature exposure within 1000°C on the phase stability of the synthesized NPs was investigated. The structure and phase composition of the samples were determined by X-ray diffraction (XRD) and Raman spectroscopy techniques. Furthermore, their optical properties were studied by analyzing spectra in the visible spectral range. Moreover, the as-prepared powders showed c-ZrO_2 phase and transformed to monoclinic phase on calcination over 500°C (m-ZrO_2) (Jithin et al., 2025). Experimental samples showed significant absorption in the middle UV range (200–250 nm), and this allowed the precise determination of the energy band gap.

DOI: 10.1201/9781003519300-3

2.2 SYNTHESIS OF ZIRCONIA NANOPARTICLES

2.2.1 Sol-Gel Method

The sol-gel method is generally used for producing zirconia NPs, which offers the advantage of controlling the size and phase composition of the particles. Zirconia NPs are synthesized using zirconium dioxide (IV) as a precursor with acid (HNO_3) or basic (NH_4OH) catalysts. The combination of methanol and isopropanol as the solvent facilitates the control of particle size and phase stability. Acid catalysis results in the formation of smaller particles and stabilizes the tetragonal phase, whereas basic catalysis favors the transformation to the monoclinic phase, as deduced from the results of a previous study (Tello et al., 2020). Zirconia-based flame retardant coatings are developed using grinding, mixing, and curing, with varying NP concentrations (0–3% w/w), as shown in Figure 2.1.

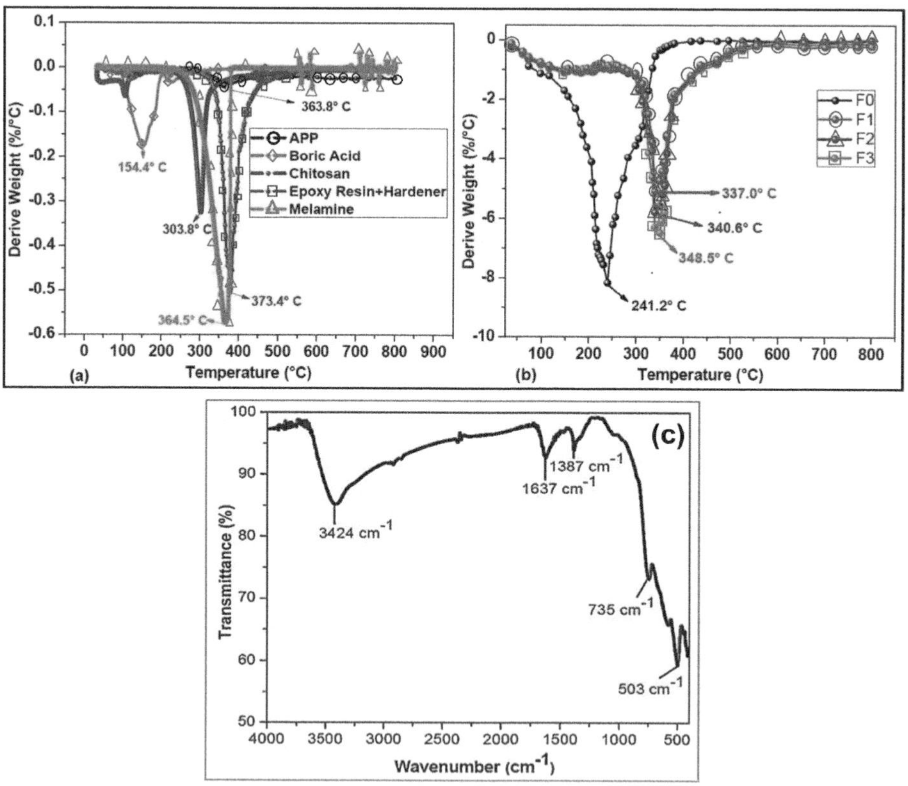

FIGURE 2.1 Differential thermogravimetry (DTG) graph of (a) individual components and (b) flame retardant samples, and (c) FTIR analysis of ZrO_2-based flame retardant coatings using multiple nanoparticles (Rao et al. 2019).

2.2.2 Effect of Catalyst and Solvent

Catalyst type and concentration substantially affect the phase changes and nanoscale sizing of ZrO_2 NPs (Zhang et al., 2006; Becker et al., 2008; Mostafa et al., 2009). Acid catalysts promote hydrolysis reactions, leading to smaller, more random atomic structure particles, whereas basic catalysts prefer condensation reactions, resulting in larger but more uniform particles. Solvents used during the refluxing process also affect particle morphology, which can be observed for methanol, owing to its high dielectric constant, thus yielding smaller submicron particles. This results in structural evolution toward the monoclinic phase, as compared to isopropanol, which stabilizes the tetragonal phase.

2.2.3 Tetragonal to Monoclinic Structure Transformation

The tetragonal (*t*)-to-monoclinic (*m*) phase transformation of zirconia is an essential mechanism for the mechanical properties of zirconia-containing materials. However, the tetragonal phase can be stabilized at nanoscale dimensions (<10 nm) and through the creation of oxygen vacancies. The transition to the monoclinic phase is preferable at high temperatures and long holding times and is facilitated by basic catalysts.

2.3 CHARACTERIZATION OF ZIRCONIA NANOPARTICLES

2.3.1 Structural and Morphological Characterization

Zirconia nanocrystals have a wide variety of functional properties depending on their crystal structure (monoclinic and tetragonal), which makes their characterization important (Tahir et al., 2007; Puigdollers et al., 2016; Manivasakan et al., 2011; Shahmiri et al., 2018). XRD analysis confirmed the predominance of the tetragonal phase, which is recognized for both its high mechanical strength and stability. The monoclinic phase was also observed, especially for larger particles. Transmission electron microscopy (TEM) morphological analysis indicated that zirconia NPs have a bimodal size distribution, where smaller tetragonal particles are spherical, while larger monoclinic particles are polyhedral (Yuan et al., 2009). This dual-phase structure allows for improved mechanical properties of zirconia, making it widely used in the fabrication of ceramics and structural composites. Energy-dispersive X-ray spectroscopy (EDS) confirmed the uniform distribution of zirconium and oxygen, ensuring the chemical purity of the synthesized NPs. Figure 2.2 presents field emission scanning electron microscopy (FESEM) images of ZnO films before and after ZrO_2 NP deposition, highlighting structural changes with and without thermal treatment. Thermal treatment at 500°C for 24 hours enhanced surface smoothness and revealed nanosheet and nanoneedle formations. Atomic force microscopy (AFM) studies were performed to further analyze the surface roughness of the synthesized NPs, showing highly uniform textures with few defects. Porosity and surface area of the obtained NPs were also examined. It was confirmed that the synthesized zirconia had a high surface area and was suitable for various catalytic applications.

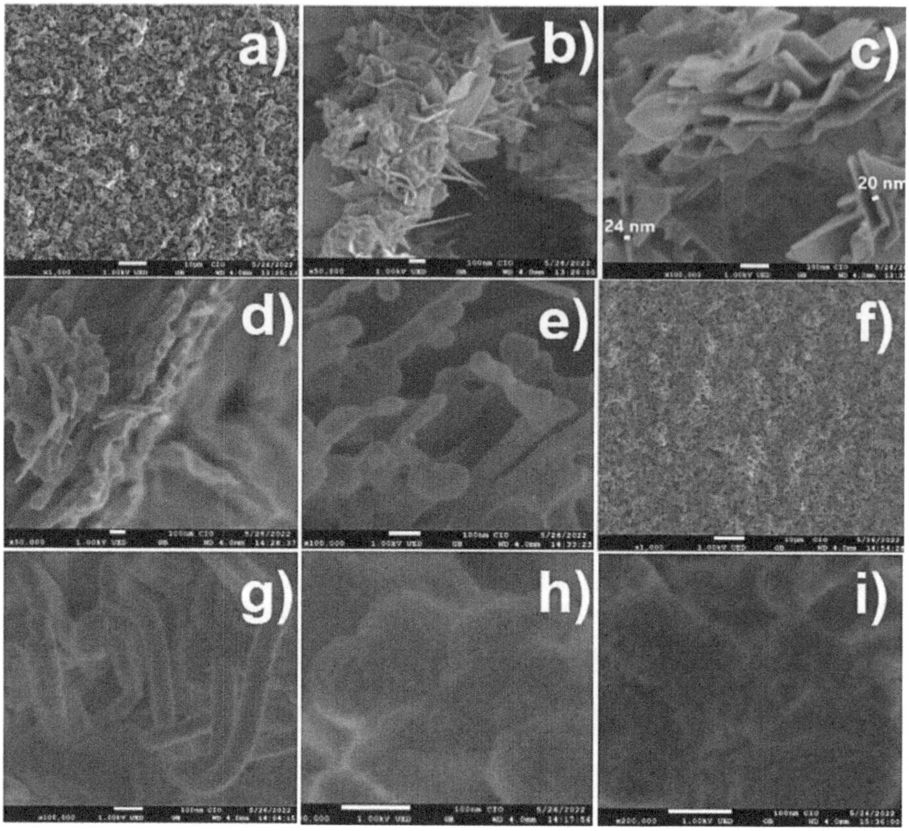

FIGURE 2.2 FESEM images of ZnO films: (a) untreated, (b) high magnification showing nanosheets and nanoneedles, (c) nanosheet thickness, (d) thermally treated at 500°C for 24 hours, (e) high magnification of the treated film, (f) smoother surface observed. After ZrO_2 NP deposition: (g) untreated, (h) high magnification of (g), (i) thermally treated at 500°C for 24 hours (Mastache et al. 2023).

2.3.2 Dispersion and Stability of Zirconia Nanoparticles

Stable dispersion of zirconia NPs in liquid media is crucial for their use in coatings, catalysis, and biomedical devices. The stability of zirconia suspensions was examined using dynamic light scattering (DLS), and an average particle size of 124 nm in an aqueous environment was measured. Ultrasonication was therefore performed to achieve much better dispersion and reduce the formation of hard agglomerates. It was also found that a suspension containing sodium chloride (NaCl) showed lower colloidal stability compared with the control, and the colloids aggregated in the presence of NaCl, thus affecting the stability of endodontic materials. Following several washing cycles to eliminate residual NaCl, the NPs showed better dispersion behavior and remained stable in the suspension for as long as 15 days. Second, the hierarchical structure of the aggregates of zirconia NPs was further explored using small-angle X-ray scattering (SAXS) analysis. Results showed the existence of a hierarchical

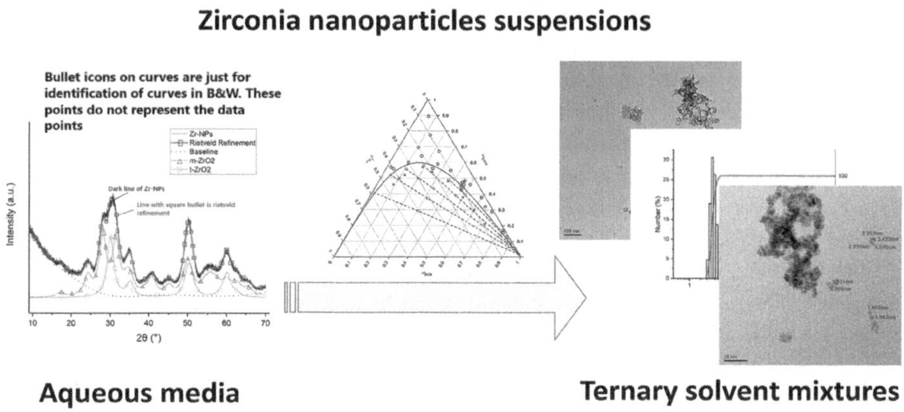

FIGURE 2.3 Zirconia NP dispersion for ternary solvent mixture (Fiorati et al. 2021).

structure, i.e., primary particles build bigger superstructures. Such results are consistent with previously documented investigations on hydrothermal ZrO_2, as controlled agglomeration improves mechanical properties while maintaining stability (Bamiduro, 2015).

A novel method for stabilizing ZrO_2 NP dispersions is using homogeneous ternary solvent mixtures. Findings suggest that this approach can be extended to other complex solvent systems for improving NP dispersion techniques, as shown in Figure 2.3. The electrostatic stability of NPs was also evaluated using zeta potential measurements. Findings showed that the zeta potential of zirconia NPs was around −35 mV in aqueous suspension and confirmed that zirconia NPs were highly dispersed with strong stability without aggregation. By incorporating various dispersing surface agents, citric acid, and polyethylene glycol (PEG), appropriate surface functionalization can enhance dispersion stability.

2.3.3 Microwave-Assisted Hydrothermal Synthesis of Zirconia Nanoparticles

Structural analysis validated the development of tetragonal and monoclinic phases (characterized by unique morphological features) (Bashir et al., 2022; Dias and Ciminelli, 2003). Stability studies revealed well-dispersed zirconia NPs with very good colloidal stability, which would be applicable for most advanced applications. These zirconia NPs can find applications in high-performance ceramics, dental implants, fuel cell electrolytes, and catalyst supports. Their enhanced mechanical properties and chemical stability make them promising candidates for reinforcement in composite materials. Optimizing the synthesis parameters should be the focus of future studies in order to fine-tune particle size distribution and phase composition to provide consistency for large-scale production (Li et al., 2024). In addition, the self-cleaning performance of zirconia NPs was evaluated based on photocatalytic experiments, which can degrade organic pollutants under UV light to achieve high self-cleaning properties (Nguyen et al., 2024; Kalpana and Rajathi, 2024).

This implies the possible use of zirconia in environmental nano-remediation and antibacterial coating. The applicability of zirconia in biomedicine as a drug delivery system and in dentistry is also investigated, highlighting its biocompatibility and its mechanical durability. Justifying the findings of this investigation, a growing body of literature demonstrates that microwave-assisted hydrothermal synthesis is an effective, scalable, and beneficial alternative to synthesize high-purity zirconia NPs. Future work will also explore the inclusion of dopants like yttrium and cerium in order to enhance the functional properties of zirconia-based materials.

The effect of dispersant has been observed with pure NP solution, as shown in Figure 2.4. The region of suspension stability has been identified when zirconia NPs are suspended with variable mass, as shown in Figure 2.4a. Zirconia NPs are synthesized using microwave-assisted and mechanochemical techniques. These synthesis methods affect the phase composition and surface properties of NPs that make them apt for diverse applications. The sol-gel method is the widely used method for the preparation of ZrO_2 nanostructures as it can control particle size and generally involves the hydrolysis of zirconium alkoxides in an acidic or basic medium with further drying and calcining (dos Santos et al., 2014; Batool et al., 2020). This method also offers the possibility to prepare ultrafine zirconia NPs of high phase purity. In contrast, the hydrothermal process can produce highly crystalline zirconia with controlled phase compositions under

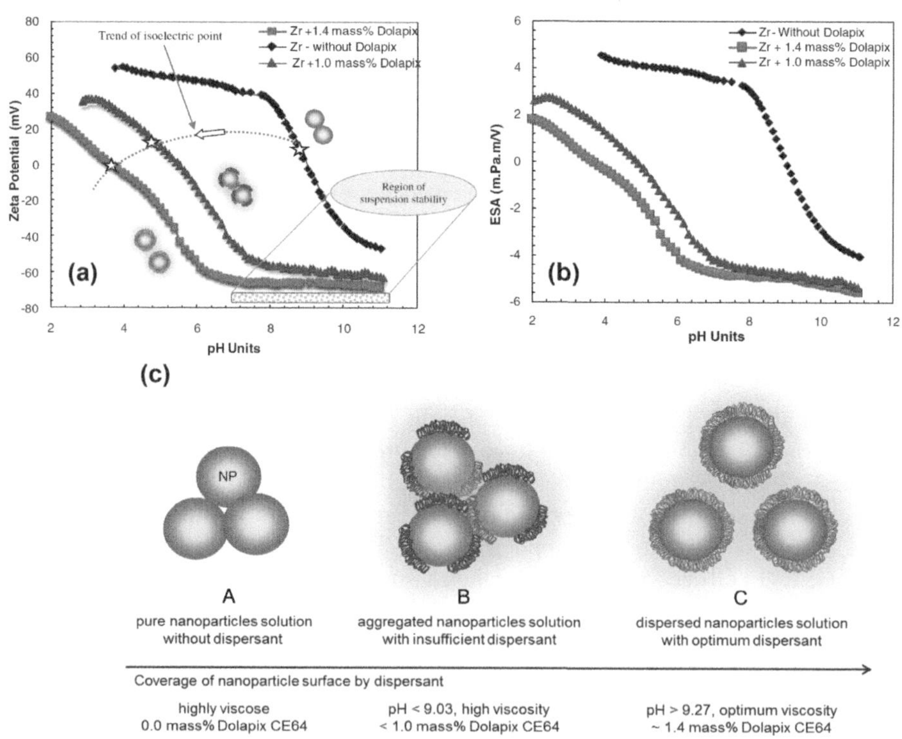

FIGURE 2.4 Zirconia particles with mass variation with isoelectric point (a), variable mass quantity of Dolapix (b) and representation of aggregated NPs (c) (Sarraf et al. 2015).

high-pressure and high-temperature conditions. Each hydrothermal synthesis method allows better control and shape of customized particles. By accelerating the kinetics of the reaction in microwave-assisted synthesis, it produces more uniform NPs with higher purity. This approach offers rapid synthesis without compromising crystallinity (Nadagouda et al., 2011). From the abundant data collected, the kilowatt-range solution combustion synthesis, a technique based on the principle of self-propagating exothermic reactions between metal precursors and fuel sources, results in the rapid production of nanosized zirconia with high surface area and catalytic efficiency. High-energy ball milling synthesizes zirconia using mechanochemical processes and tends to produce nanosized zirconia with better mechanical properties than the previously discussed method (Prasad Yadav et al., 2012; Xu et al., 2015). The characterization of zirconia NPs is usually carried out using XRD, scanning electron microscopy (SEM), XRD, Fourier-transform infrared spectroscopy (FTIR), and TEM. The corresponding as-synthesized monoclinic, tetragonal, and cubic phases were identified via XRD analysis, and TEM and SEM were used to determine particle morphology and distribution. FTIR analysis shows the presence of chemical bonding and oxygen–metal bonding in the lattice structure. Such characterization strategies are instrumental in establishing phase stability, grain size, and surface functionality.

2.3.4 Structural and Surface Modification of Zirconia Nanoparticles

The crystallographic forms of zirconia, such as monoclinic and cubic, are extremely important to its mechanical integrity, thermal stability, and chemical/oxide reactivity. Monoclinic zirconia is the most stable phase at room temperature (RT). However, it undergoes a phase transformation to the tetragonal phase at ~1170°C and finally to the cubic phase at 2370°C. This is accompanied by an expansion in volume and can result in microcracking for structural applications. Different ion dopants (Y_2O_3, CeO_2, MgO, CaO, etc.) are introduced to stabilize tetragonal and cubic phases at lower temperatures. Yttria-stabilized zirconia (YSZ) is known for its increased toughness and resistance to thermal shock and therefore is widely utilized in high-temperature applications. This is achieved by embedding dopants that hinder the unchecked phase transformation, preserving structural integrity under harsh conditions (Song et al., 2024). The functionalization of zirconia NPs is necessary for their widespread use in biomedical, catalytic, and energy applications, but it requires surface modifications. Tapped mechanical properties coupled with their reactive surface area have been used to functionalize 2D nanosheets using processes such as plasma treatment, hydrothermal etching, and coating with bioactive molecules to enhance surface reactivity and biocompatibility. Hydrothermally treated zirconia surfaces are more hydrophilic and have a higher protein adsorption rate, which together positively affect osseointegration of biomedical implants. Similarly, plasma-treated zirconia NPs exhibit higher dispersion stability and catalytic activity. In biomedical applications, zirconia surface nanotexturing has been shown to control cell adhesion and proliferation. Nanostructured patterns on zirconia surfaces can also be generated using laser ablation or anodization, and these patterns have been shown to promote selective cellular responses as well, leading to enhanced bone growth and reduced adhesion of bacteria (Siddiqui et al., 2010). Due to these alterations, zirconia appears as a promising candidate for next-generation biomaterials.

2.3.5 Actinide Separation Using Zirconia Impregnated with Oxidizing Agents

Zirconia particles are widely used in heterogeneous catalysis because of their high thermal stability, huge oxygen storage capability, and high Lewis acidity. They also serve as a catalyst support, improving the performance of industrial processes such as hydrocarbon reforming, CO_2 conversion, and pollutant degradation (Akhtar et al., 2024). A well-known use of zirconia in catalysis is in automotive catalytic converters. Zirconia-supported platinum (Pt) and palladium (Pd) catalysts are utilized for transforming toxic emissions like carbon monoxide (CO) and nitrogen oxides (NOx) into harmless materials. Zirconia has a high oxygen storage capacity that promotes oxidation and reduction reactions that are needed to control emissions.

Zirconia-supported catalysts have also been used for synthesis gas production based on either steam reforming of hydrocarbons or water–gas shift reaction. Research has presented an overview of charged zirconia composites (mainly ceria + zirconia doped ceramics) with enhanced redox characteristics and potential fuel cell performance. In photocatalysis, ZrO_2 NPs show potential in environmental remediation fields such as water purification and air pollution control. With a wide band gap (~5 eV), zirconia can absorb UV light efficiently and thus is a potential semiconductor for photocatalysis. In conjunction with other photocatalysts like titanium dioxide (TiO_2), zirconia/yellow-based nanocomposites provide high charge separation qualities, improving the degradation of organic pollutants, dyes, and pharmaceutical contaminants in wastewater treatment plants. Researchers are focusing on the expansion (using noble metal doping or composites with carbonaceous nanomaterials, i.e., graphene and carbon nanotubes) of the photocatalytic action of zirconia in the long-wave (i.e., visible light) domain (Aldeen et al., 2022). Machine learning could also refine tools for environmental clean-up to improve energy efficiency and reduce costs.

2.3.6 Analysis of Thermal Stability and Decomposition

Thermogravimetric analysis (TGA) and differential thermogravimetry (DTG) were used to analyze the thermal stability and decomposition of ZrO_2 NPs. To investigate the stages of weight loss in the precursor solution at high temperatures, the initial precursor solution was dried to RT, and the analysis was conducted for each stage. This study examined the removal of organic matter and phase transformation of ZrO_2. The TGA curve showed two main steps of weight loss. The first step, which took place between 30°C and 250°C, was linked to moisture trapped in the film, evaporating with the removal of acetate groups. The second step was an important weight loss step that occurred between 400°C and 470°C, which was attributable to the combustion of polyvinylpyrrolidone (PVP). The rest of the material over >471°C was simple carbon-free pure ZrO_2 NPs with a trace of remaining carbon content. A DTG analysis displayed precise peaks, which correlated with the degradation of polymeric components (Monteiro et al., 2012). A small peak around 62°C corresponded to moisture removal, while a larger peak around 436°C indicated PVP degradation. To obtain high-purity zirconia, the data showed that calcination should be done above 500°C. This underlined the relevance of tuning the temperature to adjust the purity and characteristics of the end NPs.

AFM images showed ZrO_2 flakes formed by compressing NPs at 18 MPa, synthesized at 400°C. The flakes were sintered at varying temperatures, revealing structural changes from 800°C to 1400°C, as shown in Figure 2.5. Moreover, thermal stability measurements showed that ZrO_2 NPs retained their crystalline structure to 900°C, suggesting their potential for high-temperature applications. Except for the decomposition of organic residues, structural degradation was negligible, which suggested good thermal stability. Accordingly, these results highlighted the appropriate features of ZrO_2 NPs to be applied in systems necessitating thermal stability, including catalysis, petroleum, and fuel cell systems. ZrO_2 nanocrystals are known for their properties such as excellent thermal stability and corrosion and mechanical resistance, which are the reasons for their widespread use in various industrial applications. Herein, we investigate the synthesis of zirconia NPs using microwave-assisted hydrothermal treatment. To illustrate how the sol-gel process operates, the sol prepared using this method is dissolved in zirconium oxychloride (typically $ZrOCl_2{\cdot}8H_2O$) in an aqueous medium, and then, sodium hydroxide (NaOH) was used for precipitation at a controlled pH range of 9–10. Nanocrystalline zirconia powders were prepared using microwave irradiation of the mixture for more

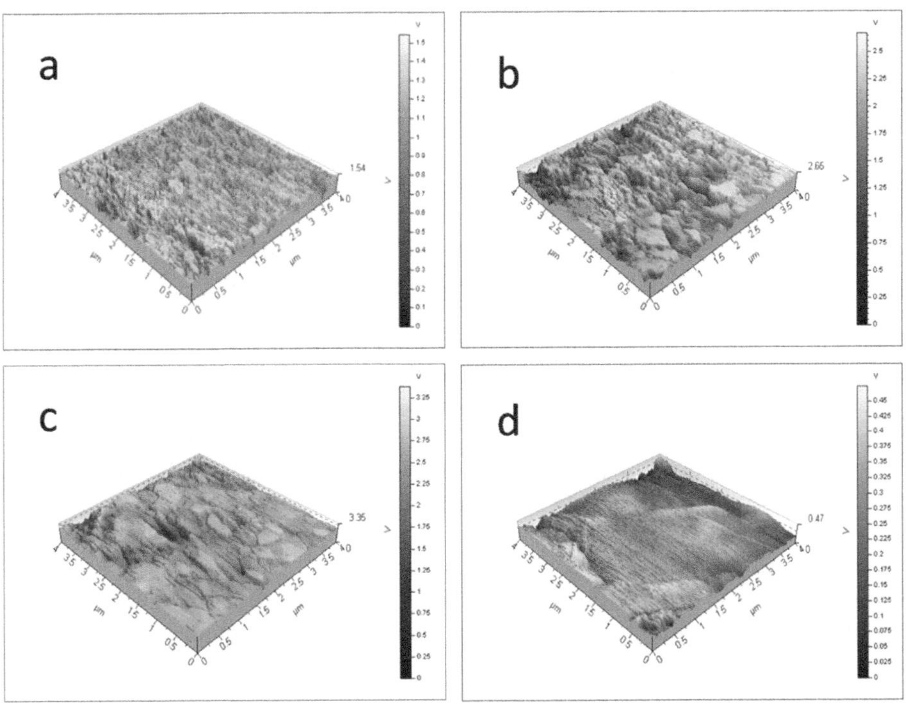

FIGURE 2.5 AFM images of ZrO_2 flakes compressed at 18 MPa from NPs synthesized at 400°C and then sintered at (a) 800°C, (b) 1000°C, (c) 1200°C, and (d) 1400°C (Wang et al. 2016).

than 2 hours. Complex tetragonal phases of zirconia were observed in TEM images, where primary particles showed a diameter between 6.9 and 13.9 nm. Analyses using small-angle X-ray scattering (SAXS) also established that the small primary particles formed roughly 50-nm agglomerates (Nirschl and Guo, 2018). DLS analysis showed that the NPs, smaller than 200 nm, were stable in aqueous suspension up to 15 days, demonstrating they were well dispersed with few agglomerates. The presence of Zr-O bonds was further verified using FTIR to validate the synthesis procedure. Microwave irradiation of crystals resulted in increasing crystallinity as a function of exposure time, as shown in the XRD data (Sari and Muzakky, 2023; Preethi et al., 2021). A separation followed by optimal dip-coating parameters facilitated effective electron transport, with the highest current density occurring at a pH of 9–10, indicating the formation of low-defect NPs. However, at the pH value of 12, locally increased defects were formed.

2.4 PHYSICAL PROPERTIES OF ZIRCONIUM OXIDE PARTICLES

2.4.1 Optical Properties and Band Gap Analysis

UV-Vis spectrophotometry was performed to analyze the optical properties of ZrO_2 NPs. The band gap energy is a key parameter that affects the behavior of semiconductors. In assessing materials for optoelectronic and photocatalytic prospects, optical transparency and electronic transitions are major factors. The energy band also decreases with high values of calcination temperature. This trend is explained by the increase in particle size since larger NPs show weaker quantum confinement effect (Keiteb et al., 2016). The trend of optical properties indicates improvement in electron mobility, which would be desirable for electronic and optical applications. The UV-Vis spectra further revealed a gradual increase in the absorbance intensity with increasing temperature, suggesting an improved ability to absorb light.

Optical reflectance spectra coefficients also verified that the synthesized NPs showed strong absorbance in the ultraviolet region with potential UV-block properties. This suggested that the band gap energy could be tuned according to desired functions by controlling calcination conditions. Combined photoluminescence (PL) measurements revealed the electronic structure and defect states of ZrO_2 NPs. PL spectra showed typical emission peaks attributed to oxygen vacancies and surface defects, affecting the recombination of charge carriers. Moreover, defect-related emission implied potential applications in photocatalysis and light-emitting devices. In a previous study, a systematic investigation of the structural, thermal, and optical properties of ZrO_2 NPs was pursued (Rizzuti et al., 2009). They successfully tuned the structural designs of such NPs despite the simplicity of the synthesis/texturing route. The versatility of the thermal treatment method was underlined by the ability to tailor structural and optical characteristics by modifying calcination temperature, laying the foundation for future developments in materials science and nanotechnology.

2.4.2 Thermal Stability and Radiation Shielding Properties

Zirconia enhances cement blends in one of the most likely ways: thermal resistance. Due to their resistance to heat-induced degradation, these composites have potential in the field of fire resistance applications and are ideal for use in high-risk conditions. Zirconia-modified composites maintain their mechanical strength even after extended exposure to high-temperature environments (>850°C) as revealed through thermal stability tests, where lower mass loss as well as retained hydration phases (as confirmed through TGA and DSC) demonstrates the resistance toward thermal decomposition for these composites (Koo and Pilato, 2020). This greatly increases the radiation shielding efficiency of cementitious mixes due to the addition of zirconia NPs. Similarly, gamma attenuation experiments were executed at GG1.G KeV and 1332.5 KeV gamma-ray energetic levels, with results indicating that 1% ZrO_2 offers better radiation-discharging ability than conventional cementitious materials. Due to their higher linear attenuation coefficient, these composites can be used in nuclear facilities, medical radiation shielding, and high-security facilities.

2.4.3 Mechanical and Thermal Properties of Zirconia Nanoparticles

Due to the excellent mechanical properties of zirconia NPs, they are applied in structural and engineering applications. Due to their high values of fracture toughness and hardness, zirconia NPs are used in machining applications and orthopedic implants. The toughening mechanism associated with their phase transformation is one of the key factors that affect the mechanical performance of zirconia NPs. The transition of phases under mechanical stress causes volume expansion and creates compressive stresses. Zirconia demonstrates good thermal stability at high temperatures. These properties make it a superlative material for use as thermal barrier coatings (TBCs) in jet engines and power turbines, where it protects metal parts from exposure to extreme heat. YSZ is the most common TBC material, which can be resistant to high temperatures (>1200°C) with no or slight degradations (Abd-Elwahed et al., 2020). Zirconia NPs also exhibit high resistance to wear, further enhancing their durability. Zirconia ceramics have low friction coefficients and high wear and abrasion resistance, being particularly useful for biomedical implants (such as femoral heads in hip prostheses) and industrial bearings. Zirconia coatings have been demonstrated to decrease the wear debris of orthopedic implants, improving performance and minimizing postsurgical complications. Despite these favorable characteristics, issues exist with optimizing the processing and sintering conditions of zirconia-based composites. Achieving similar densification typically requires higher sintering temperatures (~1400°C–1600°C), which may promote grain growth and lead to decreased mechanical strength. Advanced sintering techniques, namely spark plasma sintering (SPS) and microwave sintering, explore the potential of lowering processing temperatures while yielding fine-grained structures and improved mechanical properties.

The inclusion of t-ZrO_2 NPs into enamel and ceramic composites can enhance mechanical properties such as hardness, fracture toughness, and wear resistance (Nikkerdar et al., 2024). Studies have reported that the incorporation of nano-t-ZrO_2

into enamel coatings could decrease porosity and increase crack propagation resistance and tensile strength. Adding nano-t-ZrO2 up to 5 wt.% yields maximal strength, whereas increasing beyond that might be less effective due to agglomeration.

2.4.4 Structural Properties of Zirconia Nanoparticles

Zirconium dioxide NPs have excellent structural properties, which make them useful in different applications. There are three well-defined phases of the crystal structure of ZrO_2. The monoclinic phase is stable below 1100°C, whereas in higher temperatures, tetragonal and cubic phases are present. The structural features of ZrO_2 NPs show an important dependence on the synthesis method and thermal processing conditions. NPs were synthesized by treating zirconium (IV) acetate hydroxide thermally in the presence of PVP as a capping agent. XRD confirmed characteristic peaks corresponding to this monoclinic crystalline phase (Figure 2.2). With an increase in the calcination temperature from 600°C to 900°C, diffraction peaks became higher and sharper, revealing better crystallinity. Crystallite size, as determined from Scherrer's equation, increased with the higher temperature.

Corresponding TEM images confirmed these observations and showed the uniformity of the final morphology and gradual growth in particle size with an increase in calcination temperature (Huang et al., 2016). As such, PVP effectively prevents the agglomeration of particles, resulting in a good range of particle size distribution. Various techniques confirmed the presence of zirconium and oxygen, with only a small amount of carbon residue from PVP decomposition. The weight ratio was around 69:26 for zirconium and oxygen, as suggested by the EDX results, confirming that the synthesized NPs were pure within the experimental error. The residual carbon content observed at lower calcination temperatures may be attributed to the incomplete decomposition of organic compounds, which were progressively removed at higher temperatures. In addition, FTIR analysis confirmed the successful elimination of organic compounds after calcination. Thermal treatment efficacy was also corroborated by the disappearance of absorption peaks ascribed to acetate groups and PVP. The calcined samples showed mainly Zr-O stretching vibrations, typical of ZrO phosphors, in the spectra. Structural changes affected the stability, reactivity, and functional properties of the material synthesized.

2.4.5 Zirconia Nanoparticle-Coated Glass Fibers and Their Structural Properties

Zirconia NPs have been widely used as an additive to improve glass fiber mechanical and chemical properties. If any of these surface modifications are applied to the coating, E-glass fibers become stronger and resist deformation and breakage. The newly developed zirconia NP coating can be applied to glass fibers to improve their alkali resistance and durability, thereby enabling them to be used in harsh environments. Through inspection of SEM images, it was found that the increase in the concentration of zirconia NPs throughout the 5%–20% range led to a more homogenous coating. In addition, lower concentrations of suspensions produced unevenly dispersed NPs across the fiber substrate, while higher concentrations, such as 15% and 20%, formed a continuous layer covering the fiber surface (Samyn et al., 2018; Joshi and Bhattacharyya, 2011; Haase et al., 2015). Both elemental analyses confirmed zirconium and oxygen, verifying successful loading of NPs on glass fibers. The presence of Zr-O-Si bonds, which are key to adhesion and chemical stability, was further strengthened by FTIR analysis. Figure 2.6a describes the analysis of epoxy-based double-layer nanocomposite coatings. It presents their self-healing ability over 7 days using SEM imaging and the formation of a protective passive layer on scratched areas using EDX analysis (Figure 2.6b) (Habib et al., 2021).

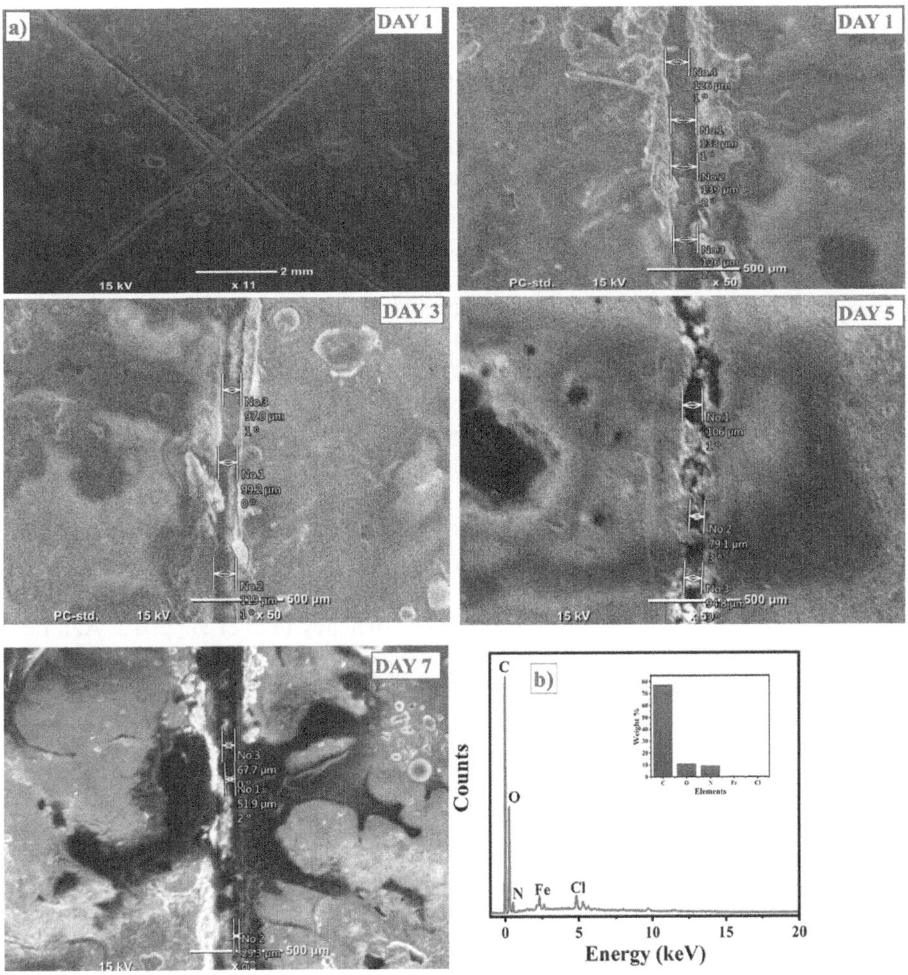

FIGURE 2.6 (a) SEM micrographs depicting the self-repairing capability of scratched epoxy-based double-layer nanocomposite coatings over a span of 7 days and (b) EDX analysis showing the formation of a passive layer on the damaged area of the synthesized epoxy-based double-layer nanocomposite coatings (Habib et al. 2021).

The TEM image shows the interaction of zirconia NPs with the silica structure of E-glass fibers. NPs also inhibited the formation of microcracks, which are one of the most common causes of mechanical failure. These free-flowing solutions of 15% and 20% zirconia were therefore used for uniform coating to reduce surface roughness for better performance under stress conditions. The crystalline phase of zirconia NPs was confirmed by XRD, as evidenced by the resulting peak intensities attributed to tetragonal and monoclinic phases of zirconia, which are necessary for structural stabilization. This structural support was evidenced by enhanced resistance against surface defects, as shown and characterized using AFM, which showed a reduction in the surface roughness of fibers compared with uncoated fibers (Liu et al., 2019; Siddiqui et al., 2010).

The effect of zirconia NP coating on the performance of E-glass fibers was studied using the tensile strength test. The tests showed that the coated fibers were stronger than bare fibers. The improvement in tensile strength with respect to untreated fibers was found to be 10.2%, 12.2%, 14.6%, and 17.4% for 5%, 10%, 15%, and 20% zirconia-solution-coated fibers, respectively. This improvement was ascribed to the nanocapsules' filling of microcracks and reinforcement of fiber integrity. The coated fibers showed a lower probability of failure according to the Weibull-based statistical analysis of tensile strength (Xu et al., 2022; Cao et al., 2022). Values of Weibull modulus higher than 1 indicated a more uniform stress distribution and lower sensitivity to defects for zirconia-coated fibers compared with uncoated ones. Fractographic analysis supported this trend, wherein SEM images of fractured fiber surfaces showed a reduction in the number of crack-initiating sites in the coated fibers. Mechanical stability results showed that 15% and 20% zirconia-solution-coated fibers had the best mechanical stability owing to denser and more uniformly coated fibers. The improved stress distribution across the fiber surface led to the increased availability of fracture toughness, confirming that high-magnification imaging shows an alteration in the failure modes.

The coated fibers also exhibited improved adhesion characteristics when used in composite materials, in addition to the aforementioned tensile strength improvements. The fiber–matrix interfacial adhesion was enhanced in the presence of zirconia NPs, resulting in the improvement of stress transfer efficiency (Wu et al., 2022; Azizi and Eslami-Farsani, 2021). Zirconia-coated glass fibers can be utilized as reinforcement in comparative applications in high-temperature industries, including aerospace, construction, and various automotive sectors, due to their high thermal tolerance. Moreover, flexural and impact resistance evaluations confirmed the versatile mechanical reinforcement capabilities of the zirconia-coated fibers beyond tensile characteristics, with superior energy absorption capacity in the coated fiber bundles.

2.5 APPLICATIONS OF ZR PARTICLES IN VARIOUS SECTORS

2.5.1 Importance of Zr Particles in Construction: Zirconia-Modified Cement Composites

The mechanical properties of zirconia-containing cement blends were examined by measuring compressive strength, evaluating flexural strength, and studying durability. The load-bearing capacity and the resistance to external stresses improved significantly with the incorporation of zirconia NPs. Compressive strength tests showed that they manifested a 12% increase in strength compared with that of control samples at 0.5% of ZrO_2 in cement pastes. The paste with 1% of ZrO_2 was found to be helpful in enhancing the strength by ~17%. The increase in strength is attributable to the ability of the NPs to fill microvoids among hydration products and to provide better interfacial adhesion. This was consistent with the findings of mechanical testing, which suggested better resistance to bending stress and demonstrated the presence of more crystalline bonds formed by the addition of zirconia, reinforcing the strength of the material structure. This mechanical performance was much better than all current widely used materials, showing their potential in heavy-load applications such as bridges, tall buildings, and factory floors. Zirconia-modified cementitious composites were gently subjected to alkaline solutions, freeze–thaw cycles, and high-temperature environments to mimic longevity tests. Samples exposed to alkaline conditions retained their strength, and similarly, samples subjected to a number of freeze–thaw cycles showed no major surface deterioration. The results indicate that the effectiveness of zirconia NPs is crucial for prolonging the service life of cement-based materials.

2.5.2 Brushing and Antimicrobial Activity of Zirconia-Based Cement Blends

Zirconia-doped cementitious materials exhibit superior mechanical and structural performance. In addition, they demonstrate unique self-cleaning and antimicrobial features. This is especially advantageous in construction applications that experience microbial contamination or demand maintenance-free durability. These materials were found to exhibit self-cleaning properties in UV-activated photocatalysis and microbial inhibition testing. The photocatalytic properties of zirconia NPs were activated by UV light, leading to the degradation of organic pollutants. This helps in sustainable urban infrastructure since it keeps buildings and roadways cleaner for longer time periods. Antimicrobial tests on composites using the agar diffusion method showed prominent resistance against microorganisms with 1% ZrO_2 composites (Nisha et al., 2025). Inhibition of *Bacillus subtilis* and *Klebsiella pneumoniae* (bacteria) and *Aspergillus niger* and *Aspergillus fumigatus* (fungi) was evident in the presence of zirconia NPs. The production of reactive oxygen species in response to UV radiation damages the microbial membrane and DNA structure.

Furthermore, the hydrophobic characteristic of zirconia allows for minimal water absorption, preventing microbial binding and biofilm formation. This property makes zirconia-modified cementitious materials suitable for use in settings that demand both hygiene and durability, such as hospitals and food-manufacturing plants. The addition of zirconia NPs to cementitious mixes causes considerable improvements in structural integrity, mechanical performance, durability, thermal stability, radiation protection, and self-cleaning. Composites with 0.5% to 1% ZrO_2 were found to yield the greatest improvements, resulting in outstanding properties that are ideal for construction applications. This is indicative of the potential for future innovations in construction technology due to the ability of zirconia-enhanced materials to withstand harsh environmental conditions, from extreme temperatures to microbial exposure. More research should be carried out on optimizing the dispersion of these NPs and hybrid formulations for achieving further performance gains. This work represents a major stride toward designing future cement-based materials that may provide the combination of permissible strength, sustainability, and multifunctionality with enormous benefits for the construction industry and beyond. The innovative use of zirconia NPs in cementitious composites has led to significant improvements in their properties, which make them suitable for construction applications. A large amount of research has been conducted on the application of industrial by-products (e.g., silica fume (SF) and marble dust (MD)) in cement matrices containing zirconia NPs. Besides improving the physical properties of the blends, this method adds to sustainable development through the recycling of industrial waste. To verify the formation of improved hydration products, a comprehensive structural analysis was performed using various techniques. SEM showed that the addition of zirconia NPs led to a denser microstructure with less porosity. The uniform dispersion of zirconia in the matrix facilitated stronger interfacial bonding. This structural integrity progressively improved with the inclusion of zirconia NPs at different concentrations (0.25%–1%). The thermal properties of these composites were investigated using TGA, which revealed decreased mass loss at higher temperatures. This is attributable to zirconia being able to catalyze hydration reactions and make the cement matrix more durable.

2.5.3 Applications of Zirconia-Based Nanomaterials in the Biomedical Field

Zirconia-based nanomaterials have received great attention as a new kind of biomaterial with excellent mechanical properties, biocompatibility, and antibacterial characteristics. These properties make them suitable for applications such as dental and orthopedic implants, drug delivery systems, and antimicrobial coatings. Zirconia NPs are widely used mostly in dental applications. Zirconia-based implants are much stronger and more biocompatible (they will not get damaged or wear out and will not affect surrounding tissues like titanium) than their titanium counterparts. Numerous studies indicate that the high fracture toughness of zirconia provides it with long-term stability in the oral environment, which contributes to a reduced risk of implant failure. Hydrothermal treatment, laser texturing, and other surface modifications can improve the osseointegration of zirconia implants and increase bone–implant contact area.

Other orthopedic uses of zirconia NPs include their application in bone scaffolds, load-bearing implants, and joint replacements. They are extremely hard and wear-resistant, which makes them suitable materials for hip and knee replacements. Moreover, hydroxyapatite- or bioactive glass-containing zirconia-based composites promote osteointegration and bone regeneration. Doped zirconia materials like YSZ have improved mechanical properties and *in vivo* inflammatory responses.

Drug delivery systems are another promising biomedical application of zirconia NPs. NPs are helpful in targeted drug delivery due to their ability to encapsulate and release therapeutic agents in a controlled manner. Surface modification by grafting polymers or conjugating ligands onto NPs results in functionalized zirconia NPs with increased bioavailability and decreased cytotoxicity. Recent progress in nanomedicine has proven that using zirconia NPs in chemotherapy drug delivery reduces systemic toxicity and improves treatment effectiveness.

Nanomaterials based on zirconia also show antimicrobial properties and can be utilized in coatings of medical devices and implants. An example of this is the strong antibacterial activity exhibited by silver-doped zirconia NPs against a range of pathogens, thus averting infections related to implanted medical devices. Various studies have proved that a combined approach with zirconia coatings can not only act as a superficial barrier that can reduce bacterial contamination, but also improve wound healing in biosensors.

2.5.4 Zirconia Nanoparticles in Catalysis and Energy Storage

Considering these great properties, zirconia NPs have attracted much attention in the energy storage and catalysis fields. They have widespread applications in fuel cells, batteries, and environmental catalytic processes. Zirconia NPs serve as a solid electrolyte in solid oxide fuel cells (SOFCs), which are one of the main types of energy storage applications. YSZ is often used as an electrolyte because of its high ionic oxygen conductivity and thermal stability (Mahato et al., 2012). Among the introduced dopants that are being investigated for improving the ionic conductivity of YSZ are scandium-423 and cerium. Enhanced electrochemical performance is achieved along with reduced operating temperatures of SOFCs. In Li-ion batteries, zirconia NPs are also used as coatings for electrodes and in solid-state electrolytes to improve charge retention and cycle stability. By limiting side reactions, zirconia-coated cathodes help prevent capacity loss and increase battery lifetime. In addition, zirconia has high thermal stability, which makes it an ideal component for enhancing the safety of Li-ion batteries by mitigating the risk of thermal runaway and electrode degradation.

Nanomaterials based on zirconia also find wide applications in catalysis. Due to their high surface area, enhanced Lewis acidity, and oxygen storage capability, these materials are well suited for various catalytic reactions such as hydrogen generation, carbon dioxide conversion, and degradation of environmental pollutants. Zirconia-supported metal catalysts are efficient in hydrocarbon reforming, catalytic conversions of automotive exhaust, and conversion of biomass to biofuels. Another application is the use of zirconia-supported palladium and platinum catalysts in automobile catalytic converters, which reduces the emissions of dangerous gases, including nitrogen oxides and carbon monoxide. Zirconia NPs have shown great promise as photocatalysts for air pollution remediation and water purification. Their potential to act as a semiconductor material demonstrates high UV light containment, which promotes the degradation of organic pollutants, thus maintaining the efficiency of wastewater treatment plants. Furthermore, with the combination of titanium dioxide (TiO_2), zirconia-based composite photocatalysts exhibit enhanced photocatalytic activities in dye degradation and environmental remediation (Chattopadhyay, 2017).

Future developments of nano zirconia focussed on specificity (as shown in Figure 2.2) lead to advanced surface engineering and the adoption of newer nano-structuring techniques. These approaches help to maintain the integrity of the structures by enhancing their suitability for energy storage and catalyst fabrication. The goal of this research is to provide innovations for better energy efficiency and sustainability, as well as addressing the environmental impact through better ionic conductivity and catalytic activities of zirconia-based hybrid materials (Hossain et al., 2023).

2.5.5 Application of Zirconia Nanoparticles in Dental Treatments

Zirconium oxide has good dispersion properties, low aggregation probability, and organic polymer biocompatibility, which increases hardness and toughness significantly. ZrO_2 is used as a filler in dental nanocomposites to enhance the mechanical properties, radiocapacity, and aesthetics of materials (Ali Sabri et al., 2021). Adding zirconium oxide NPs greatly increases hardness values in denture bases. Zirconium NPs improve the physical properties of denture bases during the establishing phase. However, this process carries some disadvantages and risk factors, so the phenomenon is highly reliant on future research (Hossain et al., 2023). The results show a high potential for the use of zirconia composites in implants with a bioactive zirconia composite layer, considering their good mechanical properties and increasingly intense implant–host tissue interaction. Recent research on zirconia NPs is summarized in Table 2.1.

TABLE 2.1
Zirconia Nanotexturized Surfaces (Hossain et al. 2023)

Treatment Method	Type	Analysis Approach	Key Findings
Hydrothermal surface modification	Subtractive	Performed physicochemical evaluations	Enhances the process of surface nanotexturization, improving material performance
Anodization	Subtractive	Studied molecular expression, morphology, biocompatibility, and surface roughness *in vitro*	Results indicate enhanced bone mineralization properties compared with untreated surfaces
Self-assembled nanostructures	Subtractive	Examined morphology, roughness, and biocompatibility *in vitro*	Leads to a substantial increase in bone mineralization on nanoscale surfaces
Nd:YAG laser treatment with silver and gold particle deposition	Combined (subtractive and additive)	Assessed morphology, atomic composition, surface functionalization, and adhesion of particles *in vitro*	Functionalization with metallic particles provides strong particle adhesion and improved surface characteristics
Sol-gel-derived titanium dioxide coating	Additive	Evaluated biocompatibility *in vitro*	Promotes effective blood coagulation and biocompatibility

(Continued)

TABLE 2.1 (*Continued*)
Zirconia Nanotexturized Surfaces (Hossain et al. 2023)

Treatment Method	Type	Analysis Approach	Key Findings
Calcium phosphate nanocoating	Additive	Conducted roughness assessments *in vivo* and studied bone–implant interactions	Nanocoatings exhibit comparable results to microsurface modifications but show superior osseointegration
Selective infiltration etching	Subtractive	Investigated surface roughness and bone–implant contact *in vivo*	Demonstrates a significant increase in bone–implant interface connection
Femtosecond laser modification	Subtractive	Analyzed surface morphology, roughness, and biological interactions *in vivo*	Cell attachment and proliferation are significantly influenced by nanotextured surfaces
Solid-state laser processing	Subtractive	Examined bone molecular expression, surface morphology, and roughness in both *in vitro* and *in vivo* conditions	This surface modification technique enhances osseointegration and improves mechanical stability

2.5.5.1 Advantages of Nanoparticles in the Field of Dental Implants

The application of NPs for enhanced osteointegration of dental implants has been clinically appreciated recently. It provides substantial benefits in terms of reducing the surface-area-to-volume ratio. Then, it reduces it until it becomes steady. Thus, these nanomaterials provide a large external surface area by retaining the volume up to a certain extent. NPs generally have the highest antibacterial potential. They are highly temperature resistant and can guarantee tough surface coating. Some particles have physical and chemical similarities with tissues like dentin and enamel such as calcium phosphate. It is applied in oral diseases, dental prosthetics, or implants. Osteoclasts are activated by NPs used in peri-implant tissues. This contributes to the strengthening of tooth tissues after concussion. It restores the force of dental displacements. NPs can provide skin that is practically similar to natural teeth. They are also used in dental fillings like root canals, providing a large surface area with antifungal properties. They also improve the mechanical properties and bonding strength through toughness.

2.5.6 Resistance to Alkali and Long-Term Durability

To test the long-term durability of NPs, exposure times were changed, ranging from 100 to 500 hours. Fibers coated with zirconia solutions of 15% and 20% did not undergo surface degradation after long standing, while almost all fibers were affected by hydroxyl ion degradation. Dissolution of the silica network, leading to the softening of fiber structure, upon exposure to alkaline solutions was identified as the degradation mechanism. Zirconia NP coating blocked the penetration of hydroxyl ions (Rawtani and Agrawal, 2012; Wang et al., 2023). Spectroscopic characterization through EDS indicated that the silicon content of uncoated fibers decreased upon treatment with alkali, while the compositions of the zirconia-coated fibers remained constant, providing additional evidence for the protective effect of the coating. In addition, an even distribution of zirconia NPs inhibited the formation of stress concentration zones that are typical initiating sites for chemical degradation.

In uncoated fibers, mechanical strength decreased rapidly after 500 hours of alkali corrosion due to the loss of tensile strength, while zirconia-coated fibers maintained their mechanical properties in the alkali environment (McKenzie, 2023; Saba et al., 2024). Tensile tests after the alkali aging procedure showed that fibers coated with 15% and 20% of the solid maintained higher than 95% of mechanical strength. This emphasizes the potential of zirconia NPs in establishing sustainable alkaline defense mechanisms. Moreover, using the X-ray photoelectron spectroscopy (XPS) technique, the stability of Zr-O-Si bonds was confirmed, ensuring the firm adherence of the coating to the fiber surface in harsh chemical environments. In particular, in areas such as marine faculties, sewage, and steel impregnation, zirconia NPs can provide a chemically strong encapsulating layer due to the formation of stable zirconia. Other thermal stability tests showed that high-temperature zirconia-coated fibers maintained high resistance to high-temperature environments, making them more suitable for extreme industrial applications.

Results showed that the usage of zirconia NP coatings on E-glass fibers can significantly improve mechanical strength, structural integrity, and alkali resistance (Ahmed, 2023; Ahmed et al., 2023; Aziz et al., 2025). The findings showed that a concentration of 15%–20% of zirconia led to the best coating in terms of functionality and mechanical characteristics. These results indicate that zirconia-coated fibers are a good candidate for composites under harsh conditions and sector utilization. The synergistic benefits of superior tensile strength, better flexural properties, long-term chemical stability, and superior high-temperature resistance make zirconia NP coatings a great candidate for next-generation glass fiber composites. Future work may progress toward better coating techniques that improve adhesion characteristics even further or hybrid NP coatings with added functions, such as antimicrobial resistance and self-healing capabilities.

2.6 MECHANISMS OF CRACK PROPAGATION AND TOUGHENING

The addition of zirconia NPs increases the toughness of the material via stress-induced phase transformation. The inclusion of zirconia NPs in enamel inhibits the development of stress concentration regions and results in lower material brittleness (2022). Zirconia NPs contributed to crack propagation through toughening by the transformation of the tetragonal M-phase. The effectiveness of zirconia NPs as a strengthening phase greatly relies on their particle size. The smaller the particles (<10 nm), the more stable the tetragonal phase, leading to improved mechanical properties of the material. Nevertheless, higher concentrations of zirconia NP aggregates subsequently diminish their mechanical properties. Generally, the presence of 0.5 wt.% zirconia NPs most effectively enhances the mechanical properties of the composite, and an increase in the population of zirconia NPs would lead to agglomeration, thus weakening the enamel structure. Mechanochemical synthesis of NPs and nanocrystalline materials is a well-characterized method for zirconia NPs and is often cited as an efficient, low-cost, and versatile process (Faure et al., 2013). High-energy milling of precursor materials results in chemical reactions occurring at RT. It has been reported that no real-time chemical reactions take place during mechanochemical milling, with crystalline CaO and amorphous $ZrCl_4$ as products. However, when the milled powder is subjected to heat treatment, hydrolysis occurs, forming zirconium hydroxide. This transformation is important as it drives particle crystallization into zirconia NPs. Thermal treatment of $Zr(OH)_4$ results in its decomposition and yields pure ZrO_2. Figure 2.7 depicts the mechanical behavior of a metal–ceramic composite for interface debonding, plastic deformation, and crack bridging. These mechanisms influence the material's strength and fracture resistance (Smirnov et al., 2017).

Nyquist plots present the electrochemical behavior of ZrO_2 in $NaNO_3$ and KCl electrolytes at different frequencies. Figure 2.8a shows the response at a high frequency by depicting charge transfer resistance. Figure 2.8b shows the low-frequency response, resulting in diffusion-related processes as shown in Figure 2.8. Zirconia exists as m-ZrO_2 at low temperatures. It has two polymorphs: above 1170°C, it is tetragonal (t-ZrO_2), and above 2370°C, it is cubic (c-ZrO_2); however, for lower temperatures, nanosized zirconia can hold the tetragonal phase due to the surface energy stabilization effect. Tetragonal zirconia is known to undergo phase transformation (monoclinic) upon crystallite size reaching a critical value (ca. 46 nm), which significantly affects its mechanical and thermal properties. Various mechanisms, such as residual stresses, surface energy effects, and the presence of dopants (e.g., yttrium (Y) and calcium (Ca)), have been proposed so far to contribute to the stabilization of tetragonal zirconia. Phase retention is sensitive to the processing of powders, requiring high-energy milling and controlled thermal treatments to be kept within desired boundary conditions to ensure appropriate mechanical and chemical properties for specific applications.

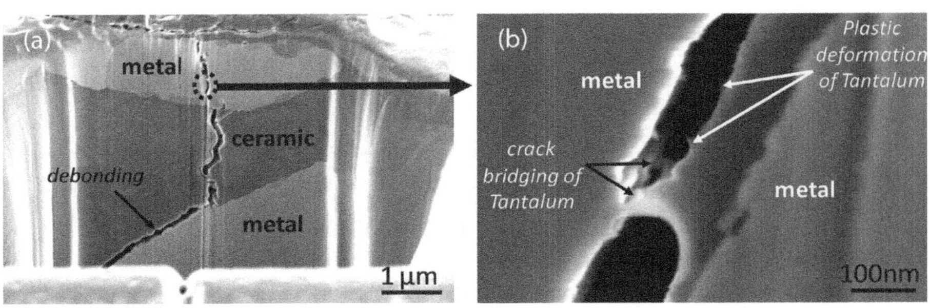

FIGURE 2.7 (a) Debonding at the interface between metal particles and ceramic and (b) plastic deformation and ligament crack bridging (Smirnov et al. 2017).

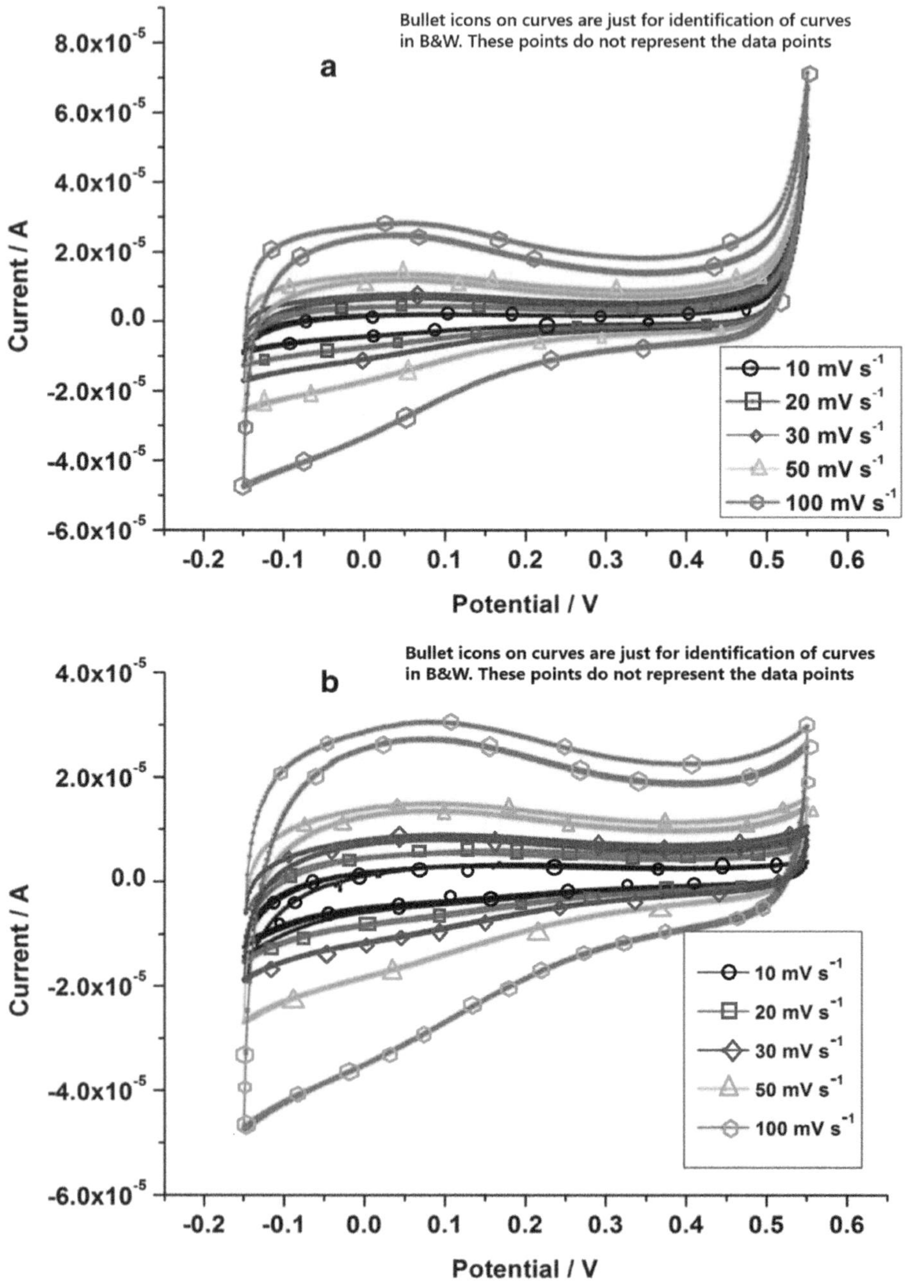

FIGURE 2.8 Nyquist plots of ZrO in Na and KCl electrolytes at (a) high and (b) low frequencies (Wang et al. 2016).

2.7 TECHNIQUES TO MODIFY THE SURFACE FOR IMPROVED BIOCOMPATIBILITY

Biocompatibility of zirconia NPs is a significant determinant in their use in biomedical implants and other applications, such as dental and orthopedic applications. Different surface modification techniques have been utilized to improve their bioactivity, osseointegration, and cellular response toward zirconia, such as laser texturing, anodization, nanocoating, and hydrothermal treatment. One common approach is using laser texturing, where a high-energy laser adds nanoscale surface roughness. Studies reveal that laser-modified zirconia surfaces enhance cellular adhesion, proliferation, and osteointegration. Higher surface roughness and surface energy promote protein adsorption and cell attachment, resulting in better osseointegration. Studies have shown that laser-textured zirconia provides better mechanical interlocking with bone tissue, reducing implant failure rates. Other promising methods include calcium phosphate nanocoatings. Such coatings imitate the natural structure of bone, which in turn creates a bioactive surface to promote the attachment and proliferation of bone cells. As reported by experimental studies, zirconia implants coated with calcium phosphate coatings can achieve significantly higher bone–implant contact compared with uncoated surfaces. An area where this technique is especially useful is in dental implantology since it allows for immediate and enhanced osseointegration, which is important for the long-term stability of a dental prosthesis.

The electrochemical mechanism modifies the properties of zirconia, resulting in the formation of a porous nanotextured layer that improves the attachment and mineralization of bone cells. Anodized zirconia surfaces have also been shown to have improved roughness, wettability, and bioactivity, making them ideal for biomedical applications (Dallali Isfahani et al., 2012). Furthermore, hydrothermal treatment has been applied to enhance the biocompatibility of zirconia surfaces. In this process, zirconia is subjected to steam or hydrothermal treatment at high temperatures, resulting in a nanoscale-structured surface layer. The hydrothermally treated zirconia demonstrated significantly increased cell adhesion, decreased bacterial colonization, and increased resistance to wear and corrosion. The synergy of such surface modification methods will revolutionize the usage of zirconia NPs in biomedical implants. Next-generation zirconia-based materials, with improved biocompatibility, mechanical strength, and longevity, can be achieved by controlling surface properties.

2.8 CHALLENGES AND FUTURE DIRECTIONS

Although nanostructured zirconia exhibits a wide range of promising applications, there are still challenges in enhancing its performance. One significant drawback is its phase instability and susceptibility to low-temperature degradation (LTD), especially in moist environments. This degradation mechanism, referred to as aging, induces surface roughening and microcracking, leading to failure of zirconia-based materials. The above concerns of LTD have led researchers to develop dopant strategies and surface treatments to prevent LTD and improve the long-term stability of zirconia ceramics. Another issue is the tendency of zirconia NPs to agglomerate during synthesis and processing. Energetic interparticle interactions result in the generation of macroscopic agglomerates, which would compromise the uniformity of dispersion and properties of the material. To enhance colloidal stability and dispersion properties, the use of surface functionalization techniques such as polymer coatings and surfactant-assisted synthesis is being investigated.

Further research directions of zirconia nanotechnology may involve the exploration of advanced multifunctional zirconia-based nanocomposites, which will provide the functionality of antimicrobial, as well as enhancing the self-healing and smart sensing characteristics. Rare-earth-doped zirconia composites may eventually find their applications in optical and electronic devices, phosphors, and transparent ceramics. Further advances in synthesis, processing, and characterization techniques will be necessary for zirconia NPs to gain ground in biomedicine, catalysis, and structural applications. Zirconia is a metal oxide, and ongoing research is focused on enhancing its performance and addressing limitations, which can lead to its applications in next-generation devices in healthcare, energy-harvesting, and environmental protection fields.

2.9 CONCLUSION

Zirconia NPs have exceptional potential as a strengthening phase in ceramic and enamel composites due to their unique mechanical, chemical, and thermal properties. In addition, their high fracture toughness, thermal stability, and biocompatibility make them suitable materials for applications across various fields, including construction, catalysis, energy storage, and biomedical sectors. The sol-gel synthesis method offers precise control over the particle size and phase stability of zirconia NPs and is suitable for producing high-purity NPs with consistent properties. Furthermore, the stabilization of the tetragonal phase at the nanoscale range significantly enhances mechanical strength and crack resistance, which is crucial for structural and dental applications.

This study also highlights the significance of surface modifications such as laser texturing and hydrothermal treatment. This improves the biocompatibility of NPs and promotes osseointegration in dental and orthopedic implants. Furthermore, zirconias role in environmental remediation, fuel cells, and radiation shielding underscores its versatility. Despite these advancements, challenges such as phase instability and NP agglomeration persist, necessitating further research into stabilizing dopants and optimized synthesis methods. Future studies should focus on multifunctional nanocomposites with self-healing, antimicrobial, and smart sensing properties.

REFERENCES

Abd-Elwahed, Mohamed Salah et al. (2020) “Effects of ZrO_2 nanoparticle content on microstructure and wear behavior of titanium matrix composite,” *Journal of Materials Research and Technology*, 9(4), pp. 8528–8534.

Ahmed, Md Kawsar (2023) “Mechanical and chemical properties of nanoparticle-coated E-glass fibers for composites applications,” (Doctoral dissertation).

Ahmed, Md Kawsar et al. (2023) “Improved mechanical properties of the E-glass fibres through TiO_2 nanoparticle coating,” *Materials Science and Technology*, 39(11), pp. 1304–1312.

Akhtar, Muhammad Saeed et al. (2024) "Recent advancements in catalysts for petroleum refining," *Catalysts*, 14(12), p. 841.

Aldeen, E.M. Sharaf et al. (2022) "Altered zirconium dioxide based photocatalyst for enhancement of organic pollutants degradation: A review," *Chemosphere*, 304, p. 135349.

Ali Sabri, Ban et al. (2021) "A review on enhancements of PMMA denture base material with different nano-fillers," *Cogent Engineering*, 8(1), p. 1875968.

Aziz, Abins et al. (2025) "Calcium-reinforced graphene oxide-biopolymer nanocomposite: A novel and sustainable sizing agent for enhancing E-glass fibers strength and adhesion in cement composites," *Composites Part A: Applied Science and Manufacturing*, 192, p. 108808.

Azizi, Hosna and Eslami-Farsani, Reza (2021) "Study of mechanical properties of basalt fibers/epoxy composites containing silane-modified nanozirconia," *Journal of Industrial Textiles*, 51(4), pp. 649–663.

Bamiduro, Faith Timilehin (2015) "Synthesis and characterisation of zinc oxide and sodium zirconate particles," (Doctoral dissertation).

Bashir, Mahwish et al. (2022) "Biodegradation of gelatin stabilized tetragonal zirconia synthesized by microwave assisted sol-gel method," *Journal of the Mechanical Behavior of Biomedical Materials*, 127, p. 105070.

Batool, Tanzeela et al. (2020) "Microwave assisted sol-gel synthesis of bioactive zirconia nanoparticles – Correlation of strength and structure," *Journal of the Mechanical Behavior of Biomedical Materials*, 112, p. 104012.

Becker, Jacob et al. (2008) "Critical size of crystalline ZrO_2 nanoparticles synthesized in near- and supercritical water and supercritical isopropyl alcohol," *ACS Nano*, 2(5), pp. 1058–1068.

Cao, Yanmei et al. (2022) "Finite element modelling of the single fibre composite fragmentation test with comparison to experiments," *Journal of Composite Materials*, 56(17), pp. 2765–2778.

Chattopadhyay, Shreyasi (2017) Titania and zirconia based nanostructured materials for energy and environmental applications. Ph.D. Thesis, Jadavpur University, India.

Chevalier, Jerome et al. (2009) "The tetragonal-monoclinic transformation in zirconia: Lessons learned and future trends," *Journal of the American Ceramic Society*, 92(9), pp. 1901–1920.

Dallali Isfahani, Taghi et al. (2012) "Mechanochemical synthesis of zirconia nanoparticles: Formation mechanism and phase transformation," *International Journal of Refractory Metals and Hard Materials*, 31, pp. 21–27.

Dias, Anderson and Ciminelli, Virginia S.T. (2003) "Electroceramic materials of tailored phase and morphology by hydrothermal technology," *Chemistry of Materials*, 15(6), pp. 1344–1352.

Faure, Bertrand et al. (2013) "Dispersion and surface functionalization of oxide nanoparticles for transparent photocatalytic and UV-protecting coatings and sunscreens," *Science and Technology of Advanced Materials*, 14(2), p. 023001.

Fiorati, Andrea et al. (2021). "Dispersions of zirconia nanoparticles close to the phase boundary of surfactant-free ternary mixtures," *Langmuir*, 37(14), 4072–4081. https://doi.org/10.1021/acs.langmuir.0c03401

Haase, Martin F. et al. (2015) "Continuous fabrication of hierarchical and asymmetric bijel microparticles, fibers, and membranes by solvent transfer-induced phase separation (STRIPS)," *Advanced Materials*, 27(44), pp. 7065–7071.

Habib, Sehrish et al. (2021) "Self-healing behavior of epoxy-based double-layer nanocomposite coatings modified with zirconia nanoparticles," *Materials & Design*, 207, p. 109839.

Hossain, Nayem et al. (2023) "Advances of plant and biomass extracted zirconium nanoparticles in dental implant application," *Heliyon*, 9(5), p. 15973. https://doi.org/10.1016/j.heliyon.2023.e15973

Huang, Sai et al. (2016) "Controlled synthesis and the effects of Gd^{3+} substitution, calcination, and particle size on photoluminescence of $(Y_{0.95-x}Gd_xTb0.05)_2O_3$ green phosphorspheres," *Chemical Engineering Journal*, 306, pp. 322–329.
Jithin, P.V. et al. (2025) "Structural and optical properties of Ce-stabilized tetragonal phase and intense blue emission of monoclinic phase in ZrO_2 nanoparticles," *Journal of Luminescence*, 277, p. 120933.
Joshi, Mangala, and Anita Bhattacharyya (2011) "Nanotechnology – A new route to high-performance functional textiles," *Textile Progress*, 43(3), pp. 155–233.
Kalpana, Krishnan and Rajathi, Kumar (2024) "Eco-friendly natural clay bentonite supported zirconium titanate (BT–$ZrTiO_4$): Synthesis, characterization and photocatalysis and self- cleaning activities," *Journal of Materials Science: Materials in Electronics*, 35(2), p. 187.
Keiteb, Aysar S. et al. (2016) "Structural and optical properties of zirconia nanoparticles by thermal treatment synthesis," *Journal of Nanomaterials*, 2016, pp. 1–6.
Koo, Joseph H. and Louis A. Pilato (2020) "Polymer nanocomposite ablatives–Part II," *International Journal of Energetic Materials and Chemical Propulsion*, 19(2), pp. 125–187.
Kumari, Nisha et al. (2022) "Zirconia-based nanomaterials: Recent developments in synthesis and applications," *Nanoscale Advances*, 4(20), pp. 4210–4236.
Li, Jinhao et al. (2024) "Recent advances in the large-scale production of photo/electrocatalysts for energy conversion and beyond," *Advanced Energy Materials*, 14(45), p. 2402441.
Liu, Jingqi et al. (2019) "Flexible printed monolithic-structured solid-state dye sensitized solar cells on woven glass fibre textile for wearable energy harvesting applications," *Scientific Reports*, 9(1), p. 1362.
Mahato, Neelima et al. (2012) "Doped zirconia and ceria-based electrolytes for solid oxide fuel cells: A review," *Nanomaterials and Energy*, 1(1), pp. 27–45.
Manivasakan, Palanisamy et al. (2011) "Synthesis of monoclinic and cubic ZrO_2 nanoparticles from zircon," *Journal of the American Ceramic Society*, 94(5), pp. 1410–1420.
Mastache, Jorge Edmundo Mastache et al. (2023). Study of the properties of zirconia nanoparticles/zinc oxide films composites: Comparation to thermally treated samples. *Results in Chemistry*, 6. https://doi.org/10.1016/j.rechem.2023.101230
McKenzie, Francesca L. (2023) *Designing Carbon Fibre-Reinforced Composites with Improved Structural Retention on Exposure to Heat/Fire*. University of Bolton, Greater Manchester, Bolton, UK.
Monteiro, Sergio N. et al. (2012) "Thermogravimetric behavior of natural fibers reinforced polymer composites-An overview," *Materials Science and Engineering:* A, 557, pp. 17–28.
Mostafa, Simon et al. (2009) "Catalytic decomposition of alcohols over size-selected Pt nanopar- ticles supported on ZrO_2: A study of activity, selectivity, and stability," *Applied Catalysis A: General*, 366(2), pp. 353–362.
Nadagouda, Mallikarjuna N. et al. (2011) "Microwave-assisted green synthesis of silver nanostructures," *Accounts of Chemical Research*, 44(7), pp. 469–478.
Nguyen, Thien Vuong et al. (2024) "Self-cleaning and antibacterial polyurethane coatings with ZrO_2- Ag hybrid nanoparticles," *Colloid and Polymer Science*, 302(2), pp. 225–236.
Nikkerdar, N. et al. (2024) "Recent progress in application of zirconium oxide in dentistry," *Journal of Medicinal and Pharmaceutical Chemistry Research*, 6(8), pp. 1042–1071.
Nirschl, Hermann and Guo, Xiaoai (2018) "Characterisation of structured and functionalised particles by small-angle X-ray scattering (SAXS)," *Chemical Engineering Research and Design*, 136, pp. 431–446.
Nisha, U. Mary et al. (2025) "Interfacial coupling effects of chitosan integrated $ZrO_2/Bi_2O_3/CeO_2$ quaternary composite for efficient wastewater treatment and antimicrobial activity," *Ionics,* 31(3), pp. 2739–2756.

Prasad Yadav, Thakur et al. (2012) "Mechanical milling: A top down approach for the synthesis of nanomaterials and nanocomposites," *Nanoscience and Nanotechnology*, 2(3), pp. 22–48.

Preethi, Jayaram et al. (2021) "Facile synthesis of Zr^{4+} incorporated chitosan/gelatin composite for the sequestration of Chromium(VI) and fluoride from water," *Chemosphere*, 262, p. 128317.

Puigdollers, Antonio Ruiz et al. (2016) "Structure and properties of zirconia nanoparticles from density functional theory calculations," *Journal of Physical Chemistry C*, 120(8), pp. 4392–4402.

Rao, Tentu Nageswara et al. (2019). Enhanced thermal properties of zirconia nanoparticles and chitosan-based intumescent flame retardant coatings. *Applied Sciences*, 9(17). https://doi.org/10.3390/app9173464

Rawtani, Deepak and Agrawal, Y.K. (2012) "Multifarious applications of halloysite nanotubes: A review," *Reviews on Advanced Materials Science*, 30(3), pp. 282–295. Available at: https://www.ipme.ru/e-journals/RAMS/no_33012/08_rawtani.pdf.

Rizzuti, Antonino et al. (2009) "Structural characterization of zirconia nanoparticles prepared by microwave-hydrothermal synthesis," *Journal of Dispersion Science and Technology*, 30(10), pp. 1511–1516.

Saba, Tasfia et al. (2024) "Precise surface engineering: Leveraging chemical vapor deposition for enhanced biocompatibility and durability in biomedical implants," *Heliyon*, 10(18), p. 37976.

Saenko, Ivan et al. (2018) "Experimental investigation of phase relations and thermodynamic properties in the ZrO_2–TiO_2 system," *Journal of the American Ceramic Society*, 101(1), pp. 386–399.

Samyn, Pieter et al. (2018) "Review: Nanoparticles and nanostructured materials in papermaking," *Journal of Materials Science*, 53(1), pp. 146–184.

dos Santos, Venina et al. (2014) "In-situ evaluation of particle size distribution of ZrO_2-nanoparticles obtained by sol–gel," *Powder Technology*, 267, pp. 392–397.

Sari, Erlin Purwita and Muzakky (2023) "The FTIR image comparison of zirconium sulfate synthesis products from the pathways of Na_2ZrO_3 and zirconium oxychloride compounds," *Journal of the Indian Chemical Society*, 100(7), p. 101029.

Sarraf, Hamid et al. (2015) "Direct probing of dispersion quality of ZrO_2 nanoparticles coated by polyelectrolyte at different concentrated suspensions," *Nanoscale Research Letters*, 10(1), pp. 1–14. https://doi.org/10.1186/s11671-015-1157-z

Shahmiri, Reza et al. (2018) "Optical properties of zirconia ceramics for esthetic dental restorations: A systematic review," *Journal of Prosthetic Dentistry*, 119(1), pp. 36–46.

Siddiqui, Naveed A et al. (2010) "Tensile strength of glass fibres with carbon nanotube-epoxy nanocomposite coating: Effects of CNT morphology and dispersion state," *Composites Part A: Applied Science and Manufacturing*, 41(4), pp. 539–548.

Smirnov, Anton et al. (2017) "Unprecedented simultaneous enhancement in damage tolerance and fatigue resistance of zirconia/Ta composites," *Scientific Reports*, 7(1). https://doi.org/10.1038/srep44922.

Song, Da et al. (2024) "Micro-structure change and crystal-structure modulated of oxygen carriers for chemical looping: Controlling local chemical environment of lattice oxygen," *Fuel*, 364, p. 131087.

Tahir, Muhammad Nawaz et al. (2007) "Facile synthesis and characterization of monocrystalline cubic ZrO_2 nanoparticles," *Solid State Sciences*, 9(12), pp. 1105–1109.

Tello, Ana C.M. et al. (2020) "Microwave-driven hexagonal-to-monoclinic transition in $BiPO_4$: An in-depth experimental investigation and first-principles study," *Inorganic Chemistry*, 59(11), pp. 7453–7468.

Wang, Annie et al. (2023) "Anticorrosive coating systems for marine propellers," *Progress in Organic Coatings*, 183, p. 107768.

Wang, Jigang et al. (2016) "Good biocompatibility and sintering properties of zirconia nanoparticles synthesized via vapor- phase hydrolysis," *Scientific Reports*, 6(1), p. 35020. https://doi.org/10.1038/srep35020

Wu, Qing et al. (2022) "Amine-caged ZrO_2@ GO multilayer core-shell hybrids in epoxy matrix for enhancing interfacial adhesion of carbon fiber composites," *Composites Part B: Engineering*, p. 110207.

Xu, Chunping et al. (2015) "Mechanochemical synthesis of advanced nanomaterials for catalytic applications," *Chemical Communications*, 51(31), pp. 6698–6713.

Xu, Ling-Yu et al. (2022) "Tailoring strain-hardening behavior of high-strength Engineered Cementitious Composites (ECC) using hybrid silica sand and artificial geopolymer aggregates," *Materials and Design*, 220, p. 110876.

Yuan, Quan et al. (2009) "Controlled synthesis and assembly of ceria-based nanomaterials," *Journal of Colloid and Interface Science*, 335(2), pp. 151–167.

Zhang, Feng et al. (2006) "Phases in ceria-zirconia binary oxide (1-x)$CeO_{2-x}ZrO_2$ nanoparticles: The effect of particle size," *Journal of the American Ceramic Society*, 89(3), pp. 1028–1036.

3 Synthesis of Zirconia-Based Materials

Moniya, Jashanpreet Singh, and
Mohamed Aboughaly

3.1 INTRODUCTION

Zirconium dioxide, or zirconia (ZrO_2), has great versatility as a ceramic material because of its high mechanical strength, chemical inertness, thermal stability, and ionic conduction capabilities [1]. This broad range of properties makes zirconia suitable for many fields such as medicine, engineering, and energy science. The synthesis of zirconia-based materials includes numerous chemical and physical techniques, all of which have distinct advantages related to the phase composition, particle size, morphology, and microstructure of the material [2]. Hydrothermal techniques, for example, can be used to treat zirconium oxyacetate aqueous solutions under pressure at high temperatures to yield size-controlled nanocrystalline zirconia powders [3]. ZrO_2 grows from the material in aqueous solutions at high temperatures and low pressures, which allows for the manufacture of well-crystallized particles. The size of the particles is controlled through the temperature applied in the process [4]. Often, to modify the properties of zirconia for certain applications, stabilizing additives such as yttrium oxide (Y_2O_3), magnesium oxide, and caladium oxide are added [5]. This is especially true for partially stabilized zirconia materials, which have started gaining increased attention of late due to higher toughness and strength compared with ordinary zirconia [6]. The properties of the colloidal suspension play a crucial role in the overall morphology and properties of the final material. To achieve unapportionable products and prevent agglomeration, suspensions should be stable and well disbursed during particle growth [7]. Zirconia nanoparticles (NPs) typically range from 1 to 100 nm in diameter. Even though the properties of NPs (chemical, mechanical, thermal, magnetic, and electrical) are outstanding, untangling their properties, fabrication, and assessment of nanostructured materials is still a challenge for scientists [8]. By adding different metal oxides of NPs to the produced material, product characteristics can be improved. Because of their diverse physical, chemical, and morphological features, metal oxide NPs find use in a wide range of industries [9]. Materials spanning from 10 to 100 nm are referred to as "nanomaterials", which emphasizes examining the features of these particles at the nanoscale. The unique features of NPs are proportional to their size, more so being microscopic [10]. Zirconium is classified as a transition metal and has the properties of titanium owing to its strong resistance to corrosion. Most zirconium metals come from minerals although it is possible to synthesize a zirconium metal or refine it using organic and inorganic

DOI: 10.1201/9781003519300-4

chlorine compounds [11]. Due to high tensile strength, hardness, and exceptional fracture toughness, zirconium NPs have found utility in a variety of applications. Five isotopes of zirconium occur naturally, of which 90Zr is the most abundant one, with a staggering incidence of 51.45%. Different approaches have been used to synthesize metal oxide and mixed metal oxide NPs [12]. Various approaches have been used to integrate many synthesis pathways. Catalytic metal complex with zirconium has catalytic applications [13]. Zirconium NPs synthesized on 2D or 3D graphene are used as catalysts for thin-walled carbon nanotubes (CNTs, tapered x-iodine growth).

Zirconium proves to be the most effective catalyst for producing carbon NPs, which boast exceptional purity and are suitable for high-quality CNTs [14]. The Kroll process was used successfully to synthesize zirconium NPs, which consists of treating zirconium tetrachloride with active metal magnesium at 800°C–900°C. Various methods for producing cubic, monoclinic, and mixed-phase zirconium NPs [15] are discussed in this chapter, including sol-gel, hydrothermal, laser ablation, wire explosion, chemical polymerization, hybrid transparent coating, and coprecipitation. Zirconia materials require further refinement of their mechanical strength optimization, manufacturing processes, and biocompatibility for biomedical uses and other areas. More focus is needed on the relationship between yttria content and zirconia's mechanical properties for applications in dental restorations. Flash sintering and 3D printing are shown to improve the microstructure optimization of zirconia, along with other aesthetic characteristics and durability in hostile environments. It is also necessary to develop hybrid materials with biocompatible components to ensure long-term safety. There are still some limitations such as low-temperature degradation (LTD), which keeps mechanical support unchanged, and ultraviolet exposure, which changes the colour and creates an issue for aesthetic durability [16]. It is harder to attain an optimal balance for clinical success in toughness, translucency, and strength because there is not an adequate composition. Moreover, inadequate long-term clinical studies pose the risk of releasing harmful NPs in the body, indicating that more research focus on zirconia materials is needed [17].

3.2 PRECURSORS AND RAW MATERIALS FOR ZIRCONIA SYNTHESIS

The selection and characteristics of precursor compounds are of utmost importance in the synthesis of zirconia materials as they influence the microstructure development, phase stability, and functional properties of the ceramic material [18]. Zirconium precursors are primary compounds used in the chemical processes aimed at synthesizing zirconia materials. They can be classified into inorganic salts, organometallic compounds, and polymeric precursors, each having specific advantages and disadvantages regarding process control, purity, and cost efficiency [19]. Some of the sources of zirconia are inorganic salts of zirconium such as zirconium chloride, zirconium nitrate (Zr_4), and zirconium sulphate (Zr_2), due to their availability and low cost. However, these inorganic precursors should be processed within specific limits of pH and temperature in order to avoid unwanted results and are often contaminated with undesirable ions that degrade the quality of the produced zirconia [20]. Zirconium alkoxides and zirconium acetylacetonates represent organometallic constituents of molecular zirconium compounds that offer a sophisticated route to zirconia synthesis due to their well-defined structures and defined reactivity, which allows for control over the kinetics and homogeneity of the reaction for the precise manipulation of the zirconia nanostructure [21]. Once zirconia is extracted, various precursors are used in different synthesis methods to regulate particle size, morphology, and crystallinity. Among them, zirconium alkoxides play a significant role in the sol-gel process, where they undergo hydrolysis in an alcoholic medium containing ammonia, forming a gel [22]. Because of the potential modification of the reactivity of these alkoxides with chelating reagents such as acetylacetone, acetic acid, and citric acid, controlling the properties of zirconia is more efficient. Another commonly used precursor is zirconium oxychloride ($ZrOCl_2$). This compound is usually derived by the chemical treatment of sodium zirconate with hydrochloric acid. In addition, zirconium nitrate and zirconyl hydroxide are frequently used in microwave hydrothermal synthesis [23]. These precursors are shown in Figure 3.1. Precursors help achieve the desired control over the shape and crystalline structure of the obtained zirconia. In terms of solution combustion synthesis, zirconium oxynitrate hydrate acts as an oxidizer when combusted with fuels such as urea or sucrose in

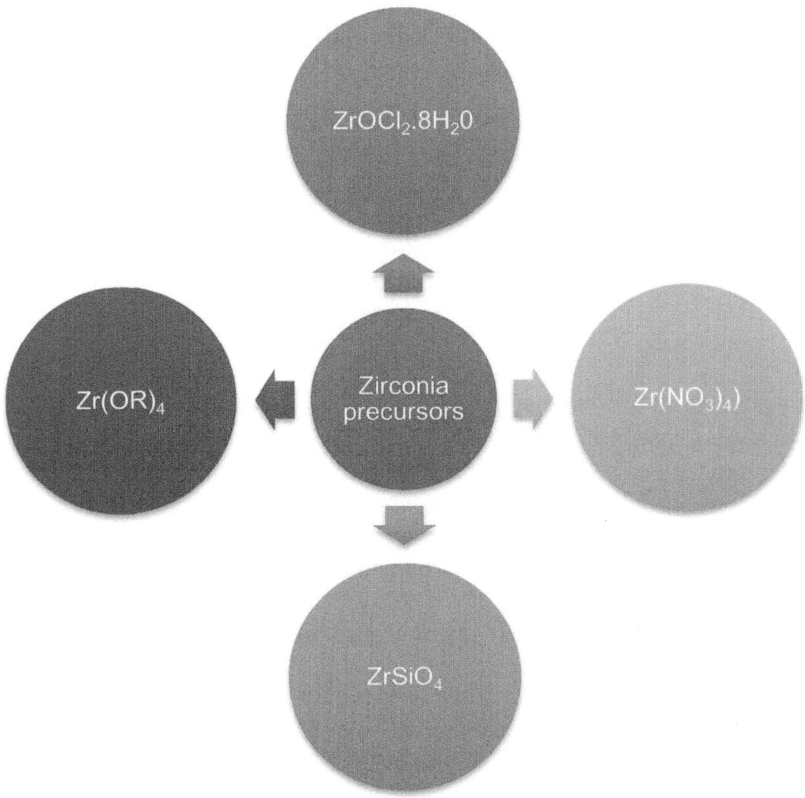

FIGURE 3.1 Various precursors for zirconia synthesis.

an exothermic reaction, yielding zirconia NPs [24]. In addition, various compounds of zirconium, such as zirconyl nitrate and zirconium alkoxides, are actively used in hydrothermal synthesis, where they come into contact with oxidizing or hydrolysing reagents in a sealed system to yield zirconia of desired structural and functional characteristics [25]. These compounds are essential in shaping the purity, stability, and operational efficiency of materials based on zirconia for numerous applications [26].

3.3 RAW MATERIALS FOR ZIRCONIA SYNTHESIS

The main source of zirconia is zircon ($ZrSiO_4$), which is a silicate mineral. It is found mostly in placer deposits and beach sands and extracted alongside rutile, ilmenite, and monazite [27]. Zircon possesses great chemical stability, requiring thermal or chemical decomposition to obtain zirconia. Australia contains the largest zircon reserves at 35% of global supply, followed by South Africa at 28% [28]. The USA and Mozambique both contribute 7% each. Zircon consists of roughly 67% zirconia (ZrO_2), 32.8% silica (SiO_2), and small quantities of hafnium oxide (HfO_2), iron oxides (Fe_2O_3), and thorium dioxide (ThO_2). The elements within zircon affect the extraction process and final purity of zirconia [29].

3.4 METHODS FOR DECOMPOSING ZIRCON TO EXTRACT ZIRCONIA

To extract zirconia from zircon, the Si–O bond in $ZrSiO_4$ should be broken. This can be done through chemical or thermal processes.

3.4.1 Chemical Methods

Using various reagents as fluxes facilitates the chemical decomposition of zircon to obtain zirconia. The most prominent method is sodium hydroxide fusion (NaOH) of zircon, where zircon is fused with caustic soda (sodium hydroxide) at a temperature of 600°C. This leads to the formation of sodium zirconate (Na_2ZrO_3) and sodium silicate (Na_2SiO_3) [30]. The latter is removable and soluble through washing, while the former undergoes HCl treatment to yield zirconium oxychloride ($ZrOCl_2$). This compound undergoes neutralization with ammonia, resulting in the precipitation of [Zr $(OH)_4$], which when heated to 900°C transforms into high-purity monoclinic zirconia (>99.5%) [31]. Another approach is sodium carbonate fusion (Na_2CO_3), in which zircon is fused with sodium carbonate at 1100°C. This process leads to the production of sodium zirconium silicate (Na_2ZrSiO_5), which transforms into sodium zirconate (Na_2ZrO_3) with excess sodium carbonate, subsequently transforming into zirconium oxychloride. The preparation of cubic zirconia with CaO and MgO involves fusion processes where zircon is melted with these oxides to form zirconia and calcium magnesium silicate {CaMgSiO4} at 1200°C. In addition, adjusting the CaO/MgO ratio can highly optimize the resulting cubic zirconia. Another efficient method of decomposition is carbon chlorination, where zircon is reacted with Cl_2 and C at 1100°C to form $ZrCl_4$ and $SiCl_4$. Then, $ZrCl_4$ is purified and hydrolysed to zirconium oxychloride ($ZrOCl_2$), which can subsequently be converted to zirconia. Each approach has its advantages when it comes to processing zircon to produce high-purity zirconia that is ideal for numerous industrial uses.

3.4.2 Thermal Plasma Dissociation

In thermal plasma decomposition, zircon is heated above 1700°C in a plasma reactor where it is decomposed into zirconia and silica. Rapid quenching is used to ensure that these oxides do not recombine, resulting in high-purity zirconia. Plasma dissociation is more efficient in regard to energy consumption compared with traditional methods and, unlike them, does not pose any environmental risks as there is no need for harmful chemicals, nor excessive waste production.

3.5 SOLVENT SELECTION AND ITS EFFECT ON ZIRCONIA SYNTHESIS

The factors influencing the choice of solvent during zirconia synthesis affect precursor dissolution, hydrolysis, and reaction kinetics, which determine the particle size, morphology, and crystallinity [7]. In wet chemical methods such as sol-gel synthesis, hydrothermal processing, and precipitation methods, different solvents are used to control these factors. In addition, water and alcohol–water mixtures as solvents in hydrothermal and coprecipitation methods aid in precursor dissolution and precipitation. Controlling pH determination alongside enhances shape-determining phase composition [26]. The remaining portion aids in the formation of homogeneous gels. Water-miscible solvents such as ethanol, isopropanol, and methanol are used in sol-gel synthesis because they facilitate the controlled hydrolysis–condensation reactions of zirconium alkoxides. Their degree of polarity and evaporation affects the pore structure and surface area of the resulting zirconia [32]. Ionic liquids and organic solvents such as acetone and dimethylformamide are also used in nonaqueous synthesis routes to reduce particle agglomeration and enhance surface energy, leading to low-bandgap dispersed nanoscale zirconia. Adjusting the remaining synthesis parameters, properties specific to advanced engineering applications can be obtained by controlling the solvent selected.

3.6 SYNTHESIS TECHNIQUES FOR ZIRCONIA-BASED MATERIAL

Zirconia synthesis involves various techniques such as sol-gel, hydrothermal, precipitation, and combustion methods, each influencing the material's phase, stability, and properties. Figure 3.2 illustrates different zirconia synthesis approaches.

3.6.1 Conventional Methods

Conventional synthesis methods for zirconia-based materials have been widely used due to their simplicity, scalability, and well-established protocols. These methods include solid-state reactions, sol-gel processes, and coprecipitation techniques [33].

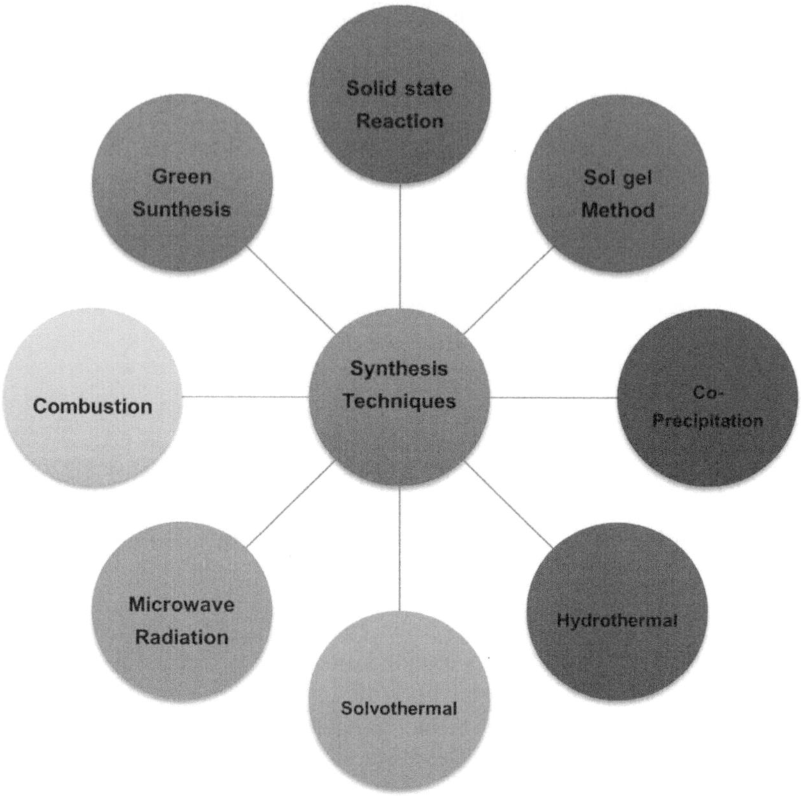

FIGURE 3.2 Synthesis techniques for zirconia-based materials.

3.6.1.1 Solid-State Reaction

The solid-state reaction method is a preferred technique for synthesizing zirconia-based materials, which uses high-temperature calcination as a chemical and phase-change-inducing method. A stoichiometric mixture of zirconium oxide and stabilizing oxides (e.g. Y_2O_3, MgO, CaO) is well ground and then heated to a temperature higher than 1200°C. This high-temperature treatment allows for ion diffusion to form a homogeneous crystalline phase, which often requires multistep calcination to promote complete reactions and phase stabilization, resulting in stepwise ion exchange. In this step, zirconia achieves its stabilizing phase. The reaction rate is dependent on temperature, particle size, and uniformity of the mixture. The calcination temperature and stabilizing oxides, which seek to preserve the high-temperature phases of the reaction, unsurprisingly dictate the transformation of monolithic zirconia into its cuboid or tetragonal shape. Its key advantages make it favourable for industrial-scale production due to ease of preparation and low cost; however, this methodology is limited due to high-temperature energy requirements and lack of control over particle shape and size.

3.6.1.2 Zirconia Production Using Sol-Gel Method

This method uses metal alkoxide precursors such as zirconium isopropoxide, bases, and acids to produce zirconia that is refined in both purity and composition. Forming a gel using hydrolysis and subsequent condensation leads to a 3D gel-like structure, which upon heating\drying at 400°C–600°C results in a crystalline form of zirconia [34]. There is a reduction in power consumption, along with better mastering of the material produced and smaller zirconia particles with greater available surface area. Expensive precursors and prolonged periods for structural forming and evaporating tend to be the challenges to overcome when using this method [35]. Regardless, other things make this optimal in synthesizing zirconia with a calibrated precision in terms of form and structure [36].

3.6.1.3 Coprecipitation Method

Coprecipitation is one of the most efficient methods for synthesizing zirconia NPs due to the ease with which high purity and phase composition are achieved. This method consists of the simultaneous precipitation of both zirconium and dopant ions from a solution. This is done through the addition of a precipitating agent, usually ammonia or urea, to an aqueous solution of $ZrOCl_2$ and dopant salts [37]. Other important parameters include pH, temperature, and precursor concentration, which greatly affect precipitation kinetics and final particle size. After precipitation, hydroxides are washed, dried, and calcined to produce crystalline zirconia. The main advantages of this method are its high-purity phase homogeneity, composition, particle size control, and scalability for industrial production [38]. However, conditions for consistent results should be carefully optimized; washing and drying steps after synthesis are tedious, and a lack of precision in controlling immediate post-synthesis conditions can lead to further complications. Regardless of these shortcomings, coprecipitation still proves to be one of the most productive methods for obtaining zirconia NPs.

3.6.2 Advanced Synthesis Approaches

Advanced synthesis techniques have been developed to overcome the limitations of conventional methods and achieve better control over the microstructure, morphology, and properties of zirconia-based materials. These methods include hydrothermal and solvothermal synthesis, microwave-assisted synthesis, combustion synthesis, and electrochemical and green synthesis approaches.

3.6.2.1 Hydrothermal and Solvothermal Synthesis

Hydrothermal and solvothermal methods are efficient in obtaining zirconia of specific morphology and particle size because they involve the use of a sealed autoclave with controlled temperature and pressure [39]. In the case of hydrothermal synthesis, zirconium precursors are put in water and subsequently heated to temperatures between 100°C and 250°C longitudinally under autogenous pressure (dissolution/recrystallization) [9]. This results in the formation of defined crystalline phases. In contrast, solvothermal synthesis uses organic solvents other than water, which offers even greater control over particle size and shape [32]. The choice of solvent is important because water-based synthesis produces spherical particles, which is not the case with organic solvents that facilitate anisotropic promotion of rod-like or plate-like structures [40]. Remarkably, these strategies provide a significantly high degree of crystallinity and control over morphology and produce NPs with a narrow size distribution. Their shortcomings, however, are the need for specialized autoclave equipment and prolonged reaction times relative to other synthesis methods. Regardless of these challenges, hydrothermal and solvothermal syntheses prove to be most useful in fabricating zirconia with properties tailored for advanced applications [41].

3.6.2.2 Synthesis by Means of Microwave Radiation

The simplicity, speed, and efficiency of microwave-assisted synthesis have drawn attention, especially with respect to the formation of zirconia nanomaterials [42]. This approach, like other techniques based on heating, suffers from the limitations of attaining high temperatures at relatively high pressures [43]. The ease with which microwaves can heat a material is referred to as dielectric heating, and this microwave heating mechanism is known as dielectric heating, where solvents and reagents used within the apparatus are exposed to microwaves and subsequently radiate it as heat. The driving energy for this process is dipolar polarization and ionic conduction [30]. Several types of zirconium salts such as nitrates, sulphates, and chlorides as well as chelating agents have been employed in this form of synthesis. Studies describe the utilization of microwave heating at the temperature of 80°C for the preparation of zirconia NPs along with its subsequent study of photoluminescence (PL) properties [44]. X-ray diffraction (XRD) analysis has shown that NPs have formed with mixed monoclinic and tetragonal phases, while microscopic analyses have indicated a particle size of less than 10 nm. The optical properties of the said NPs are outstanding, with a bandgap of 2.49 eV and a strong peak of PL emission at 414 nm [45]. Further advances in the field include the development of microwave-assisted synthesis using biological capping agents to enhance the properties of these materials. In one method, microwave synthesis was performed using honey as a capping agent for optical use in tetragonal zirconia NPs [46]. Golden honey NPs were used to cap zirconium oxides, and they had a bandgap value between 4.7 and 4.82 eV, while structural characterization showed that single-phase tetragonium oxide was formed in the high-purity level under optimized reaction conditions [47]. Moreover, a microwave-assisted sol-gel route was also utilized for the fabrication of honey-capped zirconia nanofibers. Such an approach greatly enhanced the stabilization of NPs, leading to reduced crystallite sizes and minimal harsh agglomeration. As a consequence, the stabilized tetragonal phase of zirconium oxide was retained, and its transition into the monoclinic phase was inhibited. It was observed in the study that the microwave power from 100 to 900 W has an impact in stabilizing zirconium oxide, and thus, tetragonal zirconium oxide was easily formed under a low power of about 100 W, which created crystallites of 26 nm [48]. The longstanding stability issue concerning zirconium NPs has also been dealt with, reporting that tetragonal zirconium NPs showed no significant alterations in structure over durations between 6 and 12 months under standard temperature conditions [49]. This point is crucial in relevance to practical uses that demand a high degree of stability of NPs over time. Another study showed that NPs of tetragonal zirconia could be synthesized through a microwave-assisted solvothermal method, resulting in an average crystallite size of 10 nm. Figure 3.3 depicts the synthesis of

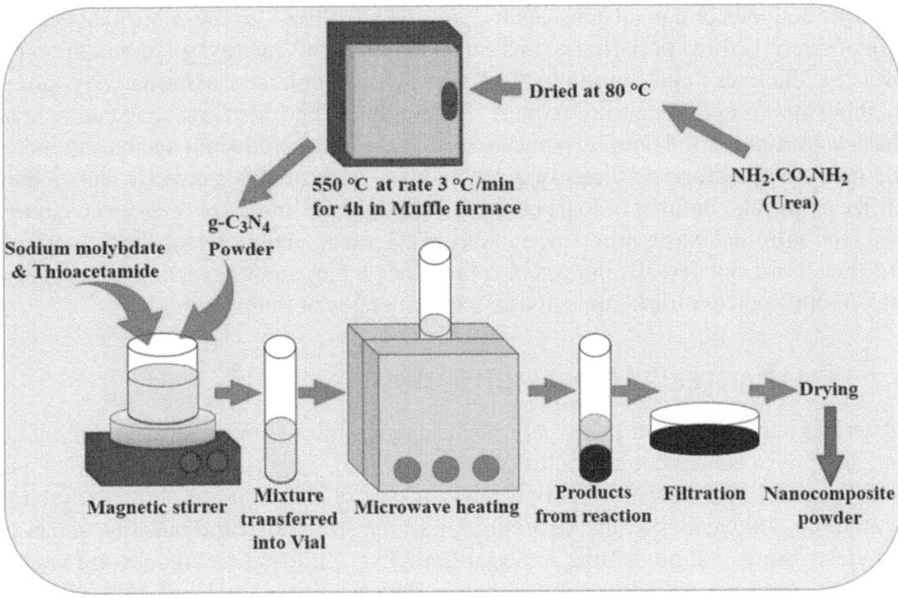

FIGURE 3.3 Synthesis of materials using microwave radiation [51].

NPs using microwave radiation. The optical analysis of the synthesized NPs revealed a bandgap of 3.67 eV, making them suitable for various photonic and electronic applications [50].

3.6.2.3 Combustion Synthesis

Combustion synthesis is a rapid and highly exothermic process used in the production of zirconia-based materials via the reaction of metal nitrates with a fuel like urea or glycine. This technique is preferred due to its simplicity and the fine powder with high surface area it generates [52]. This process commences by heating a mixture of metal nitrates and fuel, in which the fuel serves as a reducing agent, triggering an exothermic reaction that results in a temperature increase of 500°C–1000°C, aiding zirconia formation. The type and amount of fuel in relation to the oxidizer (metal nitrate) heavily influence the combustion temperature and phase composition of the end product [53]. This method is characterized by its rapid synthesis at low energy and attainment of zirconia powders with high specific surface area. Among its drawbacks are limited control over particle size and shape, risk of incomplete combustion leading to unreactive residues, or unwanted impurities. Regardless of these issues, combustion synthesis is successful and aligns with the rapid formation of zirconia-based materials.

3.6.2.4 Approaches to Electrochemical and Green Synthesis

Emphasis is being placed on the synthesis of zirconia-based materials using green and electrochemical approaches due to their sustainable characteristics, especially concerning the use of environmentally friendly reagents and energy-efficient processes. In electrochemical synthesis, zirconium salts are reduced electrochemically in either aqueous or nonaqueous electrolytes, which allows for the controlled deposition of zirconia films or particles on a substrate [54]. This method guarantees control over the thickness and morphology of the zirconia obtained. Alternatively, green methods make use of natural extracts, biotemplates, and nontoxic reagents to synthesize zirconia. For example, plant extracts that contain reducing agents can aid in the room temperature synthesis of zirconia NPs. Such methods are efficient as they utilize a minimal number of toxic chemicals. Eco-friendliness, low energy consumption, and minimal waste production of electrochemical and green synthesis methods are their most notable advantages. On the other hand, slower reaction times and a lack of optimal conditions are some of the challenges of these methods.

3.7 CHARACTERIZATION METHODS

Material characterization helps gain deep insights into the structural, morphological, and physicochemical attributes of materials [55]. This is useful in research and development in the industry. There is a wide range of techniques used to analyse the crystallographic features, surface features, particle size, molecular makeup, and even the electrostatic stability of the material [56]. The following techniques are widely used for characterization: XRD, scanning electron microscopy (SEM), transmission electron microscopy (TEM), dynamic light scattering (DLS), Fourier transform infrared (FTIR) spectroscopy, and zeta potential and surface charge analysis [57]. These techniques are critical in materials science, nanotechnology, pharmaceuticals, and biomaterials research, because they help in modifying and optimizing materials for specific applications [58].

3.7.1 X-Ray Diffraction

XRD is a powerful and nondestructive analytical technique used to determine the crystallographic structure, phase composition, and microstructural properties of solid materials. Figure 3.4 shows the XRD patterns of the as-synthesized and calcined zirconia, which is calcined at 723 K and 923 K, respectively. It is based on the constructive interference of monochromatic X-rays with a crystalline sample, governed by Bragg's Law:

$$n\lambda = 2d \sin\theta$$

where λ stands for the wavelength of the incident X-rays, d is interplanar spacing, θ is diffraction angle, and n is an integer [60]. The resulting diffraction pattern acts as a distinctive fingerprint for the material, enabling the detection of different crystalline phases and their relative proportions. XRD is widely used in the identification of phases and is capable of verifying the presence of crystalline phases in the synthesized materials to check for accuracy in composition. Moreover, peak broadening of XRD can be described by Scherrer equation in order to ascertain the crystallite size of NPs and evaluate the internal strain [61]. For thin films, XRD plays a vital role in providing information regarding crystal orientation, texture, and overall structure, which makes it an invaluable tool for characterizing coatings. In addition,

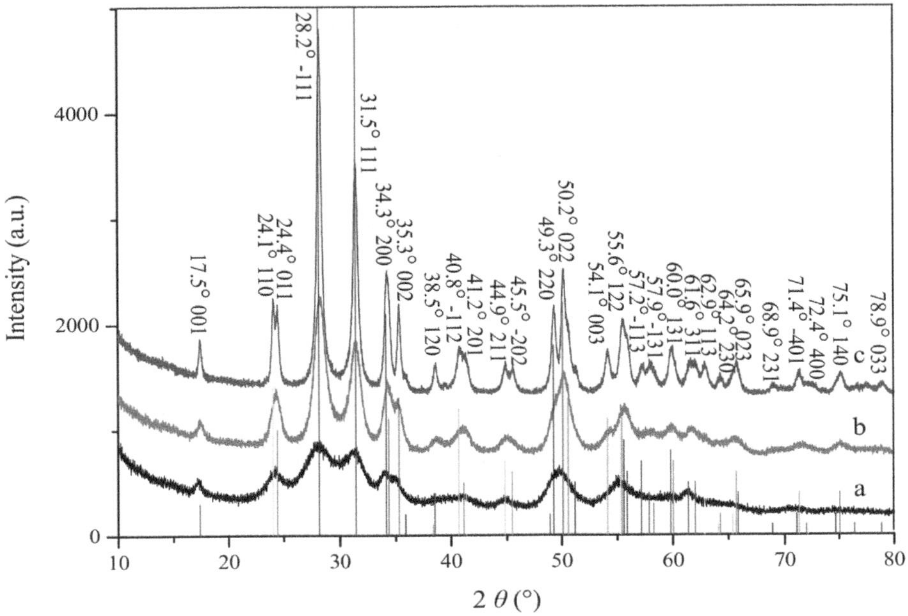

FIGURE 3.4 XRD patterns of as-synthesized (a) and calcined zirconia; (b) calcined at 723 K; (c) calcined at 923 K, vertical line [59].

it helps find lattice parameters and reveals structural imperfections such as dislocations, stacking faults, and phase transitions, which is important for understanding the material properties [62].

3.7.2 Scanning Electron Microscopy and Transmission Electron Microscopy

SEM and TEM are both imaging techniques that operate at the nanoscale and use electron beams to image the internal and external structure of different materials. SEM works by scanning the surface of the sample with a focused electron beam [63]. Along with accumulating topographical information, surface secondary electron emission provides high-resolution, three-dimensional images of the surface's topography. In addition, SEM has an energy-dispersive X-ray spectroscopy (EDS) feature to analyse the elemental composition of a sample, which makes SEM equipped for morphological analysis, engineering particle size and shape analysis, surface failure analysis, and other engineering applications. On the other hand, TEM permits imaging at the atomic level by projecting an electron beam through a very thin sample [64]. This allows the examination of nanomaterials and their defects and crystal structures. TEM is good for imaging atomic arrangements of lattices, phase identification by selected area electron diffraction (SAED) patterns, and other tomographic nanostructure imaging techniques in nanotechnology. In both nanotechnology and materials sciences, SEM and TEM serve as invaluable resources for biomedical research due to their ability to systematically characterize the structural features of

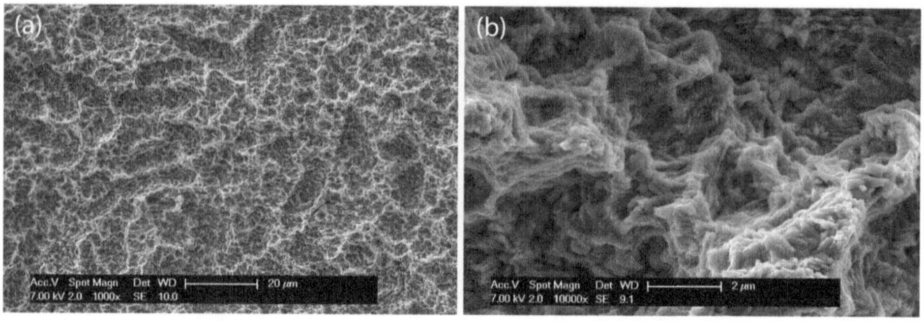

FIGURE 3.5 SEM images of an unused screw surface. (a) At 1000× magnification, irregular hollows form a fractal-like pattern. (b) At 10,000× magnification, the fine surface texture is defined by the metals grain structure, appearing clean and contamination-free [65].

sophisticated materials. Advanced material design strongly relies on the characterizing and structural analysing capabilities of these tools. Figure 3.5 shows the SEM images of an unused screw surface.

3.7.3 Dynamic Light Scattering

DLS is a powerful technique used to measure the hydrodynamic diameter of NPs in colloidal dispersions. It is based on the principle that small particles undergo Brownian motion, and when a laser is directed at them, the scattered light fluctuates due to their movement [66]. By analysing these fluctuations, particle size distribution can be determined using the Stokes–Einstein equation:

$$D = \frac{k_B T}{6\pi \eta r}$$

where D is the diffusion coefficient, k_B is the Boltzmann constant, T is the absolute temperature, η is the viscosity of the medium, and r is the particle radius. DLS is a crucial technique in nanotechnology in determining the size distribution of NPs, liposomes, and polymeric micelles. It also plays a vital role in assessing the colloidal stability of emulsions, suspensions, and drug delivery formulations [67]. Furthermore, in biophysics, DLS is used to study protein aggregation and folding behaviour. This technique is widely applied in pharmaceuticals, polymer science, and nanomaterials research, where precise control over particle size is essential for optimizing performance and functionality (Figure 3.6).

FTIR spectroscopy, a vibrational spectroscopy technique, is used to identify functional groups and molecular structures by measuring the absorption of infrared radiation by molecular bonds, generating a unique absorption spectrum for each compound. FTIR is instrumental in determining the presence of functional groups such as hydroxyl (–OH), carbonyl (C=O), and amine ($-NH_2$) groups. It is extensively used in pharmaceuticals, polymers, and biomaterials to analyse chemical compositions and ensure purity. In addition, FTIR enables real-time reaction monitoring,

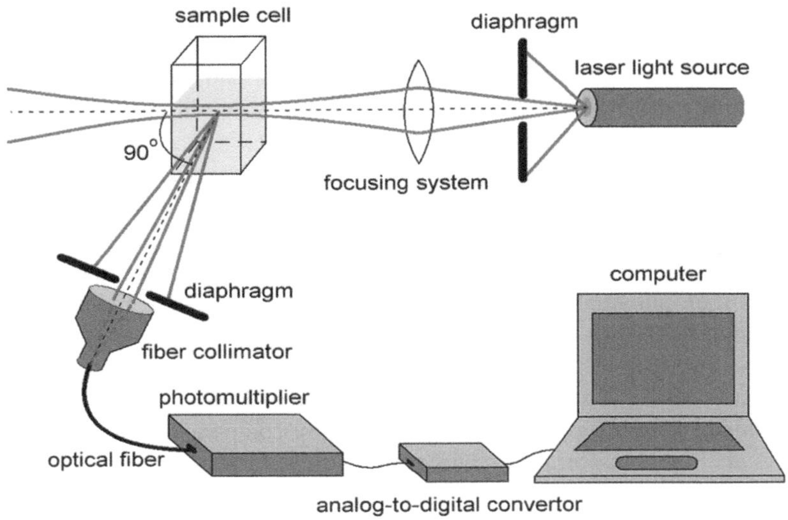

FIGURE 3.6 Scheme of a dynamic light scattering spectrometer [68].

making it invaluable in process control and material synthesis studies. Its applications span chemistry, pharmaceuticals, and material science, where it is essential for structural analysis and quality control.

3.7.4 Zeta Potential and Surface Charge Analysis

The level of electrostatic stability in colloidal dispersions may be determined using techniques such as measuring zeta potential. As it is related to NPs, emulsions, and suspensions, zeta potential or electrical potential at the stagnation point of a particle is vital in determining their stability. In a colloidal medium or fluid, stability is evaluated by zeta potential analysis. In this method, the colloid is found to be more stable when the values are higher than minus or plus at some point, whereas a value that is closer to zero means it will agglomerate [69]. It is also relevant to surface manipulations of NPs such as surface functionalization and interactions in drug delivery systems. In environmental science, zeta potential analysis contributes to the optimization of flocculation and coagulation processes in conjunction with wastewater treatment, emulsion stability, and other processes [70]. Zeta potential can be used in areas such as nanotechnology, biomedical research, and other industrial domains where performance improvement and material stability are essential.

3.8 RECENT ADVANCES IN ZIRCONIA SYNTHESIS

3.8.1 Functionalized Zirconium Nanoparticles for Targeted Therapy

The latest developments in zirconium-based nanotechnology focus on the functionalization of zirconium NPs towards targeted therapy. These modifications make it possible for NPs to selectively interact with diseased cells and reduce side effects.

Functionalization methods are surface modifications with biomolecules, polymers, or ligands that improve targeting and biocompatibility. Such functionalized zirconium NPs are being developed for drug delivery, imaging, and cancer therapy because their targeting capabilities offer significant benefits over traditional methods.

3.8.2 Hybrid Zirconium-Based Nanomaterials for Improved Drug Efficacy

Novel hybrid zirconium-based nanomaterials have been studied more thoroughly because of their potential to improve the efficacy of drugs. These materials incorporate zirconia with other nanomaterials such as silica, graphene, and metal-organic frameworks (MOFs) to enhance drug loading capacity, release profiles, and stability. Researchers seek to achieve enhanced therapeutic efficiency and reduced toxicity with sustained drug release by integrating bioactive components with zirconia. Such advances are highly important in cancer therapy, antimicrobial treatment, and regenerative medicine.

3.8.3 New Synthesis Techniques Optimizing Biocompatibility and Stability

Zirconium-based nanomaterials have experienced positive changes in biocompatibility and stability due to advances in synthesis methods. Sol-gel processing, hydrothermal synthesis, and biomimetic methods have increased the purity and shape precision and reduced the toxicity of zirconia nanomaterials. Furthermore, elements such as calcium, yttrium, and cerium have further increased the bioactivity and mechanical strength due to doping strategies. These advancements allow the better use of zirconia in medical implants, drug delivery systems, and tissue engineering.

3.9 LIMITATIONS AND CHALLENGES

3.9.1 Cytotoxicity and Biocompatibility

Cytotoxicity and biocompatibility paradoxes remain a primary focus for zirconium-based nanomaterials. While zirconia is considered biocompatible, there are some formulations and processing steps that introduce impurities or reactive species that have the potential to elicit an inflammatory reaction. Studies should be conducted to evaluate the risks of prolonged exposure and degradation byproducts in biological systems.

3.9.2 Cost-Effective Scale

The methods by which zirconia is synthesized pose issues for commercial and biomedical applications. The majority of advanced synthesis methods require high-grade precursors, elaborate devices, and intricate processes, which increases production costs. In addition, the reproducibility of large-scale synthesis, along with retaining the properties of the material, poses a barrier for its widespread use.

3.9.3 Stability and Agglomeration Challenges

Zirconia NPs have limited uses in biomedical applications because of their poor dispersion stability and performance due to their propensity to agglomerate. Various surface modifications and coating techniques have been developed to alleviate these problems, but achieving physiological long-term stability continues to be an area of research focus.

3.9.4 Regulatory Hurdles in Biomedical Applications

Zirconium-based nanomaterials have significant regulatory restrictions pertaining to biomedical use. Issues of toxicity, potential for biodegradation, and any chronic impacts should be investigated in detail before a clinical application can be considered. There are significant additional expense and time related to gaining regulatory approval due to the requisite extensive *in vivo* and clinical trials. This considerably impacts the associated cost and time for commercialization.

3.10 FUTURE PROSPECTS

3.10.1 Innovations in Synthesis for Better Control and Safety

Research in progress seeks to develop fabrication techniques that enhance the multifunctionality and safety of zirconia materials by controlling the dimensions, shape, and surface properties of particles [71]. These include approaches emerging from green synthesis, bioinspired fabrication, and nanostructuring.

3.10.2 Potential New Biomedical Applications

There is a growing body of research on the emerging uses of zirconium-based nanomaterials in drug delivery and implants, considering their surprisingly robust stability, mechanical strength, and bioinertness. These materials hold great promise in a range of healthcare applications [72].

3.10.3 Need for More Clinical Trials and Regulatory Approval Pathways

In regard to translating zirconia-based nanomaterials into clinical practice, it is important to conduct extensive preliminary and clinical trials. Research efforts should focus on the interactions of these materials within biological systems, optimizing compositions, and developing standardized safety assessment protocols. Moreover, clearer regulatory policies and guidelines need to be developed to enable the safe biomedical commercialization of zirconia products [73,74].

3.11 CONCLUSION

In summary, recent advances in zirconia synthesis have been complemented with functionalization, hybrid material development, and biocompatibility, and the achieved level is remarkable. However, concerns of cytotoxicity, economic factors, stability, and regulatory approval still need to be addressed to maximally harness the prospective capabilities of these materials in biomedical applications. The field of zirconia-based materials holds a promising future in which the further developments are expected to enhance the value of these nanomaterials by advancing zirconia-based medical devices into patient-tailored, target-oriented systems. The integration of enduring investigative work and innovative synthesis approaches, along with rigorous clinical tests, will make it possible to fully utilize the benefits zirconia-based nanomaterials have to offer in medical applications.

REFERENCES

1. S. Y. Khalaf and F. H. Ali, "Synthesis and characterization of manganese doped titanium dioxide nanoparticles," *AIP Conf Proc*, vol. 2372, no. 1, p. 160001, Nov. 2021, doi: 10.1063/5.0065981.
2. E. W. Leib et al., "Synthesis and thermal stability of zirconia and yttria-stabilized zirconia microspheres," *J Colloid Interface Sci*, vol. 448, pp. 582–592, Jun. 2015, doi: 10.1016/j.jcis.2015.02.049.
3. H. Mahdjoub, P. Roy, C. Filiatre, G. Bertrand, and C. Coddet, "The effect of the slurry formulation upon the morphology of spray-dried yttria stabilised zirconia particles." [Online]. Available: www.elsevier.com/locate/jeurceramsoc
4. M. N. Tahir et al., "Facile synthesis and characterization of monocrystalline cubic ZrO_2 nanoparticles," *Solid State Sci*, vol. 9, no. 12, pp. 1105–1109, Dec. 2007, doi: 10.1016/j.solidstatesciences.2007.07.033.
5. M. Mahdavi, M. Ramazani, and Z. Darvishi, "Synthesis and characterization of zirconium carbide nanorods at low temperature," *Int J Refract Metals Hard Mater*, vol. 56, pp. 59–62, Apr. 2016, doi: 10.1016/j.ijrmhm.2015.11.015.
6. I. L. Denry and J. A. Holloway, "Microstructural and crystallographic surface changes after grinding zirconia-based dental ceramics," *J Biomed Mater Res B Appl Biomater*, vol. 76, no. 2, pp. 440–448, Feb. 2006, doi: 10.1002/jbm.b.30382.
7. N. Kumari et al., "Zirconia-based nanomaterials: Recent developments in synthesis and applications," Aug. 23, 2022. doi: 10.1039/d2na00367h.
8. Z. Tang, X. Zhao, H. Wang, and B. Liu, "Clinical evaluation of monolithic zirconia crowns for posterior teeth restorations," *Medicine*, vol. 98, no. 40, Oct. 2019, doi: 10.1097/MD.0000000000017385.
9. S. Machmudah et al., "Hydrothermal and solvothermal synthesis of cerium-zirconium oxides for catalyst applications," *Int J Technol*, vol. 10, no. 3, pp. 582–592, May 2019, doi: 10.14716/ijtech.v10i3.2930.
10. J. P. Marković and S. K. Milonjić, "Synthesis of zirconia colloidal dispersions by forced hydrolysis," *J Serbian Chem Soc*, vol. 71, no. 6, pp. 613–619, 2006, doi: 10.2298/JSC0606613M.
11. D. Morton, S. Chen, W. Martin, R. Levine, and D. Buser, "Consensus statements and recommended clinical procedures regarding optimizing esthetic outcomes in implant dentistry," *Int J Oral Maxillofac Implants*, vol. 29, pp. 186–215, Jan. 2014, doi: 10.11607/jomi.2013.g3.
12. M. Hisbergues, S. Vendeville, and P. Vendeville, "Review zirconia: Established facts and perspectives for a biomaterial in dental implantology," Feb. 2009. doi: 10.1002/jbm.b.31147.
13. P. Fischer et al., "Bending fracture of different zirconia-based bioceramics for dental applications: A comparative study," *Materials*, vol. 14, no. 22, Nov. 2021, doi: 10.3390/ma14226887.

14. J. W. Nicholson, "Titanium alloys for dental implants: A review," Jun. 01, 2020. doi: 10.3390/prosthesis2020011.
15. C. Hu, J. Sun, C. Long, L. Wu, C. Zhou, and X. Zhang, "Synthesis of nano zirconium oxide and its application in dentistry," *Nanotechnol Rev*, vol. 8, no. 1, pp. 396–404, Jan. 2019, doi: 10.1515/ntrev-2019-0035.
16. V. Sollazzo et al., "Genetic effect of zirconium oxide coating on osteoblast-like cells," *J Biomed Mater Res B Appl Biomater*, vol. 84, no. 2, pp. 550–558, Feb. 2008, doi: 10.1002/jbm.b.30903.
17. Z. Dong et al., "Preparation and characterization of fluoride calcium silicate composites with multi-biofunction for clinical application in dentistry," *Compos B Eng*, vol. 143, pp. 243–249, Jun. 2018, doi: 10.1016/j.compositesb.2018.02.009.
18. A. Behbahani, S. Rowshanzamir, and A. Esmaeilifar, "Hydrothermal synthesis of zirconia nanoparticles from commercial zirconia," *Procedia Eng*, vol. 42, pp. 908–917, 2012, doi: 10.1016/j.proeng.2012.07.483.
19. I. H Ifijen, N. M Uzoekwe, E. B Ohiocheoya, J. A Osarobo, and S. I Omonmhenle, "Nano-zirconia synthesis methods and their pioneering applications in dentistry," *Momona Ethiop J Sci*, vol. 15, no. 2, pp. 122–152, Nov. 2023, doi: 10.4314/mejs.v15i2.1.
20. A. M. A. Moncayo, L. Peñate, M. Arregui, L. Giner-Tarrida, and R. Cedeño, "State of the art of different zirconia materials and their indications according to evidence-based clinical performance: A narrative review," Jan. 01, 2023. doi: 10.3390/dj11010018.
21. G. Pezzotti et al., "Osteogenic enhancement of zirconia-toughened alumina with silicon nitride and bioglass®," *Ceramics*, vol. 2, no. 4, pp. 554–567, Dec. 2019, doi: 10.3390/ceramics2040043.
22. I. Subuki, M. F. Mohsin, M. H. Ismail, and F. S. M. Fadzil, "Study of the synthesis of zirconia powder from zircon sand obtained from zircon minerals malaysia by caustic fusion method," *Indones J Chem*, vol. 20, no. 4, pp. 782–790, 2020, doi: 10.22146/IJC.43936.
23. A. Kristiani et al., "Sol-gel synthesis of zirconia-based nanoparticles from the side product of tin mining," in *E3S Web of Conferences*, EDP Sciences, Jul. 2024. doi: 10.1051/e3sconf/202454302013.
24. T. R. Gaydhankar, R. K. Jha, M. D. Nikalje, and K. J. Waghmare, "Influence of starting precursors and synthesis methods on the physiochemical properties of zirconia," *Mater Res Bull*, vol. 55, pp. 8–12, 2014, doi: 10.1016/j.materresbull.2014.03.010.
25. F. Rahmawati, K. Apriyani, E. Heraldy, and S. Soepriyanto, "The crystal structure and morphology of NiO-YSZ composite that prepared from local zircon concentrate of Bangka Island," in *AIP Conference Proceedings*, American Institute of Physics Inc., Mar. 2016. doi: 10.1063/1.4943447.
26. M. L. Bouchetou, J. Poirier, L. A. Morales, T. Chotard, O. Joubert, and M. Weissenbacher, "Synthesis of an innovative zirconia-mullite raw material sintered from andalusite and zircon precursors and an evaluation of its corrosion and thermal shock performance," *Ceram Int*, vol. 45, no. 10, pp. 12832–12844, Jul. 2019, doi: 10.1016/j.ceramint.2019.03.206.
27. M. L. Bouchetou et al., "Synthesis of an innovative zirconia-mullite raw material sintered from andalusite and zircon precursors and an evaluation of its corrosion and thermal shock performance," 2019, doi: 10.1016/j.ceramint.2019.03.206.
28. A. Adamczyk, "The study of the influence of ZrO_2 precursor type and the temperature of annealing on the crystallization of the tetragonal polymorph of ZrO_2 in zirconia-silica gels," *Gels*, vol. 8, no. 11, Nov. 2022, doi: 10.3390/gels8110724.
29. B. Cyriac, "Zirconia: Synthesis and characterization," in *Zirconia - New Advances, Structure, Fabrication and Applications*, IntechOpen, 2023. doi: 10.5772/intechopen.111737.

30. H. W. Park et al., "Synthesis of yttria-stabilized zirconia nanospheres from zirconium-based metal–organic frameworks and the dielectric properties," *Nanomaterials*, vol. 13, no. 1, Jan. 2023, doi: 10.3390/nano13010028.
31. R. Pokratath et al., "Mechanistic insight into the precursor chemistry of ZrO_2 and HfO_2 nanocrystals; towards size-tunable syntheses," *JACS Au*, vol. 2, no. 4, pp. 827–838, Apr. 2022, doi: 10.1021/jacsau.1c00568.
32. S. Liu et al., "Tetragonal nanosized zirconia: Hydrothermal synthesis and its performance as a promising ceramic reinforcement," *Inorganics*, vol. 11, no. 5, May 2023, doi: 10.3390/inorganics11050217.
33. S. G. Giniyatova, A. L. Kozlovskiy, R. I. Shakirzyanov, N. O. Volodina, D. I. Shlimas, and D. B. Borgekov, "Structural, dielectric, and mechanical properties of high-content cubic zirconia ceramics obtained via solid-state synthesis," *Applied Sciences*, vol. 13, no. 19, Oct. 2023, doi: 10.3390/app131910989.
34. C. J. Brinker and G. W. Scherer, *Sol-Gel Science: The Physics and Chemistry of Sol-Gel Processing*. Academic Press, 1990.
35. M. L. Matias et al., "A comparison between solution-based synthesis methods of ZrO_2 nanomaterials for energy storage applications," *Energies*, vol. 15, no. 17, Sep. 2022, doi: 10.3390/en15176452.
36. Y. Tao and P. P. Pescarmona, "Nanostructured oxides synthesised via $scCO_2$-assisted sol-gel methods and their application in catalysis," May 01, 2018. doi: 10.3390/catal8050212.
37. H. S. Kim, W. Jang, Y. G. Im, and H. P. Lim, "Antibacterial effect of zirconia nanoparticles on polyethyl methacrylate resin for provisional crowns," *Int J Nanomedicine*, vol. 17, pp. 6551–6560, 2022, doi: 10.2147/IJN.S382053.
38. H. Kim, Y. H. Lee, N. K. Kim, and I. K. Kang, "Bioactive surface of zirconia implant prepared by nano-hydroxyapatite and type I collagen," *Coatings*, vol. 12, no. 9, Sep. 2022, doi: 10.3390/coatings12091335.
39. T. Ahmad, "Hydrothermal synthesis, characterization and dielectric properties of zirconia nanoparticles," *Mater Sci Eng Int J*, vol. 1, no. 3, Oct. 2017, doi: 10.15406/mseij.2017.01.00017.
40. L. Liu et al., "From modification to mechanism: Supercritical hydrothermal synthesis of nano-zirconia," Feb. 15, 2022. doi: 10.1016/j.ceramint.2021.11.028.
41. H. Yan, J. Di, J. Li, Z. Liu, J. Liu, and X. Ding, "Synthesis of zirconia micro-nanoflakes with highly exposed (001) facets and their crystal growth," *Crystals*, vol. 11, no. 8, Aug. 2021, doi: 10.3390/cryst11080871.
42. T. Batool et al., "Microwave assisted sol-gel synthesis of bioactive zirconia nanoparticles – Correlation of strength and structure," *J Mech Behav Biomed Mater*, vol. 112, Dec. 2020, doi: 10.1016/j.jmbbm.2020.104012.
43. S. Yilmaz, S. Cobaner, E. Yalaz, and B. A. Horri, "Synthesis and characterization of gadolinium-doped zirconia as a potential electrolyte for solid oxide fuel cells," *Energies*, vol. 15, no. 8, Apr. 2022, doi: 10.3390/en15082826.
44. S. Tailor, M. Singh, and A. V. Doub, "Synthesis and characterization of yttria-stabilized zirconia (YSZ) nano-clusters for thermal barrier coatings (TBCs) applications," *J Clust Sci*, vol. 27, no. 4, pp. 1097–1107, Jul. 2016, doi: 10.1007/s10876-016-1014-y.
45. R. Dziembaj, M. Molenda, and L. Chmielarz, "Synthesis and specific properties of the ceria and ceria-zirconia nanocrystals and their aggregates showing outstanding catalytic activity in redox reactions—A review," Aug. 01, 2023. doi: 10.3390/catal13081165.
46. Y. Oh et al., "Microwave dielectric properties of zirconia fabricated using NanoParticle Jetting™," *Addit Manuf*, vol. 27, pp. 586–594, May 2019, doi: 10.1016/j.addma.2019.04.005.
47. M. A. Ja, C. L. Ca, and H. D. Kimrey, "Microwave sintering of solid oxide fuel cell materials: I, zirconia-8 mol% yttria." *J Am Ceram Soc*, vol. 75, no. 2, pp. 341–346, 1992.

48. S. S. Alsharari, M. A. Alenezi, M. S. Al Tami, and M. Soliman, “Recent advances in the biosynthesis of zirconium oxide nanoparticles and their biological applications,” *Baghdad Sci J*, vol. 20, no. 1, pp. 41–57, 2023, doi: 10.21123/bsj.2022.7055.
49. G. Negrón-Silva, C. X. Hernández-Reyes, D. Angeles-Beltrán, L. Lomas-Romero, and E. González-Zamora, “Microwave-enhanced sulphated zirconia and SZ/MCM-41 catalyzed regioselective synthesis of β-amino alcohols under solvent-free conditions,” *Molecules*, vol. 13, pp. 977–985, 2008, [Online]. Available: www.mdpi.org/molecules
50. L. Combemale, G. Caboche, D. Stuerga, and D. Chaumont, “Microwave synthesis of yttria stabilized zirconia,” *Mater Res Bull*, vol. 40, no. 3, pp. 529–536, Mar. 2005, doi: 10.1016/j.materresbull.2004.10.024.
51. M. Saxena et al., “Microwave-assisted synthesis, characterization and tribological properties of a g-C_3N_4/MoS_2 nanocomposite for low friction coatings,” *Coatings*, vol. 12, no. 12, Dec. 2022, doi: 10.3390/coatings12121840.
52. H. Ranjan and G. R. Rao, “Characterization of combustion synthesized zirconia powder by UV-vis, IR and other techniques,” *Bull Mater Sci*, vol. 23, no. 5, pp. 349–354, 2000.
53. I. Dobrosz-Gómez, M. Á. Gómez-García, J. Bojarska, M. Kozanecki, and J. M. Rynkowski, “Combustion synthesis and properties of nanocrystalline zirconium oxide,” *Comptes Rendus Chimie*, vol. 18, no. 10, pp. 1094–1105, Oct. 2015, doi: 10.1016/j.crci.2015.02.007.
54. N. Hossain, M. H. Mobarak, A. Hossain, F. Khan, J. J. Mim, and M. A. Chowdhury, “Advances of plant and biomass extracted zirconium nanoparticles in dental implant application,” May 01, 2023. doi: 10.1016/j.heliyon.2023.e15973.
55. J. M. Suominen et al., “Three-dimensional printing of zirconia: Characterization of early stage material properties,” *Biomater Investig Dent*, vol. 6, no. 1, pp. 23–31, Dec. 2019, doi: 10.1080/26415275.2019.1640608.
56. J. Rengaswamy, R. Sathasivam, P. C. Muthammal, and A. D. X. Durai, “Structural, optical, thermal, magnetic properties of zirconia nano-rods and their photocatalytic and antimicrobial properties,” *J Water Environ Nanotechnol*, vol. 6, no. 3, pp. 252–264, Jun. 2021, doi: 10.22090/jwent.2021.03.006.
57. S. Muhammad et al., “Surface charge properties of zirconium dioxide properties of zirconium dioxide,” 2012. [Online]. Available: https://www.shirazu.ac.ir/enSurfacecharge
58. Z. G. Wang et al., “Direct fabrication and characterization of zirconia thick coatings on zirconium hydride as a hydrogen permeation barrier,” *Coatings*, vol. 13, no. 5, May 2023, doi: 10.3390/coatings13050884.
59. C. Guo et al., “Morphology-controllable hydrothermal synthesis of zirconia with the assistance of a rosin-based surfactant,” *Appl Sci*, vol. 9, no. 19, Oct. 2019, doi: 10.3390/app9194145.
60. S. Pabisch, B. Feichtenschlager, G. Kickelbick, and H. Peterlik, “Effect of interparticle interactions on size determination of zirconia and silica based systems - A comparison of SAXS, DLS, BET, XRD and TEM,” *Chem Phys Lett*, vol. 521, pp. 91–97, Jan. 2012, doi: 10.1016/j.cplett.2011.11.049.
61. T. F. Alghazzawi and G. M. Janowski, “Evaluation of zirconia-porcelain interface using X-ray diffraction,” *Int J Oral Sci*, vol. 7, no. 3, pp. 187–195, Jan. 2015, doi: 10.1038/ijos.2015.20.
62. V. Gayathri and R. Balan, “Synthesis and characterisation of pure zirconium oxide (ZrO_2) nanoparticle by conventional precipitation method,” *J Environ Nanotechnol*, vol. 10, no. 4, pp. 19–21, Dec. 2021, doi: 10.13074/jent.2021.12.214447.
63. A. Scarano, S. Noumbissi, S. Gupta, F. Inchingolo, P. Stilla, and F. Lorusso, “Scanning electron microscopy analysis and energy dispersion X-ray microanalysis to evaluate the effects of decontamination chemicals and heat sterilization on implant surgical drills: Zirconia vs. steel,” *Appl Sci*, vol. 9, no. 14, Jul. 2019, doi: 10.3390/app9142837.

64. K.-F. Lin and C.-C. Lin, "Transmission electron microscope investigation of the interface between titanium and zirconia," *J Am Ceram Soc*, vol. 82, no. 11, pp. 3179–3185, 1999.
65. F. Mangano, M. Raspanti, H. Maghaireh, and C. Mangano, "Scanning Electron Microscope (SEM) evaluation of the interface between a nanostructured calcium-incorporated dental implant surface and the human bone," *Materials*, vol. 10, no. 12, Dec. 2017, doi: 10.3390/ma10121438.
66. Z. Jia, J. Li, L. Gao, D. Yang, and A. Kanaev, "Dynamic light scattering: A powerful tool for in situ nanoparticle sizing," Mar. 01, 2023. doi: 10.3390/colloids7010015.
67. M. H. Lee, J. S. Son, K. H. Kim, and T. Y. Kwon, "Improved resin-zirconia bonding by room temperature hydrofluoric acid etching," *Materials*, vol. 8, no. 3, pp. 850–866, 2015, doi: 10.3390/ma8030850.
68. E. Velichko, S. Makarov, E. Nepomnyashchaya, and G. Dong, "Molecular aggregation in immune system activation studied by dynamic light scattering," *Biology*, vol. 9, no. 6, pp. 1–13, Jun. 2020, doi: 10.3390/biology9060123.
69. H. Keramati, M. H. Saidi, and M. Zabetian, "Stabilization of the suspension of zirconia microparticle using the nanoparticle halos mechanism: Zeta potential effect," *J Dispers Sci Technol*, vol. 37, no. 1, pp. 6–13, Jan. 2016, doi: 10.1080/01932691.2015.1015077.
70. A. Serrano-Lotina, R. Portela, P. Baeza, V. Alcolea-Rodriguez, M. Villarroel, and P. Ávila, "Zeta potential as a tool for functional materials development," *Catal Today*, vol. 423, Nov. 2023, doi: 10.1016/j.cattod.2022.08.004.
71. Y. Yang, C. Hu, Q. Liu, and J. Li, "Research progress and prospects of colored zirconia ceramics: A review," *J Adv Ceram*, Oct. 2024, doi: 10.26599/jac.2024.9220941.
72. M. K. Han, "Advances and challenges in zirconia-based materials for dental applications," Sep. 01, 2024. doi: 10.1007/s43207-024-00416-7.
73. T. Chauhan et al., "Synthesis, characterization, and challenges faced during the preparation of zirconium pillared clays," *Arab J Chem*, vol. 15, no. 4, Apr. 2022, doi: 10.1016/j.arabjc.2022.103706.
74. V. Thakare, "Progress in synthesis and applications of zirconia," 2012. [Online]. Available: www.ijerd.com

4 Elaboration and Characterization of Zirconia and Its Comparison with Other Metal Oxides

Case Study

Rassim Younes

4.1 INTRODUCTION

New-generation materials use alumina–zirconia composites, which overcome both the brittleness of alumina and the susceptibility of zirconia to ageing. These composites are present in the form of alumina-toughened zirconia (ATZ, i.e. a zirconia matrix with alumina particles) or zirconia-toughened alumina (ZTA, with an alumina matrix containing zirconia particles for reinforcement). Their properties vary according to the quantity of zirconia introduced and its type (stabilized or not). The fact that stabilized zirconia is not used means that these composites are much more resistant to ageing than 3Y-TZP ceramics [1–5]. Huge research efforts are being focused on nanostructured ZTA composites. It is clear that a fine microstructure improves the mechanical properties of the ceramic. This is because intergranular defects are also smaller in size [6–8]. In addition, application of nanostructured materials generates finer wear debris and therefore is less critical during three-body friction. The transformability of zirconia can therefore be optimized by modifying the size of the zirconia particles and narrowing their distribution between their critical size for spontaneous transformation on cooling after sintering (~0.6 µm) and their critical size for transformation in the vicinity of a crack (~0.1 µm) [9–11]. There are several possible ways of obtaining micro- and/or nanostructured ZTA composites. These different routes will be discussed in the following sections.

4.2 ELABORATION OF ZTA COMPOSITES

ZTA composites are commonly obtained by mixing powders in a liquid. The resulting dispersion is poured or atomized, depending on the application. The composites obtained are generally microstructured. Various solutions have been considered in

DOI: 10.1201/9781003519300-5

TABLE 4.1
Effects of the Manufacturing Process on the Generation of Defects in the Final Product

Microstructure	Elaboration Process: Potential Effects
Initial powder	Distribution and quantity of species present, additive dosing (creation of a second phase)
Shaping (dispersion, casting)	Distribution of porosity in the green body, and therefore of final porosity
Sintering temperature (time and speed)	Density, grain size, phase ratio monoclinic
Cooling after sintering	Phase distribution, residual stresses
Hot isostatic pressing (HIP)	Density, residual stress
Bleaching	Modification of residual stresses
Machining and polishing	Surface roughness, residual stress, rate of monoclinic phase
Cleaning and sterilization	Monoclinic phase rate

the literature to minimize grain size and move towards nanocomposites [12]. First, increasingly fine and pure starting powders are used. Then, production processes are optimized (high-energy grinding, e.g. hydrothermal synthesis routes, etc.) [13,14] (see Table 4.1). Finally, heat treatment is adapted in order to optimize the structure for the application in question.

The problems most often mentioned in the literature are the presence of aggregates in the microstructure of the sintered composite and the inhomogeneity of the distribution of the second phase. These aggregates are already present during the dispersion stage and can be avoided by stabilizing and homogenizing the dispersion and optimizing its rheological properties.

4.3 DEVELOPMENT OF NANOPOWDER COMPOSITES

This is an essential step in the manufacture of nanostructured compounds, to ensure the production of a powder while also ensuring appropriate control of particle size, the presence of agglomerates, and the degree of contamination. A large number of techniques have emerged for the manufacture of ultrafine particles [15]. Most of these techniques can be classified into three types:

1. Chemical techniques (in a gas phase, liquid phase, or plasma),
2. mechanical grinding, and
3. thermophysical methods.

4.3.1 Mixture of Precursors

The first method of obtaining nanopowders involves obtaining a nanopowder in a single step, using either the sol-gel method or coprecipitation. The sol-gel method makes it possible to obtain powders with good dispersion of zirconia particles. This method involves preparing a 'sol' from inorganic salts or metal alkoxides (more generally, metal-organic compounds) [16–20].

The sol is then converted by hydrolysis and condensation reactions into a viscous fluid: the gel. The rate of the reaction can be influenced by a number of parameters, including the chemical composition of the precursors, the concentration of the reagent, pH, and temperature. The gel, containing a large mass of liquid, is dried to obtain the xerogel, which is generally amorphous and porous. The advantages of this method are the possibility of using different morphologies: films, fibres, particles, etc. [21,22], as well as controlling the chemical composition and reducing heat treatment temperatures. The major disadvantages are the price of alkoxides and their shelf life.

Coprecipitation involves preparing an aqueous solution containing the desired cations. The metal hydroxide in the solution is precipitated either by evaporation of the liquid or by the addition of a chemical reagent. This precipitation occurs through a nucleation growth phenomenon, and the characteristics of the powder are controlled by controlling the reaction parameters: temperature, concentration of salts in solution, their chemical composition, and pH. All of these factors together influence the morphology of the particles obtained. One of the major challenges posed by this method is finding the optimum conditions for the simultaneous precipitation of all the species in solution [23,24]. Simultaneous precipitation of all the species in solution is very difficult. Kikkawa [25,26] studied the two 'soft solution' preparation methods. The first procedure involves mixing aluminium chloride and zirconium (ratio Al/Zr = 0.39), while adjusting the pH by adding citric acid. The solution was stirred at 80°C to precipitate a gel. The second procedure, identical except for the presence of yttrium, was carried out with the addition of an alpha alumina suspension as a final step, in order to obtain ZTA composites. The final result shows an alumina–zirconia composite powder (with zirconia stabilized by three moles of yttrium) containing 5vol% zirconia, presenting alumina particles covered with fine zirconia particles of the order of 10nm. This latter method is referred to as post-doping.

4.3.2 Post-Doping

Post-doping consists of obtaining composite powders in several stages [25]. It involves grafting precursors of a second phase onto the surface of an existing powder [26], as shown in Figure 4.1. The heat treatment that follows breaks down the precursor, forming small particles of the second phase attached to the surface of the initial powder, thus creating a composite powder. Several techniques are available, including conventional substrate deposition (e.g. sol-gel coating) and non-conventional methods such as supercritical fluid deposition [27] (see Figure 4.2). These methods offer a compromise between standard powder mixing and sol-gel methods. In general, these techniques involve modifying the surface of the starting powder, the particles of which are then covered by the precursor (the second phase). This not only limits the inhomogeneities caused by the powder mixing technique, but also optimizes the control of the final microstructure, all in a simpler way [28,29].

Before going into detail about the reinforcement mechanisms in ZTAs, their value can be justified in a simple way. Within the composite, crack propagation takes place in the alumina matrix, which is intrinsically less sensitive to stress corrosion than zirconia. The presence of zirconia grains will lead to a certain degree of reinforcement,

FIGURE 4.1 High-resolution micrograph (SE-ESEM) of an alumina/zirconia microstructure obtained from doped and atomized powder [26]. (Permissions by ELSEVIER.)

due to either their transformation under stress or differences in the coefficient of expansion between the two phases (residual stresses).

More precisely, the mechanisms responsible for such reinforcements can be of three types, which are directly linked to the size of the zirconia grains. There are in fact two critical zirconia grain sizes (see Figure 4.3):

- Dc is the size above which zirconia is no longer retained in its phase quadratic after sintering (spontaneous transformation during cooling)
- D'c is the size below which no q-m transformation can take place, even under stress ('too stable' grains).

Only grains whose size lies between D'c and Dc can transform under stress. Two mechanisms are induced by grain transformation: microcracking and compression cracking. For zirconia grain sizes below D'c, q-m transformation is no longer possible. This leads to the third type of strengthening, which takes place as a result of residual stresses. The reinforcement mechanisms are therefore linked to grain size and their quantity and distribution. The higher the zirconia content results in the increase in the value of D'c and decrease in Dc value. The average particle size distribution lies between the critical size for spontaneous transformation (Dc) and the critical size for transformation by cracking (D'c), which means that the majority of the zirconia grains present contribute to the reinforcement of the material [30].

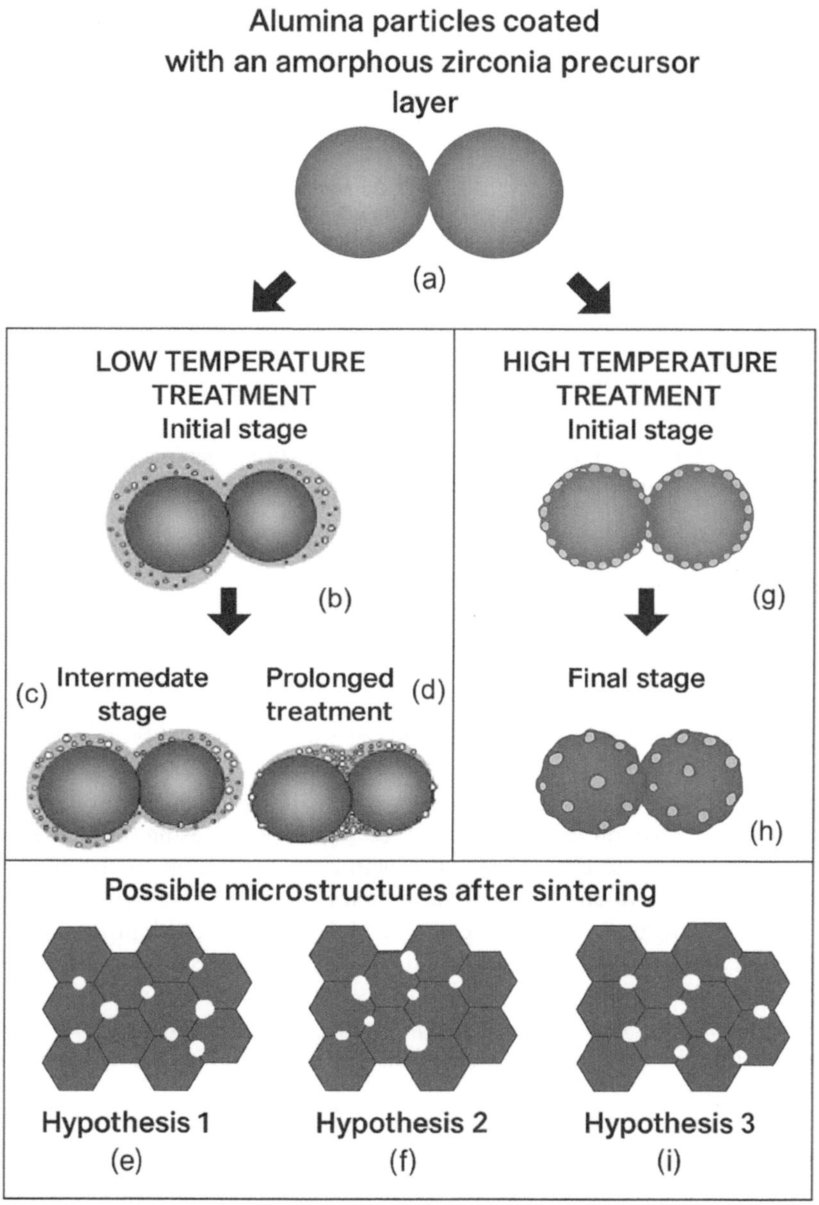

FIGURE 4.2 Diagram of the steps involved in post-doping an alumina powder with zirconia [27]. (Permissions by ELSEVIER.)

4.3.3 Reinforcement by Stress Phase Transformation

In alumina–zirconia composites, the q-m phase transformation is still responsible for the reinforcement. The literature reports toughness values for alumina–zirconia composites ranging from 5.5 to 10.5 MPa.m$^{1/2}$, for zirconia percentages varying from

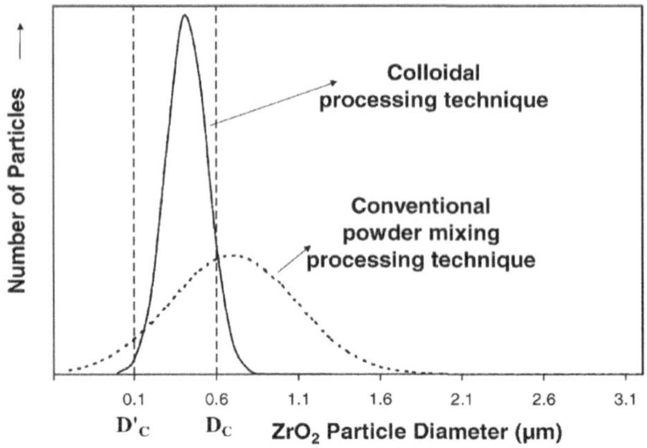

FIGURE 4.3 Distribution of zirconia grain sizes in 10-ZTA composites obtained by two processes [30]. (Permissions by ELSEVIER.)

5 to 20 vol%. The KI0–KIC ratio of the composite is much higher than that of zirconia or alumina, despite their relatively similar toughness. The composite therefore has the advantages of alumina, which is not very sensitive to propagation under high temperatures (improving the threshold), and zirconia, which strengthens the material by phase transformation (shifting the v-KI curve towards higher stress intensity factors).

The stress-induced phase transformation occurs in the presence of quadratic zirconia particles in the metastable form. The propagation of the crack is accompanied, around it, by a zone of stress under tension, which allows the release of the pressure exerted by the matrix on the dispersed zirconia particles, causing their transformation from the quadratic phase to the monoclinic phase, mainly at the crack face. The increase in the volume of zirconia particles, following this transformation, in turn generates a field of compressive stress, which opposes crack propagation. The resulting increase in fracture energy, and therefore toughness, occurs without any change in the size of the critical defects and is accompanied by an increase in fracture toughness. This mechanism depends on the volume fraction of transformable quadratic zirconia, the size and shape of the particles, and the Young's modulus of the material [31].

Lange, among others, illustrates this reinforcement mechanism on composites using yttriated and non-yttriated zirconia, in which the quadratic phase is retained after sintering. This retention of quadratic zirconia at room temperature is made possible by the compression effect of the alumina matrix, which has twice the Young's modulus of the zirconia inclusions. It is therefore possible to retain non-yttriated zirconia in the quadratic phase at room temperature when its proportion in the composite does not exceed 10 vol% (Claussen reports 15 vol%). This retention of the quadratic phase implies a very fine grain size. The increase in toughness is then due to the q-m transformation at the crack face [32–35].

The q-m transformation can occur spontaneously during cooling after sintering (particles larger than Dc). Intergranular particles seem more likely to transform than

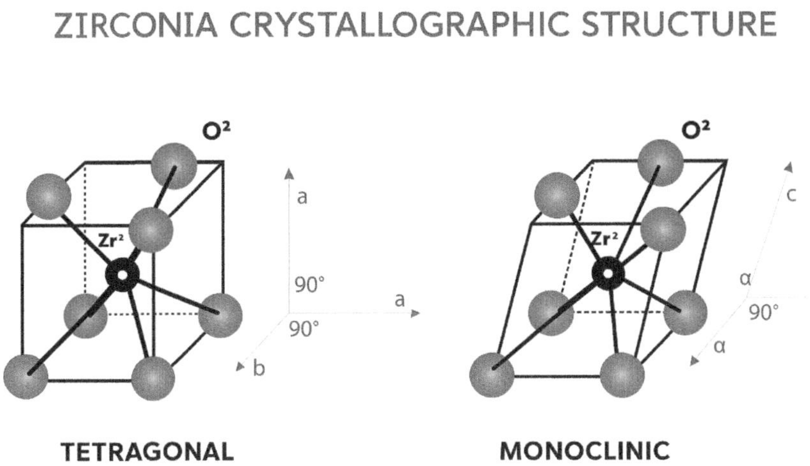

FIGURE 4.4 q-m transformation shape of zirconia [35]. (Permissions by EDP Sciences (Open Access Paper).)

intragranular particles. This would be due to a change in the germination energy of the monoclinic phase depending on the location of the particles: germs would be easier to form for intergranular particles. For Y-ZTA composites, D'c has been evaluated at 0.1 μm, while the critical size above which the quadratic phase cannot be retained at ambient phase is around 0.5–0.8 μm. The size of alumina grains has less influence on propagation resistance since the predominant strengthening mechanism is the q-m transformation (see Figure 4.4). However, this size must be small enough (<2 μm) to minimize the size of pre-existing defects and provide a better surface finish [36].

4.3.4 Reinforcement by Microcracking

Microcracks may pre-exist in the material as a result of the q-m transformation that takes place during cooling after sintering: these are known as pre-microcracked composites. During cooling after sintering, zirconia particles larger than the critical transformation size (D'c) transform from the quadratic phase to the monoclinic phase. In this configuration, zirconia grains at room temperature are no longer transformable ($d > Dc$) and do not participate in strengthening by phase transformation. There are two zones of interaction between the main crack and the microcracking network: a first zone of direct interaction, where the mechanisms of microcrack coalescence and blunting of the main crack front occur, and a second, wider zone of interaction. The latter is similar to the transformed zone discussed in the previous paragraph. It therefore helps reduce the driving force of the crack by dissipating some of the elastic energy associated with crack propagation. The increase in the surface area corresponds to an increase in fracture energy and therefore fracture toughness.

4.3.5 Internal Stress Reinforcement

When obtaining a dispersion of very small zirconia particles, sometimes present at the very heart of alumina grains, the transformation of the zirconia during cooling after sintering can be avoided. In this case, zirconia particles are retained in the metastable quadratic phase under the effect of the pressure exerted by the surrounding alumina matrix. Zirconia is then in tension in the alumina matrix, which itself is in compression. This is due to the different coefficients of expansion between zirconia and alumina. Zirconia, which has a higher coefficient of expansion, is prevented from shrinking by the alumina matrix during cooling after sintering. These particles therefore occupy a larger volume than they should (they are in traction). A crack propagating mainly in alumina will therefore see a material in compression, so its propagation speed will be lower. In this type of composite, the majority of zirconia grains are too small to be transformable ($d <$ D'c) and therefore do not participate in strengthening by phase transformation [37]. However, q-m transformation is not always responsible for strengthening. Bhaduri et al. have demonstrated high toughness (8.4MPa-m 1/2) for ZTA nanocomposites (unfortunately, the grain size is not specified). The size of these zirconia grains is assumed to be smaller than D'c, and the quadratic phase is therefore stable, even at the crack face. According to the authors, the high toughness of this composite is made possible by its plastic behaviour, which is difficult to imagine for a ceramic at room temperature (see Figure 4.5). The hardness of the material thus obtained is greatly reduced (4.95 GPa compared with 10–15 GPa for microcomposites).

4.3.6 Controlling the Size of Alumina Grains: Towards Nanocomposites

There are two ways of modifying the microstructure of alumina during sintering: the sintering method and the use of dopants. The most common sintering methods are natural sintering, flash sintering (or spark plasma sintering, SPS), microwave sintering, hot pressing, hot isostatic pressing (HIP), and all their combinations, including

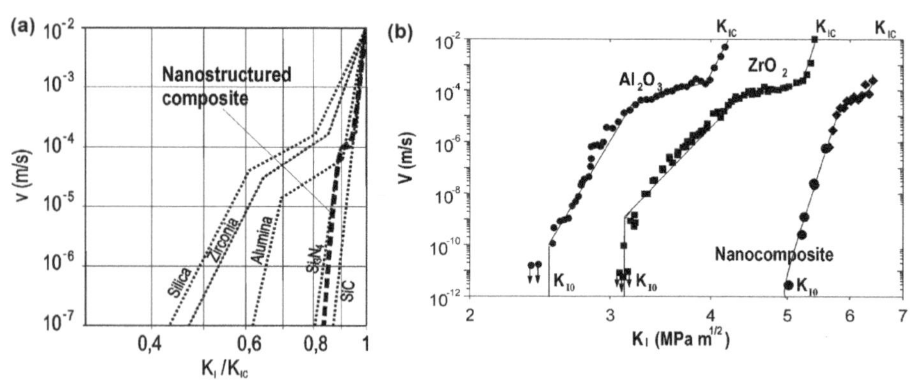

FIGURE 4.5 Comparison of slow crack propagation between alumina, zirconia, and a nanocomposite [37]. (Permissions by ACS.)

two-step sintering (TSS). In this study, only natural sintering is considered, as it appears to be the most suitable for the industrialization of THPs.

4.4 INTRODUCTION AND ROLE OF DOPING AGENTS

In ceramics, doping is defined as the deliberate introduction of foreign atoms into the material by means of additions (small quantities of impurities, mineral salts, etc.). Additives are commonly used in varying quantities in ceramic systems to improve the densification process, the mechanical and physical properties, or the aesthetics of the final ceramic. These additives are activated by modifying the sintering temperature and step times in order to influence grain size. These additions should not generate defects that could limit the strength of the material, which is why their size should be monitored so that the stresses they cause do not introduce microcracks, either during cooling or under load. The dopant can be introduced in a number of ways: dry mixing, wet mixing, wet process, or impregnation method. The aim is to distribute it homogeneously and obtain dense crucibles. The aim of this study is to understand the influence of doping on sintering and the microstructural evolution of alumina depending on the nature of the additions. Dopants may be located in the material in the form of a solid solution or may remain outside the alumina network. In the latter case, segregation occurs at grain boundaries.

4.4.1 Characteristics of Alumina without Dopants

Grain growth is necessary to achieve complete densification. This is achieved jointly by the disappearance of small grains in favour of the growth of larger adjacent grains, due to intergranular mass transport mechanisms. However, grain boundaries move easily in a solid close to its final density. Matter diffuses over short distances (of the order of the interatomic distance). As a result, the mobility of the grain boundaries will lead to a more rapid coarsening of the grains than that caused by the sintering mechanisms mentioned above and therefore will be difficult to control. This phenomenon will lead to abnormal grain growth, sometimes to the inclusion of pores within grains themselves. In the case of alumina, the abnormal growth is particularly marked and can lead to grain sizes well in excess of the expected grain size. These oversized grains are harmful because they reduce the strength of the parts by acting as stress concentrators.

4.4.2 Effects of Calcium

Cahoon demonstrated in 1956 that the addition of CaO reduced sintering kinetics and inhibited grain growth [37]. In 1987, Baik et al. studied calcium as a possible cause of exaggerated alumina grain growth. Sumita in 1988 followed Cahoon's footsteps, showing that 1% addition of CaO (116 ppm measured after sintering) reduced sintering kinetics but enabled a fine microstructure to be obtained. He partly concurred with Baik's theory, hypothesizing that the anisotropic segregation of CaO at grain boundaries is the cause of the inhomogeneous microstructures obtained and its ineffectiveness as a sintering aid. In 1989, Handwerker argued that the addition of

CaO promotes abnormal grain growth. In 2002, Ferey showed in his thesis that CaO increases the density of alumina after sintering at 1700°C. However, abnormal grain growth was observed for sintering at higher temperatures (>1800°C) and for low quantities of calcium (30 ppm). In 2003, Louet showed that calcium in the initial and intermediate stages of sintering dissolves in the regions close to the grain boundaries, creating oxygen vacancies that indirectly accelerate sintering. In addition, for a temperature close to the eutectic of the CaO–Al diagram, he noted the creation of CA6. It was not until the complexion theory of Dillon et al. in the 2000s that we really understood the behaviour of CaO in alumina during sintering. Calcium generates a complex behaviour of abnormal grain growth with multiple complexions coexisting across a wide range of temperatures. In the same work, he demonstrated that grain morphology is not a direct effect of grain growth kinetics [38].

4.4.3 Effects of Magnesium

In the 1950s, it was shown that magnesium limits the abnormal growth of grains in alumina, while making it possible to obtain a fine, dense structure, particularly for transparent ceramics, and that it reduces the mobility of grain boundaries in alumina by an order of three. These observations were supported until very recently, but could not be explained. MgO quickly became the most common and effective additive. However, its solubility in alumina is very limited (300 ppm at 1630°C), and an excess leads to the formation of an $MgAl_2O_4$ spinel phase at the grain boundaries. Analyses carried out in 1968 by Roy and Coble showed that above its solubility limit, this oxide acts as an inhibitor of abnormal grain growth by negatively influencing densification. They noted that the solubility limits of MgO in alumina for the atomic fraction of Mg/Al are as follows: 110 ppm at 1530°C, 300 ppm at 1630°C, 680 ppm at 1730°C, and 1400 ppm at 1830°C. Bae showed that the average size of alumina grains increases as a function of the MgO doping rate and that the inhibition of abnormal grain growth appears for an MgO content equal to that of CaO. In the absence of porosity, MgO reduces mobility at grain boundaries through a braking mechanism by segregation in solid solution, the effect of which is all the more marked the higher purity. In the presence of porosity, the increase in the surface diffusion rate caused by MgO in pure alumina allows the movement of pores with grain boundaries, which avoids the appearance of intragranular porosity that is very difficult to eliminate, whereas in ultrapure alumina, it is once again a reduction in grain boundary mobility that appears. Overall, MgO reduces the impact of the inhomogeneities in terms of compactness and chemistry. It limits grain coarsening in regions of high compactness, leaving regions of lower compactness free to densify. In addition, according to Handwerker, it reduces the effect of impurities (especially Ca and Si) by increasing their solubility. The distribution of dihedral angles is kept tight, pore–grain boundary separation is avoided, and abnormal growth does not occur. Many authors have studied the mechanisms of action of MgO, but different and sometimes contradictory results have been reported. A summary of these studies by Descemond (1991) shows that the effect of MgO differs according to alumina purity and density. Dispersing MgO in alumina is difficult and critical. The dispersion method has an effect on sinter densities, densification rate, grain size, grain coarsening rate, and the amount and type of porosity. Using a mechanical technique such as wet milling to dope alumina with 0.03wt% MgO leads to a low final density, a nonuniform microstructure, and an inhomogeneous dispersion of MgO in the material. An initial powder with fine grains is required, and particular attention should be paid to dispersion methods [39].

4.4.4 Effects of Silica

Abnormal grain growth is often associated with the presence of silica either in the starting commercial powder (impurities) or as an external contaminant during product development stages. One listed example suggests the influence of silica, contained in refractories in the form $MoSi_2$, during the sintering of the material. In his study, Bae showed that at least 300 ppm cationic silica is needed to obtain abnormal grain growth. In his thesis, Louet showed that its use has a negative effect on the densification of alumina during the early stages of sintering. In fact, silica is inserted in solid solutions at grain boundaries and leads to a reduction in oxygen vacancies and therefore sintering kinetics. However, at high temperatures (1600°C, defined as the temperature of onset of abnormal alumina grain growth), silica leads to the formation of intergranular liquid films, which accelerates densification. Numerous studies emphasize silica's ability to increase sintering kinetics through the formation of an intergranular glassy phase without altering the final density of the material [40]. This phase simultaneously reduces mechanical properties at high temperatures and corrosion resistance. However, the major drawback to its use is its propensity to generate heterogeneous microstructures. Dillon et al. went further to investigate the effect of silica on alumina. They observed the formation of four complexions between 1300°C and 1800°C. It would appear that at 1500°C, a transition occurs with the appearance of exaggerated grain growth, which then spreads throughout the material to the detriment of other grains. The microstructure then essentially consists of abnormal grains that block each other and re-establish a new unimodal grain size distribution. An excess of silica in solid solution would therefore have a strong influence on the kinetics and transitions between the different complexities created.

4.5 PRODUCTION OF ALUMINA

For the manufacture of dense alumina ceramics, it is necessary to transform the alumina powder into its stable α phase prior to consolidation, in order to avoid transformations during sintering. However, the formation of the stable α-Al_2O_3 phase requires a high temperature in order to achieve the activation energy necessary for its formation. Synthesis processes should enable nanometric-sized α crystallites to be obtained. To our knowledge, all studies report the formation of α-phase alumina with a grain size greater than 100 nm, i.e. about an order of magnitude larger than that of the alumina transition alumina matrix [41]. Furthermore, no isolated alpha nuclei were observed, suggesting that the growth of α-Al_2O_3 in the transition alumina matrix is 'explosive' once a critical nucleus size is reached (see Figure 4.6). The result of this growth of α-Al_2O_3 is a vermicular powder structure. An example of the morphology of α particles from a boehmite powder is shown in Figure 4.6.

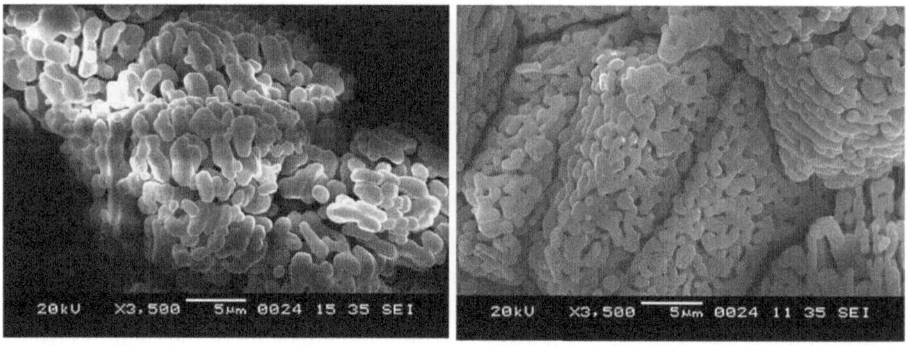

FIGURE 4.6 Vermicular structure of α-Al_2O_3 added with Gibbsite (1250°C, 2.5°C/min) (left) or by calcination of a transition alumina powder (1200°C, 5°C/min) (right) with a size distribution of less than 100nm and no agglomerates [41]. (Permissions by ELSEVIER.)

The multiconnection of α-alumina grains in a vermicular structure can be observed. These vermicules can measure several micrometres.

Currently, all the commercial α-Al_2O powders on the market have grain sizes higher than 100nm. The most widely used are TM-DAR (Taimei) and BMA15 (Baikowski). The finest powders today are necessarily transition alumina (δ, γ, θ) powders, with grain sizes below 50nm. Recently, transition alumina has been well studied because of its intrinsic nanocrystalline nature and because it can be synthesized using a variety of techniques.

4.6 MECHANISMS AND TRANSFORMATIONS DURING SINTERING

4.6.1 Nucleation Growth

The transformation from the θ phase to α-Al_2O_3 is of the nucleation growth type. It is notable, however, that the mechanism for the growth of α-Al_2O_3 crystallites appears to have different modes according to the literature [42]. Two mechanisms are proposed for the growth of the α-phase:

- Grain boundary migration between α crystallites formed and the alumina matrix of transition.
- Coalescence of α nuclei formed; then, beyond a certain critical size, growth by the migration mechanism. This coalescence mechanism necessarily involves an α-grain rearrangement mechanism. These two mechanisms are consistent with the progressive growth of vermicular α-phase single crystals.

4.6.2 Rearrangement during Transformation

This rearrangement has been mentioned in the literature. Yen et al. highlighted the importance of the rearrangement of θ particles, which, after reaching a critical diameter, should move to join other θ particles and transform into α (germination), without specifying the rearrangement mechanism. Legros et al. gave a more detailed explanation of the generation of this rearrangement created in the structure after phase transformation from γ to α. When a particle transforms from γ to α-Al_2O_3 [43], if the neighbouring particles are not in contact or arranged symmetrically, the reduction in crystallographic volume can induce interparticle forces, causing them to move. These proposed mechanisms represent a basic notion of the phase transformation mechanism. However, these studies do not show that it is possible to obtain an ultrafine (<100 nm) and monodisperse α-Al_2O_3 powder.

4.6.3 Decrease in Phase Transformation Temperature

The development of the vermicular structure, produced at high temperatures (between 1150°C and 1200°C), is a major obstacle inhibiting the densification of nanometric transition alumina at low temperatures (<1300°C). As a result, the final sintering stage of transition alumina requires a high temperature, above 1600°C, to reach a high density. This leads to grain coarsening, which is favoured at high temperatures, and slows down the sintering of the alpha alumina formed. For this reason, a reduction in the crystallization temperature has been the subject of several studies in order to control the formation of the vermicular microstructure and obtain a fine α-Al_2O_3 at lower temperatures. A number of strategies have been proposed in the literature to reduce the temperature of the crystallization of α-Al_2O_3 and thus improve densification. These studies include:

- **Effect of Grinding**: Grinding reduces the phase transformation temperature. It has thus been shown that the nucleation rate is increased. Two hypotheses are often put forward: a mechanochemical effect or a doping effect due to the contamination of the grinding media (less likely because it has not been observed).
- **Effect of Pressure during Sintering**: Studies have shown the effect of applying high pressure during sintering on reducing the phase transformation temperature. Liao showed that the application of high pressure during sintering can promote the nucleation rate and decrease the grain coarsening rate. He reported that the $\gamma \rightarrow \alpha$ transformation can occur at 460°C at a pressure of 8 GPa. Thus, by applying high pressure, a compact, finer final grain size is obtained. The same phenomenon was observed in n-TiO_2.

4.6.4 Sintering Transition of Alumina Powder

Despite the existence of certain methods for reducing the phase transformation temperature, obtaining a dense structure from nanometric alumina appears to be difficult to achieve using conventional sintering. As a result, great interest has developed in sintering techniques that use processes that enable alumina to be sintered in a short timespan. These techniques include microwave and SPS. The results of these studies are comparable to those obtained using conventional sintering. In fact, the best density obtained where a size of less than 100 nm was retained was 90%. Based on these observations, some studies have shown that high pressure can help obtain a dense nanocrystalline structure. By applying a pressure of 80 MPa to a ground powder,

Ferkel obtained a dense structure of 99% with a grain size of less than 500 nm. Mishra et al. succeeded in obtaining total densification with a grain size of 93 nm, from the same NanoTEK powder, by applying a higher pressure of 8 GPa. This is the finest dense alumina obtained in the literature. Gonzalez also highlighted the effect of pressure during compaction. Indeed, γ-alumina compacts were prepared at pressures up to 5 GPa and then sintered under pressure between 1000 and 1600°C. For compacts at pressures below 3 GPa, a porous microstructure was obtained. This result was attributed to the formation of the vermicular structure after the α-alumina phase transformation. However, compaction at pressures above 3 GPa avoided the vermicular structure and resulted in dense alumina close to the theoretical density, with a nanometric grain structure (average particle size 150 nm).

4.7 CASE STUDY RESULTS (EFFECT OF REINFORCEMENT TYPE ON PHYSICAL AND CHEMICAL CHARACTERISTICS OF ALUMINA-BASED COATINGS)

The aim of this section is to examine the influence of two types of reinforcements on the tribological behaviour of alumina-based coatings. The first type of reinforcement pertains to coatings reinforced with 3% by weight of titanium oxide (TiO_2, denoted as AT-3) [44–46], which has already undergone a microstructural, structural, and tribological study. The second reinforcement involves composite coatings of alumina alloy reinforced with 25% by weight of zirconia oxide (ZrO_2, denoted as AZ-25), deposited on low-carbon steel E335 using a thermal spray technique with oxyacetylene flame as the energy source. A tribological study was conducted between these two reinforcements based on the specified couples:

- **Couple A**: Al_2O_3-3%TiO_2 (AT-3)/Al_2O_3-13%TiO_2 (AT-13)
- **Couple B**: Al_2O_3-25%ZrO_2 (Zr-25)/Al_2O_3-3%TiO_2 (AT-13)

4.7.1 Microstructural Analyses of the Reinforcements

In Figure 4.7, the morphologies of the two filler materials in the form of beads are depicted. The microstructure of the starting material Al_2O_3–3%TiO_2 (AT-3) reveals an irregular and angular morphology in particle form, as illustrated in Figure 4.7a. In addition, the scanning electron microscopy (SEM) morphology of the second filler material Al_2O_3–25%ZrO_2 (AZ-25) (see Figure 4.7b) shows that the wire is agglomerated in a granular and more compact form than (AT-3) because the grains are closely bound together, possibly due to the sintering effect.

Observations using SEM of the Al_2O_3–3%TiO_2 (AT-3) deposits obtained through oxyacetylene flame revealed a dense, compact microstructure composed of several complex phases, with the presence of porosities and unfused particles (see Figure 4.8a). This microstructure consists of lamellae and globular pores, lacking cracks due to insufficient stresses during the formation of these stacks. In contrast, the SEM microstructure of the Al_2O_3–25%ZrO_2 (AZ-25) coating is less compact than that of AT-3 and is composed of multiple phases with a significant presence of porosities (see Figure 4.8b).

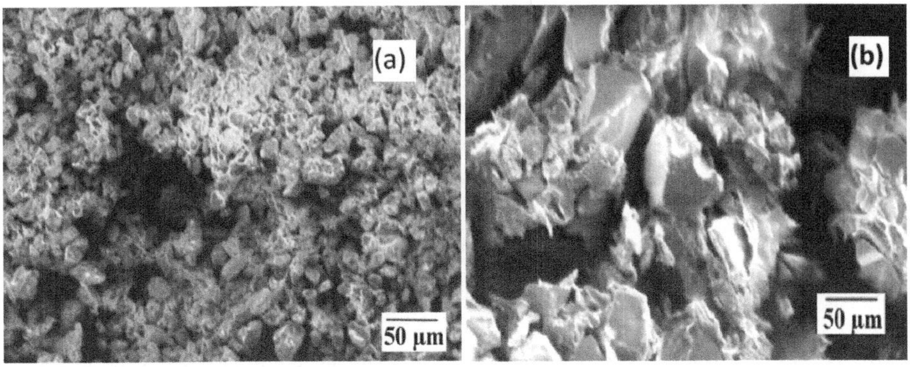

FIGURE 4.7 SEM image of alumina-based beads: (a) Al_2O_3 (Al-99), (b) Al_2O_3-25%ZrO_2 (AZ-25).

FIGURE 4.8 SEM image of alumina-based deposits: (a) Al_2O_3 (Al-99), (b) Al_2O_3-25%ZrO_2 (AZ-25).

4.7.2 X-Ray Diffraction (XRD) of the Reinforcements

Figure 4.9 displays the diffraction patterns of the alumina cord reinforced with zirconia (25% by weight of ZrO_2) and its coating obtained through flame spraying. From the peaks corresponding to zirconia-reinforced alumina (AZ-25) (see Figure 4.9), it is evident that both the starting cord material and its coating are composed of the same phases: α-Al_2O_3 matrix, monoclinic ZrO2 (JCPDS 01-078-0047), and quadratic ZrO_2 structure (JCPDS 01088-1007). The reinforcement of alumina with zirconia enhances the mechanical properties of the coating through the formation of quadratic ZrO_2. This is confirmed by the XRD spectrum of the deposit, where the proportion of quadratic ZrO_2 is more significant than that in the cord. The presence of amorphous phases is inevitable in the coating construction process, inherent to the more or less rapid cooling of sprayed particles. However, their percentages in alumina-based deposits remain limited due to background noise in the corresponding XRD diagrams. Furthermore, it is noteworthy that the use of zirconia as reinforcement promotes this tendency through the partial substitution of Al^{3+} ions by the more charged Zr^{4+} ions. Regarding the diffraction patterns obtained from the cord reinforced with 3% titanium oxide (Al_2O_3–3%TiO_2, AT-3) and its deposit, the

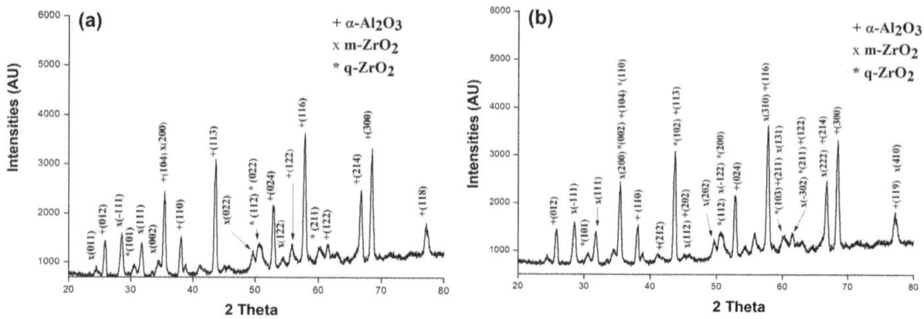

FIGURE 4.9 Diffractogram of the starting cord and its alumina-based coating Al_2O_3–25%ZrO_2 (AZ-25).

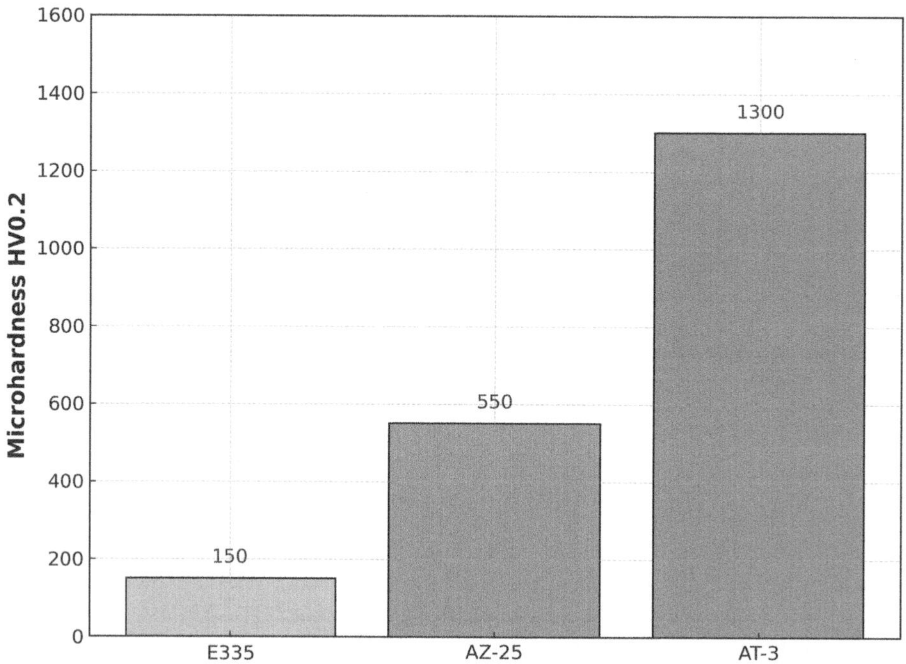

FIGURE 4.10 Vickers microhardness of the different deposits.

observed peaks in the cord reveal the presence of three phases: a major α (i.e. a major rhombohedral α-Al_2O_3 structure (JCPDS no. 046-1212)), a secondary orthorhombic $Al_6Si_2O_{13}$ structure (JCPDS no. 015-0776), and another phase in low proportion, α-TiO_2, known as rutile with an orthorhombic structure (JCPDS no. 0231446). In contrast, the spectrum of the Al_2O_3–3%TiO_2 (AT-3) deposit, as indicated in Figure 4.9, also revealed the presence of the major phase α-Al_2O_3 as the matrix and the emergence of two new alumina phases, γ-Al_2O_3 (JCPDS no. 01 074-2206) and κ-Al_2O_3 (JCPDS no. 046-1215), due to the high combustion temperature of alumina, resulting in the formation of metastable alumina (γ-κ). The processing time of the coating derived from the Al_2O_3–TiO_2 cord is not sufficient to form titanium oxides or other phases containing Ti atoms. Consequently, Ti atoms are dissolved in the γ-Al_2O_3 or κ-Al_2O_3 phase.

4.7.3 Microhardness Vickers Test

Microhardness Vickers tests are conducted on the Al_2O_3–3%TiO_2 (AT-3), Al_2O_3–25%ZrO_2 (AZ-25), and E335 steel substrate using Vickers indentation under a load of 200 g for 30 seconds. The average standard deviation calculated for each sample is based on the mean of ten measurements. The results of Vickers microhardness are illustrated in the histogram in Figure 4.10. Analysing Figure 4.10, we observe that Vickers microhardness of deposits reinforced with titanium oxide (Al_2O_3–3%TiO_2, AT-3) is nearly three times harder than that of alumina deposits reinforced with ZrO_2 particles (AZ-25) (approximately 1326 ± 66 HV0.2 for AT-3 compared to 563 ± 25 HV0.2 for AZ-25). This difference may be attributed to the influence of porosity quantity and the hardness contrast between 3%TiO_2 and 25%ZrO_2 reinforcement elements.

4.8 CASE STUDY RESULTS (EFFECT OF REINFORCEMENT TYPE ON TRIBOLOGICAL BEHAVIOUR OF ALUMINA-BASED COATINGS)

Tribological tests were conducted on Al_2O_3-based deposits with different titanium oxide and zirconia oxide reinforcements (3% TiO_2 and 25% ZrO_2) to determine friction coefficients and wear rates. The wear debris created by particle detachment during friction with the counterpart will not be eliminated from the wear tracks because, in this case, the debris contributes to friction either by acting as a third body, thus creating a protective layer and reducing the applied contact pressure on the deposit, or by acting as abrasive particles against the deposits.

4.8.1 Determination of the Friction Coefficient

Figure 4.11 illustrates the variation in the friction coefficient for the studied pairs as a function of the applied load. Similar trends in the curves depicting the variation of the friction coefficient with the normal load of the couples can be attributed to asperities generating a tangential resistant force adjacent to the normal force. The applied stress creates a connection that tends to maintain the constant ratio. The initial part of the curve (from 5N to 10N) corresponds to the accommodation phase of the disc and pin surfaces when the contact is established, followed by a gradual decrease in friction coefficients until stabilizing around a value of approximately 0.7. Overall, for all test conditions, the friction coefficients of the studied couples have a similar appearance, although the values differ. This difference can be explained by various microstructural formations and modifications during the test. However, it is interesting to emphasize the variation of the friction coefficients of the couples over time for loads of 5N and 30N. The following will present a comparative study of the

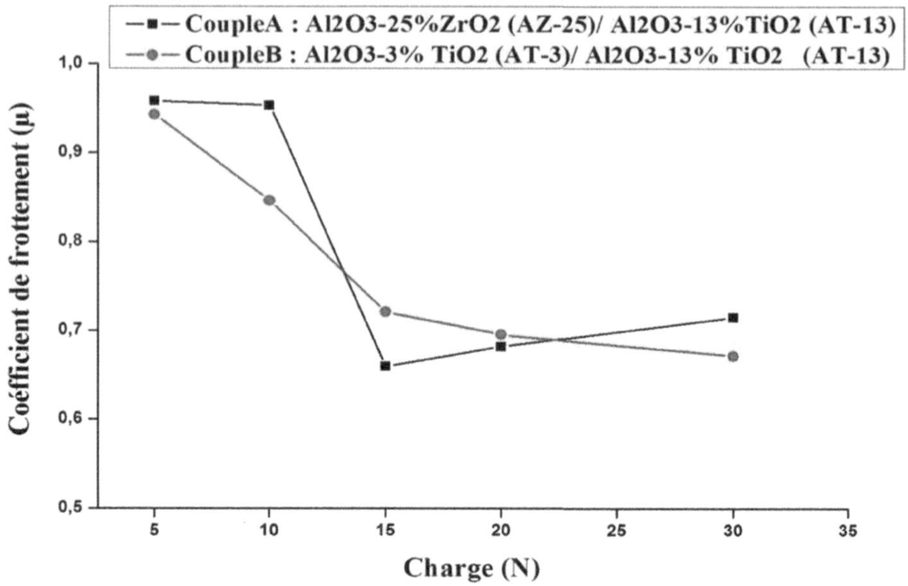

FIGURE 4.11 Variation in the coefficient of friction of couples as a function of normal load.

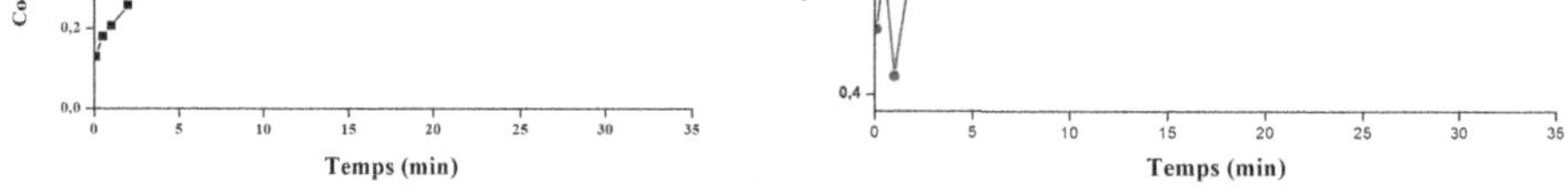

FIGURE 4.12 Variation in the friction coefficient over time: (a) couple A and (b) couple B.

friction coefficients of the couples over time for loads of 5N and 30N, as indicated in Figure 4.12a.

Figure 4.12b shows the variation in the friction coefficient over time for both couples under sliding loads of 5N and 30N. Tribological results indicate that the two curves representing the variation of the friction coefficient over time for both sliding contacts are nearly identical during the first 5 minutes under applied loads of 5N and 30N. This transition phase can be explained by the presence of asperities generating a tangential force resisting the normal force. After this transition phase, Al_2O_3 coatings reinforced with TiO_2 and ZrO_2 gave the same average friction coefficient, which is approximately 0.92 under a 5-N load. However, under a 30-N load, the friction coefficient obtained from Al_2O_3–TiO_2 coatings is higher than that of Al_2O_3–ZrO_2

coatings (0.65 for $A_{12}O_3$-TiO_2 (AT-3) compared with 0.45 for Al_2O_3-ZrO_2 (AZ-25)) (Figure 4.12a,b). From these results, we can conclude that the type of reinforcement in alumina coatings affects tribological properties. Indeed, TiO_2 reinforcement, which increases the microhardness of the alumina matrix, generates a significant shear force, leading to a high friction coefficient. On the other hand, ZrO_2 reinforcement decreases the microhardness of alumina coatings, resulting in a sliding mode and a reduction in the friction coefficient. Therefore, it is evident that pressure influences the variation of the friction coefficient over time.

4.8.2 Determination of Wear Rate

The influence of the type of reinforcement in an alumina matrix by determining wear rates is explained in this section. Figure 4.13 illustrates the variation in the wear rate as a function of the load under a sliding speed of 0.5 m/s for both couples. Tribological tests were conducted for the two previously mentioned contact couples:

- **Couple A**: Al_2O_3-3%TiO_2 (AT-3)/ Al_2O_3-3%TiO_2 (AT-13)
- **Couple B**: Al_2O_3-25%ZrO_2 (Zr-25)/ Al_2O_3-3%TiO_2 (AT-13)

In Figure 4.13, it can be observed that the wear rate is relatively stable for couple B under loads from 5N to 20N, and then, this rate increases significantly for the

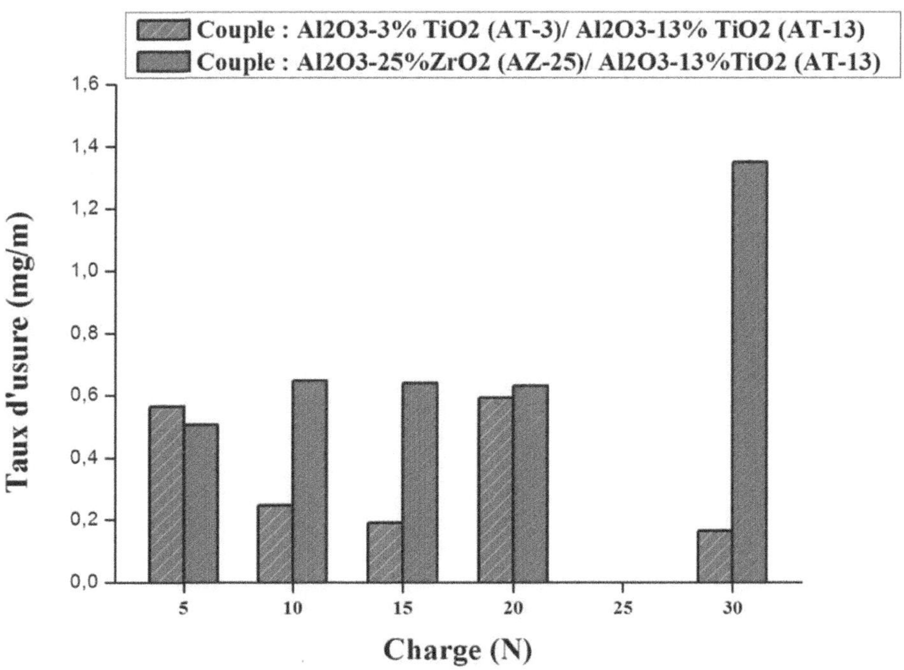

FIGURE 4.13 Variation in the wear rate of different deposits as a function of the normal load.

30-N load. This is likely due to the allotropic transformation of zirconia from the monoclinic to the quadratic form, accompanied by a volume increase of 5%–9% during thermal spraying. Furthermore, this transformation induces internal stresses that degrade the mechanical properties of couple B. Consequently, third-body particles appear, quickly ejected from the friction surface, while other particles undergo oxidation depending on the contact temperature. In contrast to couple B, there is a decrease in the wear rate after 20N for couple A. This can be explained by the similar chemical compositions of the two surfaces that do not allow for a shear mode. From this observation, it can be concluded that the load affects the tribological properties of coatings reinforced by these two types of reinforcements.

4.8.3 Roughness Profile of the Coated Pin

Figure 4.14 shows the variation in the roughness of the coated pin for different deposits as a function of the normal load. As shown in Figure 4.14, the recorded values of roughness for the studied couples show a slight difference at the beginning of the test, around a few tenths of a micron (μm). This difference increases with the applied load up to 20N. This can be explained by the emergence of a relief that promotes a roughness profile. From this observation, it can be stated that the applied load on the deposits influences the roughness profile of the contacting parts. After the tribological tests, we recorded the roughness profiles for the deposits under a load of 30 N. It varies on average from 7 μm for couple A to 12 μm for couple B.

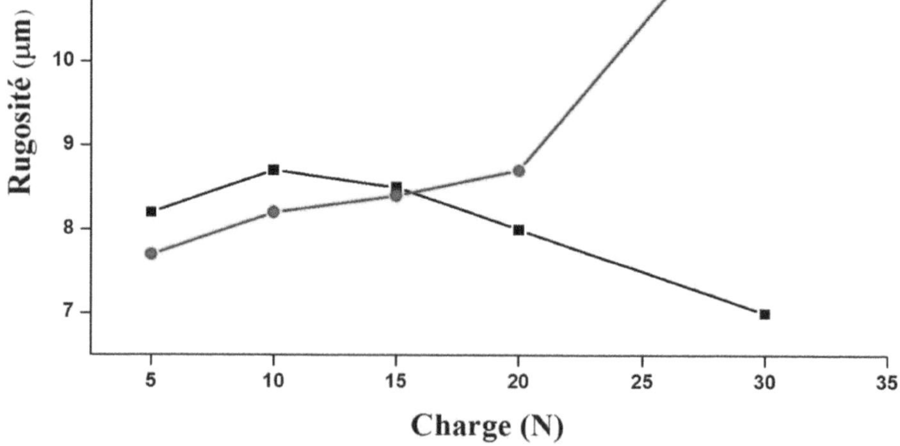

FIGURE 4.14 Variation in the roughness of the coated pin as a function of the normal load.

4.9 CONCLUSIONS

This study has successfully elaborated and characterized zirconia through a controlled sol-gel process, demonstrating its superior mechanical and thermal properties compared with other metal oxide ceramics such as alumina and titania. The synthesized zirconia exhibited excellent phase stability, hardness, and wear resistance, making it a prime candidate for applications in harsh environments, including aerospace, electronics, and biomedical engineering.

The characterization techniques revealed that the crystal structure and grain size of zirconia are highly influenced by synthesis conditions, with finer grain sizes contributing to improved mechanical properties. In addition, zirconia's higher fracture toughness and thermal stability distinguish it from other metal oxides, which are more prone to phase transformations under extreme thermal conditions. When compared to alumina, zirconia shows a marked improvement in toughness, though alumina retains better wear resistance in some applications. While titania and ceria exhibit lower hardness and thermal stability, they still hold advantages in specific catalytic and environmental applications.

In conclusion, zirconia is a promising material with unique properties that make it suitable for a wide range of industrial applications. However, further optimization of its synthesis processes and exploration of composite materials may further enhance its performance, opening new avenues for innovation in high-performance engineering.

REFERENCES

1. Wang, L., Jiao, Y., Yao, L., Sheng, Y., Hao, Z., Tang, W., & Dou, R. (2023). Investigation of mechanical properties and low-temperature degradation of dental 3Y-TZP ceramics fabricated by stereolithography in combination with microwave sintering. *Journal of the Mechanical Behavior of Biomedical Materials*, *148*, 2023. https://doi.org/10.1016/j.jmbbm.2023.106211.
2. Bermejo, R., Torres, Y., Baudin, C., Sanchez-Herencia, A. J., Pascual, J., Anglada, M., & Llanes, L. (2007). Threshold strength evaluation on an Al_2O_3–ZrO_2 multilayered system. *Journal of the European Ceramic Society*, *27*(2–3), 1443–1448.
3. Cattani-Lorente, M., Scherrer, S. S., Ammann, P., Jobin, M., & Wiskott, H. A. (2011). Low temperature degradation of a Y-TZP dental ceramic. *Acta Biomaterialia*, *7*(2), 858–865.
4. Chevalier, J., Deville, S., Münch, E., Jullian, R., & Lair, F. (2004). Critical effect of cubic phase on aging in 3 mol% yttria-stabilized zirconia ceramics for hip replacement prosthesis. *Biomaterials*, *25*(24), 5539–5545.
5. Mukhopadhyay, A. K., Sarapure, S., & Maiti, P. (2023). 18 - Advanced tape cast multilayer thin ceramics and composites with inelastic failure behaviors for damage-resistant applications. In *Advanced Flexible Ceramics* (pp. 391–410). Elsevier. https://doi.org/10.1016/B978-0-323-98824-7.00019-1.
6. Guo, L. (2023). 6—Ceramic coatings in turbine applications. In *Advanced Ceramic Coatings for Emerging Applications* (pp. 103–131). Elsevier. https://doi.org/10.1016/B978-0-323-99624-2.00001-2.
7. Tyona, M. D., & Itodo, A. (2023). 7 - Effects of surface modification (surface treatment) on friction and surface abrasion of ceramic composites. In *Surface Modification and Functionalization of Ceramic Composites* (pp. 91–114). Elsevier. https://doi.org/10.1016/B978-0-323-85883-0.00002-8.

8. Ruys, A. J. (2023). Chapter 9 - SiC-fibre reinforced SiC composites (SiC–SiC). In *Silicon Carbide Ceramics* (pp. 491–548). Elsevier. https://doi.org/10.1016/B978-0-323-89869-0.00011-3.
9. Chevalier, J., & Gremillard, L. (2008). 11- Zirconia ceramics. In *Bioceramics and Their Clinical Applications* (pp. 243–265). Woodhead Publishing. https://doi.org/10.1533/978 1845694227.2.243.
10. Danzer, R., Lube, T., Morrell, R., Supancic, P. (2013). Chapter 9.1- Mechanical properties of ceramics. In *Handbook of Advanced Ceramics* (Second Edition) (pp. 609–632). Academic Press. https://doi.org/10.1016/B978-0-12-385469-8.00033-2.
11. Telle, R. (2015). Chapter 1- Properties of ceramics. In *Handbook of Ceramics Grinding and Polishing* (pp. 1–49). William Andrew Publishing. https://doi.org/10.1016/B978-1-4557-7858-4.00001-7.
12. Anglada, M. (2017). Assessment of mechanical properties of ceramic materials. In *Advances in Ceramic Biomaterials* (pp. 83–109). Woodhead Publishing. https://doi.org/10.1016/B978-0-08-100881-2.00003-8.
13. Jorand, Y., Dubois, J., & Fantozzi, G. (1993). Elaboration by acid dispersion of yttria-stabilized zirconia-alumina composites with good mechanical properties. *Journal of Materials Science Letters*, 12(20), 1579–1581. https://doi.org/10.1007/BF00627019.
14. Exare, C., Kiat, J. M., Guiblin, N., Porcher, F., & Petricek, V. (2015). Structural evolution of ZTA composites during synthesis and processing. *Journal of the European Ceramic Society, 35*(4), 1273–1283. https://doi.org/10.1016/j.jeurceramsoc.2014.10.031.
15. Magnani, G., & Brillante, A. (2005). Effect of the composition and sintering process on mechanical properties and residual stresses in zirconia–alumina composites. *Journal of the European Ceramic Society, 25*(15), 3383–3392. https://doi.org/10.1016/j.jeurceramsoc.2004.09.025.
16. Bach, J. P., & Thevenot, F. (1989). Fabrication and characterization of zirconia-toughened alumina obtained by inorganic and organic precursors. *Journal of Materials Science, 24*(8), 2711–2721. https://doi.org/10.1007/BF02385615.
17. Rojas, A. R., Elguezabal, A. A., Porporati, A. A., Bernal, M. B., & Ponce, H. E. E. (2023). *Performance of Metals and Ceramics in Total Hip Arthroplasty.* Springer. https://doi.org/10.1007/978-3-031-25420-8.
18. Liang, Y. M., & Zhao, J. H. (1999). Effect of zirconia particle size distribution on the toughness of zirconia-containing ceramics. *Journal of Materials Science, 34*, 2175–2181. https://doi.org/10.1023/A:1004588516105.
19. Malode, S. J., & Shetti, N. P. (2022). 18- ZrO_2 in biomedical applications. In *Metal Oxides for Biomedical and Biosensor Applications* (pp. 471–501). Elsevier. https://doi.org/10.1016/B978-0-12-823033-6.00016-8.
20. Wallace, J. S., Claussen, N., Rühle, M., & Petzow, G. (1981). Development of phases in in situ-reacted mullite-zirconia composites. In *Surfaces and Interfaces in Ceramic and Ceramic — Metal Systems.* Springer. https://doi.org/10.1007/978-1-4684-3947-2_13
21. Savina, D. L., Tokiy, V. V., Tokiy, N. V., & Konstantinova, T. E. (2007). Modeling of dehydration and dehydrogenation in zirconia with anion impurity. In *Hydrogen Materials Science and Chemistry of Carbon Nanomaterials.* Springer. https://doi.org/1 0.1007/978-1-4020-5514-0_63.
22. Shi, J. L., Lin, Z. X., & Yan, D. S. (1990). Phase transformation charactheristic of nano-sized zirconia-alumina composite powder. In *Transactions of the Materials Research Society of Japan.* Springer. https://doi.org/10.1007/978-94-009-0789-8_23.
23. Shoja Razavi, R., & Loghman-Estarki, M. R. (2017). Advance techniques for the synthesis of nanostructured zirconia-based ceramics for thermal barrier application. In *Sol-Gel Based Nanoceramic Materials: Preparation, Properties and Applications.* Springer. https://doi.org/10.1007/978-3-319-49512-5_2.

24. Srinivasan, R., & Davis, B. H. (1997). Zirconia: A review of a super ceramic. In *Materials Synthesis and Characterization*. Springer. https://doi.org/10.1007/978-1-4899-0145-3_8.
25. Kikkawa, S., Kijima, A., Hirota, K., & Yamaguchi, O. (2002). Soft solution preparation methods in a ZrO_2–Al_2O_3 binary system. *Solid State Ionics, 151*(1–4), 359–364.
26. Palmero, P., Naglieri, V., Chevalier, J., Fantozzi, G., & Montanaro, L. (2009). Alumina-based nanocomposites obtained by doping with inorganic salt solutions: Application to immiscible and reactive systems. *Journal of the European Ceramic Society, 29*(1), 59–66.
27. Naglieri, V., Joly-Pottuz, L., Chevalier, J., Lombardi, M., & Montanaro, L. (2010). Follow-up of zirconia crystallization on a surface modified alumina powder. *Journal of the European Ceramic Society, 30*(16), 3377–3387.
28. Rozov, K., Curtius, H., & Bosbach, D. (2015). Preparation, characterization and thermodynamic properties of Zr-containing Cl-bearing layered double hydroxides (LDHs). *Radiochimica Acta, 103*(5), 369–378.
29. Vahid, B. R., Saghatoleslami, N., Nayebzadeh, H., & Toghiani, J. (2018). Effect of alumina loading on the properties and activity of SO_4^{2-} /ZrO_2 for biodiesel production: process optimization via response surface methodology. *Journal of the Taiwan Institute of Chemical Engineers, 83*, 115–123.
30. De Aza, A. H., Chevalier, J., Fantozzi, G., Schehl, M., & Torrecillas, R. (2002). Crack growth resistance of alumina, zirconia and zirconia toughened alumina ceramics for joint prostheses. *Biomaterials, 23*(3), 937–945.
31. Vautrot, L., Geandier, G., Mourot, M., Dehmas, M., Aeby-Gautier, E., Denand, B., & Denis, S. (2014). Internal stresses in metal matrix composites in relation with matrix phase transformations. *Advanced Materials Research, 996*, 944–950. https://doi.org/10.4028/www.scientific.net/amr.996.944.
32. Geandier, G., Dehmas, M., Mourot, M., Aeby-Gautier, E., Denis, S., Martin, O., & Karnatak, N. (2014). In situ structural evolution of steel-based MMC by high energy X-ray diffraction and comparison with micromechanical approach. *HTM Journal of Heat Treatment and Materials, 69*(1), 54–59.
33. Cho, S., Lee, J., Shin, S., Lee, D., Kim, M., Kwon, H., ... & Kim, J. (2023). Enhancing high-temperature properties of stainless steel composite with titanium carbide reinforcement: A study on coefficient of thermal expansion, thermal conductivity, and strength. *Journal of Materials Research and Technology, 25*, 7241–7253. https://doi.org/10.1016/j.jmrt.2023.07.162.
34. Fournier, S., Chevalier, J., Baeza, G. P., Chaput, C., Louradour, E., Sainsot, P., ... & Reveron, H. (2023). Ceria-stabilized zirconia-based composites printed by stereolithography: Impact of the processing method on the ductile behaviour and its transformation features. *Journal of the European Ceramic Society, 43*(7), 2894–2906.
35. CASAS, T. (2017). Zircone translucide & CFAO directe: Une réponse pertinente pour de nombreuses situations cliniques. *Actualités Odonto-Stomatologiques, 283*, 4. https://doi.org/10.1051/aos/2017034.
36. Shishido, S., Inagaki, R., Kanno, T., Svanborg, P., Barkarmo, S., Örtengren, U., & Nakamura, K. (2023). Residual stress associated with crystalline phase transformation of 3–6 mol% yttria-stabilized zirconia ceramics induced by mechanical surface treatments. *Journal of the Mechanical Behavior of Biomedical Materials, 146*, 106067. https://doi.org/10.1016/j.jmbbm.2023.106067.
37. Chevalier, J., Deville, S., Fantozzi, G., Bartolomé, J. F., Pecharroman, C., Moya, J. S., ... & Torrecillas, R. (2005). Nanostructured ceramic oxides with a slow crack growth resistance close to covalent materials. *Nano Letters, 5*(7), 1297–1301.
38. Baik, S., & White, C. L. (1987). Anisotropic calcium segregation to the surface of Al_2O_3. *Journal of the American Ceramic Society, 70*(9), 682–688. https://doi.org/10.1111/j.1151-2916.1987.tb05739.x.

39. Hezaveh, H., & Muhamad, I. I. (2013). Effect of MgO nanofillers on burst release reduction from hydrogel nanocomposites. *Journal of Materials Science: Materials in Medicine*, *24*(6), 1443–1453. https://doi.org/10.1007/s10856-013-4914-5.
40. Laby, L., Klein, L. C., Turniansky, A., & Avnir, D. (1997). Synthesis of alumina and silica-containing alumina xerogel hosts. *Journal of Sol-Gel Science and Technology*, *10*, 177–184. https://doi.org/10.1023/A:1018351630652.
41. Chen, H., Ren, B., Chen, L., Zhao, F., Bai, Q., Chen, J., & Li, B. (2021). Preparation of equiaxed α-Al_2O_3 by adding oxalic acid. *Ceramics International*, *47*(22), 31512–31517. https://doi.org/10.1016/j.ceramint.2021.08.029.
42. Gürel, S. B., & Altun, A. (2009). Reactive alumina production for the refractory industry. *Powder Technology*, *196*(2), 115–121.
43. Bekale, V. M., Legros, C., Haut, C., Sattonnay, G., & Huntz, A. M. (2006). Processing and microstructure characterization of ceria-doped yttria-stabilized zirconia powder and ceramics. *Solid State Ionics*, *177*(37–38), 3339–3347.
44. Younes, R., Bradai, M. A., Sadeddine, A., Mouadji, Y., Bilek, A., & Benabbas, A. (2016). Effect of TiO_2 and ZrO_2 reinforcements on properties of Al_2O_3 coatings fabricated by thermal flame spraying. *Transactions of Nonferrous Metals Society of China*, *26*(5), 1345–1352.
45. Younes, R., Bradai, M. A., Sadeddine, A., Mouadji, Y., Bilek, A., & Benabbas, A. (2015). Microstructural and tribological properties of Al_2O_3–13pctTiO_2 thermal spray coatings deposited by flame spraying. *Metallurgical and Materials Transactions B*, *46*, 2394–2403.
46. Issam, B., Rassim, Y., Abdelhek, I., Amokrane, B. M., Nadia, S., & Eugenia, D. (2023). The effect of post heat treatment and TiO_2 addition on tribological and electrochemical performances of thermally sprayed alumina. *Canadian Metallurgical Quarterly*, *63*(4), 1298–1311.

Section II

Industrial and Engineering Applications

5 Applications of Zirconia and Zirconia-Based Materials

Moniya, Jashanpreet Singh, Vikas Sharma, and Abdul Wahab Hashmi

5.1 INTRODUCTION

Zirconia (ZrO_2) is a versatile ceramic material with multi-industry applications owing to its unparalleled mechanical, thermal, and chemical properties [1,2]. Zirconia was initially developed for structural applications, but is currently being implemented in aerospace, dentistry, biomedical implants, gas sensors, cutting tools, and medical refractories. One of its most prominent features is its polymorphic nature, which allows it to exist in three different phases: monoclinic, tetragonal, and cubic [3–6]. It is possible to stabilize the tetragonal form with dopants such as yttria (Y_2O_3), magnesia (MgO), and calcium oxide (CaO), which increases its mechanical toughness and enables it to endure demanding applications [7,8]. Zirconia-based bioceramics find extensive use in dental restorations, prosthetic implants, and orthopaedic applications due to their biocompatibility and exceptional fracture resistance [9,10]. Dental zirconia has high aesthetic value and is mechanically durable and resistant to wear; thus, it is the most preferred material for crowns and bridges [11]. In orthopaedic applications, zirconia-based implants can be used instead of traditional metal implants as they are less likely to cause allergic reactions and are better tolerated by tissues [12,13]. Because of zirconia's extreme temperature and corrosive resistance, it is also useful in energy applications such as solid oxide fuel cells (SOFCs). The high ionic conductivity of yttria-stabilized zirconia (YSZ) enhances efficiency in SOFCs and sustainable energy production [14]. Furthermore, zirconia is widely used in the manufacturing of oxygen sensors due to high-temperature controlled oxygen levels in industrial and automotive applications that require precise measurement and control. Zirconia's unmatched applications range from its use in thermal barrier coatings (TBCs) for the aerospace and automotive industries to the manufacture of optical instruments and jewellery. Zirconia-based coatings enhance the service life of engines by preventing overheating. At the same time, zirconia's high refractive index, along with optical clarity, makes it a cheaper alternative to diamonds, and thus, it is used in cubic zirconia (CZ) and optical tools [15,16]. Though facing challenges with phase transformation at low temperatures, potential brittleness, and fabrication complexity, numerous advantages of zirconia make it an ever-researched material [17]. Advanced processing techniques in nanotechnology and composite formulations are bound to improve the toughness and stability of zirconia-based materials [18].

DOI: 10.1201/9781003519300-7

5.2 MATERIAL PROPERTIES RELEVANT TO APPLICATIONS

Due to its remarkable mechanical, thermal, electrical, and biocompatible features, zirconia is a highly versatile material. As for its mechanical properties, zirconia has high hardness (~8.5 Mohs), excellent wear resistance, and superior fracture toughness, which makes it ideal for cutting tools, crowns in dentistry, and structural applications. The ability of zirconia to withstand crack propagation through transformation toughening further enhances its durability under high-stress environments such as biomedical implants and aerospace components. Zirconia has a high melting point (~2715°C), making it thermally critical as it can withstand extreme temperatures without structural degradation. Moreover, its low thermal conductivity allows it to be an excellent insulator that is applied widely in TBCs for jet engines and refractory linings. YSZ is a widely known solid electrolyte with high ionic conductivity, strengthening its position in SOFCs as well as oxygen sensors in automobile and industry appliances. Adding biocompatibility enhances the importance of zirconia further since it is nontoxic, corrosion resistant, and highly compatible with biological systems, which makes it the primary choice in dental and orthopaedic implants. Altogether, these properties have made zirconia an essential material in diverse industries ranging from aerospace, energy, and healthcare to advanced engineering.

5.3 INDUSTRIAL CUTTING TOOLS AND WEAR-RESISTANT COMPONENTS

Industries such as paper and packaging, textiles, woodworking, automotive, medical, and food production use zirconium oxide in cutting tools because of its unique durability, resistance to fatigue, and toughness [19–21]. These tools are essential where electrical insulation and edge strength are required. Cutting edges in yarns used in textiles have important applications where yttrium zirconium ceramic (Y-TZP) is known for exceptional cutting-edge strength [22]. Zirconium oxide ceramics are resistant to corrosion, which enable acids and even alkalis to be used in cleaning, thus significantly prolonging service life. They are now widely used in the industry; for instance, all-in-one thread guides and splicer shears used in winding machines can optically detect yarn that is too thin or too thick and cut it out [23]. Tests show that the lifetime of ceramic components is about four times longer than that of those made with hard metals. Furthermore, the introduction of zirconium oxide in cutting edges of weaving machines has proven to be successful as it can withstand up to ten cuts per second, as opposed to the few cuts per minute that winding machines manage. The unique combination of hardness, wear, and toughness makes zirconium oxide suitable for use in kitchen knives. Apart from cutting instruments, zirconium oxide ceramics are primarily used in protective and wearing parts, particularly in forming technology, where tools should resist very high forces and loads. One of the important properties is its low friction when rubbing against metals, which helps to further decrease wear. A remarkable example is the manufacture of spindles, which are devices that transform rotary motion into linear motion. Even under the most demanding conditions, ceramic spindle units can operate for long periods without suffering any significant wear, even after millions of cycles. Furthermore, the incorporation of zirconium oxide completely removes the possibility of galling, which is the adhesion of sliding surfaces due to mechanical wear. In pumps and engines, the use of zirconium oxide is increasing, especially in pump impellers. Because these elements are usually not subjected to dynamic or tensile loads, over-fracture toughened zirconium oxide is ideal for such critical components, which need to be reliable under harsh industrial conditions.

5.4 BIOMEDICAL AND DENTAL APPLICATIONS

5.4.1 Zirconium Oxide in Dentistry

Zirconium oxide is widely known for its use in dental technology because of its strength, impressive fracture toughness, biocompatibility, and good aesthetics [24]. The increasing importance of zirconia in dental restorations is attributed to its ability to resemble natural dentition. This substance finds application in orthodontic brackets and dental prosthetics for root posts, crowns, and bridges [25]. Furthermore, zirconia is used for basic materials in surgical abutments and implants. More sophisticated mixed ceramics based on zirconium oxide, like alumina-toughened zirconia (ATZ), are also used in medicine for the production of hearing aids and finger endoprostheses. Zirconia is not an entirely new material in medicine; it has been investigated as an implant material for hip joint prosthetic devices with good results in long-term studies [26].

5.4.2 Requirements for Dental Crowns and Bridges

Dental applications use zirconium oxide ceramics for a range of procedures, from conservative restorations and crowns to bridges and implant prostheses. These materials should meet the crucial requirements of surgical implant materials; they should be low in radioactivity and hence safe enough to use in the human body [27]. The key advantages of zirconium oxide as dental materials are its biocompatibility, low solubility in water, and aesthetic appeal, especially its ability to look like natural teeth. However, initially, this opacity precluded its application elsewhere apart from posterior teeth without ceramic veneering [28]. However, the advent of more translucent forms of zirconia has made it possible to apply them to other more visible areas as they allow light to pass through them better than their predecessors. Dental zirconium oxide has three primary colours: light, medium, and intensive. Extra matching of shades can be done with staining techniques. Newer shades of traditional zirconium oxide frameworks, which require ceramic veneering for perfect aesthetics, enable monolithic crowns and bridges without secondary layering. As much as this improved the aesthetics, translucency is lost, which brings forth the concern regarding mechanical properties like fracture toughness of ZrO_2 [29]. An advantage that stands out with this material is its poor heat conductivity, which is nearly equal to the properties of natural teeth, hence reducing thermal irritations to improve comfort during treatment. To make dental treatments easier, unlike most other ceramics, zinc phosphate or glass ionomer cements may also be bonded with zirconium oxide. In addition, its mechanical features such as hardness, bending strength, and elastic

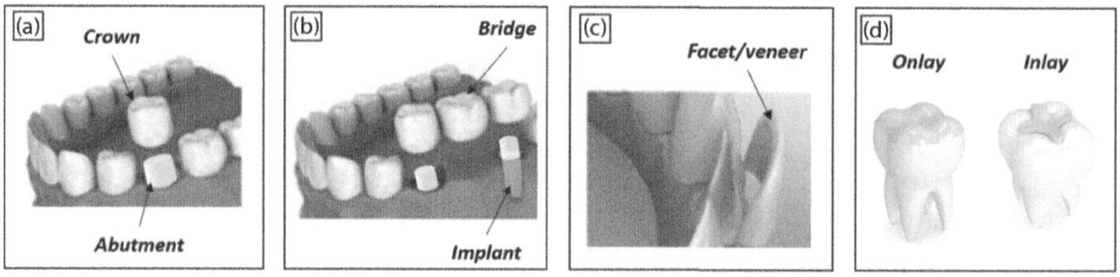

FIGURE 5.1 Types of dental restorations: (a) crown and abutment; (b) bridge and implant; (c) facet/veneer; (d) onlay and inlay [30].

modulus are very appropriate for dental use. Furthermore, an important method to improve the strength of zirconium oxide is transformation toughening. When tensile stress is applied, the tetragonal phase of the material transforms into the monoclinic phase, and this particular transformation increases the volume by about 5%, which helps in closing cracks and preventing them from worsening. Although wearing down the opposing teeth is possible, especially for a patient who tends to grind their teeth (bruxism), the associated longevity makes it difficult to ignore because of its superior hardness. Another possible disadvantage is chipping, where ceramic veneers separate from the underlying zirconium oxide infrastructure lattice. Opposing teeth may get worn out due to high levels of abrasion resistance inherent in monolithic zirconium oxide crowns. Figure 5.1 shows the different types of dental restorations.

Notably though in dental jargon, these are referred to as "zircon" or "zircon oxide" crowns rather than "zirconium oxide" crowns, which is actually the more formal term scientifically. This is because people commonly believe that "zircon" refers to the metal, hence thinking that such crowns are metallic not ceramic.

5.4.3 Fabrication of Zirconium Oxide Ceramic Dentures

The conventional techniques of sintering, which use direct powder processing, are poorly implemented for the fabrication of zirconium oxide dental restorations within the laboratory. Thus, prefabricated forms of zirconium oxide ceramics, such as green and white compacts, tend to be used instead. These materials are now available in a presintered state where they are porous and chalk-like, hence easier to machine. These blanks are used to form crowns and bridge frameworks, which are further processed by carbide or diamond grinding tools. Due to material shrinkage, which occurs during final sintering, workpieces are made about 25% larger than the desired dimensions upfront. Material properties are then improved by the densification of porosity and compaction of the ceramic, which occurs during final sintering [12,31–33].

The field of dental prosthetics has been profoundly affected by developments in computer-aided design (CAD) and computer-aided manufacturing (CAM), with automation enabling the milling of dental blanks, which are available in bulk and have relatively low shrinkage and high edge stability. Such blanks undergo precompaction and presintering processes that increase their machinability, which makes them easier, faster, and more precise to manufacture. The speed and accuracy with which CAD/CAM techniques make dental restorations both before and after the final sintering

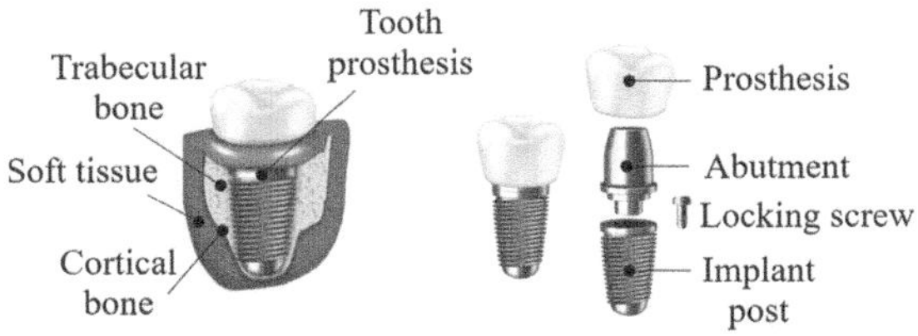

FIGURE 5.2 Dental implant with components [35].

stage is unrivalled. Although most dental technicians are switching to CAD/CAM methodologies, there are still some who prefer traditional methods as they allow for the creation of one-of-a-kind, custom restorations as opposed to the more generic, mass-produced custom restorations. It is likely that these techniques will be adopted more widely due to the CAD and CAM features that significantly ease the process.

5.4.4 Zirconium Oxide Implants

Zirconium oxide has also been shown to be a rather effective material for endosseous implants, which are placed within the jawbone. Figure 5.2 shows the dental implant with components, and their use is achievable due to the material's combination of strength, biocompatibility, and aesthetic properties. For example, alumina ceramics have not become widely accepted as dental implant materials, but zirconium oxide, especially yttrium-stabilized TZP-A, has considerable benefits. This material undergoes further processing with the hot isostatic pressing (HIP) method, where sintered ceramics are treated under high pressure after they are formed in order to increase their density and enhance mechanical attributes. Unlike titanium implants, zirconium oxide implants are more attractive because they are white and blend well with natural teeth. They are ideal for patients with thick gums. Like titanium implants, zirconium oxide implants also osseo-integrate well with bone, but the former require a longer healing time. One of the strongest advantages of zirconium oxide is that it has less bacterial retention around the gum line, which is helpful for oral health. On the negative side, such implants have some downsides such as increased pricing and insufficient data on long-term clinical use. The ageing issue of zirconium oxide is also a problem, adding to the lack of knowledge for optimal surface features for osseo-integration. Further studies must be done to create surface features analogous to etched titanium implants in order to enhance bone bonding. Apart from crowns, bridges, and implants, zirconium oxide ceramics are incorporated into various dental products. For instance, drills used in dentistry and manufactured from zirconium oxide have a slender, three-sided paradigm and great resistance to wear, making them indispensable in contemporary dentistry [34].

5.5 THERMAL BARRIER COATINGS AND HIGH-TEMPERATURE APPLICATIONS

Because zirconium oxide is highly resistant to heat, has low thermal conductivity, and has high phase stability, it is widely used in high-temperature applications [36,37]. It is actively used in TBCs in gas turbines, jet engines, and industrial kilns, where it improves the efficiency and durability of the components at high temperatures. In kiln furniture, ZrO_2 acts as an inert thermal shock support for ceramics, metals, and glass during their firing processes while withstanding over 1500°C [38,39]. As a result, it is essential in the ceramics, glass, and metal industries. In the same vein, ZrO_2 is used in TBCs for gas turbines and jet engines due to its low thermal conductivity, high melting point, and thermal expansion compatibility with Ni-based superalloys, which aids in reducing the metal temperature by 200°C–300°C and avoiding thermal fatigue, CMAS corrosion, and oxidation [40,41]. It is often used in turbine blades, combustion chambers, and exhaust nozzles, where it increases lifespan [42]. This enables engines like those in the Rolls-Royce Trent series to function at high temperatures over 1400°C, which reduces emissions and fuel consumption while increasing efficiency [43]. Aside from TBCs, zirconium oxide is broadly used within the aerospace, automotive, and power generation industries in relation to engine components that operate at high temperatures [44]. It is often used in aerospace nozzles, hypersonic aircraft exhaust components, and rocket engines because of their high wear resistance coupled with low thermal conductivity (11×10^{-6}/K) [45]. These components can survive aerodynamic heating. In the automotive industry, they are incorporated in turbine blades of turbocharger turbines, as well as in pistons to improve heat endurance and efficiency [46–48]. Moreover, in reactors such as very high-temperature gas-cooled reactors (VHTRs) with ZrO_2 deployed, fuel rods and reactor cores are protected from thermal degradation over 1000°C [45,46,49,50]. Subsequent developments focusing on the utilization of zirconium oxide aim towards next-generation TBCs with dual-phase zirconates to improve self-insulation and self-mending ceramics that heal microfractures through thermal stress. Figure 5.3 shows the schematic diagram of a gas turbine, blade of the turbine, and cross-section TBC system.

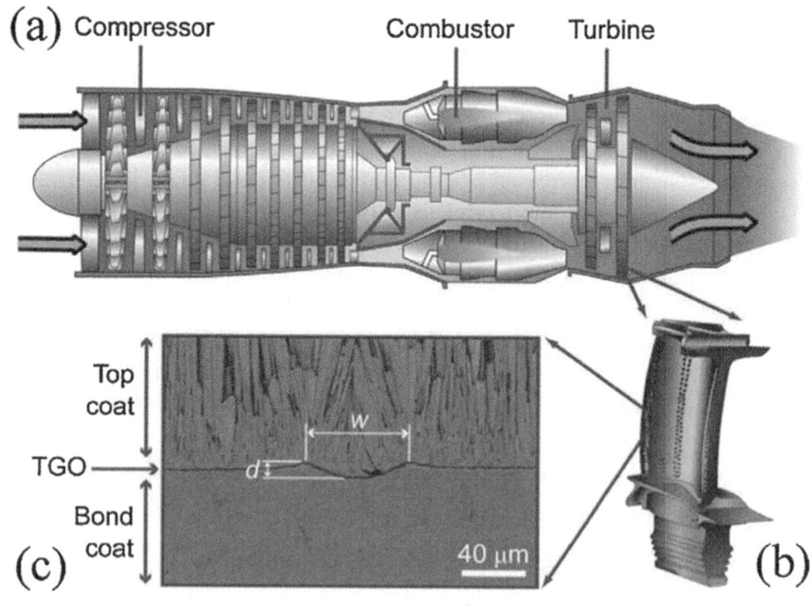

FIGURE 5.3 (a) Schematic diagram of a gas turbine, (b) blade of the turbine, and (c) cross-section TBC system [51,52].

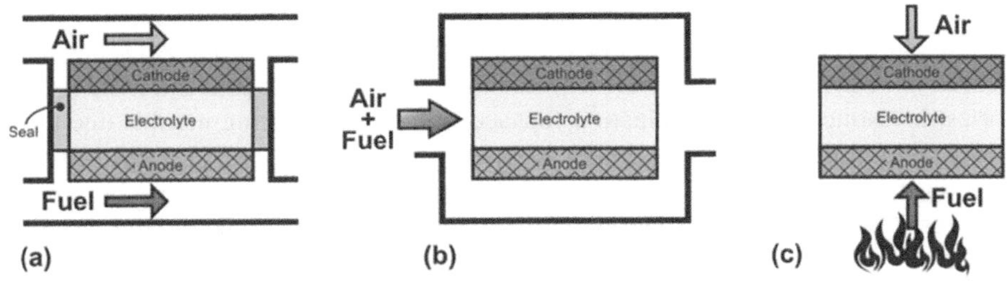

FIGURE 5.4 Schematics of (a) dual-chamber SOFC, (b) single-chamber SOFC, and (c) no-chamber SOFC [64].

In addition, there are plans for 3D-printed TBCs for specific turbine parts [36,53]. Further reasons for the advanced thermal protection capabilities of ZrO_2 are ongoing research aimed at novel coatings for hypersonic vehicles and re-entry spacecraft, alongside self-insulating TBCs. Ongoing efforts in tuning ZrO_2 TBCs for modular turbine components enhance cutting-edge technologies [54–56].

5.6 OXYGEN SENSORS AND SOLID OXIDE FUEL CELLS

The use of ZrO_2 in oxygen sensors and SOFCs oxygen ions is due to its conductive, thermally stable, and chemically resistant properties [57–59]. Oxygen ZrO_2 sensors enable precise fuel monitoring and emission control. They are widely used in automotive exhaust systems, industrial combustion monitoring, and medical devices [60]. The stability of ZrO_2 at high temperatures (600°C–1000°C) makes it ideal for lambda sensors, where it ensures proper response time (TPRH) to control the air–fuel ratio within strict emission standards of Euro 6 and EPA. Furthermore, YSZ is a solid electrolyte used in zirconia SOFCs. These batteries are used in stationary power supplies, military equipment, and newer vehicles powered by hydrogen because of their efficiency, high fuel flexibility, and long lifespan [61–63]. SOFCs require a temperature of 700°C–1000°C [18]. Figure 5.4 shows the schematics of a dual-chamber SOFC, single-chamber SOFC, and no-chamber SOFC. Further developments are aimed at sensor miniaturization with the possibility of using GDC (gadolinium-doped ceria) or SDC (samarium-doped ceria) as a material to decrease the operational temperature of SOFC and improve monitoring capabilities in ZrO_2 hydrogen production and storage. As a key material in clean energy and emission control technologies, zirconium oxide continues to drive innovations in sustainable power generation and environmental monitoring.

5.7 REFRACTORY AND METALLURGICAL APPLICATIONS

Because of its melting point, which is higher than 2700°C, zirconium oxide is extensively applied in refractors and metallurgical purposes. Besides, its exceptional chemical resistance and low wettability from molten metals make ZrO_2 even more important. In high-temperature industries such as steel, ceramics, and petrochemicals, zirconium oxide is used in the production of crucibles, as well as furnace linings and protective coatings. This ensures thermal stability and resistance to chemical attacks from molten metal oxides, acids, and alkalis. In addition, its superior fracture toughness enables thermal cycling endurance without cracking. ZrO_2 has been proven to be more effective as a low-wettability component for tundish, ladle refractories, investment casting, induction furnace linings, and casting moulds due to its lower contamination on metal products. Besides, it improves the efficiency of metal refining by extending the lifespan of furnace components and improving corrosion resistance, thermal insulation, and casting precision. Further research is being pushed towards nanoengineered refractors, hybrid ceramic metal composites, and 3D-printed refractor structures to improve durability and thermal shock resistance even further. Zirconium oxide strongly delivers when utilized in high-performance refractor systems, industrial heat treatment processes, and metallurgical casting, making it essential to these applications.

5.8 ELECTRICAL AND ELECTRONIC COMPONENTS

Due to the superior dielectric strength, poor electrical conductivity, and excellent thermal stability of ZrO_2, it is extensively used in electrical and electronic systems. Its ability to sustain high voltages without breakdown enables its ideal applications in high-power electronics, capacitors, and circuit components. In addition, because of the resistance of ZrO_2 to thermal shock and its high melting point, it can be used in induction and microwave heating systems as a high-frequency heating element. Its mechanical strength and chemical inertness give stability in microelectronics, which makes it suitable for use as dielectric layers, insulating substrates, and semiconductors. Moreover, its non-magnetic nature prevents interference with sensitive electronic devices such as systems for MRI and airplane electronics, making these unique properties extremely useful features of ZrO_2 in advanced electrical and electronic systems. Figure 5.5 shows the schematic diagram of the MLCC (multilayer ceramic capacitor) fabrication process in which zirconium is used.

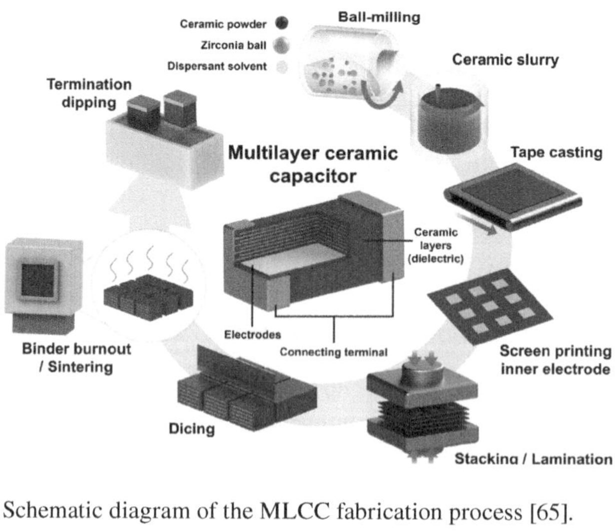

FIGURE 5.5 Schematic diagram of the MLCC fabrication process [65].

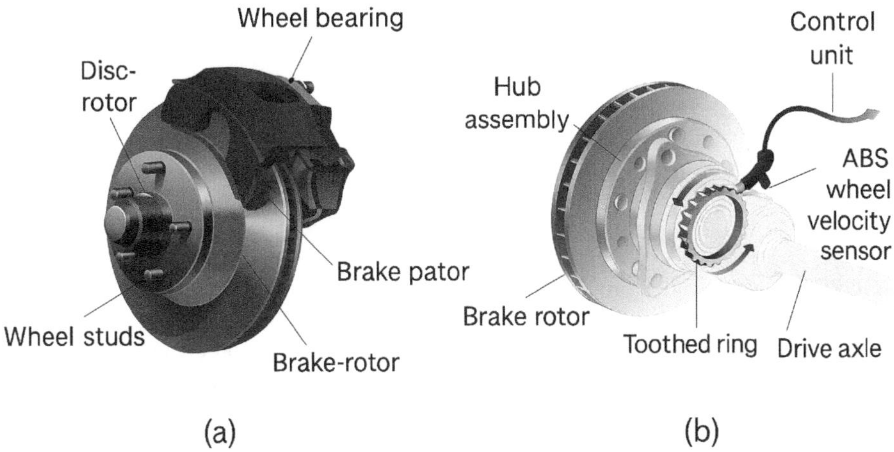

FIGURE 5.6 Braking systems: (a) pads with disc; (b) antilock braking system (ABS) [16].

5.9 AUTOMOTIVE INDUSTRY

Due to their outstanding toughness, mechanical resistance, and thermal insulation properties, zirconia ceramics exhibit extensive use in the automotive industry. Their low thermal conductance, in combination with a relatively high thermal expansion coefficient, allows them to act as insulators in engine combustion chambers and enhances thermal productivity by minimizing heat loss. While zirconia is used in the production of critical engine components, such as cylinder head bottom plates, valve seat rings, and piston crowns, large-scale commercial adoption remains limited as these ceramic components tend to exhibit differing degrees of strength under extreme engine operating conditions. Zirconia ceramics possess higher resistance to strength and high temperatures in comparison with alumina ceramics, which makes them ideal materials to encapsulate combustion chambers, specifically in diesel engines [66]. There are ongoing studies to replace zirconia with lightweight ceramic materials in piston pins, piston rings, and other moving parts to improve engine efficiency. Moreover, in some automobiles, especially luxury cars, zirconia ceramic is used in sensors or shock absorbers due to their piezoelectric and electrorestrictive properties, which enables adaptive vibration damping on rough road surfaces. Figure 5.6 shows the cross-section of a ceramic brake disc with zirconia reinforcement.

5.10 COMPOSITE MATERIALS AND ADVANCED CERAMICS

Zirconia-toughened alumina (ZTA) improves the fracture toughness of ceramic materials due to the increased toughness accompanied by the high hardness of zirconia and alumina. This gives ZTA wide applications in industries and bioengineering fields that require high durability and wear resistance [67,68]. On the other hand, ATZ has superior mechanical strength and toughness, which renders it highly efficacious for utilization in dental and orthopaedic biocompatible implants, where longevity

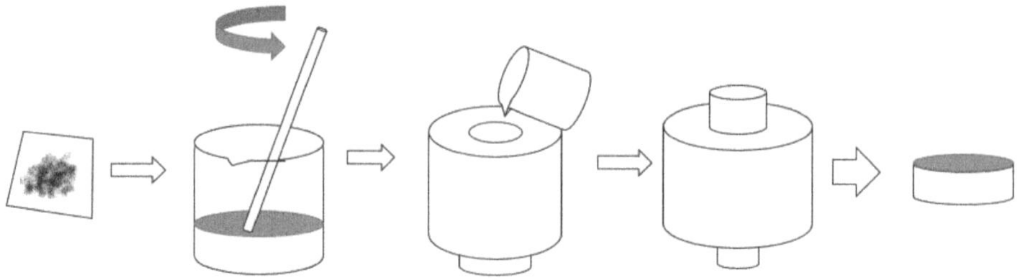

FIGURE 5.7 Preparation process diagram of zirconia coating.

is crucial. Furthermore, the use of particle- and fibre-reinforced zirconium oxide ceramic composites has shown a significant increase in fracture toughness, which enables their application in aerospace and mechanical engineering, where highly increased strength, thermal stability, and resistance to extreme operating conditions are prerequisites [68,69].

5.11 PIGMENTS AND COATINGS

ZrO_2 is widely used in the production of white pigments in ceramics due to its unparalleled brightness and opacity. These features give enhanced value and strength to ceramic products. Due to its high refractive index, light scattering is superlative, which makes it a useful whitening agent in glazes, tiles, and sanitary ware. Moreover, ZrO_2 is also used as a coating in automotive paints, varnishes, and electronic devices, where it greatly increases resistance to scratches, thus improving the durability and lifetime of coatings. Aside from its application as a white pigment, ZrO_2 also produces a yellow compound when combined with vanadium oxide, from which a highly vibrant yellow pigment is obtained, which withstands a large range of conditions and is therefore used in the production of ceramics, coatings, and decorative surfaces that require vivid colour with high stability and wear resistance. Figure 5.7 shows the preparation process diagram of zirconia coating.

5.12 USE IN CHEMICAL AND CORROSIVE ENVIRONMENTS

For its excellent wear resistance and chemical stability, ZrO_2 is utilized in many industrial applications where aggressive chemicals are present. Its corrosion resistance, along with its ability to mechanically sustain stress, makes it reliable in demanding fields such as the pharmaceutical and chemical processing industries. ZrO_2 is used in valve parts, pump impellers, and press dies, which are some of the most severely corroded components in chemical industries. To maximize efficiency and longevity, these components should have high durability and resistance to chemical degradation. In addition, ZrO_2 is used as glue rollers in cigarette-producing machines as it offers smooth, low-wear operation with prolonged life due to its corrosion resistance against adhesives. Such properties make ZrO_2 a crucial material in precision-driven industrial processes.

5.13 JEWELLERY AND SYNTHETIC GEMSTONES

CZ is now widely used as a diamond simulant because of its great hardness, optical beauty, and similarity to natural diamonds. Due to its high refractive index and great dispersion, which makes CZ reflect light like a real diamond and beautifully sparkle in its true form, CZ is one of the best jewellery adornments for those who wish to achieve a luxurious appearance at a reasonable cost. Diamonds are known to dazzle and mesmerize onlookers as a result of the unique way in which they scatter light. CZ is far more accessible in terms of affordability than natural diamonds, which makes it most appealing. In contrast to diamonds, which are expensive and hard to come by due to how they are created and subsequently extracted, CZ is synthetically produced, making it less expensive. Its accessibility enables the public to appreciate the beauty of exquisite jewellery while still being economical, quickly turning CZ into the most popularly chosen stone for modern-age rings, necklaces, earrings, bracelets, and other adornments. Along with its beauty, its durability and resistance to scratches make it appealing as well, having a Mohs hardness rating of 8–8.5. This means CZ is softer and is only surpassed by diamonds and a few others. This quality means jewellery made from CZ can withstand the daily wear and tear from use while retaining its polish and lustre for a long period. Furthermore, unlike some gemstones, CZ does not need special maintenance to retain its brilliance, making CZ products moisture- and heat-friendly. CZ might have many advantages, but is different from diamonds. The lack of natural "blemishes" provided by diamonds results in CZ being too perfect under examination. Without these imperfections, zirconia becomes denser and softer, making it easier to suffer minor abrasions over time. Thus, people in search of brilliant-looking jewellery without steep prices can always opt for CZ, making it a great option in the world of fashion and fine jewellery.

5.14 CONCLUSION

With regard to the electronics, aerospace, automotive, biomedical, and other industries, zirconium oxide ceramic is extensively used across these sectors because of its mechanical strength, thermal resistance, chemical stability, and exceptional optical properties. Due to its ability to withstand high temperatures and highly corrosive environments, ZrO_2 is an ideal candidate for structural components of thermal barriers and industrial coatings. For microelectronics, high-power insulators, and high-frequency heating elements, zirconium oxide serves as a stable insulating material for electronic devices and is therefore a critical component for semiconductor insulation. In the medical sector, zirconia-based ceramics are being used in surgical instruments, medical implants, and plastic prostheses. In the fields of automotive and aerospace engineering, ZrO_2 is crucial for improving the durability and efficiency of engines. The application of zirconium oxide in combustion chamber insulation, cylinder head components, and valve seat rings results in enhanced performance alongside providing higher fuel economy. The biomedical field benefits greatly by the application of zirconia as well: in dental and orthopaedic implants, biocompatible, wear-resistant, and mechanically strong ZrO_2 ensures reliable long-term use.

The jewellery industry's preference for zirconium oxide lies within its optical brilliance and hardness, especially in the use of CZ as an alternative to diamond because it is economical and easy to use. On top of its uses in jewellery, ZrO_2 is integrated in the production of coatings, pigments, and ceramics, where it enhances resistance to wear, colour fidelity, and durability in automobile paints, shiners, and ceramic glazes. In the pharmaceutical, chemical, and cigarette-manufacturing industries, its use in valve and pump parts, press dies, and glue rollers is made possible by its corrosion and wear resistance. As strength and hardness requirements increase, reinforced zirconia ceramics with added fibres and particles have greater fracture toughness, which is ideal for aerospace and mechanical engineering applications. With the progression of technology, this will change with the application of zirconium oxide in SOFCs, biomedical engineering, and advanced coatings, which will increase its use in protective structural materials. Its versatility and blend of strength and stability render zirconium oxide an essential material in advances in modern engineering, science, and industries.

REFERENCES

1. F. Murillo-Gómez et al., "Mechanical, adhesive and surface properties of a zirconia-reinforced lithium silicate CAD/CAM ceramic exposed to different etching protocols," *Materials*, vol. 17, no. 20, Oct. 2024, doi: 10.3390/ma17205039.
2. Y. Han and J. Zhu, "Surface science studies on the zirconia-based model catalysts," *Top Catal*, vol. 56, no. 15–17, pp. 1525–1541, 2013, doi: 10.1007/s11244-013-0156-5.
3. N. Kumari et al., "Zirconia-based nanomaterials: Recent developments in synthesis and applications," Aug. 23, 2022, doi: 10.1039/d2na00367h.
4. G. Albarghouti, S. Mahmoud, and F. Darwish, "Processing of zirconia." [Online]. Available: www.intechopen.com
5. L. P. Webber, H. L. Chan, and H. L. Wang, "Will zirconia implants replace titanium implants?," *Appl Sci*, vol. 11, no. 15, Aug. 2021, doi: 10.3390/app11156776.
6. G. Magnani, P. Fabbri, E. Leoni, E. Salernitano, and F. Mazzanti, "New perspectives on zirconia composites as biomaterials," *J Compos Sci*, vol. 5, no. 9, Sep. 2021, doi: 10.3390/JCS5090244.
7. T. D. Vulić, M. V. Carević, N. D. Abazović, T. B. Novaković, Z. D. Mojović, and M. I. Čomor, "Application of zirconia/alumina composite oxide ceramics as photocatalysts for removal of 2,4,6-trichlorophenol from water," *Photochem*, vol. 2, no. 4, pp. 905–917, Dec. 2022, doi: 10.3390/photochem2040058.
8. J. Xie et al., "Research progress of high dielectric constant zirconia-based materials for gate dielectric application," Jul. 01, 2020, doi: 10.3390/coatings10070698.
9. S. Lamnini, D. Pugliese, and F. Baino, "Zirconia-based ceramics reinforced by carbon nanotubes: A review with emphasis on mechanical properties," Sep. 01, 2023, doi: 10.3390/ceramics6030105.
10. S. Liu et al., "Tetragonal nanosized zirconia: Hydrothermal synthesis and its performance as a promising ceramic reinforcement," *Inorganics*, vol. 11, no. 5, May 2023, doi: 10.3390/inorganics11050217.
11. A. Y. Alqutaibi et al., "Revolution of current dental zirconia: A comprehensive review," Mar. 01, 2022, doi: 10.3390/molecules27051699.
12. L. N. Khanlar, A. S. Rios, A. Tahmaseb, and A. Zandinejad, "Additive manufacturing of zirconia ceramic and its application in clinical dentistry: A review," Sep. 01, 2021, doi: 10.3390/DJ9090104.

13. A. M. A. Moncayo, L. Peñate, M. Arregui, L. Giner-Tarrida, and R. Cedeño, "State of the art of different zirconia materials and their indications according to evidence-based clinical performance: A narrative review," Jan. 01, 2023, doi: 10.3390/dj11010018.
14. A. Arena et al., "Nanostructured zirconia-based ceramics and composites in dentistry: A state-of-the-art review," *Nanomaterials*, vol. 9, no. 10, Oct. 2019, doi: 10.3390/nano9101393.
15. G. Pezzotti et al., "Osteogenic enhancement of zirconia-toughened alumina with silicon nitride and bioglass®," *Ceramics*, vol. 2, no. 4, pp. 554–567, Dec. 2019, doi: 10.3390/ceramics2040043.
16. F. Ilie and A. C. Cristescu, "Tribological behavior of friction materials of a disk-brake pad braking system affected by structural changes—A review," *Materials*, vol. 15, no. 14, Jul. 2022, doi: 10.3390/ma15144745.
17. B. Arnold, *Zircon, zirconium, zirconia - similar names, different materials*. Springer, Berlin, Heidelberg, 2022. doi: 10.1007/978-3-662-64269-6.
18. N. Donker, D. Schönauer-Kamin, and R. Moos, "Mixed-potential ammonia sensor based on a dense yttria-stabilized zirconia film manufactured at room temperature by powder aerosol deposition," *Sensors*, vol. 24, no. 3, Feb. 2024, doi: 10.3390/s24030811.
19. P. Sokola et al., "Comprehensive study of stereolithography and digital light processing printing of zirconia photosensitive suspensions," Dec. 01, 2024, doi: 10.3390/ceramics7040104.
20. T. N. Rao, I. Hussain, J. E. Lee, A. Kumar, and B. H. Koo, "Enhanced thermal properties of zirconia nanoparticles and chitosan-based intumescent flame retardant coatings," *Applied Sciences*, vol. 9, no. 17, Sep. 2019, doi: 10.3390/app9173464.
21. C. I. Resende-Gonçalves et al., "Porous zirconia blocks for bone repair: An integrative review on biological and mechanical outcomes," Mar. 01, 2022, doi: 10.3390/ceramics5010014.
22. K. Jitwirachot, P. Rungsiyakull, J. A. Holloway, and W. Jia-Mahasap, "Wear behavior of different generations of zirconia: Present literature," 2022, doi: 10.1155/2022/9341616.
23. T. D. Vulić, M. V. Carević, N. D. Abazović, T. B. Novaković, Z. D. Mojović, and M. I. Čomor, "Application of zirconia/alumina composite oxide ceramics as photocatalysts for removal of 2,4,6-trichlorophenol from water," *Photochem*, vol. 2, no. 4, pp. 905–917, Dec. 2022, doi: 10.3390/photochem2040058.
24. A. M. Barbasz and B. Dyba, "Direct interaction of zirconia nanoparticles with human immune cells," *Biophysica*, vol. 4, no. 1, pp. 83–91, Mar. 2024, doi: 10.3390/biophysica4010006.
25. C. Larsson and P. V. Von Steyern, "Implant-supported full-arch zirconia-based mandibular fixed dental prostheses. Eight-year results from a clinical pilot study," *Acta Odontol Scand*, vol. 71, no. 5, pp. 1118–1122, Sep. 2013, doi: 10.3109/00016357.2012.749518.
26. P. Pittayachawan, "Comparative study of physical properties of zirconia based dental ceramics," 2014. [Online]. Available: https://www.researchgate.net/publication/32897236
27. D. Morena et al., "Comparative clinical behavior of zirconia versus titanium dental implants: A systematic review and meta-analysis of randomized controlled trials," Aug. 01, 2024, doi: 10.3390/jcm13154488.
28. M. F. R. P. Alves, S. Ribeiroa, P. A. Suzuki, K. Strecker, and C. dos Santosc, "Effect of Fe_2O_3 addition and sintering temperature on mechanical properties and translucence of zirconia dental ceramics with different Y_2O_3 content," *Materials Research*, vol. 24, no. 2, Mar. 2021, doi: 10.1590/1980-5373-MR-2020-0402.
29. Y. Zhang and B. R. Lawn, "Novel zirconia materials in dentistry," *J Dent Res*, vol. 97, no. 2, pp. 140–147, Feb. 2018, doi: 10.1177/0022034517737483.
30. A. C. Branco, R. Colaço, C. G. Figueiredo-Pina, and A. P. Serro, "Recent advances on 3D-printed zirconia-based dental materials: A review," Mar. 01, 2023, doi: 10.3390/ma16051860.

31. W. Burger and G. Kiefer, "Alumina, zirconia and their composite ceramics with properties tailored for medical applications," *Journal of Composites Science*, vol. 5, no. 11, p. 306, Nov. 2021, doi: 10.3390/jcs5110306.
32. E. Borie, E. Rosas, R. F. de Souza, and F. J. Dias, "Zirconia implants: A brief review and surface analysis of a lost implant," Aug. 01, 2024, doi: 10.3390/coatings14080995.
33. T. M. Alnassar, "Color stability of monolithic zirconia in various staining liquids: An in vitro study," *Applied Sciences*, vol. 12, no. 19, Oct. 2022, doi: 10.3390/app12199752.
34. P. Fischer et al., "Bending fracture of different zirconia-based bioceramics for dental applications: A comparative study," *Materials*, vol. 14, no. 22, Nov. 2021, doi: 10.3390/ma14226887.
35. M. Martinez-Mondragon et al., "Numerical analysis of zirconium and titanium implants under the effect of critical masticatory load," *Materials*, vol. 15, no. 21, Nov. 2022, doi: 10.3390/ma15217843.
36. D. R. Clarke and S. R. Phillpot, "Thermal barrier coating materials," *Materials Today*, vol. 8, no. 6, pp. 22–29, 2005, doi: 10.1016/S1369-7021(05)70934-2.
37. I. T. Yilmaz, İ. T. Yilmaz, and M. G. Mehmet Akçay, "Thermal barrier coatings for diesel engines," 2010. [Online]. Available: https://www.researchgate.net/publication/315576118
38. D. Ye et al., "Research on modified thermal barrier coatings against CMAS corrosion driven by mechanism–data hybrid model," *Coatings*, vol. 14, no. 12, Dec. 2024, doi: 10.3390/coatings14121513.
39. F. Ye, Y. Yao, F. Meng, and T. Luo, "CMAS corrosion behavior of mid-entropy rare-earth hafnate ($Y_{0.3}Gd_{0.3}Yb_{0.4}$)$4Hf_3O_{12}$ as thermal barrier coating candidate," *Materials*, vol. 17, no. 23, Dec. 2024, doi: 10.3390/ma17235892.
40. C. Jiang, R. Li, F. He, Z. Cheng, W. Li, and Y. Zhao, "Anti-sintering behavior of GYYSZ, thermophysical properties, and thermal shock behavior of thermal barrier coating with YSZ/composite/GYYSZ system by atmospheric plasma spraying," *Nanomaterials*, vol. 14, no. 22, Nov. 2024, doi: 10.3390/nano14221787.
41. M. Zhang, G. Wang, J. Wang, and Z. Zhao, "Thermo-physical properties of hexavalent tungsten W^{6+}-doped ta-based ceramics for thermal/environmental barrier coating materials," *Metals*, vol. 14, no. 12, Dec. 2024, doi: 10.3390/met14121368.
42. A. Kumar, J. Moledina, Y. Liu, K. Chen, and P. C. Patnaik, "Nano-micro-structured 6%–8% YSZ thermal barrier coatings: A comprehensive review of comparative performance analysis," Dec. 01, 2021, doi: 10.3390/coatings11121474.
43. J. Smialek, "Phase stability in plasma-sprayed partially stabilized zirconia yttria," 1981. [Online]. Available: https://www.researchgate.net/publication/23589949
44. M. Zhang, Y. Wang, Y. Feng, Y. Shang, S. Gong, and Y. Cheng, "Failure mechanism analysis of thermal barrier coatings under a service simulation environment," *Coatings*, vol. 15, no. 1, Jan. 2025, doi: 10.3390/coatings15010078.
45. I. Barwinska, M. Kopec, D. Kukla, C. Senderowski, and Z. L. Kowalewski, "Thermal barrier coatings for high-temperature performance of nickel-based superalloys: A synthetic review," Apr. 01, 2023, doi: 10.3390/coatings13040769.
46. M. Bogdan and I. Peter, "A comprehensive understanding of thermal barrier coatings (TBCs): Applications, materials, coating design and failure mechanisms," May 01, 2024, doi: 10.3390/met14050575.
47. W. Fan, Y. Wang, Y. Liu, Y. Bai, Y. Wang, and Q. Liu, "Mechanical properties durability of Sc_2O_3-Y_2O_3 Co-stabilized ZrO_2 thermal barrier materials for high temperature application," *Coatings*, vol. 12, no. 2, Feb. 2022, doi: 10.3390/coatings12020155.
48. S. Stecura, "Effects of compositional changes on the performance of a thermal barrier coating system," In *Ann. Conf. on Composite and Advanced Mater.*, No. NASA-TM-78976, 1978.

49. S. M. Lakiza et al., "Thermal barrier coatings: Current status, search, and analysis," *Powder Metallurgy and Metal Ceramics*, vol. 57, no. 1–2, pp. 82–113, May 2018, doi: 10.1007/s11106-018-9958-0.
50. A. G. Evans, D. R. Mumm, J. W. Hutchinson, G. H. Meier, and F. S. Pettit, "Mechanisms controlling the durability of thermal barrier coatings." [Online]. Available: www.elsevier.com/locate/pmatsci
51. A. Mehta et al., "Processing and advancements in the development of thermal barrier coatings: A review," Sep. 01, 2022, doi: 10.3390/coatings12091318.
52. W. U. de Goes, N. Markocsan, and M. Gupta, "Thermal swing evaluation of thermal spray coatings for internal combustion engines," *Coatings*, vol. 12, no. 6, Jun. 2022, doi: 10.3390/coatings12060830.
53. C. Leyens, "Advanced materials and coatings for future gas turbine applications." In *Proceedings of the 24th International Congress of the Aeronautical Sciences*, Yokohama, Japan, vol. 29. 2004.
54. R. J. Takahashi, J. M. K. de Assis, L. H. Fazan, L. A. A. Rodríguez, A. G. Capella, and D. A. P. Reis, "TBC development on Ti-6Al-4V for aerospace application," *Coatings*, vol. 15, no. 1, Jan. 2025, doi: 10.3390/coatings15010047.
55. A. Niaz, A. F. Al Fuhaid, and M. I. Faraz, "Understanding corrosion degradation processes of a multi-component CoNiCrAlY-coating system," *Coatings*, vol. 12, no. 10, Oct. 2022, doi: 10.3390/coatings12101396.
56. A. Kussaiyn-Murat et al., "Application of discrete exterior calculus methods for the path planning of a manipulator performing thermal plasma spraying of coatings," *Sensors*, vol. 25, no. 3, p. 708, Jan. 2025, doi: 10.3390/s25030708.
57. J. Wang, X. Li, Z. Wang, J. Feng, W. Lin, and J. Peng, "Research on application characteristics of zirconia-based high-temperature NOx sensors," *Energies*, vol. 15, no. 8, Apr. 2022, doi: 10.3390/en15082919.
58. S. Yilmaz, S. Cobaner, E. Yalaz, and B. A. Horri, "Synthesis and characterization of gadolinium-doped zirconia as a potential electrolyte for solid oxide fuel cells," *Energies*, vol. 15, no. 8, Apr. 2022, doi: 10.3390/en15082826.
59. H. W. Park et al., "Synthesis of yttria-stabilized zirconia nanospheres from zirconium-based metal–organic frameworks and the dielectric properties," *Nanomaterials*, vol. 13, no. 1, Jan. 2023, doi: 10.3390/nano13010028.
60. A. Beketova et al., "Zirconia nanoparticles as reinforcing agents for contemporary dental luting cements: Physicochemical properties and shear bond strength to monolithic zirconia," *Int J Mol Sci*, vol. 24, no. 3, Feb. 2023, doi: 10.3390/ijms24032067.
61. M. A. Alam, U. A. Samad, A. Anis, E. S. M. Sherif, H. S. Abdo, and S. M. Al-Zahrani, "The effect of zirconia nanoparticles on thermal, mechanical, and corrosion behavior of nanocomposite epoxy coatings on steel substrates," *Materials*, vol. 16, no. 13, Jul. 2023, doi: 10.3390/ma16134813.
62. S. G. Giniyatova, A. L. Kozlovskiy, R. I. Shakirzyanov, N. O. Volodina, D. I. Shlimas, and D. B. Borgekov, "Structural, dielectric, and mechanical properties of high-content cubic zirconia ceramics obtained via solid-state synthesis," *Appl Sci*, vol. 13, no. 19, Oct. 2023, doi: 10.3390/app131910989.
63. S. Tailor, M. Singh, and A. V. Doub, "Synthesis and characterization of yttria-stabilized zirconia (YSZ) nano-clusters for thermal barrier coatings (TBCs) applications," *J Clust Sci*, vol. 27, no. 4, pp. 1097–1107, Jul. 2016, doi: 10.1007/s10876-016-1014-y.
64. K. A. Kuterbekov et al., "Classification of solid oxide fuel cells," Apr. 01, 2022, doi: 10.3390/nano12071059.
65. S. Pattipaka, Y. Lim, Y. H. Son, Y. M. Bae, M. Peddigari, and G. T. Hwang, "Ceramic-based dielectric materials for energy storage capacitor applications," May 01, 2024, doi: 10.3390/ma17102277.

66. C. Y. Lee, S. Lee, J. H. Ha, J. Lee, I. H. Song, and K. S. Moon, "Effect of the zirconia particle size on the compressive strength of reticulated porous zirconia-toughened alumina," *Applied Sciences*, vol. 12, no. 5, Mar. 2022, doi: 10.3390/app12052316.
67. I. Žmak, D. Ćorić, V. Mandić, and L. Ćurković, "Hardness and indentation fracture toughness of slip cast alumina and alumina-zirconia ceramics," *Materials*, vol. 13, no. 1, Jan. 2020, doi: 10.3390/ma13010122.
68. N. Song et al., "Evaluation of phase transformation and mechanical properties of metastable yttria-stabilized zirconia by nanoindentation," *Materials*, vol. 12, no. 10, 2019, doi: 10.3390/MA12101677.
69. U. Betke, M. Scheunemann, and M. Scheffler, "Refitting of zirconia toughening into open-cellular alumina foams by infiltration with zirconyl nitrate," *Materials*, vol. 12, no. 12, Jun. 2019, doi: 10.3390/ma12121886.

6 ZrO_2 Used for Thermal Protection Systems of Ultra-High Temperature Automobile and Aerospace Applications

Amrinder Mehta, Hitesh Vasudev, and Che Hua Yang

6.1 INTRODUCTION

The rapid increase in automation and mechanization has resulted in an escalating demand for electrical energy. This surge in demand can be attributed to the substantial utilization of electricity in the residential, industrial, and mechanical sectors. Various electricity generation strategies are employed to fulfil India's full-size power needs, including thermal, hydroelectric, gasoline, solar, and biomass-based energy plants. As the United States continues to modernize and industrialize, the demand for reliable and sustainable energy resources is increasing. The diversification of strength-era techniques not only meets the growing demand but also contributes to power protection and environmental sustainability. India's strength landscape is evolving, with a growing emphasis on renewable resources such as solar and biomass, in addition to traditional thermal and hydroelectric plants, to create a balanced and resilient energy infrastructure that can assist the country's economic growth and development [1,2].

Zirconium dioxide (ZrO_2), also referred to as zirconia, has emerged as a promising material for thermal protection systems in ultra-high-temperature applications. Its superthermal properties, such as low thermal conductivity and high melting point (2715°C), make it an excellent candidate for defensive components exposed to intense heat. ZrO_2 exhibits excellent thermal resistance and maintains its structural integrity at high temperatures, which is critical for aerospace and hypersonic automobile programs [3]. The polymorphic nature of the protection, with three distinct crystal structures (monoclinic, tetragonal, and cubic) depending on the temperature, contributes to its specific thermal behavior. Stabilized zirconia, frequently obtained through the addition of dopants such as yttria, further enhances its high-temperature stability and mechanical properties. These traits, combined with its chemical inertness and resistance to oxidation, have led to the expanded research and development of

DOI: 10.1201/9781003519300-8

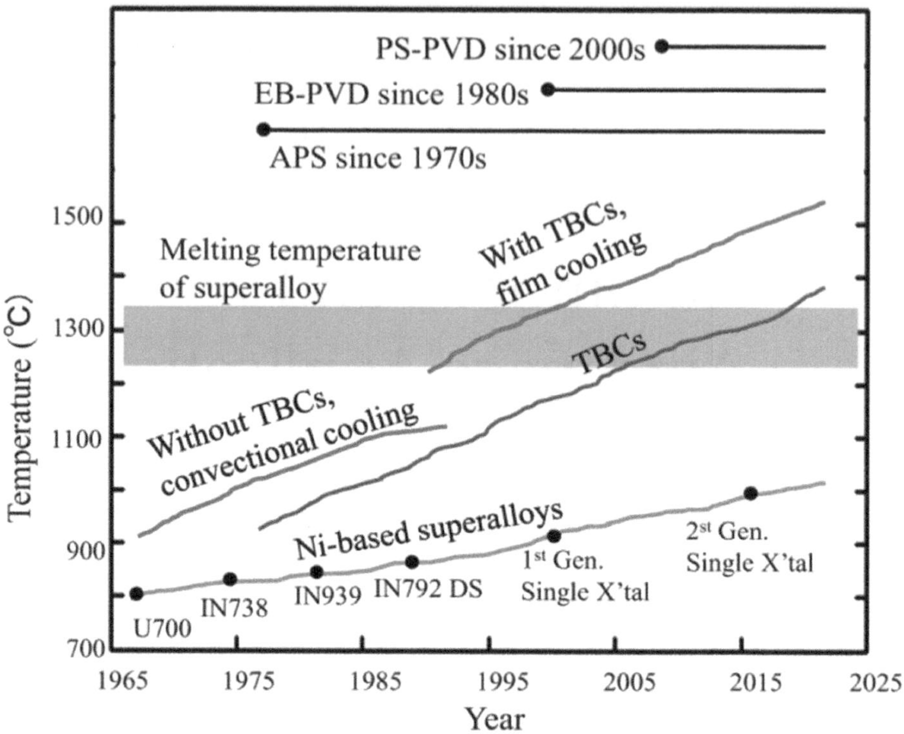

FIGURE 6.1 Modification of gas turbine engine from the 1960s to the current progression [5]. (Creative Commons Attribution 4.0.)

thermal protection structures based primarily on ZrO2 for use in high-temperature environments [4]. To address this shortcoming, Figure 6.1 suggests an increase in operating temperatures over time for nickel-based superalloys utilized in high-temperature applications, such as fuel generators. Advancements in materials and cooling structures have resulted in high operating temperatures for these components, which in turn enhances their performance [5]. The degree Celsius (°C) on this axis indicates the temperature of the tool when in use. The decrease in the temperature range shown in the diagram is approximately 800°C, even in the direction of the high end, where the temperatures exceed 1500°C. As can be seen below, the *x*-axis (year) represents the year from the middle of the 1960s to around 2025 [5].

Its thermal balance enables it to withstand high temperatures without degradation, making it suitable for additives exposed to excessive warmth, such as engine parts and thermal barrier coatings (TBCs). The mechanical energy of ZrO_2 contributes to its sturdiness and resistance to wear, enabling its use in essential structural elements and high-pressure applications [6]. Furthermore, the tribological properties of zirconia, including low friction and excessive wear resistance, make it a first-rate material for bearings, seals, and different moving components that require minimum friction and long-term reliability. These mixed characteristics have led to the considerable adoption of ZrO_2 in numerous applications, including fuel injection structures

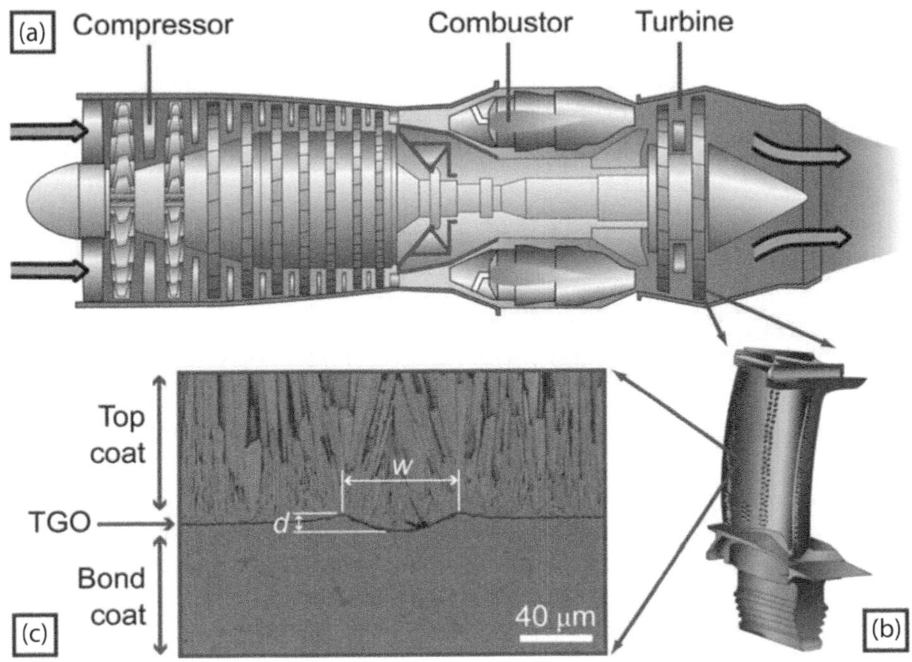

FIGURE 6.2 (a) Thermal spray coatings for gas turbines concerning turbine locations, (b) blade of the turbine, and (c) cross-section TBC system [8]. (Reproduction permissions.)

and exhaust gasoline sensors in automobiles, turbine blades, and aerospace coatings, where performance under worrying conditions is critical. The ability of ZrO_2 to resist excessive temperatures is crucial for the development of thermal safety systems in the automotive and aerospace industries [7].

Figure 6.2 illustrates the inner workings and material compositions of jet engines. Figure 6.2a depicts the go-phase of the engine, highlighting the compressor, combustor, and turbine. Air enters the compressor, where it is pressurized, and then flows into the combustor, where it mixes with the gas and ignites. The resulting hot gases power the turbine, which, in turn, powers the compressor and generates thrust. Figure 6.2b presents a turbine blade, an important component exposed to high temperatures and stresses. Finally, Figure 6.2c offers a microscopic view of a coating gadget on the blade, consisting of a bond coat, a thermally grown oxide (TGO) layer, and a top coat. This coating protects the blade from oxidation and corrosion at high temperatures, with measurements indicating the dimensions of the coating [8].

It plays a crucial role in the TBCs used in engine pistons. TBCs are crucial for reinforcing the overall performance and sturdiness of internal combustion engines, imparting thermal insulation and safety against high-temperature exposure. ZrO_2 is especially valued in TBC packages because of its excellent thermal properties, such as low thermal conductivity and high thermal expansion coefficient [9]. When applied to engine pistons, ZrO_2-based TBCs create a defensive layer that reduces heat transfer from the combustion chamber to the piston body, thereby improving

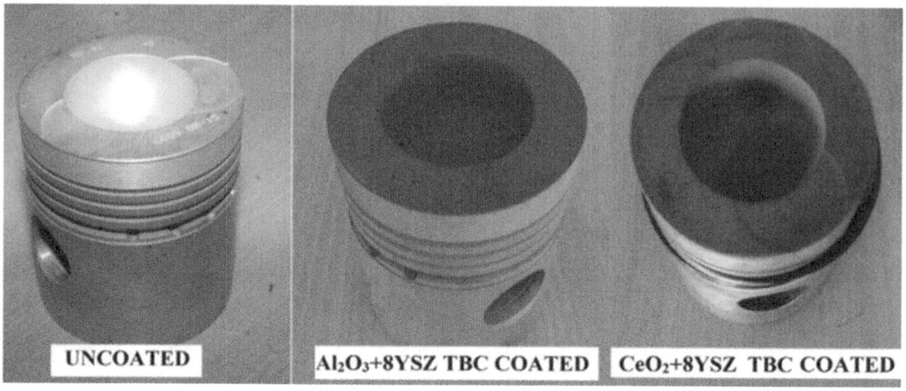

FIGURE 6.3 Coated and uncoated pistons [11]. (Reproduction permissions.)

the thermal performance and reducing the thermal pressure on the piston material. Additionally, the high-temperature stability of ZrO_2 allows it to withstand extreme situations in the engine, maintaining its structural integrity and protective characteristics over extended durations of operation [10]. This thermal barrier effect allows for higher combustion temperatures, potentially leading to improved engine efficiency and decreased emission.

Figure 6.3 presents three engine pistons side by side, contrasting their respective surface treatments. The first piston was uncoated and exhibited a regular metallic appearance. The second piston functioned as a blue coating of Al_2O_3+8YSZ TBC, imparting thermal barrier properties. The 0.33 piston was lined with a dark grey layer of CeO_2+8YSZ TBC, which was also designed for thermal insulation. The purpose of the coating is to improve the overall performance and sturdiness using a temperature-lowering switch [11].

Figure 6.4 shows the relationship between the brake power (kW) and brake-specific fuel consumption (BSFC) (kg/kW-hr). The overall performance of three exclusive engine setups was compared: an uncoated engine, an engine coated with Al_2O_3/8YSZ, and an engine lined with CeO_2/8YSZ. The graph shows that as the brake electricity increases, the BSFC decreases for all three setups. This shows that the engine becomes more fuel efficient at higher energy outputs. The lined engines, especially the CeO_2/8YSZ-lined engine, exhibited lower BSFC values than the uncoated engine across the range of brake energy tested, suggesting improved gasoline efficiency with these coatings. The largest distinction in BSFC was observed at lower brake strength degrees, with the traces converging as the brake power increased [11].

ZrO_2 (zirconia) plays a critical role in TBCs for extremely high-temperature packages in motors and aerospace applications. Its low thermal conductivity and high melting point make it ideal as a defensive engine additive against excessive heat. Typically applied as yttria-stabilized zirconia (YSZ), ZrO_2-based TBCs can withstand temperatures exceeding 1200°C while maintaining structural integrity [12]. This allows engines to operate at higher temperatures, thereby improving their efficiency and reducing emissions. The relationship between the brake electricity and the brake-particular gas intake (BSFC) demonstrates this performance gain. As the

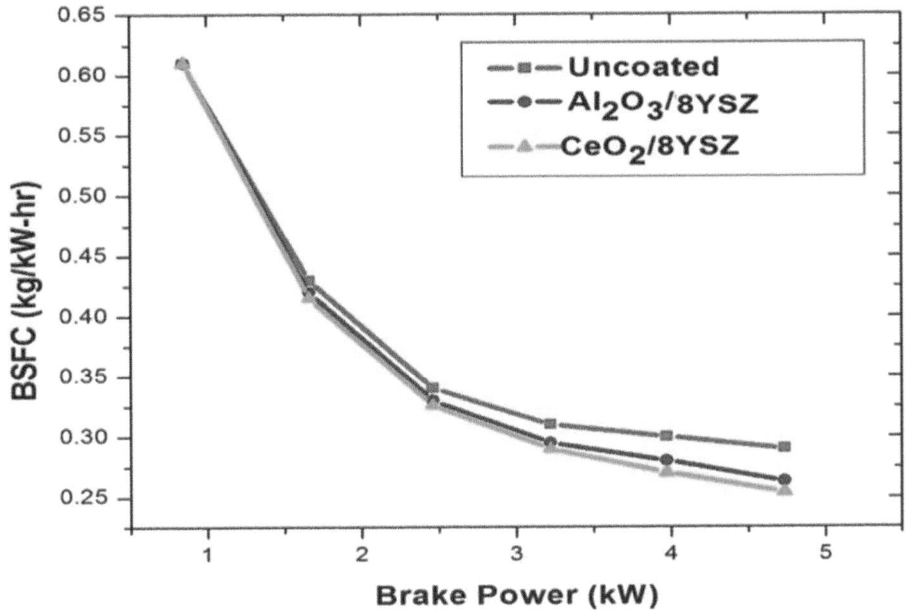

FIGURE 6.4 Brake power vs. brake-specific fuel consumption [11]. (Reproduction permissions.)

brake energy increased, the BSFC generally decreased, indicating improved gasoline performance at higher-power outputs [13]. Engines lined with ZrO_2-based TBCs often exhibit decreased BSFC values compared to uncoated engines throughout several brake energy tiers, specifically at decreased outputs. This improved performance is due to the capacity of the TBCs to reduce heat loss from the combustion chamber, allowing additional electricity to be transformed into useful power instead of being wasted as heat. Plasma-sprayed deposition is a widely used approach for developing ZrO_2-based TBCs for ultra-high-temperature packages in the automobile and aerospace industries [14]. This procedure involves injecting ZrO_2 powder into an excessive-temperature plasma jet, which melts the debris and propels it toward the substrate at high velocities. Consequently, the molten particles solidified unexpectedly and formed a layered, splat-like microstructure in the plasma-sprayed coatings. This precise microstructure, along with the flattened splats, pores, and microcracks, contributes to the low thermal conductivity and strain tolerance of the ZrO_2 TBCs [15]. Figure 6.5 illustrates the plasma-spray coating process. The process began with the feeding of YSZ powder into a plasma spray gun. Inside the gun, the powder is subjected to a high-temperature plasma flame that melts the YSZ debris. These molten particles are then propelled in the direction of the surface to form a coating. As the molten YSZ particles impacted the surface, they rapidly solidified and formed a layered structure, building up the deposited coating. The image presents a coating pattern, showcasing the quality of the coating manufactured using the plasma spray process. This method is used to apply protective or functional coatings to various materials, thereby improving their properties [16].

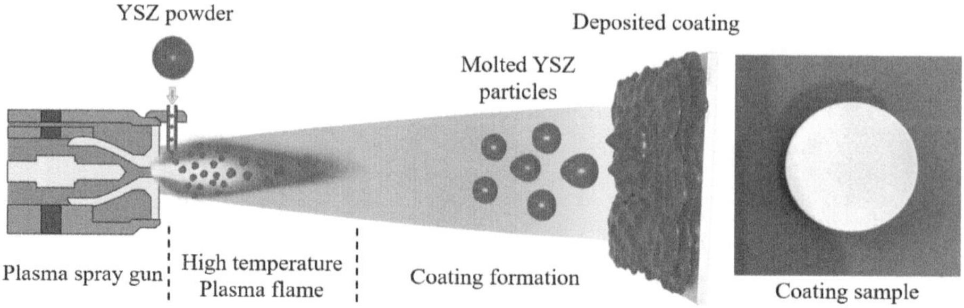

FIGURE 6.5 Schematic of the YSZ coating deposited by the plasma-spraying process and images of the prepared coating samples [16]. (Creative Commons Attribution 4.0.)

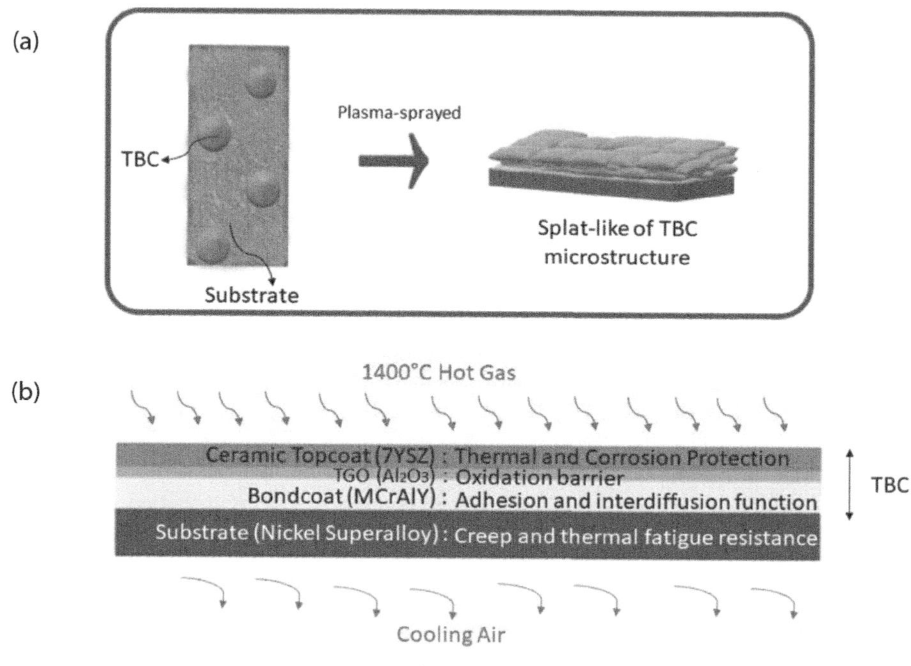

FIGURE 6.6 Plasma-sprayed deposition-based microstructures mainly consist of (a) splat-like morphology and (b) functional aspects of TBC multilayer coatings [17]. (Creative Commons Attribution 4.0.)

Figure 6.6 shows a general illustration of TBCs applied to parts. These coatings are typically applied under high-temperature operating conditions, such as in gas turbines. In Figure 6.6a, the application and microstructure of the TBC are investigated. Here, an example of the application of TBC onto a substrate using a plasma-spraying technique is presented. Figure 6.6b shows the functionality and organization of the TBC system layers. This segment offers a top-level view of the several layers of the general TBC gadget, which can be broadly classified as follows: the 7% yttria-partially stabilized zirconia (7YSZ) usually forms the top layer [17].

Fast solidification additionally results in the formation of metastable levels, together with a suitable tetragonal t' segment, which enhances the thermal stability and mechanical properties of the coating. By controlling plasma spray parameters, such as gasoline glide charges, powder feed price, and spray distance, the microstructure can be tailored to optimize thermal insulation, mechanical electricity, and durability for unique high-temperature programs in engines, mills, and aerospace components [18].

Previous research has established that ZrO_2 is well known for its excessive thermal balance and occasional thermal conductivity, which are suitable for TBCs. The low thermal conductivity of ZrO_2 is essential for its function as a thermal barrier, effectively decreasing the temperature difference between warm gases and the underlying metal additives [19]. This aggregate of houses enables ZrO_2-based coatings to protect crucial additives from thermal harm, extend their operational lifespan, and improve the normal performance of devices. Furthermore, the flexibility of ZrO_2 allows for its change via doping or the advent of composite structures, potentially improving its thermal barrier properties and adapting it to unique software requirements [20]. However, there is limited knowledge regarding the long-term performance and sturdiness of ZrO_2 under extremely high-temperature conditions, which are standard in automotive and aerospace environments. Addressing this knowledge gap is essential for developing greener and more reliable thermal safety systems that can withstand the stressful conditions of modern high-temperature programs [21]. This study aimed to assess the long-term thermal performance and durability of ZrO_2 when used in thermal protection systems for extremely high-temperature packages in the automotive and aerospace sectors.

6.2 METHODS USED IN THERMAL PROTECTION SYSTEMS

Zirconium dioxide (ZrO_2) has emerged as a promising material for thermal protection systems owing to its excellent thermal properties and high-temperature balance. Various strategies have been employed to increase and optimize ZrO_2-based coatings for thermal barrier applications [22]. These include plasma spraying, electron beam physical vapor deposition (EB-PVD) [23], sol-gel processing, and chemical vapor deposition (CVD) [24]. Plasma spraying is extensively used because of its versatility and cost-effectiveness, allowing the deposition of thick coatings with controlled porosity. EB-PVD produces columnar microstructures that enhance pressure tolerance and thermal cycling resistance. Sol-gel techniques offer specific control over composition and nanostructure, even as CVD enables the introduction of dense, uniform coatings with incredible adhesion. Researchers have also explored doping ZrO_2 with rare-earth elements to enhance its thermal and mechanical properties. These strategies, which are regularly used in combination, have contributed to the improvement of advanced ZrO_2-based thermal protection structures that can withstand extreme temperatures and harsh environmental conditions [25]. Figure 6.7 shows a comparative analysis of various YSZ TBCs fabricated using distinct thermal spray techniques. These coatings are crucial for protecting components in high temperature environments. The array consists of conventional air plasma-sprayed (APS) coatings, which exhibit a medium stage of porosity, and APS coatings engineered for

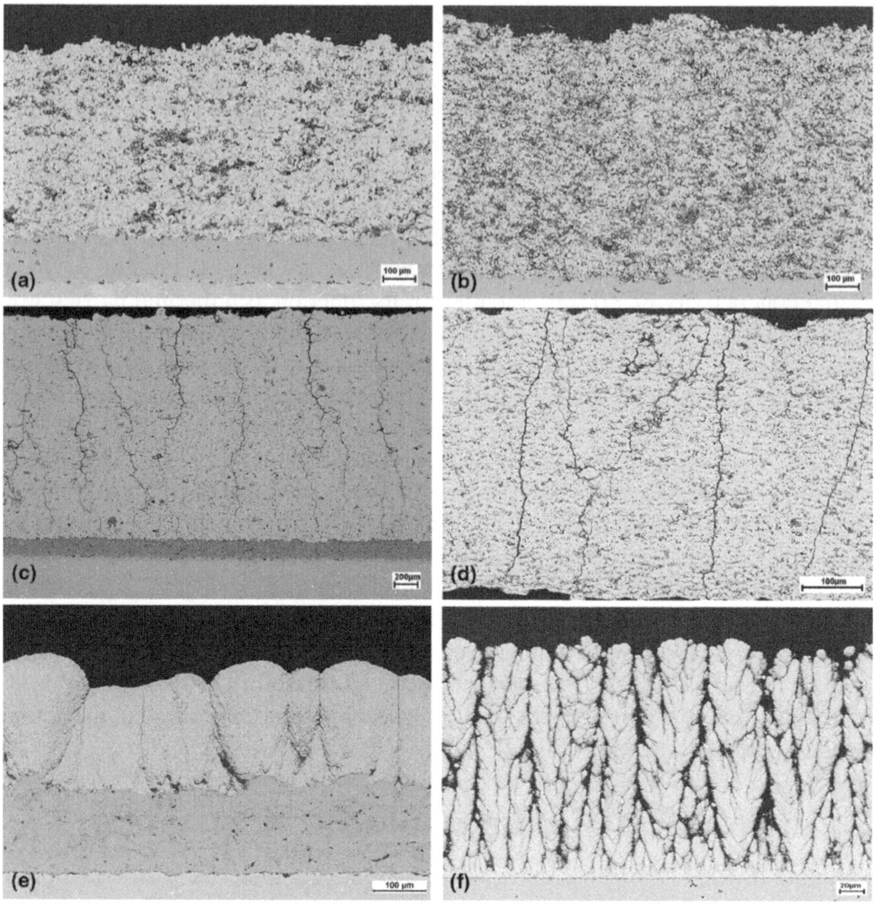

FIGURE 6.7 A variety of YSZ TBCs that are thermally sprayed include (a) conventional APS (medium porosity level), (b) very porous APS, (c) thick segmented APS, (d) segmented SPS, (e) columnar SPS, and (f) columnar PS-PVD [26]. (Creative Commons Attribution 4.0)

extremely high porosity. Thick segmented APS and segmented suspension plasma spray (SPS) coatings were also featured. A large portion of the photograph depicts columnar microstructures along with both SPS and plasma spray-physical vapor deposition (PS-PVD) methods. Each method resulted in distinct morphological characteristics that influenced the thermal insulation and strain control of the coating. The porosity, segmentation, and columnar structure of elements contribute to their overall performance and strength [26].

6.3 CHALLENGES IN TBCS

Zirconia (ZrO_2) is widely used in TBCs because of its low thermal conductivity and high toughness. However, it faces several challenges in TBC applications, including:

6.3.1 Phase Transformation

At elevated temperatures, ZrO_2 undergoes a phase transformation from tetragonal to monoclinic, resulting in cracking inside the coating. Phase transformation in ZrO_2 coatings is a vital phenomenon that occurs at multiple temperatures. Zirconia (ZrO_2) exists in three crystallographic structures: monoclinic, tetragonal, and cubic [27]. At room temperature, the strong phase was monoclinic. However, as the temperature increases, ZrO_2 undergoes a phase transformation, that is, it transforms to the tetragonal phase and then to the monoclinic phase upon cooling. This transformation is followed by a tremendous volume enlargement of approximately 3–5%, which induces massive stress inside the coating [28]. Figure 6.8 presents a comprehensive analysis of YbYSZ (ytterbium-doped yttria-stabilized zirconia) materials with varying Yb^{3+} contents. It contains four part-figures (a, b, c, and d), each illustrating the distinctive components of the material traits. Figure 6.8a shows the X-ray diffraction (XRD) patterns of the YbYSZ samples. The XRD patterns revealed remarkable peaks corresponding to the crystallographic stages within the materials. Reference styles (PDF#70-4430 and PDF#70-4426) were included for phase identification. Figure 6.8b shows a zoomed-in view of a particular region of the XRD patterns,

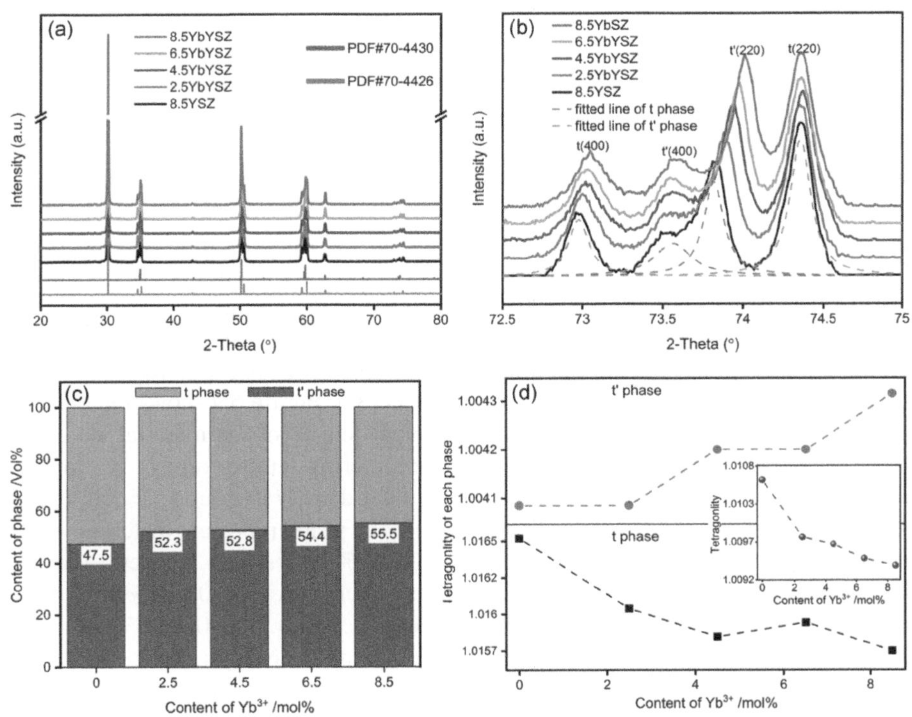

FIGURE 6.8 XRD patterns of xYbYSZ ($x = 0$, 2.5, 4.5, 6.5, and 8.5): (a) $2\theta = 20°–80°$ at a scan rate of 5°/min, (b) $2\theta = 72.5°–75°$ at a scan rate of 1°/min, (c) phase content, and (d) tetragonality of t phase and t' phase [29]. (Creative Commons Attribution 4.0.)

highlighting the height splitting related to the tetragonal (*t* and *t′*) phases. The fitted strains were used to estimate the contributions from each section. Figure 6.8c indicates the quantity possibilities of the *t* and *t′* levels as a function of Yb^{3+} content. The graph indicates how the segment composition is modified at various dopant concentrations. Figure 6.8d shows the tetragonality of each segment against the Yb^{3+} content. Tetragonality is a measure of the distortion of the crystal lattice from the best cubic shape of the lattice. The inset graph provides a better view of the tetragonality of the *t-section*. The trends observed in these plots provide insights into the relationship between the Yb^{3+} content and the resulting crystal shape and segment distribution of the YbYSZ materials [29].

These stresses can result in the formation of microcracks and, in intense instances, cause complete failure of the coating through spallation. The segment transformation and subsequent cracking can seriously compromise the protective properties of the ZrO_2 coating, reducing its effectiveness as a thermal barrier or corrosion-resistant layer. To mitigate this difficulty, numerous techniques have been developed, such as doping ZrO_2 with stabilizing dealers like yttria (Y_2O_3) to keep the tetragonal section at better temperatures and reduce the chance of unfavorable section modifications [30].

6.3.2 Sintering Effect

The sintering effect plays an essential role in the overall performance and longevity of yttria partly stabilized zirconia (Y-PSZ) coatings when exposed to high temperatures. At temperatures exceeding 1100°C, Y-PSZ undergoes sintering, which results in considerable adjustments in its material properties [31]. This phenomenon increases both the Young's modulus and thermal conductivity of the coating. The expanded Young's modulus results in reduced stress tolerance, making the coating extra vulnerable to cracking and spallation under thermal and mechanical stresses [32]. Simultaneously, the elevated thermal conductivity compromises the coating's capability to provide effective thermal insulation, which is usually the primary characteristic of Y-PSZ in high-temperature applications. These blended sintering results ultimately contribute to a shortened coating lifespan, limiting its effectiveness and sturdiness in extreme-temperature environments [33]. Figure 6.9 shows the relationship between the sintering temperature and bulk density of the material at various weight percentages. The *x*-axis represents the sintering temperature in °C, ranging from 1250 to 1450, while the *y*-axis shows the bulk density in grams per cubic centimeter, ranging from 5.5 to 6.1. The four lines on the graph represent different weight percentages: 0.0, 0.3, 0.5, and 1.0 wt%. The 0.0 wt% line shows a general increase in the bulk density as the sintering temperature increases. The 0.5 wt% line shows an increase up to 1400°C, after which it decreases slightly. The 0.3 wt% line shows an increase until 1350°C, after which it remains almost constant. The 1.0 wt% line was almost constant. The graph provides a visual representation of how different weight percentages affect the bulk density of the material as the sintering temperature changes [34].

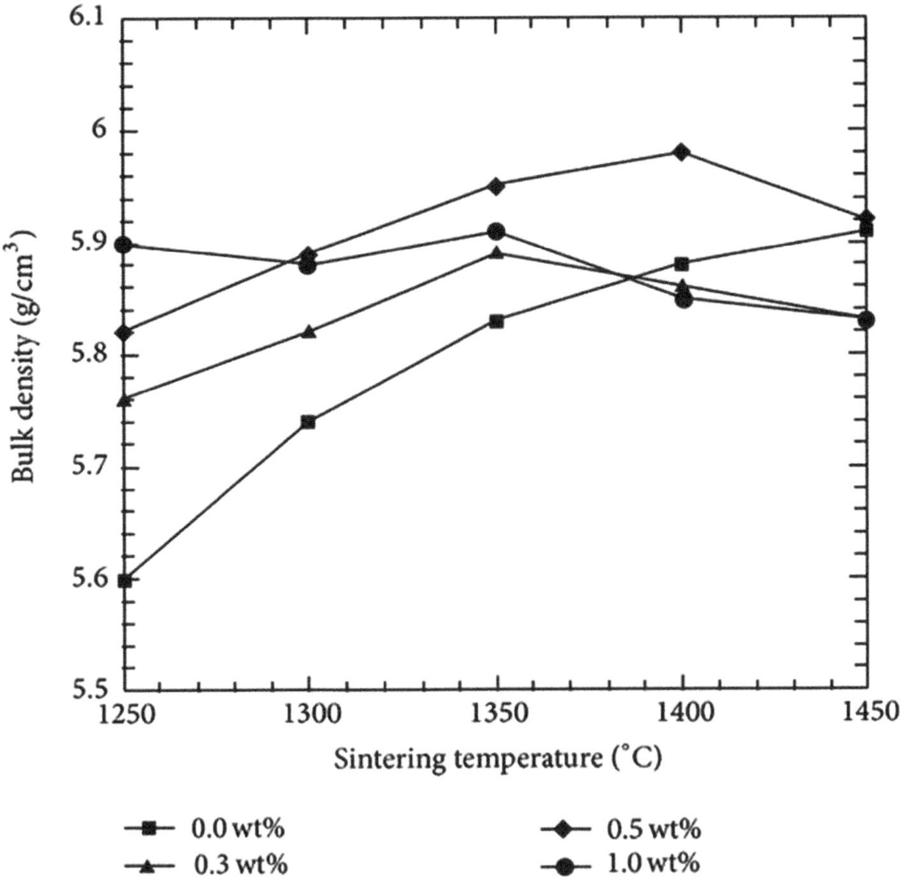

FIGURE 6.9 Effect of sintering temperature and CeO_2 addition on the bulk density of Y-TZP [34]. (Creative Commons Attribution 3.0.)

6.3.3 Thermal Oxidation

Thermal oxidation poses a significant challenge in the high-temperature packaging of zirconium dioxide (ZrO_2) coatings. As the temperature increases, ZrO_2 undergoes a transformation that increases its oxygen permeability, allowing oxygen molecules to pass through more easily. This phenomenon, called oxygen transparency, compromises the protective characteristics of the ZrO_2 layer, exposing the underlying bond coat to oxidative environments [35]. Consequently, the bond coat becomes vulnerable to elevated oxidation, potentially leading to degradation of its structural integrity and adhesive properties. This oxidation can result in the formation of unfavorable oxide scales, spallation, and, ultimately, failure of the coating device. Understanding and mitigating this thermal oxidation is crucial for enhancing the durability and overall performance of ZrO_2-based TBCs in extreme temperature environments [36].

Figure 6.10 presents a microscopic view of the material cross-section, likely a coating or layered structure, as revealed by a scanning electron microscope. As shown in Figure 6.10a, a significant crack was observed at the interface between the top layer and the underlying material. Pores are scattered at some points of the shape, and dark, elongated areas are classified as "oxide." A vast portion of the material was marked as "TGO," likely indicating a thermally grown oxide layer. Figure 6.10b suggests a similar shape; however, with extra labels: "YSZ" (likely yttria-stabilized zirconia)

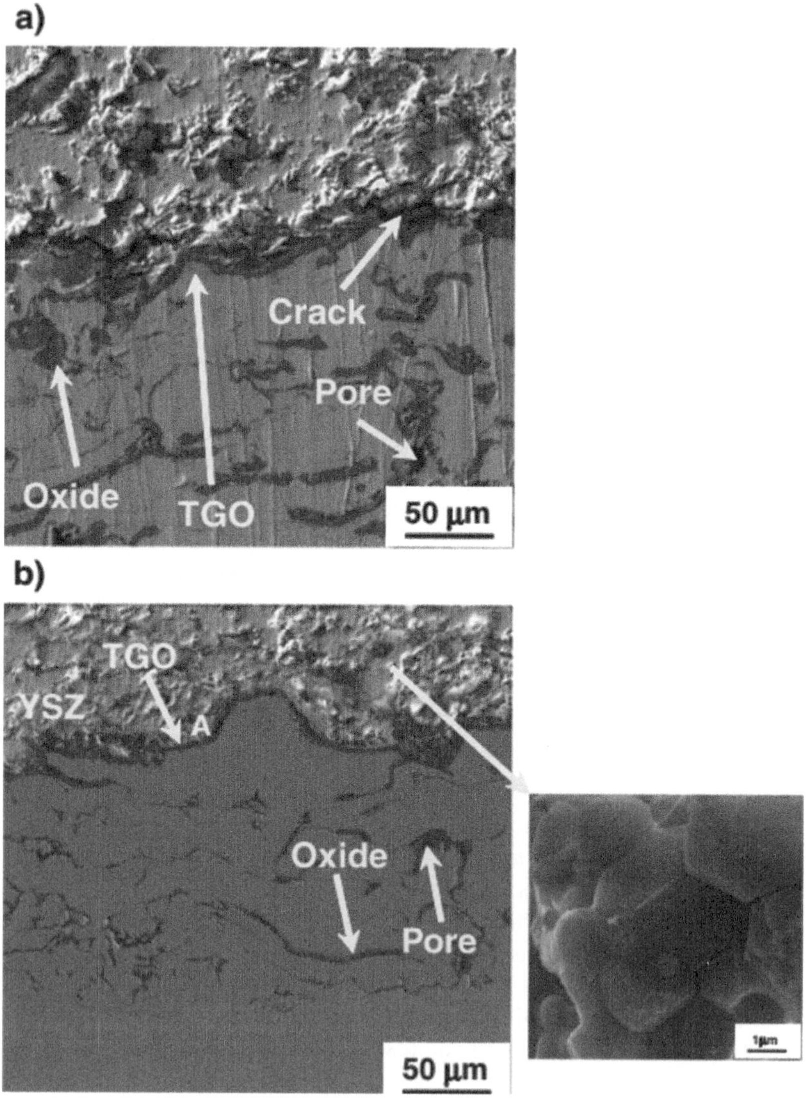

FIGURE 6.10 Cross-sectional SEM images of TBC coatings after oxidation test: (a) conventional YSZ after 192 hours and (b) nanostructured YSZ after 256 hours [37]. (Reproduction permissions.)

and "A," the TGO place is also visible on this diagram. The inset presents a better magnification view, revealing the microstructure in more detail and highlighting grain limitations or other microstructural functions. The scale bars suggest a resolution of 50 μm within the foremost photos and 1 μm inside the inset [37].

The limited reduction in thermal conductivity is closely related to the crystal structure of the protection. Zirconia exists in extraordinary crystalline levels, with the tetragonal *t*′-phase being the most suitable for TBC packages because of its stability at high temperatures and resistance to section transformations. To maintain this vital *t*′-segment shape, yttria is commonly added as a stabilizing agent [38]. While the addition of yttria first facilitates a reduction in the thermal conductivity by introducing factor defects and oxygen vacancies within the crystal lattice, there may be a restriction on the amount that can be introduced without compromising the stability of the *t*′-phase. This constraint efficiently limits the extent to which the thermal conductivity can be reduced via yttria doping, as excessive addition could lead to the formation of less desirable levels, potentially compromising the overall performance and sturdiness of the coating [39]. Poor adhesion to metal substrates is a critical issue that affects the lifetime and reliability of zirconia coatings. The enormous difference in the thermal expansion coefficients between the ceramic zirconia coating and the underlying steel substrate can cause thermal stresses during the heating and cooling cycles [40]. These stresses may cause the coating to delaminate or spall off, exposing the steel substrate to excessive temperatures and causing catastrophic failures. Additionally, chemical incompatibility between the ceramic coating and metal substrate can further exacerbate adhesion problems. This poor bonding not only reduces the coating's effectiveness in shielding the underlying metal but also shortens the overall lifespan of the element, necessitating extra common protection or an alternative [41].

6.3.4 Hot Corrosion

Hot corrosion is a substantial concern for zirconia-based TBCs exposed to environments containing sodium sulfate (Na_2SO_4) and vanadium pentoxide (V_2O_5). This degradation mechanism is particularly problematic in high-temperature programs, fuel turbines, and aerospace engines. When these contaminants are present, they can react with the protective zirconia layer, compromising its structural integrity and thermal insulation [42]. The procedure generally starts with the formation of a molten salt at the coating surface, which then penetrates the porous TBC. As the molten salt infiltrates, it reacts with zirconia, causing section transformations and the formation of new compounds. This chemical interaction results in quantity modifications, cracking, and spallation of the coating [43].

Figure 6.11 presents three scanning electron microscopy (SEM) micrographs, labeled (a), (b), and (c), showcasing the surface morphology of the material at a high magnification. The scale bar in each photo represents 20 μm. All three micrographs show a hard, granular texture with irregularly formed particles dispersed across the floor. The arrows highlight precise features in the microstructure, specifically pointing towards what appears to be dendritic systems. These dendritic systems, observed in all three images, suggest a solidification process wherein crystal growth occurs

FIGURE 6.11 Low-magnification FE-SEM images of the surfaces of (a) YSZ, (b) Al_2O_3/YSZ, and (c) CeO_2/YSZ coatings after the hot corrosion test [44]. (Creative Commons Attribution 4.0.)

in a branching, tree-like manner. The differences in particle size, distribution, and dendritic formation throughout the three micrographs may imply differences in the composition or processing conditions of the material [44].

The presence of vanadium compounds is detrimental because they can destabilize the tetragonal phase of zirconia and accelerate the degradation process. Over time, warm corrosion can significantly reduce the effectiveness of TBC, potentially leading to premature failure of the underlying factor. To mitigate this issue, researchers have explored alternative coating compositions and floor remedies that provide improved resistance to high-temperature corrosion in harsh environments.

6.4 FUTURE SCOPE

Research into superior TBCs has significantly improved the overall performance of zirconia-based materials. One key technique involves doping zirconia with rear-earth elements or transition steel oxides. This method complements the phase balance and decreases the thermal conductivity, which are essential elements for TBC effectiveness. By incorporating these dopants, researchers can manage the crystal shape and thermal residence of zirconia, resulting in coatings that maintain their integrity under high-temperature conditions while providing advanced insulation [45]. Another promising avenue is the development of the latest compositions, particularly rear-earth zirconates, including $La_2Zr_2O_7$. These materials exhibit higher temperature capabilities than standard yttria-stabilized zirconia (YSZ). Rare-earth zirconates have a unique crystal structure that contributes to their notable high-temperature stability and thermal conductivity. This makes them the best candidates for subsequent-era TBCs, which can resist increasingly demanding situations in advanced gasoline turbines and other high-temperature programs [46].

Composite materials have emerged as effective solutions for enhancing the mechanical properties of TBCs. Zirconia-toughened alumina (ZTA) is an excellent example of this technique. By combining zirconia and alumina, researchers have created a material that exhibits improved hardness and wear resistance compared to its male or female components. ZTA leverages the transformation toughening mechanism of zirconia and, at the same time, capitalizes on the high hardness of alumina. This results in a coating that can withstand the tough mechanical and thermal stresses encountered in turbine environments [47]. The exploration of alternative stabilizers for zirconia is an active area of research. Cerium oxide (CeO_2) has garnered attention as a capability stabilizer owing to its potential to improve corrosion resistance. Unlike traditional yttria stabilization, ceria-stabilized zirconia has shown promising results in resisting degradation in corrosive environments, which is particularly important in positive business programs where chemical assault is a subject [48].

These ongoing studies aim to address the constraints of traditional zirconia-based TBCs and expand extra sturdy coating structures for extremely high-temperature environments. By first tuning the composition, structure, and properties of these materials, scientists and engineers are working toward developing TBCs that may better resist temperatures, resist degradation, and offer superior thermal insulation. The last goal is to allow the development of greener and sturdier high-temperature systems, including advanced fuel mills for energy and aerospace programs [49].

Yttria-stabilized zirconia (YSZ) coatings are widely utilized in thermal barrier programs because of their excellent thermal insulation and high-temperature stability. The segmentation of YSZ coatings is an important aspect of their microstructural characterization and overall performance evaluation. This procedure involves analyzing the internal structure of the coating, along with pores, cracks, and splat boundaries, which considerably affect the thermal and mechanical properties of the coating [50]. Advanced imaging strategies, including scanning electron microscopy (SEM) and X-ray computed tomography (CT), are generally employed to visualize the coating's pass segment and its three-dimensional shape. Image evaluation algorithms are then applied to segment the diverse capabilities, allowing researchers to quantify the porosity, crack density, and splat morphology. Understanding the segmentation of YSZ coatings is critical for optimizing their deposition parameters, predicting their sturdiness, and developing better green thermal barrier systems for applications in fuel generators, aerospace components, and different high-temperature environments [51]. Figure 6.12 shows a rectangular solid object with a rough, cracked surface. The object appears to be somewhat worn or aged. The cracks run across the surface, suggesting that it may be dry or brittle. A small, dark spot was also visible in one area of the object. The background was plain white [52].

Figure 6.13 shows scanning electron microscopy (SEM) micrographs revealing the microstructural characteristics of the material. Several capabilities were

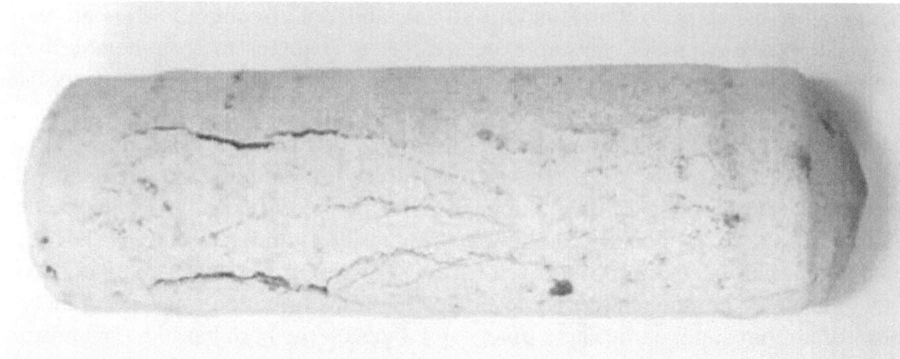

FIGURE 6.12 Thermal cycling test of as-sprayed YSZ coating (without segmentation) after the first cycle [52]. (Reproduction permissions.)

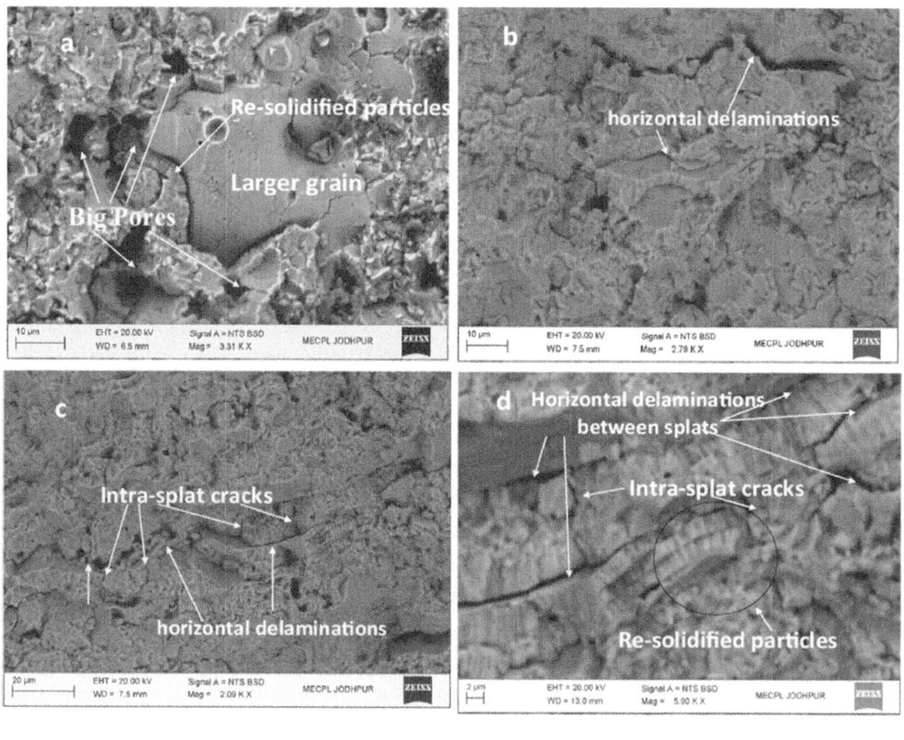

FIGURE 6.13 Fracture surfaces of the cross-section of the segmented YSZ coating [52]. (Reproduction permissions.)

highlighted, including "big pores," "re-solidified debris," and variations in grain size. "Horizontal delaminations" are routinely observed, indicating layer separation inside the material. Additionally, "intra-splat cracks" were observed, suggesting cracking within individual splats or layers. The presence of "segmentation crack penetration" was also noted, showing how cracks propagate through the material's structure.

This microstructural information provides insights into the protection's integrity, ability to fail, and usual first-rate, which can be critical for packages in which the overall performance of the protection is crucial. Together, these photos provide a comprehensive view of the internal structure of the protection at the microscopic level [52].

Convolutional neural networks (CNNs) have emerged as effective devices for modeling complex techniques in substance technology, including the atmospheric plasma-spraying method. The implementation of CNNs in this context allows for the analysis and prediction of numerous parameters that impact the roughness and properties of plasma-sprayed coatings. By leveraging the ability of CNNs to extract hierarchical features from facts, researchers can broaden the methods that capture the complex relationships among system variables, such as plasma fuel composition, spray distance, powder feed charge, resulting coating microstructure, and overall performance [53]. These CNN-based models can be trained on large datasets of experimental results and manner parameters, enabling them to analyze the underlying styles and correlations that govern the plasma spraying procedure. As a result, the implementation of CNNs in this area can optimize procedure parameters, except for coating homes, and ultimately improve the efficiency and reliability of atmospheric plasma-spraying strategies in commercial programs [54].

Figure 6.14 illustrates a method for implementing CNNs in modeling the atmospheric plasma-spraying (APS) process. The method begins with statistical guidance, where control parameters are fed into experiments or simulations to generate

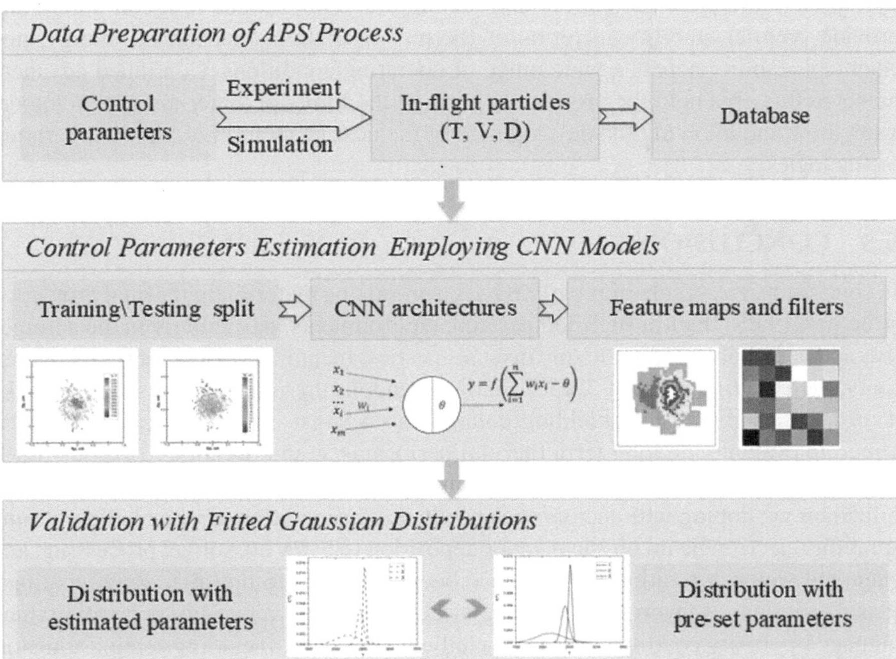

FIGURE 6.14 Method for implementing CNNs in modeling the atmospheric plasma spraying process [55].

statistics on in-flight debris, characterized by temperature (T), velocity (V), and diameter (D). The data were then compiled into a database. Next, the control parameters are expected to use CNN fashions. The statistics are divided into training and testing sets, which can be used to broaden the CNN architectures. These architectures produce characteristic maps and filters that are vital for extracting relevant information. Finally, the method involves validating the use of outfitted Gaussian distributions. The distributions generated with the predicted parameters were compared with distributions that had pre-set parameters to evaluate the accuracy and reliability of the CNN fashions in predicting and controlling the APS process [55].

The methodology employed in this study has a look at both experimental and computational processes to evaluate the overall performance of ZrO_2 in high-temperature environments. The primary protection used in this study was zirconium dioxide (ZrO_2), sourced from a reputable provider and characterized for its purity and thermal stability [56]. Data were changed into amassed via a series of controlled laboratory experiments designed to simulate the ultra-high temperature conditions encountered in automobile and aerospace programs. The thermal properties and performance metrics of ZrO_2 were calculated using advanced computational methods and confirmed by experimental data. Statistical evaluation was performed to assess the reliability and importance of the experimental results using ANOVA and regression analysis [57]. As research progresses, the combination of these approaches can lead to the introduction of hybrid or multi-layered coating systems that integrate the strengths of different substances and techniques. This should result in TBCs that provide premier stability in terms of thermal insulation, mechanical energy, and chemical stability across a wide range of operating conditions. Continued advancements in this area hold the promise of pushing the limits of materials' technological know-how and allowing the development of the next generation of high-temperature technologies.

6.5 CONCLUSION

In conclusion, zirconium dioxide (ZrO_2) is a promising material for thermal protection structures (TPS) in ultra-high-temperature environments, particularly in the automotive and aerospace sectors. Its intrinsic properties, including its low thermal conductivity, high melting point of 2715°C, and tremendous thermal shock resistance, make it an ideal candidate for shielding components against excessive heat. This study aimed to evaluate the long-term thermal performance and durability of ZrO_2-based TPS, acknowledging the polymorphic nature of the protection and the benefits of stabilization via doping with factors such as yttria. Various strategies, including plasma spraying, electron beam physical vapor deposition (EB-PVD), sol-gel processing, and chemical vapor deposition (CVD), have been employed to optimize ZrO_2 coatings. Plasma spraying is favored for its value-effectiveness and versatility in creating thick managed-porosity coatings. However, challenges remain, including section transformation, sintering at high temperatures, oxygen transparency, restricted thermal conductivity, poor adhesion to metal substrates, and hot corrosion in harsh environments. Future research should focus on addressing these limitations using superior TBCs.

Doping with rare-earth elements or transition metal oxides, developing new compositions such as rear-earth zirconates (e.g., $La_2Zr_2O_7$), and exploring stabilizers such as cerium oxide (CeO_2) are promising avenues. Composite substances, such as zirconia-toughened alumina (ZTA), also offer superior mechanical properties. Computational methods, including CNNs, are increasingly utilized to optimize plasma spraying techniques, predict coating homes, and enhance method efficiency. The technique used in this study integrates experimental and computational strategies to evaluate the performance of ZrO_2 under high-temperature conditions, paving the way for the development of next-generation high-temperature materials. By combining different materials and strategies, hybrid or multilayered coating structures can be developed to provide excellent stability in terms of thermal insulation, mechanical energy, and chemical balance, ensuring long-term durability and reliability in challenging operational environments.

REFERENCES

1. Majid M. Renewable energy for sustainable development in India: Current status, future prospects, challenges, employment, and investment opportunities. *Energy, Sustainability Society*. 2020;10:1–36.
2. Aboushelib MN, Kleverlaan CJ, Feilzer AJ. Selective infiltration-etching technique for a strong and durable bond of resin cements to zirconia-based materials. *The Journal of Prosthetic Dentistry*. 2007;98:379–88.
3. AbuShanab WS, Moustafa EB, Taha MA, Youness RA. Synthesis and structural properties characterization of titania/zirconia/calcium silicate nanocomposites for biomedical applications. *Applied Physics A*. 2020;126:1–12.
4. Afzal A. Implantable zirconia bioceramics for bone repair and replacement: A chronological review. *Materials Express*. 2014;4:1–12.
5. Zhang X, Deng Z, Li H, Mao J, Deng C, Deng C, et al. Al_2O_3-modified PS-PVD 7YSZ thermal barrier coatings for advanced gas-turbine engines. *J npj Materials Degradation*. 2020;4:31.
6. Lin J, Stinnett TC. Development of thermal barrier coatings using reactive pulsed dc magnetron sputtering for thermal protection of titanium alloys. *Surface Coatings Technology*. 2020;403:126377.
7. Manimaran S, Dhas AAG, Joy N, Raja K, Saikia K, Kumar S. Analysis of surface texture of Ni-based super alloy with thermal barrier coating system (TBC) for gas turbine blade. *AIP Conference Proceedings*: AIP Publishing; 2020.
8. Hille TS, Nijdam TJ, Suiker AS, Turteltaub S, Sloof WG. Damage growth triggered by interface irregularities in thermal barrier coatings. *Acta Materialia*. 2009;57:2624–30.
9. Fang H, Wang W, Yang Z, Yang T, Wang Y, Huang J, et al. Phase stability, thermal shock behavior and CMAS corrosion resistance of Yb_2O_3-Y_2O_3 co-stabilized zirconia thermal barrier coatings prepared by atmospheric plasma spraying. *Surface Coatings Technology*. 2021;427:127864.
10. Gell M, Jordan EH, Teicholz M, Cetegen BM, Padture NP, Xie L, et al. Thermal barrier coatings made by the solution precursor plasma spray process. *Journal of Thermal Spray Technology*. 2008;17:124–35.
11. Venkadesan G, Muthusamy J. Experimental investigation of Al_2O_3/8YSZ and CeO_2/8YSZ plasma sprayed thermal barrier coating on diesel engine. *Ceramics International*. 2019;45:3166–76.
12. Devoino O, Okovity V. Research of plasma spraying of the ZrO_2–Y_2O_3 electrolyte coatings. *Mokslas–Lietuvos Ateitis/Science–Future of Lithuania*. 2014;6:621–5.

13. Bragadeshwaran A, Kasianantham N, Ballusamy S, Tarun KR, Dharmaraj AP, Kaisan MU. Experimental study of methyl tert-butyl ether as an oxygenated additive in diesel and Calophyllum inophyllum methyl ester blended fuel in CI engine. *Environmental Science Pollution Research.* 2018;25:33573–90.
14. Mrdak M, Lačnjevac Č, Bajić D. Characterization of zirconium oxidecoating stabilized with cerium and yttrium oxide deposited on the bonding coating nickel chromium aluminum cobalt yttrium oxide. *Materials Protection.* 2023;64:5–12.
15. Omran K, Ahmed MS, Elmeniawi M, Sallam H, Naga S. A Review on Thermal Barrier Coatings. *The Egyptian International Journal of Engineering Sciences Technology.* 2021;36:1–6.
16. Huang J, Sun W, Huang R, Ma W. Cracking behavior of atmospheric plasma-sprayed 8YSZ thermal barrier coatings during thermal shock test. *Coatings.* 2023;13:243.
17. Rafiq NM, Wang S. Thermal barrier coatings: An insight into conventional plasma spray and water-stabilized plasma spray. *Coatings.* 2022;12:1916.
18. Ataie A, Mostaghimi J, Pershin L, Xu P. Fabrication of nanostructured cobalt ferrite coatings using suspension plasma spraying (SPS) technique. *Surface Coatings Technology.* 2017;328:451–61.
19. Bakan E, Vaßen R. Crack healing mechanisms in atmospheric plasma sprayed Yb-silicate coatings during post-process heat treatment. *Journal of the European Ceramic Society.* 2023;43:3684–93.
20. Bakhsheshi-Rad H, Hamzah E, Ismail A, Daroonparvar M, Kasiri-Asgarani M, Jabbarzare S, et al. Microstructural, mechanical properties and corrosion behavior of plasma sprayed NiCrAlY/nano-YSZ duplex coating on Mg–1.2 Ca–3Zn alloy. *Ceramics International.* 2015;41:15272–7.
21. Bobzin K, Öte M, Knoch M. Surface pre-treatment for thermally sprayed ZnAl15 coatings. *Journal of Thermal Spray Technology.* 2017;26:464–72.
22. Brandon JR, Taylor R. Thermal properties of ceria and yttria partially stabilized zirconia thermal barrier coatings. *Surface and Coatings Technology.* 1989;39:143–51.
23. Chavez JJG, Naraparaju R, Mikulla C, Mechnich P, Kelm K, Ramana C, et al. Comparative study of EB-PVD gadolinium-zirconate and yttria-rich zirconia coatings performance against Fe-containing calcium-magnesium-aluminosilicate (CMAS) infiltration. *Corrosion Science.* 2021;190:109660.
24. Behera GC, Sahoo SP, Kumari S, Datta S. Study on wear morphology of uncoated and MT-CVD TiCN-Al_2O_3 coated carbide inserts during dry machining of Inconel 825 superalloy. *Materials Today: Proceedings.* 2023.
25. Sabzi M, Mousavi Anijdan S, Shamsodin M, Farzam M, Hojjati-Najafabadi A, Feng P, et al. A review on sustainable manufacturing of ceramic-based thin films by chemical vapor deposition (CVD): Reactions kinetics and the deposition mechanisms. *Coatings.* 2023;13:188.
26. Vaßen R, Bakan E, Mack DE, Guillon O. A perspective on thermally sprayed thermal barrier coatings: Current status and trends. *Journal of Thermal Spray Technology.* 2022;31:685–98.
27. Ahmed S, Atia NN. Fast and efficient zirconia-based reversed phase chromatography for selective determination of triptans in rat plasma. *Journal of Pharmaceutical Biomedical Analysis.* 2017;143:241–51.
28. Almeida V, Balzaretti N, Costa T, Machado G, Gallas M. Surfactants for CNTs dispersion in zirconia-based ceramic matrix by sol–gel method. *Journal of Sol-Gel Science Technology.* 2013;65:143–9.
29. Cao Z, An S, Song X. Effect of thermal treatment at high temperature on phase stability and transformation of Yb_2O_3 and Y_2O_3 co-doped ZrO_2 ceramics. *Scientific Reports.* 2022;12:9955.

30. Cañas E, Orts M, Boccaccini A, Sánchez E. Solution Precursor Plasma Spraying (SPPS): A novel and simple process to obtain bioactive glass coatings. *Materials Letters.* 2018;223:198–202.
31. Arantes V, Coutinho R, Martins S, Huang S, Vleugels J. Solid state sintering behavior of zirconia-nickel composites. *Ceramics International.* 2019;45:22120–30.
32. Bacchelli B, Giavaresi G, Franchi M, Martini D, De Pasquale V, Trirè A, et al. Influence of a zirconia sandblasting treated surface on peri-implant bone healing: An experimental study in sheep. *Acta Biomaterialia.* 2009;5:2246–57.
33. Baig M, Khalid F, Khan F, Rehman K. Properties and residual stress distribution of plasma sprayed magnesia stabilized zirconia thermal barrier coatings. *Ceramics International.* 2014;40:4853–68.
34. Ragurajan D, Satgunam M, Golieskardi M. The effect of cerium oxide addition on the properties and behavior of Y-TZP. *International Scholarly Research Notices.* 2014;2014:828197.
35. Chen H, Fan M, Zhu W, Norton A. High temperature oxidation behaviour of combustion flame sprayed CoNiCrAlY coatings. *Surface Coatings Technology.* 2020;385:125431.
36. Choudhary S, Islam A, Mukherjee B, Richter J, Arold T, Niendorf T, et al. Plasma sprayed lanthanum zirconate coating over additively manufactured carbon nanotube reinforced Ni-based composite: Unique performance of thermal barrier coating system without bondcoat. *Applied Surface Science.* 2021;550:149397.
37. Keyvani A, Saremi M, Sohi MH. An investigation on oxidation, hot corrosion and mechanical properties of plasma-sprayed conventional and nanostructured YSZ coatings. *Surface Coatings Technology.* 2011;206:208–16.
38. Cinca N, Lima CRC, Guilemany JM. An overview of intermetallics research and application: Status of thermal spray coatings. *Journal of Materials Research Technology.* 2013;2:75–86.
39. Di Girolamo G, Blasi C, Brentari A, Schioppa M. Microstructural, mechanical and thermal characteristics of zirconia-based thermal barrier coatings deposited by plasma spraying. *Ceramics International.* 2015;41:11776–85.
40. Di Girolamo G, Blasi C, Pagnotta L, Schioppa M. Phase evolution and thermophysical properties of plasma sprayed thick zirconia coatings after annealing. *Ceramics International.* 2010;36:2273–80.
41. Di Girolamo G, Blasi C, Schioppa M, Tapfer L. Structure and thermal properties of heat treated plasma sprayed ceria–yttria co-stabilized zirconia coatings. *Ceramics International.* 2010;36:961–8.
42. Ahmadi-Pidani R, Shoja-Razavi R, Mozafarinia R, Jamali H. Evaluation of hot corrosion behavior of plasma sprayed ceria and yttria stabilized zirconia thermal barrier coatings in the presence of Na_2SO_4+ $_{V2}O_5$ molten salt. *Ceramics International.* 2012;38:6613–20.
43. Di Iorio G, Paglia L, Pedrizzetti G, Genova V, Marra F, Bartuli C, et al. Ytterbium Disilicate/Monosilicate multilayer environmental barrier coatings: Influence of atmospheric plasma spray parameters on composition and microstructure. *Coatings.* 2023;13:1602.
44. Mukherjee B, Islam A, Pandey KK, Rahman OA, Kumar R, Keshri AK. Impermeable CeO_2 overlay for the protection of plasma sprayed YSZ thermal barrier coating from molten sulfate-vanadate salts. *Surface Coatings Technology.* 2019;358:235–46.
45. Doleker K, Ozgurluk Y, Özkan D, Mesekiran N, Karaoglanli AC. Comparison of microstructures and oxidation behaviors of ytria and magnesia stabilized zirconia thermal barrier coatings (TBC). *Materiali in Tehnologije.* 2018;52:315–22.
46. Doleker KM, Ozgurluk Y, Karaoglanli AC. Isothermal oxidation and thermal cyclic behaviors of YSZ and double-layered YSZ/$La_2Zr_2O_7$ thermal barrier coatings (TBCs). *Surface Coatings Technology.* 2018;351:78–88.

47. Doleker KM, Odabas O, Ozgurluk Y, Askerov H, Karaoglanli AC. Effect of high temperature oxidation on Inconel 718 and Inconel 718/YSZ/$Gd_2Zr_2O_7$. *Materials Research Express*. 2019;6:086456.
48. Doleker KM, Ozgurluk Y, Ahlatci H, Karaoglanli AC. Evaluation of oxidation and thermal cyclic behavior of YSZ, $Gd_2Zr_2O_7$ and YSZ/$Gd_2Zr_2O_7$ TBCs. *Surface Coatings Technology*. 2019;371:262–75.
49. Doleker KM, Ozgurluk Y, Karaoglanli AC. TGO growth and kinetic study of single and double layered TBC systems. *Surface Coatings Technology*. 2021;415:127135.
50. Doleker KM, Ozgurluk Y, Parlakyigit AS, Ozkan D, Gulmez T, Karaoglanli AC. Oxidation behavior of NiCr/YSZ thermal barrier coatings (TBCs). *Open Chemistry*. 2018;16:876–81.
51. Fan W, Bai Y. Review of suspension and solution precursor plasma sprayed thermal barrier coatings. *Ceramics International*. 2016;42:14299–312.
52. Tailor S, Upadhyaya R, Manjunath S, Dub A, Modi A, Modi S. Atmospheric plasma sprayed 7%-YSZ thick thermal barrier coatings with controlled segmentation crack densities and its thermal cycling behavior. *Ceramics International*. 2018;44:2691–9.
53. Cheng G, Yang C, Yao X, Guo L, Han J. When deep learning meets metric learning: Remote sensing image scene classification via learning discriminative CNNs. *IEEE Transactions on Geoscience Remote Sensing*. 2018;56:2811–21.
54. Imrana Y, Xiang Y, Ali L, Noor A, Sarpong K, Abdullah MA. CNN-GRU-FF: A double-layer feature fusion-based network intrusion detection system using convolutional neural network and gated recurrent units. *Complex Intelligent Systems*. 2024:1–18.
55. Zhu J, Wang X, Kou L, Zheng L, Zhang H. Prediction of control parameters corresponding to in-flight particles in atmospheric plasma spray employing convolutional neural networks. *Surface Coatings Technology*. 2020;394:125862.
56. Hospach A, Mauer G, Vaßen R, Stöver D. Characteristics of ceramic coatings made by thin film low pressure plasma spraying (LPPS-TF). *Journal of Thermal Spray Technology*. 2012;21:435–40.
57. Jamali H, Mozafarinia R, Razavi RS, Ahmadi-Pidani R. Comparison of thermal shock resistances of plasma-sprayed nanostructured and conventional yttria stabilized zirconia thermal barrier coatings. *Ceramics International*. 2012;38:6705–12.

7 Zirconia-Based Thermal Barrier Coatings

Advancements in Doped, Nanostructured, Multilayered, and Gradient-Structured Systems

Amrinder Mehta, Hitesh Vasudev, and Jeyaprakash Natarajan

7.1 INTRODUCTION

Zirconia-based ceramics are known for their high heat resistance and are used in various enhanced applications. All these materials are based on zirconium dioxide (ZrO_2), often known as zirconia, which has exceptional mechanical, chemical, and thermal properties. The most promising feature of zirconia-based materials is their high mechanical strength and fracture resistance. This is attributed to the fact that the crystals in the material undergo a characteristic phase transformation when subjected to stress, which is referred to as transformation toughening [1]. Figure 7.1 shows the different phases of the crystal structure of zirconia (ZrO_2), which are familiar in material science. These structures are shown as large spheres for zirconium elements and small spheres for oxygen, connecting lines are illustrated as

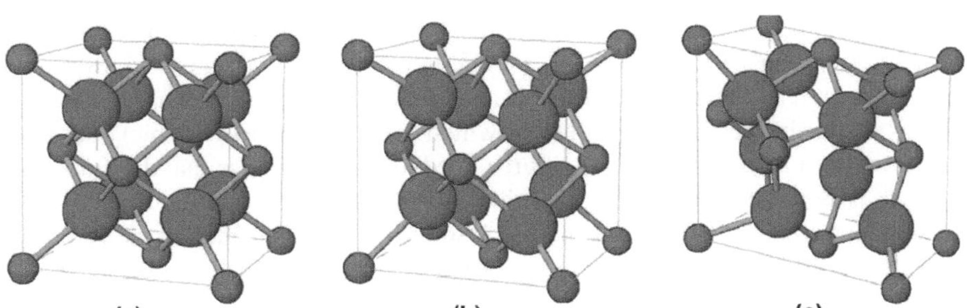

FIGURE 7.1 Pictorial representation of the three polymorphic forms of zirconia: (a) cubic, (b) tetragonal, and (c) monoclinic [2]. (Reproduction permissions.)

DOI: 10.1201/9781003519300-9

plain lines for clarity. As seen in the three separate regions (a, b, c) of the figure, zirconia is not a single variant, but shows polymorphism. In Figure 7.1a, the monoclinic phase, which is stable at room temperature with a rather more complex crystal arrangement, is depicted. Structure (b) indicates the tetragonal phase, where the atomic order is slightly better than the random state, and it is observed at intermediate temperatures. Structure (c) corresponds to the cubic phase, in which the atomic arrangement is the most symmetric, and it is observed at high temperatures. The areas of stability of each phase and their properties are different, so zirconia can be used in a wide range of applications. The transition between these phases is rather dramatic and is at the very center of the material's transformation toughening behavior [2].

This property renders zirconia-based ceramics highly prone to fracture and crack propagation rather than initiation, making them suitable for applications in which reliability and endurance are critical. In dentistry, materials based on zirconia ceramics have become popular as replacements for conventional metal–ceramic restorations. The primary advantage of zirconia-based ceramics over other widely used materials is their excellent appearance because they resemble teeth; they can be employed in constructing crowns, bridges, and implants. Various experimental investigations of the biocompatibility of zirconia have revealed that there are no allergic reactions when it is implanted in bone tissues [3]. Zirconia materials are also used in orthopedics, especially in the analysis of total hip replacements. Owing to the favorable wear characteristics of these materials, such as their ability to withstand wear and low friction, they are suitable for use in femoral heads in hip replacement systems. Furthermore, the use of zirconia in this field can be attributed to its chemical stability and biocompatibility. With the progress of zirconia-based materials, different types of materials are available, such as yttria-stabilized zirconia (YSZ) and partially stabilized zirconia (PSZ) [4]. These variations offer different properties that can be designed to include various characteristics. Yttrium-stabilized zirconia (YSZ) combined with electronics has triggered the use of electronics, particularly in the development of ferroelectric systems with ultralow power systems [5]. The phase diagram of temperature–composition for a material system with YO1.5 in mole percent is shown in Figure 7.2. What can be observed in the diagram is different phase regions and phase transitions throughout the different forms of crystal structures. A liquid phase (L) exists at high temperatures of approximately 2700°C. The diagram features three main solid phases: the cubic form F, the transformable tetragonal form T, and the monoclinic form T. Two substantial tie lines are represented by two colors at 1600°C and 1400°C. Several two-phase regions are clearly shown in the diagram, some of which are L + F (liquid + cubic) and T + F (tetragonal + cubic). A feature that has received much attention is that the tetragonal T phase is nontransformable in the middle composition part of the diagram. All phase boundaries are depicted using black solid lines, transformation pathways are indicated with dashed lines for clarity. The composition ranges from 0 to 20 mol% YO1.5, illustrating how and to what extent the phases, which are at equilibrium with the referent phases, depend on temperature and the concentration of roaming elements [6].

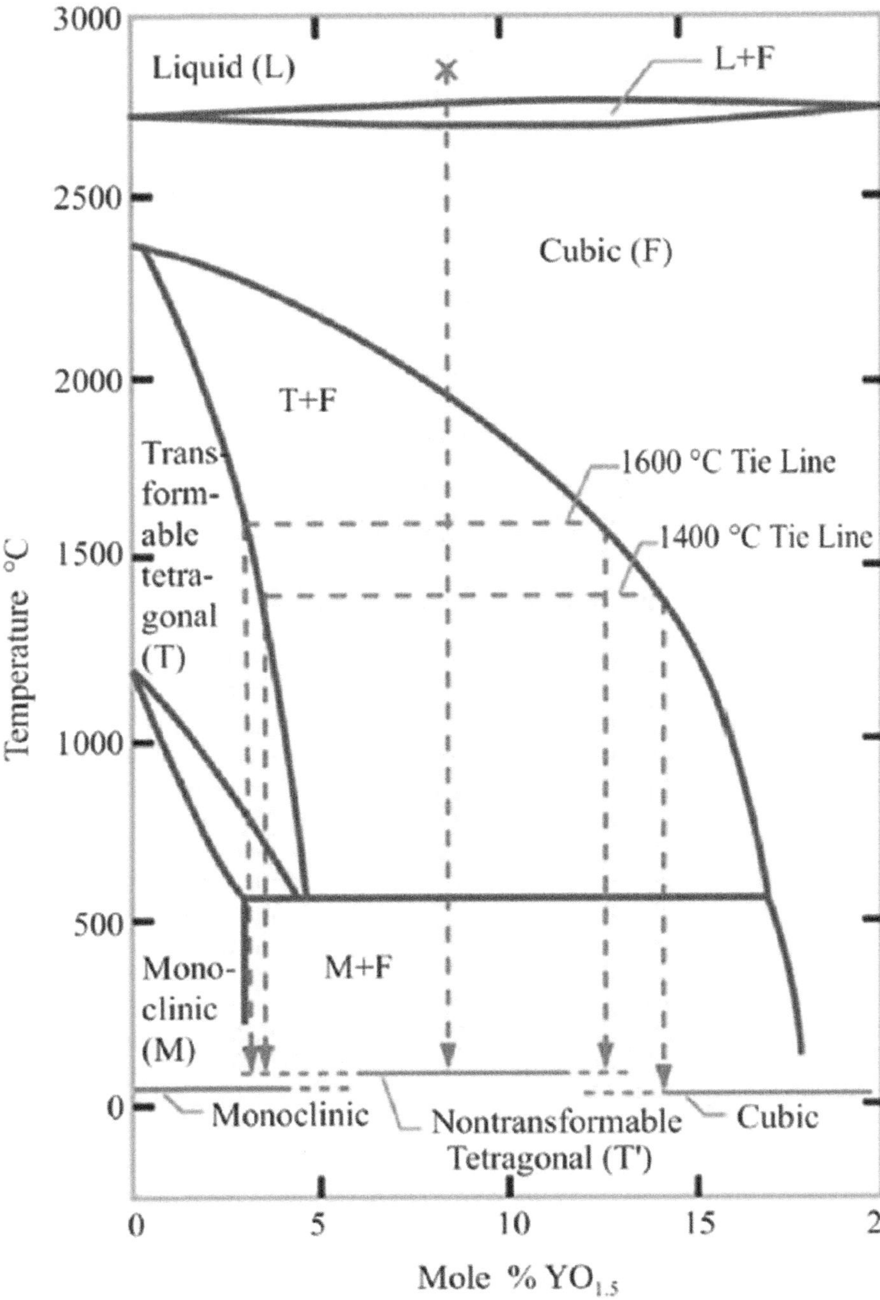

FIGURE 7.2 Phase diagram of the ZrO_2 – Y_2O_3 system [6]. (Creative Commons Attribution 4.0.)

The production of zirconia materials has also advanced. Additive manufacturing and spark plasma sintering have been used to produce geometrically complex shapes and enhance properties. The adopted complex manufacturing techniques enable better control of microstructure and composition of the material, resulting in better application performance.

However, some problems are encountered with zirconia-based materials, and some researchers are working hard to solve them [7]. These include enhancing their stamina and performance and lowering steep switching voltages in electronics. Dental preclinical studies and trials are underway to improve the osteointegration of zirconia implants and establish protocols for the fabrication of these implants and their use in clinical practice. They are also considered for TBCs because of their low thermal conductivity and good chemical stability when used to protect components at high temperatures [8]. With the increasing development of research in this field, new opportunities for the use of zirconia-based materials have emerged. Further changes and enhancements to the properties of the zirconia-based materials are currently being investigated through the synthesis of nanostructured, multilayered, and gradient-structured materials [9]. Improvements have been made to overcome the current drawbacks of zirconia-based materials, and their potential in science and technology continues to expand.

7.2 ROLE OF ZIRCONIA IN HIGH-TEMPERATURE ENVIRONMENTS

Zirconia-based ceramics have recently been established as important materials for high-temperature components, such as thermal barrier coatings (TBCs) for gas turbine engines. These materials, produced from zirconium dioxide (ZrO_2), possess excellent thermal, mechanical, and chemical stabilities and are highly recommended for shielding metallic parts from heat and corrosive conditions. One of the most vital uses of zirconia in high-temperature application is in the form of YSZ in TBCs [10,11]. Figure 7.3 represents major requirements for the new TBC materials using a circular diagrammatic method with eight key properties. The core potentiality is identified as "requirements for alternative TBC materials," which is enclosed in a circle of related characteristics. These essential requirements encompass phase stability in beta, oxidation resistance to the right, and low thermal conductivity in the bottom right corner. Thermal shock resistance is also highlighted in the diagram, which is an important area of concern with regard to temperature variation. A high value of the coefficient of thermal expansion (CTE) is suggested at the bottom. High mechanical and thermal properties and sintering resistance are suggested at the left side. Corrosion resistance emerged in the lower left box to meet the minimum requirements. Each property is shown in a differentiated box with arrows pointing to the central concept, thus giving a graphic representation of the combined characteristics that are vital for TBC materials [12].

YSZ coatings are critical for protecting metallic components in gas turbine engines from the corrosive effects of high-temperature combustion gases. The thermal conductivity of the coatings is an important characteristic that determines the design and performance of a gas turbine and the life span of the said apparatus. New developments in temperature gradient measurement, for instance, the laser temperature

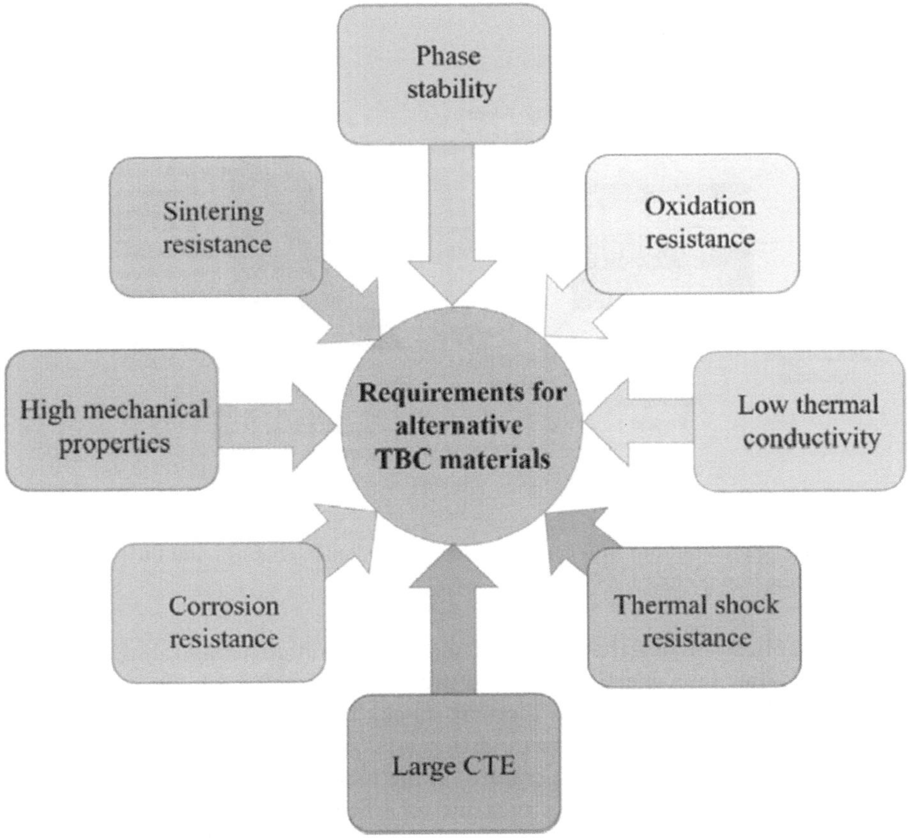

FIGURE 7.3 Alternative TBC materials must satisfy certain fundamental prerequisites [12]. (Reproduction permissions.)

gradient technique, have made it possible to determine the thermal conductivities of porous YSZ coatings in a much closer approximation to operating conditions [13]. Figure 7.4 depicts the photonic device coated through the thin film process, and the resulting structure consists of a top-coat layer, bond coat layer, and substrate layer. The following deposition techniques are practiced around this structure: plasma spray process, high velocity oxy fuel (HVOF), high velocity air fuel (HVAF), magnetron sputtering, ion beam sputtering, chemical vapor deposition, physical vapor deposition, and flame spray process. Stress transfer, internal stress release, and the reduction of microcrack propagation are essential functions of the bond coat layer. On the right side, there is a triangle where the main advantages of duplex coatings are described, including protection against corrosion, less maintenance, better appearance, increased life span, cheaper replacement, economic efficiency, and better durability for steel. This multistep coating system shows that surface preparation and treatment are best done through many processes that will complement one another to give the best result to the target material [14].

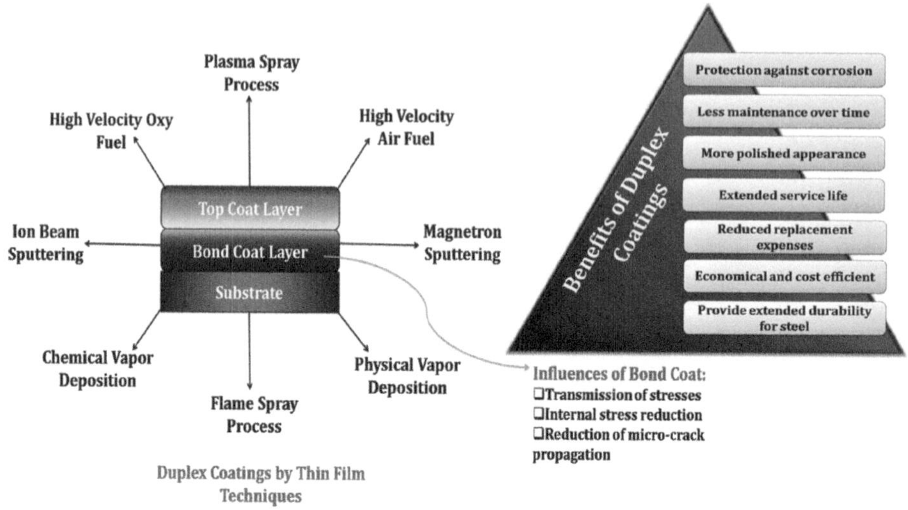

FIGURE 7.4 Potential thin film coating methods for duplex coatings and their advantages [14]. (Reproduction permissions.)

Such application of zirconia-based materials at high temperatures depends on some attainable properties. YSZ features low thermal conductivity, chemical stability, and good resistance to thermal shock. These characteristics enable the establishment of a large temperature differential across the thickness of the coating to minimize the actual temperature on the metal parts below [15]. This temperature reduction is a necessary measure as it helps to increase the life cycle of the turbine parts, as well as the efficiency of the entire engine. Investigative work has consistently aimed at ameliorating the performance of zirconia-based materials for demanding high-temperature applications. One way is to code the YSZ with other rare earth oxides, such as Scandia, in order to enhance the characteristics of the TBC material. Furthermore, the utilization of other substances, such as ceria and thoria, in YSZ has been shown to guarantee enhanced high-temperature operation. Another area of study of zirconia in recent years has been on the behavior of the material under characteristic temperature gradients [16]. It has been observed, through molecular dynamic simulations, that oxygen ion diffusion in YSZ can be influenced by sharp temperature gradients, and depending on the gradients, the diffusion coefficients may rise. It is very important for previewing as well as for enhancing the efficiency of YSZ in TBCs and solid oxide fuel cells (SOFCs) applications. SOFCs have emerged as promising energy sources for both stationary and portable applications. To achieve optimal performance, the design of SOFCs, including their structure and operation, must be optimized. The importance of zirconia is not only limited to some applications of TBC. It has also been applied in the design of functionally graded ceramics for HI-TEMP applications. Materials such as functionally graded thermal barrier zirconia-mullite/alumina composites have displayed better durability characteristics and thus remain applicable in several important high-temperature engineering

TABLE 7.1
Ceramic Zirconia Possesses Mechanical Properties [18]

S. No.	Mechanical Properties	Amount
1	Density	6.05 g/cm^3
2	Hardness	1200 HV
3	Bend strength	900–1200 MPa
4	Compressive strength	2000 MPa
5	Fracture toughness	7–10 $MPa{\cdot}m^{1/2}$
6	Young's modulus	210 GPa
7	Thermal expansion coefficient	11×10^{-6} 1/K

Source: Permissions CC BY 4.0.

applications [17]. In Table 7.1, the mechanical parameters of commercially available zirconia that has been stabilized with yttria are presented [18].

Further advancements in processing and characterization of zirconia-based functional materials for high-temperature applications are still ongoing. The laser cladding method is being adopted to incorporate thermocouple sensors into TBCs so that real-time temperature readings in critical interface regions can be taken. Regarding hostile environments, this innovation has created other prospects for optimizing the activity of gas turbines and their protection. It should be noted that each work is performed with a certain composition and structure of zirconia-based materials, which have current restrictions [19]. The YSZ composites based on neodymium and ytterbium oxides, with the addition of titanium dioxide to suppress the relevant phase transformations and increase the thermal degradation resistance at temperatures above 1200°C, proved effective. Zirconia is used in high-temperature applications, mainly as TBC in gas turbine engines. Its non-conductive, highly stable, high-temperature, and chemically inert features make it an enabling material for the development of high-temperature technologies. The molecular development work currently performed on zirconia-based materials effectively points to the prospects of further enhancements in the thermal cycling, durability, and general efficiency of high-temperature operating systems [20].

7.3 APPLICATIONS OF ZIRCONIA IN AEROSPACE AND AUTOMOTIVE INDUSTRIES

This ceramic material, called zirconia, is excellent and has been used to replace other materials in the aerospace and automotive industries. Zirconia is widely used in aerospace materials to improve the efficiency of aircraft engines. One of its applications is as a thermal barrier coating on turbine blades, where it shields the parts from high-temperature operation and permits engines to run at hotter temperatures, enhancing fuel economy and diminishing emissions [21]. The lightweight and wear resistance of the zirconia material also enhance the performance of parts used in critical engine locations, bearings, and seals, thereby reducing the cyclic variations, enhancing the reliability, and increasing the service intervals. In the automotive industry, zirconia has been used in several ways to improve vehicle performance and make the environment more friendly. Its most important application is in oxygen sensors, where the high stability of zirconia in high-temperature applications allows monitoring of exhaust components, thereby enhancing combustion efficiency of engines and lowering pollutant emissions. Zirconia is also applied in spark plug insulators where stable ignition is needed in harsh environments. In high-performance engines, zirconia has been used in turbocharger rotors where it contributes to efficiency instead of inertia. The wear-resistant and high-temperature stability of the material make it fit for use in fuel injectors so that fuel flow can be accurately controlled and, in turn, contributes to the overall performance of the engine [22]. Figure 7.5 suggests four of the most critical areas of emerging advanced technology and transportation annually related to completely different photographs. There is a jet airplane on a runway at sunset in the "Aviation" section, which gives a feeling of elegance and efficiency of present-day luxury. The "Power Generation" section is represented by an industrial complex where tall structures and pipework complexity refer to energy production facilities. The "Marine" section includes a segment of a large cruise liner cruising on the sea – transport and travel. In the middle of these four images, there is a drawing illustrating a typical blade of the turbine labeled TBC, which implies that this is a unifying technology of these four disciplines. This central element suggests that turbine technology is indispensable

FIGURE 7.5 Industrial applications of zirconia using the TBCs method [23]. (Reproduction permissions.)

in the broad context of the industrial economy and may be used in aviation and navigation, in industrial and transportation machinery, in electric power engineering, and even in space research [23].

Zirconia possesses some unique characteristics that make it suitable also in brake systems; high-performance brake pads and discs made of zirconia are characterized by high operational stability of the friction coefficients at high temperatures. Both industries use zirconia in numerous sensors where its data are properly fed to various engine management systems and other important necessities. In its support for exhaust gas recirculation valves as well as in the catalytic converters, it plays a significant role in reducing undesirable emissions. In fulfilling technological demands, the application of zirconia in the aerospace and automotive sectors rises and subsequently boosts the efficiency and performance relevant to environmental consciousness of the said industries [24].

7.4 FUTURE TRENDS IN THE MANUFACTURING OF ZIRCONIA-BASED MATERIALS

The subsequent deterioration of zirconia TBCs is essential to making components more durable in high-temperature applications, such as in turbines. The current developments are aimed at doped, nanostructured, multiple-layered, and gradient TBCs based on zirconia, which exhibit significantly better thermal properties and stability under severe operating conditions compared to YSZ coatings. These innovations are designed to overcome the shortcomings of conventional TBCs, such as low thermal conductivity and stability to oxidation and erosion [25]. Figure 7.6 highlights the future trends of the TBCs. This allows readers to understand the different areas of TBC development by improving the mind map structure. The conceptual core is called the "Future scope of Thermal Barrier Coatings (TBCs)," and

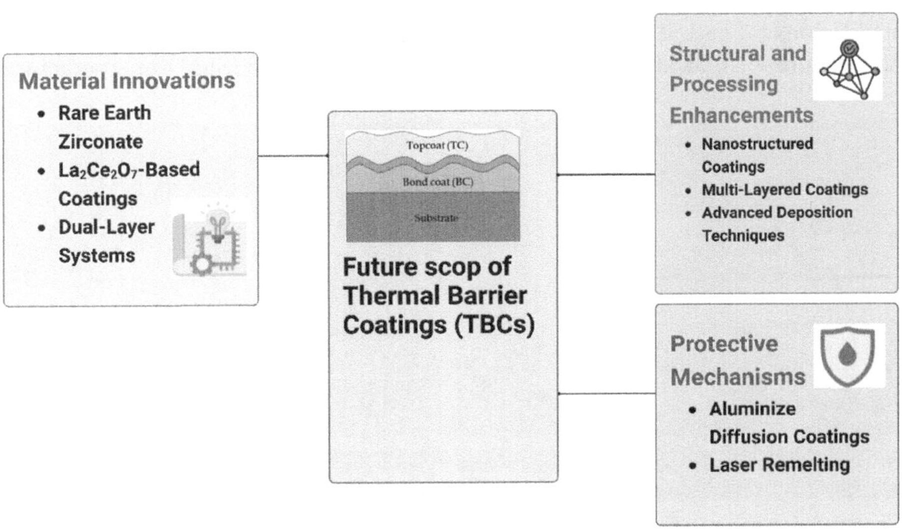

FIGURE 7.6 Future scope of the Zr-based TBCs.

this section is linked to three areas of advancement. The first section, "Material Innovations," consists of materials such as rare-earth zirconate, protective coating based on $La_2Ce_2O_7$, and combined two-layer systems [26]. The second section covers "Structural and Processing Enhancements," including nanostructured coatings, multiple coated coatings, and advanced deposition technologies. The third section, "Protective Mechanisms," focuses on aluminized diffusion coatings and laser remelting. It also illustrates a cross-sectional view of a conventional TBC system, which includes the topcoat TC, bond coat BC, and the substrate. This overall flowchart presents all the main directions in TBC research and development and underlines the complex nature of enhancing the performance of thermal barriers through material science, structural design, and protection [27]. The following are the principal features of these advanced TBCs:

7.4.1 Doped Zirconia-Based TBCs

Doped zirconia-based TBCs have attracted much attention in recent years due to improvements in their performance parameters. The addition of rare earth elements, specifically gadolinium oxide (Gd_2O_3) and yttria-stabilized zirconia (YSZ), seems to improve the overall properties of these coatings [28].

The incorporation of Gd_2O_3 in YSZ was found to be useful for stabilizing the tetragonal phase of zirconia, which is required for the strength of the coating at elevated temperatures. This stabilization is due to the larger ionic radius of Gd^{3+} than Zr^{4+}, which leads to lattice distortions and oxygen vacancies, thus suppressing the phase transformations that lead to coating failure. From the above results and discussion, we can infer that the introduction of Gd_2O_3 into the YSZ coatings enhanced their thermal characteristics. The thermal conductivity of Gd_2O_3-doped YSZ (GdYSZ) was reduced as the Gd_2O_3 volume fraction increased, with phonon scattering observed at the atomic level. The reduction in the thermal conductivity is desirable for TBC applications because it results in improved thermal insulation of the metal substrate [29]. Figure 7.7 indicates two scanning electron microscope (SEM) micrographs of

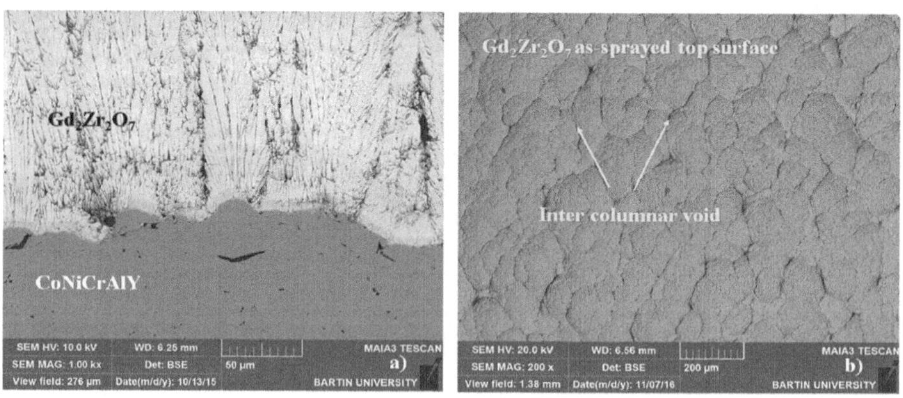

FIGURE 7.7 SEM micrographs: (a) a cross-sectional view of a two-layer structure and (b) a top-down view of the as-sprayed $Gd_2Zr_2O_7$ surface [30]. (Reproduction permissions.)

Bergamer labeled with (a) and (b) from Bartin University. Figure 7.7a shows a section micrograph where two layers are observed: the $Gd_2Zr_2O_7$ layer on top of the CoNiCrAlY layer. The $Gd_2Zr_2O_7$ layer was characterized by a vertical columnar microstructure, whereas CoNiCrAlY was identified as an adherent, uniform bottom layer. The right micrograph (Figure 7.7b) shows the as-sprayed $Gd_2Zr_2O_7$ surface at lower magnification, indicating a porous network of inter-columnar voids between the crystallites. These voids are illustrated in the image, where they are indicated by arrows for clarity. From the SEM parameters, the pictures show that Figure 7.7a was obtained at 10.0 kV with 100x magnification and Figure 7.7b was obtained at 20.0 kV with 200x magnification. For a better description of the size of the microstructural features, the scale bars for Figure 7.7a and b are 50 and 200 μm, respectively [30].

Additionally, the incorporation of Gd_2O_3 results in the improvement of the mechanical characteristics of the designed coating. The experimental results indicate that GdYSZ has a higher value of hardness, elastic modulus, and fracture toughness than YSZ. These improvements are because of solid solution strengthening and the creation of a more refined cellular structure that does not readily allow the crack to spread. The sintering resistance of zirconia-based TBCs is also enhanced when Gd_2O_3 is added into the TBC system. The presence of Gd^{3+} ions at the grain boundaries limits the surface energy and grain boundary energy, establishing a low driving force for sintering reactions [31]. This improved sintering resistance assists in preserving the porous structure of the coating, which is pivotal to the insulation properties. In addition, Gd_2O_3-doped YSZ coatings have higher phase stability when subjected to high temperatures, when compared with non-sophisticated YSZ. This improved stability is important for long-term use in environments of high temperatures, which are experienced in gas turbine engines. The integration of Gd_2O_3 in zirconia-based TBC gives several facet solutions to improve coating characteristics. Hence, suppressing the tetragonal phase stability, their low thermal conductivity, enhancement of mechanical properties, as well as sintering resistance and phase stability in GdYSZ coatings can pave the way for a new era in TBC systems in high-temperature applications [32].

7.4.2 Nanostructured TBCs

Recent complex nanostructured TBCs are promising next-generation coating materials for high-temperature applications, such as gas turbine engines and other thermal insulation fields. Employment of these coatings results in enhanced performance as compared to conventional TBCs in terms of thermal conductivity and thermal shock resistance. However, for nanostructured TBCs, there stands the exciting prospect that microstructures govern such performance [33]. These coatings are generally constituted by yttria-stabilized zirconia (YSZ) and generally possess a grain form of between 30 and 70 nm. This type of nanostructure results in piled-up grain boundaries that constitute a great obstacle to heat flow. A higher number of interfaces triggers multiple phonon collisions, resulting in lower thermal conductivity than simple coatings with larger grain sizes [34].

Figure 7.8 presents a set of SEM micrographs of the cross-sectional view of a multilayer coating system. Figure 7.8a presents a cross-sectional view with three distinct

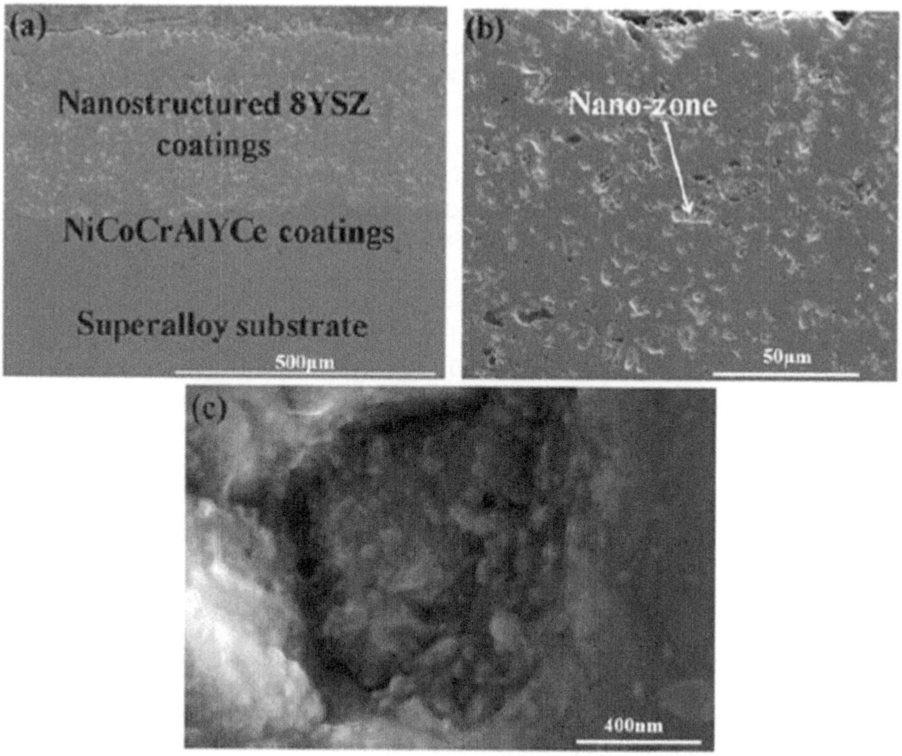

FIGURE 7.8 (a) Cross-sectional scanning electron microscopy images of as-sprayed 8YSZ coatings, (b) enlarged picture of (a), and (c) enlarged image of unmelted feedstocks in (b) [35]. (Creative Commons Attribution 4.0.)

layers: a thin nanostructured 8YSZ layer at the top, NiCoCrAlYCe in the middle layer, and a superalloy substrate at the base with a scale bar of 500 μm. On the same scale, Figure 7.8b shows a surface view at 50 μm whereby a portion of the area, specifically the nanozone, represents a unique surface morphology with small particles randomly distributed throughout the zone. Figure 7.8c is another high-power image captured at 400 nm scale to show that the nanostructure morphology of the coating surface has a cloudy and irregular surface with contrast variation, confirming the surface topography of the coating. The micrographs provided in this work show the coating system from the macro to the nanoscale, and this type of information is useful for evaluating the coating properties and its behavior in high-temperature environments [35].

Retaining porosity during high-temperature operation is perhaps the most important benefit of nanostructured TBCs. The microstructure of the coating was formed by nano melts partially sintered and embedded in a fully molten matrix. These partially melted regions, nanozones, are of great importance to the performance of the entire coating. Nanozones are regions that contain pores that are filled with a large number of grains of the same dimension as nanoseconds and microseconds, which start to prevent the extension of the crack and improve the characteristics of fracture toughness. This feature of the structure is instrumental in increasing the thermal shock and general toughness of the coating [36].

These cracks give nanostructured TBCs a tri-modal pore size distribution of micro, fine, and nano-scale pores on the TBC substrate. The formation of these structures is critical for reducing thermal conduction while promoting the flexibility of the coating. The pores also slow down the heat transfer processes and allow for thermal stresses to be absorbed and distributed over the coating surface, thus enhancing thermal cycle and thermal shock characteristics of the coating. Another interesting characteristic of nanostructured TBCs is their lesser value of elastic modulus than that possessed by normal coatings. This lesser stiffness enables the coating to effectively withstand thermal stress and strain, thus minimizing the chances of spallation or delamination during thermal cycling. Predominantly, nanostructured TBCs offer lower thermal conductivity, and their enhanced mechanical properties make them suitable for severe thermally gradient regions and thermal cycling [37]. Different experimental analyses have shown the improved thermal shock resistance in nanostructured TBCs. The results indicated that these coatings could undergo more thermal cycles up to failure than conventional coatings. This enhanced performance is due to the improved distribution of and reduction of thermal stresses by the optimized porosity of the coating nanostructure [38]. Nevertheless, the fabrication and application of nanostructured TBCs have several issues. One issue is high-temperature reactivity and the ability to sinter and densify, which can raise thermal conductivity and reduce flexibility over time. Nevertheless, the precise engineering of nanostructures, the relevant part of the herein described coatings, such as the percentage of nanozones and optimum porosity, can prevent such consequences and sustain the overall performance for a long time [39].

Nanostructured TBCs are indeed a new generation of TBCs that define the state of the art in the engineering of thermal protection materials. Possessing nanozones, a developed system of porosity and submicron crystalline grains, they, however, have special thermal coefficients of conductivity and thermal-stress coefficients that allow them to demonstrate comparatively low thermal conductivity and high-temperature staking capability as compared to other coatings. With further studies in this area, nanostructured TBCs have a strong potential to enhance all aspects of using high-temperature components across industries.

7.4.3 Multilayered TBCs

A multilayered TBC system is a highly developed concept for improving the service capacity and durability of high-temperature parts and components in various uses, including gas turbine engines and combustion chambers. These complex coating systems utilize a precisely tailored multilayer system so as to optimize thermal stress and enhance the material's general service life. The major benefit of multilayered coatings is that they reduce both thermal and mechanical stresses in regions where they are most concentrated in the interfaces between different layers. This stress distribution mechanism reduces the likelihood of early coating failure and delamination, both of which are characteristics of standard single-layer TBCs. It is in this way

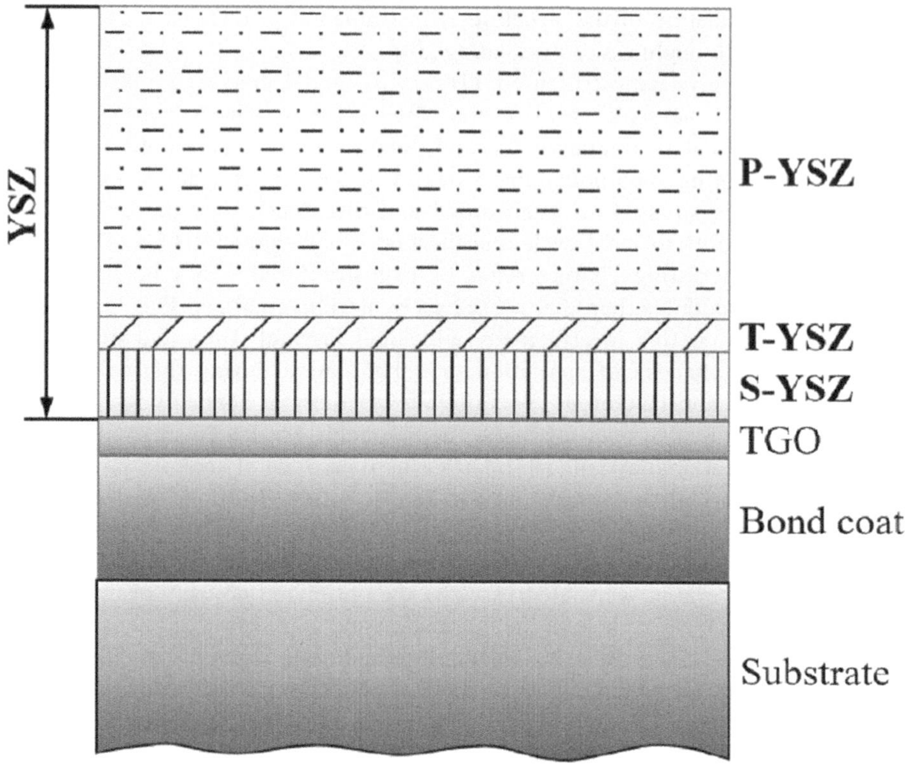

FIGURE 7.9 Schematic diagram of the structure of the multi-layer TBCs [41]. (Reproduction permissions.)

that the diverse layers with dissimilar composition and properties can make the coating system flexible, which will provide thermal insulation, mechanical compliance, and chemical stability [40].

In Figure 7.9, we see a schematic of the cross-sectional view of a multilayer ceramic coating system. A substrate layer acts as a base material layer from the bottom of the mold. Above it is a layer of bond coat, which is commonly used to improve the bond between the structure and overlay, as represented in the figure. Lying on the bond coat is a very slender thermally grown oxide (TGO) layer marked in gold. The main coating structure consists of three distinct YSZ layers: S-YSZ at the bottom represents a vertical columnar style, T-YSZ in the middle exhibits diagonal patterns, and P-YSZ at the top exhibits a horizontal dotted-line pattern. This type of layered structure is widely employed in all TBC systems, especially in applications at high temperatures. Based on the different patterns obtained in each layer, the microstructural features of the YSZ layers are believed to be arranged in different ways to tailor properties such as thermal insulation and strain tolerance. This includes the development of an overall protection for the whole system against high temperatures and other rigorous working conditions of the substrate [41].

As is well-known, one of the main function of multilayered TBCs is to prevent corrosive species from interacting with the material. The orientation of the layers can reduce the access of undesirable phases, such as sulfur and vanadium, which are normally observed under severe working conditions. This barrier effect greatly minimizes the hot corrosion, a phenomenon that affects the underlying substrate material in a very terrible manner. Reducing the ability of corrosive agents to penetrate the material layers results in increased overall use time of the coated parts by multilayered TBCs [42]. Materials that have similar properties are used in many designs of the multilayered TBCs. For instance, we could expect a thin, chemically stable bond coat on top of which we may have layers of ceramics with progressively varying chemical and physical characteristics. This gradual transition reduces the TEML and is effective in reducing or preventing further thermal expansion variation between the substrate and the outermost ceramic layer of the coating, thus increasing the resistance of the coating to thermal cycling fatigue [43]. Functional gradients could also be incorporated within the layers of the TBC or even across the overall coating system in other high-level multilayer TBCs. These gradients can be controlled by adjusting either the composition or porosity or microstructure gradually across the thickness of the coating to obtain the best of thermal insulation and mechanical properties. In some studies, new materials have been introduced to improve the performance and lifetime, such as rare earth-doped ceramics or nanocomposite materials [44]. Multilayer TBCs provide an elaborate solution for the protection of high-temperature components under severe conditions. Indeed, by exploiting the possibilities offered by new advanced coating systems to manage thermal stresses, offer protection against corrosive species, and establish property gradients, the above high-temperature applications considerably enhance the operational life and efficiency of critical components.

7.4.4 Gradient-Structured TBCs

Advanced thermal coating designs known as gradient thermal barrier coatings (TBCs) are applied to high-temperature components in a variety of applications, such as combustion chambers and turbine engines. These state-of-the-art coating systems embed a precisely controlled transition in either composition or microstructure from topping to substrate, thus giving rise to a desired gradient of thermal/mechanical properties across the entire coating. It therefore becomes apparent that gradient-structured TBCs offer the core benefit of reducing thermal mismatch and stress concentrations between various layers and the substrate [45]. Therefore, these highly specialized coatings reduce the shear stress at the interface and the thermal stress. Hence, they enhance thermal and mechanical property compatibility, making the probability of coating failure by delamination or spallation extremely low. Consequently, the change can be gradual and can be controlled by the porosity distribution, chemical composition, or microstructure of the coating [46].

The present functional graded TBCs may contain a mixture of materials with the necessary characteristics of the composite material. For example, a standard design might feature a close-tolerance, chemically stable bond coat adjacent to the substrate, and a less-dense ceramic wear-resistant top coat with strong thermal conductivity.

This transition process also assists in adapting between the thermal expansion coefficients of the metallic substrate and the ceramic top coat; otherwise, when exposed to cycles of heating and cooling, the coating will fail under thermal cycling fatigue. Here, the particularity of TBC designs based on gradients means that the properties of the protective films can be adjusted to meet the requirements for certain operating conditions [47].

As shown in Figure 7.10, substantial information about functionally graded materials (FGMs) is provided. In the top left diagram (Figure 7.10a), the gradient change of the material properties and volume percentage between two materials, one being material A and the other being material B, is illustrated across the thickness. The top right diagram (Figure 7.10b) categorizes FGMs into three main types.

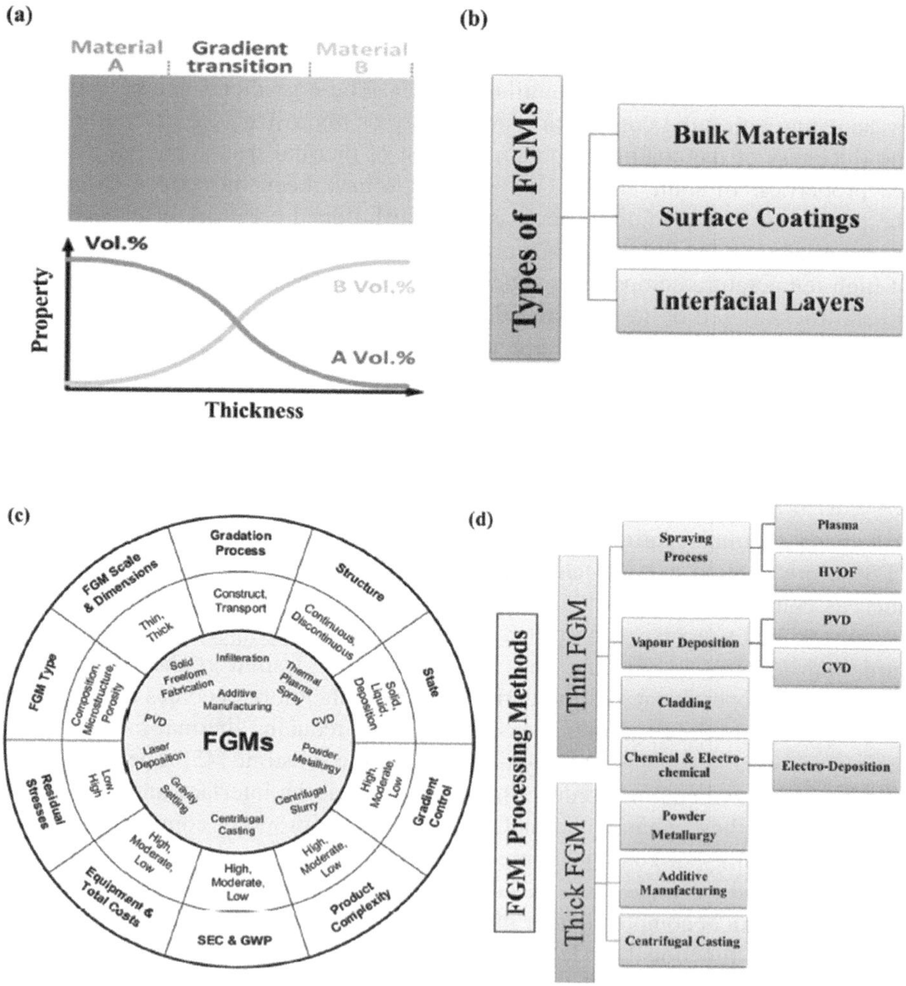

FIGURE 7.10 Basic concept, classifications for production methods, and types of FGMs [48]. (Reproduction permissions.)

The three areas considered here are bulk materials, surface coatings, and interfacial layers. A detailed classification wheel of FGMs for different aspects, such as gradation processes, structure, scale, fabrication methods, and environment, is shown in the bottom left circular diagram (Figure 7.10c). The flowchart at the bottom right (Figure 7.10d) shows FGM processing techniques classified by thin and thick FGMs. Thin FGM processes include spraying, plasma, high-velocity oxygen fuel (HVOF), vapor deposition (PVD), chemical vapor deposition (CVD), cladding, and chemical/electrochemical methods. Thick FGM processes include powder metallurgy, additive manufacturing, and centrifugal casting. This single picture integrates excellently the extensive and polymorphic applications of FGM technology with its manufacturing techniques [48].

There is an adequate control of the thermal conductivity whereby the value is reduced from the bond coat to the top coat while constantly achieving adequate adhesion onto the substrate. Likewise, the mechanical constants like Young's modulus and fracture strength can also be controlled to design a compliant-coated system, which is capable of tolerating thermal and mechanical loads. These gradient structures can be produced through more sophisticated fabrication methods, including plasma spraying with varying feedstock composition or electron beam physical vapor deposition with controlled deposition conditions [49]. These methods enable strict control over the composition and microstructure of the coating material, thus enabling the production of precisely designed TBC systems for targeted applications. Gradient-structured TBCs are a complex solution for maintaining high-temperature parts under elevated conditions and severe applications. These newest systems also show an extended range of thermal and mechanical properties through the full coating thickness, allowing for an improved lifetime of the most demanding parts in a range of high-temperature applications, which results in increased efficiency and protection of the coated equipment [50]. However, some discrepancies remain, for instance, TGO layers at the interface; such development creates a weakened layer that can compromise the strength of the entire system. Further work is required to fine-tune these coatings for precise applications, such as addressing the problems of thermal expansion and chemical stability.

7.5 CONCLUSION

Finally, after briefly reviewing the themes of this review, it is clear that current studies have mainly focused on doped, nanostructured, multilayered, and gradient structures in zirconia-based TBCs. These innovations overcome the drawbacks of standard YSZ coatings and find numerous applications in high-temperature systems. New-generation-doped zirconia-based TBCs, such as those containing gadolinium oxide, exhibit higher TBC thermal stability, lower thermal conductivity, and better mechanical performance. These dopants stabilize the tetragonal phase of zirconia, thereby creating a more heterogeneous microstructure, which retards crack advance and degrades the overall strength. A nanostructured TBC possesses low thermal conductivity and high thermal shock resistance, which are attributed to the composite microstructure, nanozones, and porosity distribution. Features such as a large number of grain boundaries and trimodal pore size distribution lead to better insulation and mechanical characteristics. Multilayered TBCs incorporate a designed layering concept that covers the monitoring of thermal stresses and acts as a barrier against potential corrosive species. These systems increase the inherent carrying capacity of the laminates by distributing stress across different layers with propitious material attributes, thereby further increasing their total quotient of duration and efficiency. Cross-sectional graded TBCs are characterized by a controlled number of layers and a stepwise change in composition and structure, which effectively reduces or eliminates the thermal and mechanical stresses at the interface between the layers and the substrate. This enables the generation of optimum property profiles in the coating to match the particular operating conditions, leading to improved thermal cycling fatigue characteristics of the coating. These advanced TBC systems provide enhanced performance and durability on the TBCs; however, issues like the formation of thermally grown oxide layers continue. Further studies or even enhancements to implement these innovative coating systems for high-temperature applications are continuously required. With the development of the field, these more advanced zirconia-based TBCs are expected to play a significant role in improving the efficiency and life of components in aerospace, automotive, and electrical power generation.

REFERENCES

1. Cao X, Vassen R, Stöver D. Ceramic materials for thermal barrier coatings. *Journal of the European Ceramic Society*. 2004;24:1–10.
2. Sengupta P, Bhattacharjee A, Maiti HS. Zirconia: A unique multifunctional ceramic material. *Transactions of the Indian Institute of Metals*. 2019;72:1981–98.
3. Afrasiabi A. Aplication of plasma spraed YSZ-AL_2O_3 composite TBC on Ni-based Super Alloy (IN-738) and evaluation of the microestructure and its thermall stability (Doctoral dissertation, PhD Thesis. Under supervisión of Dr. M. Saremi. University of Tehran). 2008.
4. Afrasiabi A, Saremi M, Kobayashi A. A comparative study on hot corrosion resistance of three types of thermal barrier coatings: YSZ, YSZ+ Al_2O_3 and YSZ/Al_2O_3. *Materials Science Engineering: A*. 2008;478:264–9.
5. Ahmadi MS, Shoja-Razavi R, Valefi Z, Jamali H. Laser surface modification of air plasma-sprayed Al_2O_3/YSZ multilayer thermal barrier coating to improve hot corrosion resistance in V_2O_5–Na_2SO_4 salts. *Journal of Thermal Spray Technology*. 2019;28:1906–18.
6. Devoino O, Okovity V. Research of plasma spraying of the ZrO_2–Y_2O_3 electrolyte coatings. *Mokslas–Lietuvos ateitis/Science–Future of Lithuania*. 2014;6:621–5.
7. Aniołek K, Barylski A, Kupka M, Leszek I. The tensile properties, scratch behaviors and sliding wear of oxide scale formed on titanium Grade 2. *Materials*. 2020;13:3048.
8. Mehta A, Singh G. Consequences of hydroxyapatite doping using plasma spray to implant biomaterials. *Journal of Electrochemical Science Engineering Failure Analysis*. 2023;13:5–23.
9. Antunes FC, Goulart CA, Andreeta MR, de Souza DP. YSZ/Al_2O_3 multilayer thick films deposited by spin coating using ceramic suspensions on Al_2O_3 polycrystalline substrate. *Materials Science Engineering: B*. 2018;228:60–6.
10. Bakhsheshi-Rad H, Hamzah E, Ismail A, Daroonparvar M, Kasiri-Asgarani M, Jabbarzare S, Medraj M. Microstructural, mechanical properties and corrosion behavior of plasma sprayed NiCrAlY/nano-YSZ duplex coating on Mg–1.2 Ca–3Zn alloy. *Ceramics International*. 2015;41:15272–7.

11. Liu L, Wang S, Zhang B, Jiang G, Liu H, Yang J, Wang J, Liu W. Present status and prospects of nanostructured thermal barrier coatings and their performance improvement strategies: A review. *Journal of Manufacturing Processes*. 2023;97:12–34.
12. Esmaeilkhanian AH, Sharifianjazi F, Ahmadi E, Ijadi A, Meskher H, Zarei R, Nili-Ahmadabadi M, Irandoost M, Karimi N, Ghiasvand A. Thermal barrier coating with improved durability: An overview of doped, nanostructured, multilayered, and gradient-structured zirconia-based thermal barrier coatings. *Materials Today Communications*. 2023;37:107514.
13. Bernard B, Quet A, Bianchi L, Joulia A, Malié A, Schick V, Rémy B. Thermal insulation properties of YSZ coatings: Suspension plasma spraying (SPS) versus electron beam physical vapor deposition (EB-PVD) and atmospheric plasma spraying (APS). *Surface Coatings Technology*. 2017;318:122–8.
14. Sathish M, Radhika N, Saleh B. Duplex and composite coatings: A thematic review on thermal spray techniques and applications. *Metals Materials International*. 2023;29:1229–97.
15. Carpio P, Rayón E, Pawłowski L, Cattini A, Benavente R, Bannier E, Salvador M, Sánchez E. Microstructure and indentation mechanical properties of YSZ nanostructured coatings obtained by suspension plasma spraying. *Surface Coatings Technology*. 2013;220:237–43.
16. Mehta A, Vasudev H, Singh S. Sustainable manufacturing approach with novel thermal barrier coatings in lowering CO_2 emissions: Performance analysis with probable solutions. *International Journal on Interactive Design Manufacturing*. 2023:1–13.
17. Chen H, Zhang C, Xuan J, Liu B, Yang G, Gao Y, Luo H. Effect of TGO evolution and element diffusion on the life span of YSZ/Pt–Al and YSZ/NiCrAlY coatings at high temperature. *Ceramics International*. 2020;46:813–23.
18. Madfa AA, Al-Sanabani FA, Al-Qudami NH, Al-Sanabani JS, Amran AG. Use of zirconia in dentistry: An overview. *The Open Biomaterials Journal*. 2014;5.
19. Chen L, Shan L, Zhang M, Li H, *Modified Dynamic Viscosity Formula for YSZ Droplet Solidification Based on CEL Method*. 2022. Bellingham, Washington: SPIE.
20. Chen X, Zhao Y, Gu L, Zou B, Wang Y, Cao X. Hot corrosion behaviour of plasma sprayed YSZ/$LaMgAl_{11}O_{19}$ composite coatings in molten sulfate–vanadate salt. *Corrosion Science*. 2011;53:2335–43.
21. Cheng Z, Yang J, Shao F, Zhong X, Zhao H, Zhuang Y, Ni J, Tao S. Thermal stability of YSZ coatings deposited by plasma spray–physical vapor deposition. *Coatings*. 2019;9:464.
22. Dai M, Song X, Lin C, Liu Z, Zheng W, Zeng Y. Investigation of microstructure changes in Al_2O_3-YSZ coatings and YSZ coatings and their effect on thermal cycle life. *Journal of Advanced Ceramics*. 2022;11:345–53.
23. Song J-B, Wang L-S, Dong H, Yao J-T. Long lifespan thermal barrier coatings overview: Materials, manufacturing, failure mechanisms, and multiscale structural design. *Ceramics International*. 2023;49:1–23.
24. De la Roche J, Alvarado-Orozco JM, Gomez PA, Cano IG, Dosta S, Toro A. Hot corrosion behavior of dense CYSZ/YSZ bilayer coatings deposited by atmospheric plasma spray in $Na_2SO_4 +{}_{V2}O_5$ molten salts. *Surface Coatings Technology*. 2022;432:128066.
25. de la Roche J, Gómez PA, Alvarado-Orozco JM, Toro A. Hot corrosion and thermal shock resistance of dense-CYSZ/YSZ bilayer thermal barrier coatings systems applied onto Ni-base superalloy. *Journal of the European Ceramic Society*. 2020;40:5692–703.
26. Ariharan S, Parchoviansky M, Singh P, Rani P, Maurya R, Sekar A, Keshri AK, Pakseresht A. Hot corrosion behavior of $La_2Ce_2O_7$-based plasma-sprayed coating. *High Temperature Corrosion of Materials*. 2024:1–10.

27. Lashmi PG, Ananthapadmanabhan PV, Chakravarthy Y, Unnikrishnan G, Balaji N, Aruna ST. Hot corrosion studies on plasma sprayed bi-layered YSZ/$La_2Ce_2O_7$ thermal barrier coating fabricated from synthesized powders. *Journal of Alloys and Compounds*. 2017;711:355–64.
28. Bahamirian M, Hadavi S, Farvizi M, Keyvani A, Rahimipour M. ZrO_2 9.5Y_2O_3 5.6Yb_2O_3 5.2Gd_2O_3; a promising TBC material with high resistance to hot corrosion. *Journal of Asian Ceramic Societies*. 2020;8:898–908.
29. Bobzin K, Zhao L, Öte M, Linke TF. Deposition and characterization of thermal barrier coatings of ZrO_2–4 mol.% Y_2O_3–1 mol.% Gd_2O_3–1 mol.% Yb_2O_3. *Surface Coatings Technology*. 2015;268:205–8.
30. Karaoglanli AC, Doleker KM, Ozgurluk Y. Interface failure behavior of yttria stabilized zirconia (YSZ), $La_2Zr_2O_7$, $Gd_2Zr_2O_7$, YSZ/$La_2Zr_2O_7$ and YSZ/$Gd_2Zr_2O_7$ thermal barrier coatings (TBCs) in thermal cyclic exposure. *Materials Characterization*. 2020;159:110072.
31. Guo L, Zhang C, Li M, Sun W, Zhang Z, Ye F. Hot corrosion evaluation of Gd_2O_3-Yb_2O_3 co-doped Y_2O_3 stabilized ZrO_2 thermal barrier oxides exposed to Na_2SO_4+ $_{V2}O_5$ molten salt. *Ceramics International*. 2017;43:2780–5.
32. Sun L, Guo H, Peng H, Gong S, Xu H. Influence of partial substitution of Sc_2O_3 with Gd_2O_3 on the phase stability and thermal conductivity of Sc_2O_3-doped ZrO_2. *Ceramics International*. 2013;39:3447–51.
33. Mehta A, Vasudev H, Singh S, Prakash C, Saxena KK, Linul E, Buddhi D, Xu J. Processing and advancements in the development of thermal barrier coatings: A review. *Coatings*. 2022;12:1318.
34. Keyvani A, Saremi M, Sohi MH, Valefi Z. A comparison on thermomechanical properties of plasma-sprayed conventional and nanostructured YSZ TBC coatings in thermal cycling. *Journal of Alloys Compounds*. 2012;541:488–94.
35. Zhou F, Liu M, Wang Y, Wang Y, Deng C. Depth-sensing indentation creep behavior of nanostructured thermal barrier coatings from as-synthesized t'-8YSZ feedstocks. *Nanomaterials*. 2019;10:38.
36. Keyvani A, Saremi M, Sohi MH, Valefi Z, Yeganeh M, Kobayashi A. Microstructural stability of nanostructured YSZ–alumina composite TBC compared to conventional YSZ coatings by means of oxidation and hot corrosion tests. *Journal of Alloys Compounds*. 2014;600:151–8.
37. Lima RS, Marple BR. Toward highly sintering-resistant nanostructured ZrO_2–7wt.% Y_2O_3 coatings for TBC applications by employing differential sintering. *Journal of Thermal Spray Technology*. 2008;17:846–52.
38. Mehta A, Vasudev H, Thakur L. Applications of numerical modelling techniques in thermal spray coatings: A comprehensive review. *International Journal on Interactive Design Manufacturing*. 2024;18:3525–45.
39. Sezavar A, Sajjadi SA, Babakhani A, Peng RL, Yuan K. Oxidation behavior of a nanostructured compositionally graded layer (CGL) thermal barrier coating (TBC) deposited on IN-738LC. *Surface Coatings Technology*. 2019;374:374–82.
40. Dokur MM, Goller G. Processing and characterization of CYSZ/Al_2O_3 and CYSZ/ Al_2O_{3+} YSZ multilayered thermal barrier coatings. *Surface Coatings Technology*. 2014;258:804–13.
41. Li C-J, Li Y, Yang G-J, Li C-X. A novel plasma-sprayed durable thermal barrier coating with a well-bonded YSZ interlayer between porous YSZ and bond coat. *Journal of Thermal Spray Technology*. 2012;21:383–90.
42. Lashmi P, Ananthapadmanabhan P, Unnikrishnan G, Aruna S. Present status and future prospects of plasma sprayed multilayered thermal barrier coating systems. *Journal of the European Ceramic Society*. 2020;40:2731–45.

43. Mahade S, Curry N, Björklund S, Markocsan N, Nylén P. Thermal conductivity and thermal cyclic fatigue of multilayered $Gd_2Zr_2O_7$/YSZ thermal barrier coatings processed by suspension plasma spray. *Surface Coatings Technology*. 2015;283:329–36.
44. Alkunte S, Fidan I, Naikwadi V, Gudavasov S, Ali MA, Mahmudov M, Hasanov S, Cheepu M. Advancements and challenges in additively manufactured functionally graded materials: A comprehensive review. *Journal of Manufacturing Materials Processing*. 2024;8:23.
45. Chakrabarty R, Song J. A modified Johnson-Cook material model with strain gradient plasticity consideration for numerical simulation of cold spray process. *Surface Coatings Technology*. 2020;397:125981.
46. Deng Z, Liu D, Liu G, Liu J, Li C, Li S, Chen TJO, Technology L. Numerical simulation study on the formation mechanism of hydroxyapatite-silver gradient bioactive ceramic coatings under wide-band laser cladding. *Optics Laser Technology*. 2023;163:109412.
47. Deng Z, Liu D, Liu G, Xiong Y, Xin S, Li S, Li C, Chen T. Study on the corrosion resistance of SiC particle-reinforced hydroxyapatite-silver gradient bioactive ceramic coatings prepared by laser cladding. *Surface Coatings Technology*. 2023:129734.
48. Sathish M, Radhika N, Saleh B. A critical review on functionally graded coatings: Methods, properties, and challenges. *Composites Part B: Engineering*. 2021;225:109278.
49. Ghadami F, Aghdam ASR, Ghadami S. Preparation, characterization and oxidation behavior of CeO_2-gradient NiCrAlY coatings applied by HVOF thermal spraying process. *Ceramics International*. 2020;46:20500–9.
50. Goudarzi ZM, Valefi Z, Zamani P, Taghi-Ramezani S. Comparative investigation of the effect of composition and porosity gradient on thermo-mechanical properties of functionally graded thick thermal barrier coatings deposited by atmospheric plasma spraying. *Ceramics International*. 2022;48:28800–14.

8 Modified Zirconia as an Efficient Thermal Barrier Coating

A Case Study on ZrO_2-Based Coatings

Raghavendra Sagar and Nacer Badi

8.1 INTRODUCTION TO ZO_2 AS TBC MATERIAL

Thermal barrier coatings (TBCs) find widespread applications in aerospace and ground-based turbine engines to protect hot section components.[1] The outstanding performance of TBCs is attributable to the low thermal conductivity, significant fracture toughness, and robust corrosion resistance of TBCs.[2] A key function of ceramic coating is its ability to safeguard against thermal cycles. TBCs enhance the overall engine efficiency through their thermal protection capabilities. TBCs help carefully design the engine and manage cooling airflow to minimize metal surface temperatures, and it is feasible to achieve efficiency enhancements while also extending the creep and oxidation durability of blades and vanes.[3–5] In addition, the development of TBCs imparts enhanced thermal insulation in demanding conditions and high temperatures while also extending the lifespan of coated components by safeguarding the underlying metallic and ceramic materials. Furthermore, using TBCs contributes to improved efficiency by reducing the energy consumption of cooling systems.[6,7]

The pursuit for new materials aimed at producing TBCs with superior thermomechanical properties has attracted considerable attention, resulting in advances in diverse ceramic-based coatings.[8] Some oxide materials have emerged as the most thoroughly investigated materials for these coatings, although their application is hindered by the high costs associated with raw materials and processing.[9] These materials, however, exhibit the necessary characteristics for TBC applications, including excellent thermal stability, high fracture toughness, and a coefficient of thermal expansion (CTE) that is comparable to that of superalloys. These TBCs contribute to reducing the stresses that arise from consecutive thermal cycles.[10]

Zirconia (ZrO_2) stands out as a prominent oxide in the turbine sector and is commonly applied in TBCs for several components, such as the combustion chamber and fixed and rotating blades in gas turbines and aero-engines.[11] A key factor for the efficacy of a TBC is its phase stability across the operational temperature range and during thermal cycling. This characteristic prevents the TBC from undergoing early

DOI: 10.1201/9781003519300-10

degradation caused by phase changes and fluctuations in crystallographic allotropies. The conversion of the monoclinic phase (*m*-ZrO_2) to the tetragonal phase (*t*-ZrO_2) is a reversible process, occurring at a temperature of 1170°C during heating, while the reverse transformation to the monoclinic phase takes place at approximately 950°C during cooling.[12]

In the past few decades, yttria-stabilized zirconia (YSZ) has been widely employed.[13] However, the performance of YSZ coatings is substantially reduced in high-temperature and severe environments beyond 1200°C, due to challenges such as the transition from tetragonal to monoclinic phases and significant sintering.[14] Thick zirconia-based TBCs are capable of withstanding high gas temperatures exceeding 1250°C.[15] However, extended exposure to thermal conditions leads to pronounced sintering in YSZ coatings, which significantly influences their microstructure, phase composition, and mechanical characteristics. Research into the effects of heat treatment on thick ZrO_2-based TBCs reveals that pores and microcracks gradually heal during thermal processing. With increasing temperatures and prolonged durations, the initial metastable tetragonal phase (*t′*-ZrO_2) evolves into stable tetragonal (*t*-ZrO_2) and yttrium-enriched cubic (*c*-ZrO_2) phases as a result of yttrium diffusion.[16] Upon cooling, the newly formed *t*-ZrO_2 transforms into the monoclinic phase (*m*-ZrO_2). This transformation is generally associated with a volume expansion of 3%–5%, and the resulting thermal stress may lead to the peeling of YSZ TBCs. Consequently, it is essential to investigate phase stability and microstructural evolution at high temperatures and their effects on the performance of these coatings. Therefore, it is crucial to investigate the stability of these phases and the evolution of microstructure at high temperatures, as well as their implications for the performance of the coatings. In addition, it is imperative to investigate approaches that allow for engine operation at higher gas temperatures.[17,18]

Unlike micron-sized powders, the unique size effect of nanopowders significantly contributes to the improvement in coatings. The Hall–Petch empirical relationship indicates that the mechanical strength of materials can be markedly enhanced when grain size is decreased from conventional levels to the nanoscale level.[19] The increased wear rate can be attributed to the enhanced microhardness and structural integrity resulting from improved cohesion. At the same time, an increase in hardness contributes to improved wear resistance. Furthermore, high-purity nano-TBCs are characterized by lower porosity, which contributes to a higher elastic modulus. Regarding thermal performance, the thermal diffusivity of nanostructured coatings is reported to be between 60% and 90% of that of traditional coatings, and their thermal conductivity is approximately 40% less than that of microcoatings.[20]

Furthermore, YSZ is the predominant ceramic composition that is recognized for its outstanding physical and chemical attributes, which encompass high fracture toughness, swift ionic conductivity, and minimal thermal conductivity, and it has been used as a TBC in various engines.[21] The maximum operating temperature for 8YSZ is around 1200°C, taking into account the phase transitions from the *t′* phase to the tetragonal (*t*) and cubic (*c*) phases, followed by the transition from the *t* phase to the monoclinic (*m*) phase. This transformation results in a significant volumetric expansion of approximately 3% or higher, leading to substantial stresses and ultimately causing spallation of the coating.[22] It is essential that the material remains stable and does not experience phase transformations during cycling between ambient and high temperatures.

The tetragonal phase of zirconia, which does not undergo transformation, is essential to address the increasing demands for the thermophysical and mechanical properties of TBCs.[23] Therefore, research focused on advancements in new TBC materials for high-temperature applications has demonstrated that specific dopants in zirconia–yttria, such as Ta_2O_5, Nb_2O_5, and RE_2O_3, play a crucial role in stabilizing the tetragonal phase.[24] These dopants enable the material to function effectively at temperatures exceeding 1200°C while simultaneously improving its properties by reducing the thermal conductivity.[25]

Apart from YSZ, trivalent and pentavalent oxides are emerging as effective stabilizers for zirconia. Studies have shown that a formulation comprising 16.6% $YO_{1.5}$ and 16.6% $TaO_{2.5}$ can stabilize 7YSZ, which points to its potential durability in TBC applications.[26] Coatings of single-phase tetragonal niobia- and yttria-co-doped zirconia, deposited using electron beam physical vapor deposition (EBPVD), have been found to have lower thermal conductivity than the standard 6–8 mol% YSZ coatings. In addition, the addition of niobia, at levels up to 6 wt.%, as a co-dopant in yttria partially stabilized zirconia coatings does not significantly impact the microstructure.[27]

8.2 PREPARATION OF PRISTINE ZRO_2 AND YSZ TBC FROM VARIOUS TECHNIQUES

This section discusses all reported techniques from which pristine ZrO_2 and YSZ are developed as TBCs on various substrates. The classification of the development of TBCs based on solid and liquid starting materials is illustrated in Figure 8.1.

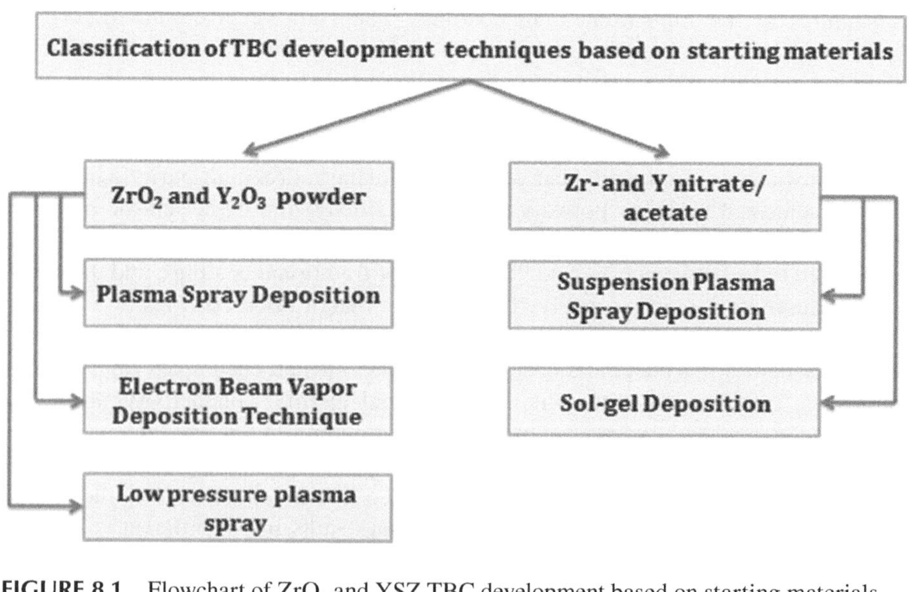

FIGURE 8.1 Flowchart of ZrO_2 and YSZ TBC development based on starting materials.

There are as many as seven ways from which TBCs can be developed using both solid-state and wet chemical approaches. The overview of each technique is as follows:

8.2.1 Plasma Spray Deposition

The plasma spray deposition (PSD) technique is one of the extensively used TBC deposition processes mainly used for materials like ZrO_2 and YSZ.[28–30] These coatings are developed onto metal substrates that require high-temperature resistance like gas turbines and jet engines. In this technique, high-energy plasma is used to jet, melt, and propel powder particles onto a substrate, where they swiftly cool and solidify, forming a coating. The important components present in the PSD unit are substrate, plasma feedstock, and plasma torch. The substrate belongs to the deposition unit on which materials are coated. The plasma feedstock is a part where the powder is fed to propel on the substrate. The plasma torch is a device that generates plasma using argon or nitrogen, or both, mixed with hydrogen to produce high-temperature plasma jet. Figure 8.2 shows the schematic representation of a PSD unit.

First, Ni-based alloys that are normally used as substrates are cleaned using sandpapers or chemical etching, followed by washing with organic solvents such as isopropyl alcohol. Then, the surface of the cleaned substrates is roughened using a grinding machine, which helps achieve a better adhesion of the thermal barrier with the substrate material. Second, ZrO_2 powder or YSZ is sieved to obtain particles

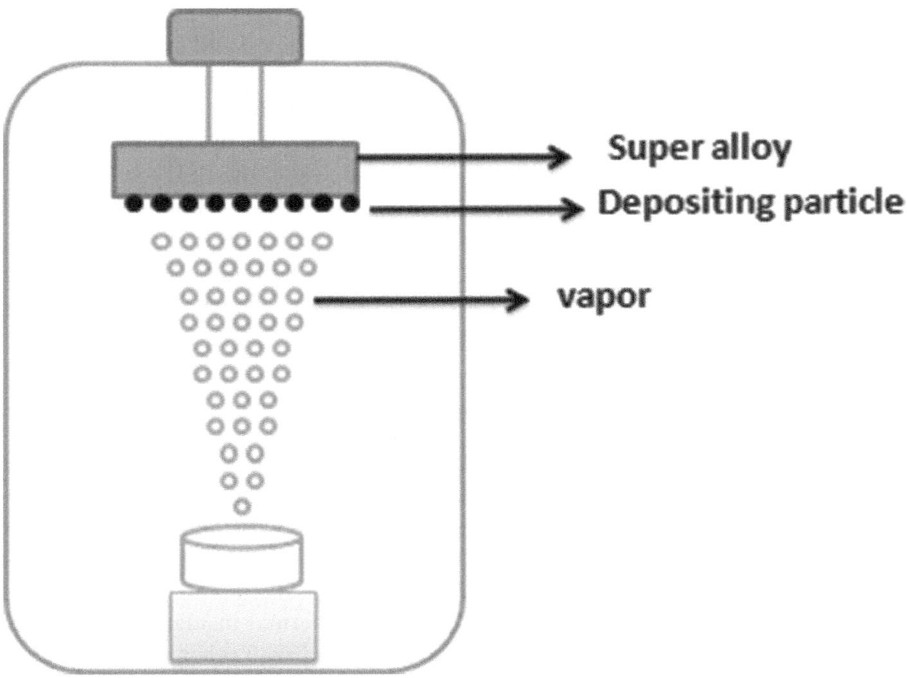

FIGURE 8.2 Schematic representation of PSD for the development of TBCs.

TABLE 8.1
Comparison between Thermal, Structural, Microstructure, and Mechanical Features of ZrO_2 and YSZ Prepared Using the PSD Technique

Characteristic	ZrO_2	YSZ
Thermal conductivity	1.7 W/(m.K)	2.2–2.9 W/(m.K)
Thermal expansion	$2.3\ 10^{-6}K^{-1}$	$(7.1 \pm 1.9) \cdot 10^{-6}K^{-1}$
Crystal structure	Monoclinic	Stabilized cubic phase
Porosity	Higher porosity	Dense microstructure
Bonding strength	Moderate	High for metallic substrates
Used for	General thermal insulation	Gas turbines
Toughness	Lower toughness	High toughness and durability

with sizes ~50 μm, which are then fed in the propeller. This powder is propelled in a plasma stream, which produces an electrical arc between two electrodes. Because of the high temperature produced due to the plasma arc, the powder melt moves toward the substrate. As soon as the plasma encounters the substrate, it is cooled, forming a thin coating on the Ni-based alloy substrate. This can be continued for many cycles to achieve the desired thickness of ZrO_2 and/or YSZ on the substrate. The cooling of the substrate facilitates to form a dense microstructure and hard TBC layer. Nevertheless, the rate of cooling also influences the microstructure and thermal conductivity of the alloy with TBCs.

Many parameters affect the coating of TBC layers in the PSD technique.[31] For example, an elevated plasma jet temperature can significantly improve the adhesion strength of TBC, however, this advantage is often offset by an accelerated rate of oxidation. If the distance between the substrate and the plasma torch is higher, the porosity and roughness of the coating will increase. Using the PSD technique, one can easily develop TBCs with a thickness of around 500 μm.[32] Many reports have shown that YSZ deposited using the PSD technique exhibits low thermal conductivity, stable cubic crystal phase, high density, and superior adhesion with the substrate, which is in sharp contrast with ZrO_2 properties.[13] Table 8.1 shows the characteristics of ZrO_2 and YSZ prepared using the PSD technique.

8.2.2 Electron Bean Physical Vapor Deposition

It is a vacuum-based technique, where a high-power electron beam is used to vaporize the atoms or molecules of the target material and coat them over the desired substrate. This process is mainly used to develop thin films with dense microstructure that could be used in applications ranging from photovoltaic to aerospace industries. This process yields columnar structured high-quality TBCs with an almost least porous microstructure with enhanced adhesion and thermal insulation.[33] However, the process of producing TBCs is too slower compared with other deposition techniques, and this technique is mainly used to protect the components used in turbine blades of jet engines.

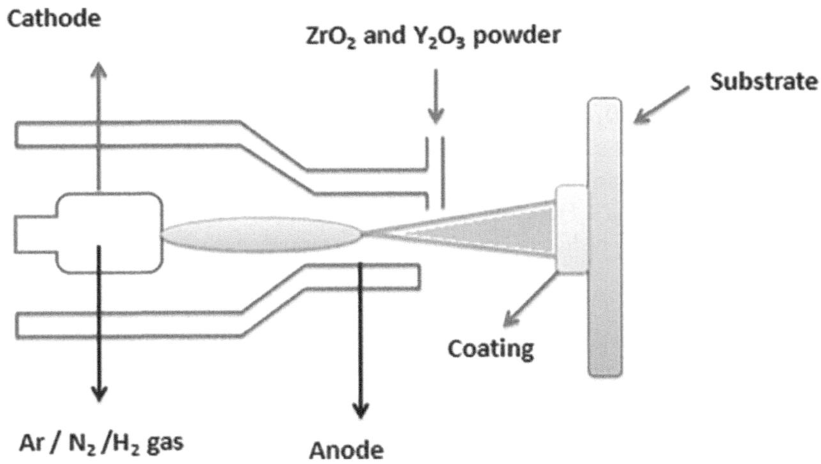

FIGURE 8.3 Illustration of the EBPVD setup used for the deposition of TBCs.

It is one of the most preferred techniques to prepare YSZ with a stable crystal structure. This technique involves three major components: vacuum chamber, electron beam generator, and target. First, the substrate is placed in a vacuum chamber, and YSZ and ZrO_2 targets are placed in the path of the electron beam. Then, the vacuum chamber is evacuated till 10^{-6} to 10^{-7}Torr. This vacuum is necessary to reduce possible contamination and achieve hindrance-free path for the electron beam to the target. A beam generator is heated with high voltage so that a beam with 5–10 keV is generated and directed toward the target material. Once this high-power electron beam hits the target, it vaporizes the target material, and these vaporized atoms travel through the vacuum chamber and get deposited on the substrate material to form a coating of desired thickness. Factors such as the power of the electron beam, gap between the substrate and YSZ, and the pressure of the chamber influence the thickness of the coating/film. The rate of coating is influenced by several factors. Proper deposition of YSZ on the substrate can be achieved by controlling these parameters. In addition, microstructure, crystallinity, and quality of the film can be controlled by substrate temperature.[34] After YSZ deposition, annealing of the coated substrate in the appropriate atmosphere at phase formation temperature facilitates the stabilization of the crystal phase. Several reports are also available suggesting the heat treatment of the deposited substrates under oxygen atmosphere to form proper stoichiometric YZS or ZrO_2 coatings. However, the synthesis of TBCs using EBPVD is more expensive than other conventional techniques. Figure 8.3 displays the EBPVD setup used in TBC deposition.

8.2.3 Low Pressure Plasma Spray (LPPS)

The LPSS deposition technique is one of the extensively used techniques where a low-pressure plasma torch is used to melt YSZ or ZrO_2 and accelerate it toward the superalloys. The use of low pressure facilitates an uninterrupted flow of atoms toward the substrate and helps achieve controlled deposition of TBCs with excellent adherence with the substrate.[35]

In this technique, first, YSZ and ZrO_2 powder is pressed typically in a spherical shape of grain size ~50 µm to facilitate the deposition process. The substrates that are normally used such as metallic or nickel-based alloys on which YSZ and ZrO_2 is deposited are cleaned and heated between 300°C and 500°C to impart good adhesiveness with TBC layers. Here, substrate cleaning is performed by either sandblasting or grit-blasting to produce a rough surface to improve the adhesiveness of YSZ and ZrO_2 TBCs.[36]

In order to deposit TBCs, the vacuum chamber is maintained with the appropriate pressure recommended by previous studies, between 10 and 100 mbar. Moisture and probable contaminants are removed to obtain highly pure TBCs on the substrate. Then, Ar or a mixture of Ar and H_2 gas is ionized using the plasma torch. Typically, high-voltage electrodes are used to create electric discharge and ionize the gas.[37] Here, high-temperature plasma could be achieved, and a staggering 10,000°C can be reached. When ZrO_2 and YSZ are introduced into this plasma stream, because of the high temperature, they easily melt and are direct toward metallic or Ni-based alloy substrates. In this powder injection process, the density, porosity, and surface morphology of TBCs are important parameters that determine the properties of coatings. Molten TBC particles are impinged on the substrate and get solidified to form a top bond layer (TBL). This results in the layer-by-layer formation of TBCs with the feeding of ZrO_2 and YSZ in the plasma jet. Often, the substrate is rotated to obtain uniform thickness of coating on the surface. The temperature within the vacuum chamber should be carefully monitored to evade substrate overheating and ensure accurate blending of sprayed TBC material. Since the thickness of the TBC affects how well it works, it is closely monitored during deposition. The normal thickness of TBCs is between 100 and 500 µm. Spraying is frequently done in several passes with strict control over the deposition rate to obtain the required thickness. Following the application of the coating, the substrate is allowed to cool. The rate of cooling is crucial to prevent coating delamination or cracking.[38] To improve the qualities of the coating and further stabilize the YSZ phase, post-treatment procedures such as thermal cycling or sintering may be employed. The coating may also be polished or ground to create a smooth surface, which is crucial for enhancing the thermal performance and longevity of the coating. Figure 8.4 demonstrates the LPPS process used in TBC deposition.

Many factors affect TBC coatings. The strength of the plasma torch controls the temperature of the plasma and the amount of energy available to melt the feedstock powder. The rate of injection of ZrO_2 or YSZ powder into the plasma controls the rate of deposition of the coating. In addition, it is possible to improve particle adhesion and reduce the risk of thermal stress by preheating the substrate. Oxidation is reduced, and a more regulated deposition process is made possible by lower pressure. The plasma-forming gas, which is usually argon or a combination of argon and hydrogen, determine the coating quality and plasma temperature. TBCs typically have a coating thickness of 100–500 µm, depending on the application.[39]

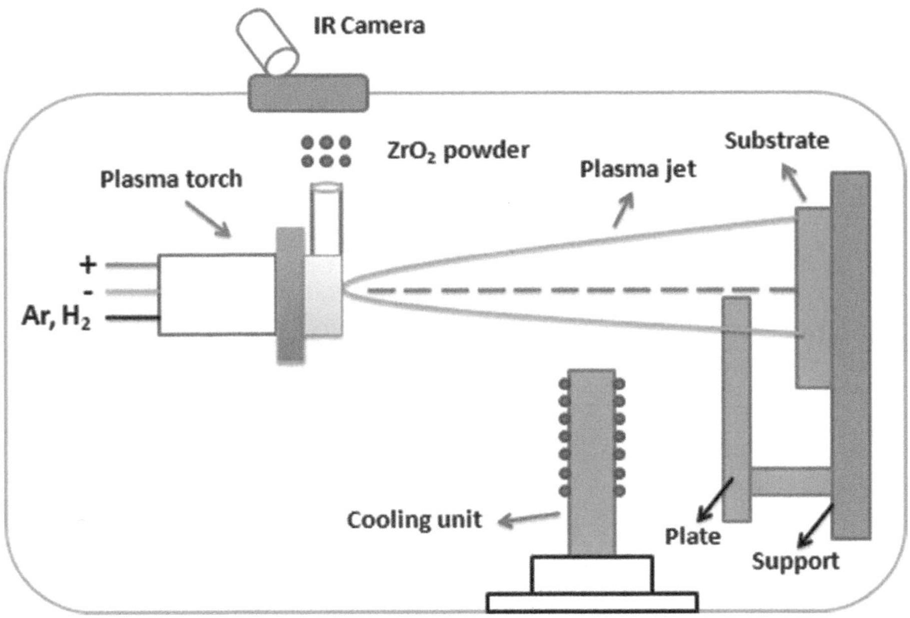

FIGURE 8.4 Schematic representation of the LPPS unit used for the deposition of TBCs.

There are several advantages to LPSS. It can yield high-quality coating due to the bonding strength, control over oxidation, and composition, but the cost of LPSS equipment and complexity in the coating process are the primary challenges.

8.2.4 Suspension Plasma Spray (SPS)

SPS is a novel thermal spraying method used to deposit coatings like YSZ and zirconia. It is particularly effective for applications requiring TBCs as it enables the production of coatings with a fine microstructure, controlled porosity, and dense morphology. For TBCs to operate effectively in high-temperature environments, such as gas turbines and aerospace engines, these characteristics are essential. Instead of solid powder feedstock, SPS uses a suspension of fine powder particles in a liquid medium, unlike the conventional plasma spray technique.[40] Usually, this suspension is made up of ZrO_2 or YSZ particles that have been distributed throughout a solvent, like alcohol or water. The coating is formed when the liquid is injected into the plasma jet and evaporates, which causes the particles to heat up quickly and accelerate toward the substrate. Because of their fine particle size and quick liquid-medium evaporation, SPS coatings have special properties.

To create the suspension, zirconia or YSZ powder is combined with a liquid medium, which may include water, ethanol, or isopropanol.[41] The concentration of the powder within the suspension is generally maintained between 10% and 30% by weight, depending upon the specific properties desired in the coating. In addition, a dispersing agent is often incorporated into the suspension to mitigate particle agglomeration and ensure a stable and homogeneous dispersion. Whether it is

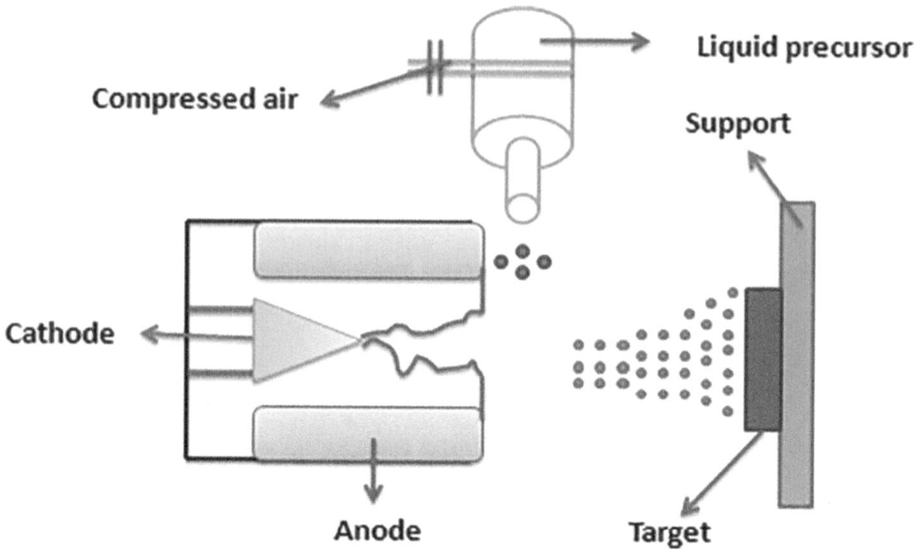

FIGURE 8.5 Schematic representation of the SPS setup used to develop TBCs.

a metal or a ceramic, such as a superalloy or a ceramic component, the substrate should be thoroughly cleaned to remove impurities, oils, and other residues in order to guarantee the best adhesion. Grit-blasting, sandblasting, and acid cleaning are common surface preparation techniques that roughen the substrate and promote better coating adhesion. In order to minimize oxidation and precisely regulate plasma parameters, the SPS procedure is typically carried out in a vacuum or controlled atmosphere chamber. It is a common practice to maintain a vacuum environment between 10 and 100 mbar, which helps avoid contamination during the deposition process. The gas inside the chamber—typically argon, hydrogen, or a combination of the two—is ionized using a plasma torch. By using a high voltage to ionize the gas, a high-temperature plasma jet is produced. Fine ZrO_2 or YSZ powder particles can be melted and accelerated by plasma temperatures of up to 10,000°C. The liquid suspension is injected and evaporated in the plasma jet that the torch creates, which is the focal point of the procedure. The particles are quickly heated by the high-energy plasma stream, melting or partially melting them. Using a nozzle or injector system, a suspension of ZrO_2 or YSZ powder is introduced into the high-temperature plasma jet in this procedure. While fine powder particles are heated by the plasma stream to their melting point, the liquid component of the suspension is evaporated almost instantly.[42] When the powder particles are heated, they change from a solid to a molten state, causing the solvent to evaporate quickly. The plasma jet then accelerates and directs the molten particles toward the substrate. These molten or partially molten particles solidify upon impact with the substrate surface, forming a layer-by-layer coating. In order for the plasma spray to evenly cover the entire surface, the substrate is frequently tilted, rotated, or moved during the spraying process. The molten particles gradually form a dense and consistent coating as they bond and solidify upon coming into contact with the substrate. The injected material's fine particle size allows for high deposition rates without sacrificing microstructure refinement. The structure of the deposited coating is essential for its effectiveness as a thermal insulator. Typically, coatings created through SPS feature fine, compact structures with low porosity levels. Key factors that influence the properties of the final coating include the dimensions of the droplets, their solidification behavior, and the number of splats that form on the surface. After the coating application, it is subjected to annealing, which is a heat treatment process utilized to enhance the crystallinity of the YSZ phase and the thermal stability of the coating, followed by polishing or grinding. For applications that require a high-quality surface finish, the coating may be polished or ground to achieve a smooth and consistent surface. Figure 8.5 shows the schematic representation of SPS for developing TBCs.

The benefits of SPS encompass fine microstructure, regulated porosity, enhanced adhesion, and superior coating characteristics. However, it has challenges related to process complexity, financial implications, quality of the powder used, and compatibility with substrates.[43]

8.2.5 Sol-Gel Method

The sol-gel process is a chemical methodology used to create materials from a liquid solution, known as sol, which subsequently undergoes gelation to establish a solid framework. This technique has gained popularity for producing thin, dense, and uniform coatings. This approach allows for meticulous control over both chemical composition and microstructure, which is essential for the development of high-performance TBCs.[44] In the realm of TBCs, ZrO_2 and YSZ are preferred due to their low thermal conductivity, exceptional thermal stability, and capacity to undergo phase transformations that ensure effective thermal insulation in high-temperature environments. YSZ, when doped with 6%–8% yttria, stabilizes the tetragonal or cubic phases of zirconia at high temperatures, thereby preventing adverse phase transformations that could undermine the efficacy of the coating.

The sol is generally a liquid precursor solution comprising metal alkoxides or metal salts. For the synthesis of ZrO_2, zirconium alkoxides are predominantly utilized as the precursor, whereas for YSZ, a combination of zirconium alkoxide and yttrium alkoxide (or yttria salt) is employed. The preparation of the ZrO_2 sol typically involves dissolving zirconium alkoxides, such as zirconium isopropoxide or zirconium n-propoxide, in an alcohol solvent, commonly ethanol or isopropanol. Hydrolysis is initiated by the addition of a small quantity of water along with an acid (like nitric acid or acetic acid) or a base to promote the development of a gel network. In the case of YSZ, the sol is created by blending zirconium alkoxide with yttrium alkoxide (for instance, yttrium nitrate) in a similar alcohol medium. The yttria concentration is generally maintained at 6%–8% by weight to ensure the stabilization of the tetragonal or cubic phase of ZrO_2. Subsequently, hydrolysis is performed using water and an acid/base to yield a stable gel precursor. It is essential to regulate the molar ratios of the precursors (ZrO_2 and Y_2O_3) to achieve the desired stoichiometric balance and facilitate appropriate phase formation in the resultant YSZ material. The sol-gel process encompasses several key reactions:[45]

- **Hydrolysis:** In this initial stage, alkoxide or salt precursors interact with water present in the solvent. Hydrolysis is essential for the transformation of the precursor into metal hydroxides, which subsequently initiate the polymerization process to establish the sol network.
- **Polymerization:** Following hydrolysis, the solution experiences polymerization. The hydrolyzed metal ions engage in bonding with adjacent molecules, resulting in a gel-like formation. The conditions of polymerization can be fine-tuned by modifying the concentrations of water, solvent, and acid/base, leading to the creation of a sol, which is a colloidal suspension of fine particles that can later evolve into a gel.
- **Gelation:** During gelation, sol particles, which consist of metal hydroxides or oxides, interconnect to form a three-dimensional network. This process yields a viscous, semisolid structure. The gel network is subsequently allowed to mature, enhancing its stability and promoting a more uniform distribution of oxide particles within the gel.

After the sol is prepared, it can be applied to the substrate using various coating methods, with dip-coating, spray-coating, and spin-coating being the most prevalent techniques. In dip-coating, the substrate, usually made of metal or ceramic, is immersed in the sol and then withdrawn at a controlled speed, where the withdrawal rate influences the coating thickness: slower speeds yield thicker layers as the sol forms a thin film that undergoes gelation. In spin-coating, the sol is deposited onto the substrate, which is subsequently rotated at high speeds. This high-speed rotation facilitates the even distribution of the sol over the surface, resulting in a uniform coating. Spin-coating is especially advantageous for the production of thin, consistent films. Spray-coating involves applying the sol onto a heated substrate through an aerosol spray technique, allowing sol droplets to create a thin layer upon contact with the substrate surface.

Then, films are subjected to drying, during which they are subjected to conditions that facilitate the removal of excess solvent and water. It is crucial to monitor this stage carefully to prevent cracking or deformation of the coating. A gradual drying approach is generally favored, allowing the sol to evaporate slowly and uniformly. After the drying phase, the coating undergoes pyrolysis, where it is subjected to high temperatures in a furnace, typically ranging from 400°C to 600°C. This thermal treatment serves multiple functions: it eliminates any residual organic materials, such as solvents and dispersing agents, and promotes the crystallization of remaining metal hydroxides or oxides, resulting in a robust oxide network, such as ZrO_2 and YSZ. In addition, for YSZ, this process aids in stabilizing the desired tetragonal or cubic phases of zirconia, with the yttria content ensuring the durability of the coating in thermal barrier applications. Figure 8.6 shows the flowchart depicting the sol-gel process.

The sol-gel process offers several advantages, including fine microstructure, high uniformity, and low porosity with controlled crystallinity. However, it also has some drawbacks, such as issues with coating thickness, potential for cracking during the drying phase, and time-consuming nature.[46] Table 8.2 shows a summary of TBCs developed using the methods explained in Section 8.2.

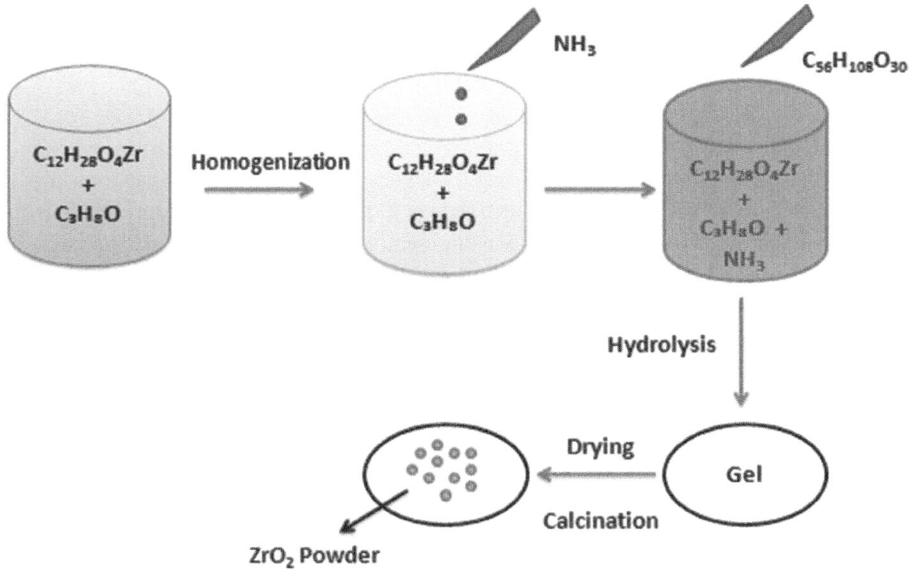

FIGURE 8.6 Flowchart of the sol-gel process illustrating ZrO_2 powder preparation.

TABLE 8.2
Summary of TBCs Developed Using the Discussed Methods

Method	Advantages	Disadvantages	Applications
PSD	Superior deposition rates, commercially viable	Highly porous with low thermal conductivity	Gas turbines and aerospace engines
EBPVD	High-quality coatings, dense, better thermal cycling	Slow, expensive	Turbine blades of aerospace engines
LPPS	Microstructure tunable	Slower, expensive	Aerospace engines
SPS	Least porous	Requires precise control	Aerospace engines
Sol-gel	Composition controllability and good adhesion	Slow and difficult to have dense coatings	Gas turbines

8.3 RARE EARTH ION-SUBSTITUTED ZRO_2-BASED TBCS

Rare earth (RE) elements, including Y, Ce, La, and Sm, are utilized within the ZrO_2 lattice to stabilize the tetragonal or cubic phases at high temperatures.[47] This incorporation is essential for enhancing the high-temperature phase stability of zirconia, which is crucial for the sustained performance of TBCs. The presence of RE dopants aids in stabilizing the tetragonal and cubic phases, ensuring that the coating maintains its desired microstructure even in extreme temperature environments. Such increased phase stability is vital in preventing cracking due to phase transformations, thus contributing to the long-term durability of the coating. Moreover, the addition of RE elements can significantly enhance the mechanical properties of zirconia, including its hardness, fracture toughness, and wear resistance.[48] Some of the most common RE ions doped with ZrO_2 are discussed below:

8.3.1 Yttria-Stabilized Zirconia

YSZ is extensively researched and utilized as a TBC in high-temperature environments, particularly in gas turbines, jet engines, and industrial heating systems. Its widespread use can be attributed to its superior thermal insulation capabilities, commendable thermal stability, and relatively low thermal conductivity.[49] These characteristics are essential for safeguarding the underlying metal substrates from extreme temperatures and oxidative conditions. However, despite its prevalent applications, YSZ faces challenges, such as its mechanical performance at high temperatures and potential for phase transformations that may compromise its long-term stability.[50] YSZ is generally produced by doping ZrO_2 with yttria, which stabilizes its tetragonal or cubic phase at high temperatures. At room temperature, pure zirconia exists in a monoclinic phase, which, while thermodynamically stable, is susceptible to cracking due to phase changes that occur upon heating. By doping with yttria, typically ~6–10 mol% of Y_2O_3, the tetragonal phase is preserved at high temperatures, thereby enhancing the mechanical properties of the material. The primary benefit of YSZ lies in its ability to maintain a stable tetragonal or cubic structure under high-temperature conditions. This structure exhibits higher resistance to thermal stress and cracking compared with the monoclinic phase, which could otherwise result in coating failure. The high-temperature stability of YSZ is vital for the performance of TBCs as they should withstand significant temperature gradients without degradation.

The CTE of YSZ is relatively well suited for metallic substrates utilized in turbine blades, such as nickel-based superalloys. This alignment helps minimize the likelihood of thermal stresses and delamination that occur between the coating and the substrate during thermal cycling. However, it is important to note that the CTE of YSZ is still higher than that of some metals, potentially leading to challenges when coatings face extreme temperature variations. The important advantages of YSZ as TBCs are as follows:[51]

1. High-Temperature Durability: YSZ exhibits exceptional resistance to degradation at high temperatures, making it particularly suitable for applications where metal components are subjected to temperatures exceeding 1,000°C. Its stability during thermal cycling, resistance to oxidation, and capacity to maintain structural integrity under thermal stress are crucial for prolonging the lifespan of turbine blades and other components in high-temperature settings.
2. Good Adhesion and Bonding: YSZ demonstrates strong adhesion to metal substrates, attributable to the comparable CTE of YSZ and nickel-based superalloys commonly utilized in gas turbines. The bond between the bond coat (typically MCrAlY or other metallic alloys) and YSZ can be improved through the use of intermediate layers or by fine-tuning the microstructure of the YSZ coating, thereby enhancing the overall durability of the TBC system.

3. Low Thermal Conductivity: The low thermal conductivity of YSZ stands out as one of its most significant benefits. By functioning as a thermal insulator, YSZ minimizes the heat transfer from combustion gases to the metal substrate, thereby preventing substrate softening and thermal degradation. This characteristic contributes to the efficiency and durability of gas turbines operating under extreme thermal conditions.
4. Phase Transformation Toughening: YSZ has a distinctive capability to undergo a phase transformation from the tetragonal to monoclinic phase when subjected to stress or damage. This transformation results in a volume expansion that can assist in "self-healing" microcracks, thus enhancing fracture toughness, although this property is more pronounced in partially stabilized zirconia systems.

However, YSZ also faces several challenges in its long-term performance. High temperatures can cause phase instability, leading to reduced toughness and increased brittleness. Sintering and densification can also affect YSZ coatings, reducing porosity and insulating properties. Cracking and spalling are common issues due to the mismatch in thermal expansion between the coating and the substrate.

8.3.2 Cerium-Stabilized Zirconia (CeSZ)

Historically, YSZ has been the leading material to be used in TBCs. Nevertheless, the quest for alternative materials to enhance TBC performance has led to the investigation of CeSZ. CeSZ has gained prominence due to cerium's exceptional chemical properties, which can significantly improve the thermal stability, sintering resistance, and mechanical performance of the zirconia matrix.[52] The production of CeSZ involves substituting a portion of zirconium with cerium in the ZrO_2 lattice. As a RE element, cerium is instrumental in stabilizing the tetragonal and cubic phases of zirconia at high temperatures, which is vital for maintaining the mechanical strength and thermal performance of the coating. Cerium doping enhances the stability of these phases at temperatures above 1,200°C, thereby augmenting the thermal stability of the material. Furthermore, cerium introduces oxygen vacancies into the zirconia lattice, which improves oxygen ion conductivity and enhances the stability of high-temperature phases.

The thermal conductivity of CeSZ plays a pivotal role. CeSZ shows lower thermal conductivity than pure zirconia and is comparable to YSZ, rendering it an effective thermal insulator. The presence of oxygen vacancies due to cerium doping diminishes phonon propagation, which in turn reduces thermal conductivity and protects the substrate from thermal damage. This decrease in thermal conductivity is particularly significant in high-temperature scenarios, where regulating heat flux to the substrate is crucial for optimizing the performance and durability of turbine blades and other essential components.[53]

The mechanical properties of CeSZ coatings are notably enhanced, particularly fracture toughness and resistance to sintering at high temperatures. The presence of oxygen vacancies, resulting from cerium doping, plays a crucial role in improving the material's fracture toughness by hindering the propagation of cracks.

This enhancement renders CeSZ coatings more resilient to cracking under both mechanical and thermal stress conditions. Furthermore, the superior mechanical properties of CeSZ contribute to enhanced wear resistance compared with undoped zirconia, which is vital for TBCs exposed to erosion and abrasion. The CTE of CeSZ typically aligns well with that of metal substrates, such as nickel-based superalloys used in turbine blades, thereby reducing the risk of delamination and cracking due to thermal cycling. However, it is worth mentioning that the CTE of CeSZ may still be higher than that of metal substrates, which could pose challenges in scenarios involving significant temperature gradients.[54]

The advantages of CeSZ as TBCs include enhanced phase stability that prevents the degradation of the coating's mechanical properties, such as toughness, and prevents cracking due to phase transformations in pure zirconia. CeSZ coatings also show better sintering resistance, preserving the coating's insulating properties over longer periods of use. Oxygen vacancies in CeSZ enhance the material's toughness and durability, promoting toughening mechanisms like transformation-induced plasticity and crack bridging. Its inherent oxidation resistance prevents degradation due to reactive oxygen species in high-temperature environments. CeSZ also shows low thermal conductivity, essential for effective thermal insulation in TBCs. However, challenges such as cost and availability, phase stability at very high temperatures, thermal expansion mismatch with the substrate, and long-term durability under harsh conditions require further investigation.

8.3.3 Lanthanum-Stabilized Zirconia (LaSZ)

Zirconia is doped with lanthanum oxide (La_2O_3) to yield LaSZ.[55] The tetragonal and cubic phases of zirconia are stabilized at high temperatures by the RE element lanthanum. By stabilizing zirconia's tetragonal and cubic phases at high temperatures, lanthanum doping inhibits phase transformation and improves the material's mechanical and thermal stability. Up to temperatures above 1,200°C, lanthanum stabilizes the phases, thus improving the material's mechanical strength and thermal insulation qualities. LaSZ is a suitable thermal insulator because it has comparable, or even slightly lower, thermal conductivity than other zirconia-based TBCs like YSZ. By inhibiting the propagation of cracks, especially in the presence of thermal shock, LaSZ increases fracture toughness. The material becomes less brittle due to transformation toughening caused by oxygen vacancies produced by lanthanum doping. The superior hardness and wear resistance of LaSZ coatings are critical for halting erosion and preserving the coating's integrity in turbines with high-velocity gas flow. The CTE of LaSZ is compatible with metal substrates, like nickel-based superalloys, which is crucial for reducing thermal stresses and avoiding TBC delamination during thermal cycling. In contrast to other zirconia-based materials, LaSZ and the substrate may still have some mismatched CTE, which could cause stress buildup and coating failure. Figure 8.7 shows the XRD pattern of LaSZ prepared using the sol-gel technique.

LaSZ offers several advantages in TBCs, including enhanced high-temperature stability, which prevents phase transformation and degradation at temperatures above 1,200°C. This stability contributes to increased resistance to cracking and

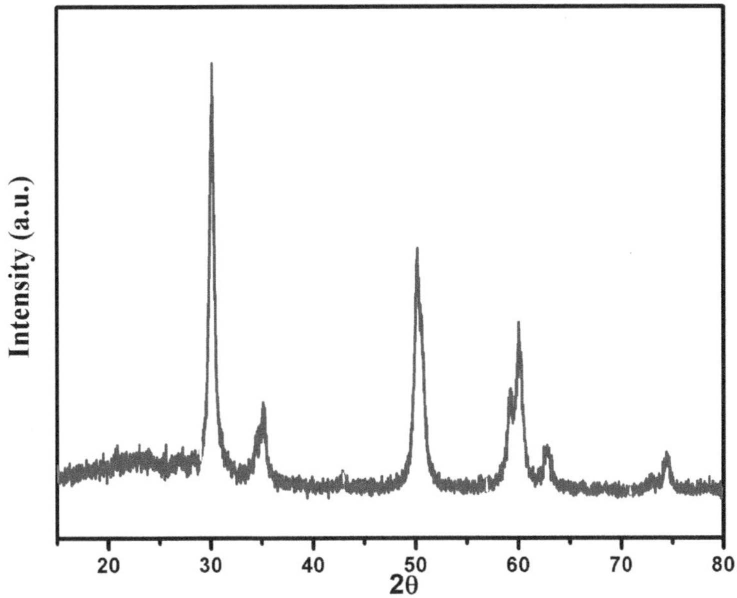

FIGURE 8.7 Powder XRD pattern of LaSZ prepared by the sol-gel technique.

TABLE 8.3
Comparison between Structure and Thermal Properties of Some Rare-Earth-Ion-Stabilized Zirconia

Properties	YSZ	CeSZ	LaSZ
Structure at R_T	Cubic	Cubic	Cubic
Lattice parameter	5.13Å	5.09–5.18Å	10.8Å
Thermal conductivity	2.2–2.9 W/(m K)	1.88–1.33 W/(m K)	1.75 W/(m K)

failure during thermal cycling. LaSZ also demonstrates improved sintering resistance, maintaining porosity and low thermal conductivity even under prolonged high temperatures, making it valuable for turbine engine applications.[56] Its superior mechanical properties, such as fracture toughness and wear resistance, enhance durability under mechanical stresses and thermal shocks. In addition, LaSZ provides excellent thermal insulation, reducing heat transfer to the substrate and improving the lifespan of components. However, challenges in using LaSZ as TBCs include the lack of availability of lanthanum, thermal expansion mismatch, and the need for further research on long-term durability under extreme conditions. Processing challenges and advancements in deposition techniques are necessary to enhance the quality and reproducibility of LaSZ coatings. Table 8.3 shows a comparison between the structure and thermal properties of RE-ion-stabilized zirconia discussed in Section 8.3.

8.4 TBCs IN COMBUSTION ENGINE COMPONENTS OF AUTOMOBILES

TBC is a crucial component of combustion engine parts, particularly those found in contemporary automobile engines. The following describes the purpose and importance of TBCs:

- **Heat Insulation**: TBCs are used to insulate engine parts from excessive heat, including exhaust manifolds, turbine blades, cylinder heads, and pistons. They offer a layer that stops excessive heat transfer, preventing the components from overheating and wearing out too soon, especially in combustion engines, where high temperatures are common.[57]
- **Increased Efficiency**: By insulating engine components, TBCs aid in regulating the combustion chamber's temperature, which maximizes the combustion of the fuel–air mixture. Better thermal efficiency enables the engine to draw more energy from the fuel, which reduces fuel consumption and enhances engine performance in general.[31]
- **Longer Component Life**: High temperatures can cause material deterioration, oxidation, and thermal fatigue. TBCs increase components' resilience and prolong the life of engine parts by shielding them from these impacts.[58]
- **Emissions Reduction**: TBCs can help improve the combustion process, which can lead to complete fuel combustion and less production of harmful emissions like particulate matter and nitrogen oxides (NOx), thus providing a way to adhere to more stringent environmental regulations.[59]
- **Improved Turbocharger and Exhaust System Performance**: TBCs are frequently used in exhaust and turbocharger turbines.[60]

8.5 CONCLUSION

The development of TBCs has led to the advancement of various ceramic-based coatings, with oxide materials like zirconia being particularly studied due to their thermal stability, high fracture toughness, and CTE. Zirconia is commonly used in TBCs for combustion chambers and fixed and rotating blades in gas turbines and aero-engines. Nanopowders significantly improve coatings by enhancing mechanical strength, wear rate, and thermal performance. High-purity nano-TBCs have lower porosity and higher elastic modulus, with thermal diffusivity between 60% and 90% that of traditional coatings and thermal conductivity 40% less than that of microcoatings. Dopants like Ta_2O_5, Nb_2O_5, and RE_2O_3 stabilize the tetragonal phase, allowing the material to function at temperatures exceeding 1200°C while lowering thermal conductivity.

One of the primary benefits of RE-element-doped zirconia lies in its superior thermal insulation capabilities. The replacement of certain elements with RE elements leads to the formation of oxygen vacancies, which in turn decreases the thermal conductivity of the substance. This property is especially useful in TBCs, where the essential function is to shield the underlying substrate from high temperatures. With lower thermal conductivity, coatings made from RE-element-doped zirconia can deliver enhanced thermal protection and improved efficiency in high-temperature applications. In addition, these coatings are characterized by better resistance to cyclic thermal fatigue. The presence of RE elements modifies the microstructure, which helps reduce the propensity for crack propagation during repeated thermal cycling. This is particularly significant for TBCs that endure thermal cycles in operational environments. The improved durability against mechanical and thermal stresses results in a longer operational life of TBCs.

TBC is a crucial component in modern combustion engine parts, insulating components from excessive heat. It increases efficiency by regulating the combustion chamber's temperature, reducing fuel consumption, and enhancing overall performance. TBCs also prolong component life by shielding them from deterioration and oxidation caused by high temperatures. They also reduce emissions by promoting complete fuel combustion and reducing harmful emissions, making them essential for stringent environmental regulations.

REFERENCES

1. Avcı, A., Karabaş, M., Eker, A. A., & Akman, E. (2025). Optimization of laser process parameters and improved corrosion behaviour of LZ/YSZ thermal barrier coating. *Journal of the Australian Ceramic Society*. https://doi.org/10.1007/s41779-025-01149-4
2. Karthikeyan, K., & Vidyavathy, S. M. (2024). Synthesis of lanthanum pyrochlore–lanthanum phosphate composite powders for thermal barrier coating applications. *Powder Metallurgy and Metal Ceramics*, *62*(11–12), 712–721. https://doi.org/10.1007/s11106-024-00429-1
3. Kebriyaei, A., Rahimipour, M. R., & Razavi, M. (2024). Solution medium suspension plasma spray (SMSPS): A microstructure and high-temperature properties analysis of YSZ thermal barrier coating. *Journal of Thermal Spray Technology*, *33*(7), 2395–2407. https://doi.org/10.1007/s11666-024-01848-7
4. Kumar, S., & Pandey, S. M. (2024). Phase composition, microstructural characterization and mechanical properties of plasma-sprayed samarium niobate: A possible replacement of YSZ for thermal barrier coating applications. *Journal of Materials Engineering and Performance*. https://doi.org/10.1007/s11665-024-10168-4
5. Nayak, H., Kollur, S., Prasad, K., Erannagari, S., Prasad, C. D., & Nagabhushana, N. (2024). Development of equal proportional YSZ + Al_2O_3 thermal barrier coating and effect of coating thickness on the corrosion behaviour on cast iron substrate. *Deleted Journal*, *101*(2), 291–308. https://doi.org/10.1007/s11085-024-10222-5
6. Parthiban, K., & Ghosh, S. (2024). Hot corrosion study of novel glass–ceramic-based thermal barrier coating system. In *Springer Proceedings in Materials* (pp. 129–140). https://doi.org/10.1007/978-981-97-2969-2_8
7. Zhang, Z., Zeng, W., Zhang, X., & Zeng, Y. (2023). Comparative study on different methods for prediction of thermal insulation performance of thermal barrier coating used on turbine blades. *Journal of Thermal Science*, *33*(1), 172–189. https://doi.org/10.1007/s11630-023-1904-x
8. Cheng, C., Lv, T., Song, B., Chai, Y., Luo, J., Li, G., Yang, L., & Zhou, Y. (2024). Thermal cycling property of the novel $Hf_6Ta_2O_{17}$/YSZ TBCs prepared by atmospheric plasma spraying technology. *Surfaces and Interfaces*, *53*, 105050. https://doi.org/10.1016/j.surfin.2024.105050
9. Wang, X., Zhen, Z., Huang, G., Mu, R., He, L., & Xu, Z. (2021). Thermal cycling of EB-PVD TBCs based on YSZ ceramic coat and diffusion aluminide bond coat. *Journal of Alloys and Compounds*, *873*, 159720. https://doi.org/10.1016/j.jallcom.2021.159720

10. Yin, Q., Shen, K., Wang, Y., & Xing, Y. (2025). Research progress of rare earth zirconate thermal barrier coating ceramic materials. *Materials Today Communications*, *42*, 111579. https://doi.org/10.1016/j.mtcomm.2025.111579
11. Nesbitt, J. A., Zhu, D., Miller, R. A., & Barrett, C. A. (2003). Failure morphologies of cyclically oxidized ZrO_2-based thermal barrier coatings. *Materials at High Temperatures*, *20*(4), 507–517. https://doi.org/10.1179/mht.2003.059
12. Vakilifard, H., Shahbazi, H., Liberati, A. C., Saraswathy, R. B. N., Lima, R. S., Pugh, M. D., & Moreau, C. (2024). High entropy oxides as promising materials for thermal barrier topcoats: A review. *Journal of Thermal Spray Technology*, *33*(2–3), 447–470. https://doi.org/10.1007/s11666-024-01744-0
13. Shi, X., Han, X., Sun, C., Wang, D., Li, G., Deng, L., Wu, S., Zhang, X., & Wang, Y. (2024). The effect of two raw YSZ materials with different morphology on TBCs via atmospheric plasma spraying: Microstructures and high-temperature performance. *Surface and Coatings Technology*, *483*, 130723. https://doi.org/10.1016/j.surfcoat.2024.130723
14. Bursich, S., Morelli, S., Bolelli, G., Cavazzini, G., Rossi, E., Mecca, F. G., Petruzzi, S., Bemporad, E., & Lusvarghi, L. (2023). The effect of ceramic YSZ powder morphology on coating performance for industrial TBCs. *Surface and Coatings Technology*, *476*, 130270. https://doi.org/10.1016/j.surfcoat.2023.130270
15. Dong, T., Kong, L., Fu, B., Li, J., & Li, G. (2022). Effect of CeO_2 doping on high temperature oxidation resistance of YSZ TBCs. *Ceramics International*, *48*(24), 36450–36459. https://doi.org/10.1016/j.ceramint.2022.08.205
16. Dou, M., Gao, W., Xue, T., Han, J., Wu, D., Zou, Y., & Zhang, Y. (2024). CMAS corrosion characteristics of YSZ and Al_2O_3-YSZ thermal barrier coatings at 1250°C–1350°C. *Journal of the European Ceramic Society*, *44*(15), 116789. https://doi.org/10.1016/j.jeurceramsoc.2024.116789
17. Feng, Y., Dong, T., Li, G., Wang, R., Ma, G., Zhao, X., & Liu, Q. (2020). The roles of stress in the thermal shock failure of YSZ TBCs before and after laser remelting. *Journal of Alloys and Compounds*, *828*, 154417. https://doi.org/10.1016/j.jallcom.2020.154417
18. Gong, J., Du, J., Wang, J., Gao, P., Xue, B., Wang, X., & Yang, B. (2024). Microstructural evolution and high-temperature oxidation behavior of plasma sprayed SiC-YSZ TBCs. *Materials Characterization*, *215*, 114211. https://doi.org/10.1016/j.matchar.2024.114211
19. Karaoglanli, A. C., Doleker, K. M., & Ozgurluk, Y. (2019). Interface failure behavior of yttria stabilized zirconia (YSZ), $La_2Zr_2O_7$, $Gd_2Zr_2O_7$, YSZ/$La_2Zr_2O_7$ and YSZ/$Gd_2Zr_2O_7$ thermal barrier coatings (TBCs) in thermal cyclic exposure. *Materials Characterization*, *159*, 110072. https://doi.org/10.1016/j.matchar.2019.110072
20. Lei, X., Wang, Y., Wang, Q., Li, L., & Ning, X. (2021). A case study on CMAS corrosion of an in-situ microstructure modification YSZ TBCs fabricated by EB-PVD. *Surface and Coatings Technology*, *425*, 127738. https://doi.org/10.1016/j.surfcoat.2021.127738
21. Lai, Q., Chen, J., Chang, F., Pei, J., Liang, Y., Chen, X., Feng, Q., Cen, Z., & Luo, N. (2023). Cold sintering process assisted sintering for 8YSZ ceramic: A way of achieving high density and electrical conductivity at a reduced sintering temperature. *Ceramics International*, *49*(9), 14744–14749. https://doi.org/10.1016/j.ceramint.2023.01.070
22. Mahmoud, M. M., Arif, M., & Gasem, Z. M. (2024). Hybrid microwave sintering of yttria stabilized zirconia. In *Lecture Notes in Mechanical Engineering* (pp. 267–278). https://doi.org/10.1007/978-3-031-64362-0_27
23. Xu, X., Feng, Y., Zhong, Z., Guo, X., Zhang, Z., Li, J., Zhao, S., Wu, S., & Sun, H. (2022). Modified room temperature solid-state synthesis of yttria-stabilized zirconia (YSZ) nano-powders for solid oxide fuel cells. *Journal of Rare Earths*, *41*(9), 1385–1391. https://doi.org/10.1016/j.jre.2022.07.012

24. Kim, D. (1990). Effect of Ta_2O_5, Nb_2O_5, and HfO_2 alloying on the transformability of Y_2O_3-stabilized tetragonal ZrO_2. *Journal of the American Ceramic Society, 73*(1), 115–120. https://doi.org/10.1111/j.1151-2916.1990.tb05100.x
25. Raghavan, S., Wang, H., Dinwiddie, R. B., Porter, W. D., Vaβen, R., Stöver, D., & Mayo, M. J. (2004). Ta_2O_5/Nb_2O_5 and Y_2O_3 co-doped zirconias for thermal barrier coatings. *Journal of the American Ceramic Society, 87*(3), 431–437. https://doi.org/10.1111/j.1551-2916.2004.00431.x
26. Chen, L. B. (2006). Yttria-stabilized zirconia thermal barrier coatings — A review. *Surface Review and Letters, 13*(05), 535–544. https://doi.org/10.1142/s0218625x06008670
27. Yin, J., Lai, C., Zhang, X., Zhang, L., Wang, X., Wang, X., Deng, Z., Zhuo, X., He, B., & Deng, L. (2021). Microstructural and columnar growth characteristics of 7YSZ thermal barrier coatings fabricated by plasma spray physical vapor deposition. *Journal of the European Ceramic Society, 41*(16), 315–323. https://doi.org/10.1016/j.jeurceramsoc.2021.09.008
28. Amaya, C., Aperador, W., Caicedo, J., Espinoza-Beltrán, F., Muñoz-Saldaña, J., Zambrano, G., & Prieto, P. (2009). Corrosion study of alumina/yttria-stabilized zirconia (Al_2O_3/YSZ) nanostructured thermal barrier coatings (TBC) exposed to high temperature treatment. *Corrosion Science, 51*(12), 2994–2999. https://doi.org/10.1016/j.corsci.2009.08.028
29. Gao, Z., Zhang, X., Chen, Y., Chalk, C., Nicholls, J., Brewster, G., & Xiao, P. (2023). Strain tolerance evolution of EB-PVD TBCs after thermal exposure or CMAS attack. *Journal of the European Ceramic Society, 44*(1), 426–434. https://doi.org/10.1016/j.jeurceramsoc.2023.08.047
30. Jia, Z., Wang, W., Qi, W., Li, W., Song, K., & Xie, L. (2024). Effective thermal conductivity and thermal insulation performance of plasma-sprayed Mo/YSZ functionally graded thermal barrier coatings. *Surface and Coatings Technology*, 131698. https://doi.org/10.1016/j.surfcoat.2024.131698
31. Reddy, D. K. S., Barmavatu, P., Das, M. K., & Aepuru, R. (2022). Mechanical properties evaluation and microstructural analysis study of ceramic-coated IC engine cylinder liner. *Materials Today Proceedings, 76*, 518–523. https://doi.org/10.1016/j.matpr.2022.11.157
32. Mehta, A., Vasudev, H., & Singh, S. (2023). Development and characterization of YSZ/Al_2O_3/CeO_2 plasma spray coatings. *Materials Today Proceedings*. https://doi.org/10.1016/j.matpr.2023.10.161
33. Tahir, M., Qasim, M., Ahmed, N., Satti, A. N., Malik, A. E., Khan, Z. S., & Anwar, M. (2024). Impact of atmospheric plasma spraying parameters on microstructure, mechanical properties and thermal cycling performance of YSZ coatings. *Ceramics International*. https://doi.org/10.1016/j.ceramint.2024.10.253
34. Yetim, A., Tekdir, H., Turalioglu, K., Taftali, M., & Yetim, T. (2023). Tribological behavior of plasma-sprayed yttria-stabilized zirconia thermal barrier coatings on 316L stainless steel under high-temperature conditions. *Materials Letters, 336*, 133873. https://doi.org/10.1016/j.matlet.2023.133873
35. Franco, D., Vargas, F., López, E., & Ageorges, H. (2023). Wear behavior at high temperature of ZrO_2–Y_2O_3 (YSZ) plasma-sprayed coatings. *Journal of Materials Science, 59*(1), 20–37. https://doi.org/10.1007/s10853-023-09204-w
36. He, P., Yin, S., Song, C., Lapostolle, F., & Liao, H. (2015). Characterization of yttria-stabilized zirconia coatings deposited by low-pressure plasma spraying. *Journal of Thermal Spray Technology, 25*(3), 558–566. https://doi.org/10.1007/s11666-015-0375-3

37. Song, C., Xie, S., Fan, X., He, P., Liu, M., Zhou, K., Deng, C., & Liao, H. (2020). Very low-pressure plasma-sprayed dense yttria-stabilized zirconia coatings using an axial bi-cathode plasma torch. *Surface and Coatings Technology*, *402*, 126281. https://doi.org/10.1016/j.surfcoat.2020.126281
38. Wang, X., Wang, Y., Song, X., Li, W., Zheng, W., & Zeng, Y. (2022). Analysis of thermal shock performance of Y_2O_3 stabilized ZrO_2 (YSZ) coating based on residual stress and micro-morphology. *Ceramics International*, *48*(13), 19081–19089. https://doi.org/10.1016/j.ceramint.2022.03.198
39. Zhao, K., Huang, W., Deng, P., Zhong, R., Tan, Z., Hu, Y., Li, J., & Mao, W. (2024). Mechanical properties, thermal shock resistance and stress evolution of plasma-sprayed 56 wt% Y_2O_3-stabilized ZrO_2 thick thermal barrier coatings. *Surface and Coatings Technology*, 131352. https://doi.org/10.1016/j.surfcoat.2024.131352
40. Kavaliauskas, Ž., Kėželis, R., Milieška, M., Marcinauskas, L., Valinčius, V., Aikas, M., Uscila, R., Baltušnikas, A., & Žunda, A. (2021). Influence of different plasma spraying methods on the physical properties of YSZ coatings. *Surfaces and Interfaces*, *24*, 101120. https://doi.org/10.1016/j.surfin.2021.101120
41. Miranda, F. S., Caliari, F. R., Campos, T., Leite, D., Pessoa, R., Essiptchouk, A., & Petraconi, G. (2020). High-velocity plasma spray process using hybrid SiO_2+ZrO_2 precursor for deposition of environmental barrier coatings. *Surface and Coatings Technology*, *404*, 126447. https://doi.org/10.1016/j.surfcoat.2020.126447
42. Sun, Y., Wu, X., Tan, Z., Yang, L., & Zhou, Y. (2022). Boosting the strain tolerance of Ta_2O_5 stabilized ZrO_2 TBCs through prefabricated cracks by in-situ reaction. *Journal of Alloys and Compounds*, *935*, 168131. https://doi.org/10.1016/j.jallcom.2022.168131
43. Wang, J., Bai, L., Ma, F., Wan, S., Yi, G., Sun, J., Tian, X., & Yang, Z. (2021). Evaluation of microstructure evolution of thermal barrier YSZ coating after thermal exposure. *Ceramics International*, *48*(5), 6681–6690. https://doi.org/10.1016/j.ceramint.2021.11.218
44. Iqbal, A., & Moskal, G. (2023). Recent development in advance ceramic materials and understanding the mechanisms of thermal barrier coatings degradation. *Archives of Computational Methods in Engineering*, *30*(8), 4855–4896. https://doi.org/10.1007/s11831-023-09960-7
45. Keyvani, A., Bahamirian, M., & Esmaeili, B. (2020). Sol-gel synthesis and characterization of ZrO_2_25wt.%CeO_2-2.5wt.%Y_2O_3 (CYSZ) nanoparticles. *Ceramics International*, *46*(13), 21284–21291. https://doi.org/10.1016/j.ceramint.2020.05.219
46. Saemi, H., Rastegari, S., Sarpoolaky, H., & Latifi, R. (2021). Oxidation resistance of double-ceramic-layered thermal barrier coating system with an intermediate Al_2O_3-YAG layer. *Journal of Thermal Spray Technology*, *30*(4), 1049–1058. https://doi.org/10.1007/s11666-021-01191-1
47. Bahamirian, M., Hadavi, S., Farvizi, M., Keyvani, A., & Rahimipour, M. R. (2020). ZrO_2 9.5Y_2O_3 5.6Yb_2O_3 5.2Gd_2O_3; a promising TBC material with high resistance to hot corrosion. *Journal of Asian Ceramic Societies*, *8*(3), 898–908. https://doi.org/10.1080/21870764.2020.1793474
48. Chen, L., Li, B., & Feng, J. (2024). Rare-earth tantalates for next-generation thermal barrier coatings. *Progress in Materials Science*, *144*, 101265. https://doi.org/10.1016/j.pmatsci.2024.101265
49. Zhao, D., Chen, M., Pi, Z., & Zhang, F. (2023). Phase equilibria and non-transformable tetragonal zirconia in ZrO_2-$RETaO_4$ systems and their stabilization mechanism: Experiments and calculations. *Ceramics International*, *49*(10), 15969–15978. https://doi.org/10.1016/j.ceramint.2023.01.193
50. Zhou, Y., Gan, G., Ge, Z., Feng, J., & Peng, S. (2021). Investigation of thermophysical properties of ZrO_2-Sm_3TaO_7 ceramics. *Journal of Asian Ceramic Societies*, *9*(2), 629–638. https://doi.org/10.1080/21870764.2021.1907025

51. Peng, Z., Wang, Y., Wang, S., Yao, J., Zhao, Q., Xie, E., Chen, G., Wang, Z., Liu, Z., Wang, Y., & Ouyang, J. (2024). Improvement strategy on thermophysical properties of $A_2B_2O_7$-type rare earth zirconates for thermal barrier coatings applications: A review. *International Journal of Minerals Metallurgy and Materials*, *31*(5), 1147–1165. https://doi.org/10.1007/s12613-024-2853-4
52. Qu, C., Chen, L., Lv, L., Wang, Y., Ji, X., Yun, H., Su, C., & Feng, J. (2022). Low thermal conductivity and anisotropic thermal expansion of ferroelastic $(Gd_{1-x}Y_x)TaO_4$ ceramics. *Journal of Advanced Ceramics*, *11*(11), 1696–1713. https://doi.org/10.1007/s40145-022-0641-z
53. Chen, Z., Cui, X., Fang, Y., Yan, C., Wang, R., Jing, Y., Wang, X., Jin, G., & Wang, X. (2022). A comparative study on the hot corrosion behavior of plasma sprayed Sm/Gd doped $La_2Zr_2O_7$/YSZ and Sm and Gd co-doped $La_2Zr_2O_7$/YSZ coatings. *Surface and Coatings Technology*, *452*, 129077. https://doi.org/10.1016/j.surfcoat.2022.129077
54. Guo, L., Li, M., & Ye, F. (2016). Phase stability and thermal conductivity of RE_2O_3 (RE=La, Nd, Gd, Yb) and Yb_2O_3 co-doped Y_2O_3 stabilized ZrO_2 ceramics. *Ceramics International*, *42*(6), 7360–7365. https://doi.org/10.1016/j.ceramint.2016.01.138
55. Kurapova, O. Y., Glukharev, A. G., Borisova, A. S., Golubev, S. N., & Konakov, V. G. (2019). Phase formation, stability and heat capacity of ternary TiO_2–CeO_2–ZrO_2 solid solutions. *Materials Chemistry and Physics*, *242*, 122547. https://doi.org/10.1016/j.matchemphys.2019.122547
56. Saini, A., Singh, G., Kumar, H., Sehgal, S., & Singh, A. (2018). Investigations on the use of lanthanum strontium zirconate coating in maintaining temperature distributions of gas turbine blades. *Materials Today Proceedings*, *5*(14), 28408–28412. https://doi.org/10.1016/j.matpr.2018.10.126
57. Wang, J., Sun, J., Jing, Q., Liu, B., Zhang, H., Yongsheng, Y., Yuan, J., Dong, S., Zhou, X., & Cao, X. (2018). Phase stability and thermo-physical properties of ZrO_2-CeO_2-TiO_2 ceramics for thermal barrier coatings. *Journal of the European Ceramic Society*, *38*(7), 2841–2850. https://doi.org/10.1016/j.jeurceramsoc.2018.02.019
58. Shanmugam, M., Sathiyamurthy, S., Rajkumar, G., Saravanakumar, S., Prabakaran, S. T., & Shaisundaram, V. (2020). Effect of thermal Barrier coating in CI engines fueled with Citrus Medica (Citron) peel oil biodiesel dosed with cerium oxide nanoparticle. *Materials Today Proceedings*, *37*, 1943–1956. https://doi.org/10.1016/j.matpr.2020.07.485
59. Thiruselvam, K., Saravanan, R., & Gift, M. M. (2022). Experimental analysis of the combined effect of thermal barrier coating material and the nanoparticle augmented fuel usage on CI engine characteristics. *Materials Today Proceedings*, *72*, 2351–2357. https://doi.org/10.1016/j.matpr.2022.09.403
60. Zhou, F., Yu, J., Wu, C., Fu, J., Liu, J., & Duan, X. (2023). The application prospect and challenge of the alternative methanol fuel in the internal combustion engine. *Science of the Total Environment*, *913*, 169708. https://doi.org/10.1016/j.scitotenv.2023.169708

9 Wear Performance of Plasma-Sprayed Nanostructured Zirconia Coatings on Stainless Steel under Distilled Water Conditions

Praveen Kumar Verma, Hitesh Vasudev, and Yanhai Cheng

9.1 INTRODUCTION

Stainless steel, especially austenitic grades like 316L, is widely used in biomedical implants, marine technology, and industrial equipment because of its high corrosion resistance, high mechanical strength, and biocompatibility. All these characteristics render stainless steel an ideal candidate for load-bearing applications where both structural integrity and environmental stability are paramount. Despite its better corrosion resistance in aqueous and severe environments, stainless steel has relatively poor tribological behavior under mechanical contact, sliding, or erosion conditions. This is mostly due to its lower hardness and tendency to wear mechanisms such as adhesive wear, abrasive wear, and fretting wear when subjected to cyclic loading or corrosive conditions. As such, surface treatment methods have extensively been pursued for the enhancement of wear resistance in stainless steel with its innate corrosion resistance and mechanical strength intact [1].

To improve the tribological performance of stainless steel components, several approaches have been used in surface engineering, one of them being plasma spraying of nanostructured ceramic coating. Plasma-sprayed nanostructured zirconia (ZrO_2) coatings are particularly attractive due to their unique combination of high hardness, excellent chemical inertness, and low thermal conductivity [2]. These coatings have demonstrated remarkable potential in mitigating surface wear by reducing friction, improving load-bearing capacity, and enhancing surface toughness. The application of zirconia-based coatings is particularly valuable in biomedical implants, where generation of wear debris should be kept at a minimum to avoid inflammation, and in marine or industrial applications, where long service life under aggressive operating conditions is a critical requirement [3].

DOI: 10.1201/9781003519300-11

Zirconia is a polymorphic ceramic material that can exist in three different crystalline phases: monoclinic (m-ZrO_2), tetragonal (t-ZrO_2), and cubic (c-ZrO_2). Monoclinic zirconia occurs naturally at room temperature, transforming reversibly into the tetragonal phase around 1170°C and further into the cubic phase at temperatures above 2370°C. The monoclinic-to-tetragonal ($m \rightarrow t$) phase transformation is also associated with large volume expansion (~4%–5%), which creates tensile stresses, microcracking, and ultimately coating failure. To counter this built-in restriction, zirconia is often stabilized with dopant oxides such as yttria (Y_2O_3), calcium oxide (CaO), and magnesia (MgO), which support the stability of the tetragonal and/or cubic phases under room temperatures. Yttria-stabilized zirconia (YSZ) has attracted widespread interest for its enhanced toughness, improved phase stability, and improved wear resistance, making it the most suitable applicant for high-performance protective coatings [4]. Figure 9.1 shows the parameters that influence nanostructured coating formation.

The excellent mechanical and tribological qualities of zirconia coatings are further augmented when nanostructured zirconia powders are utilized in the process of plasma spraying. Nanostructured coatings have an enhanced microstructure with a better developed grain boundary density, and these grain boundaries play an extremely important role in resisting crack movement and enhancing mechanical integrity. In comparison with traditional microstructured coatings, nanostructured zirconia coatings have increased microhardness, higher fracture toughness, and better adhesion strength because of their smaller grain size and homogeneous

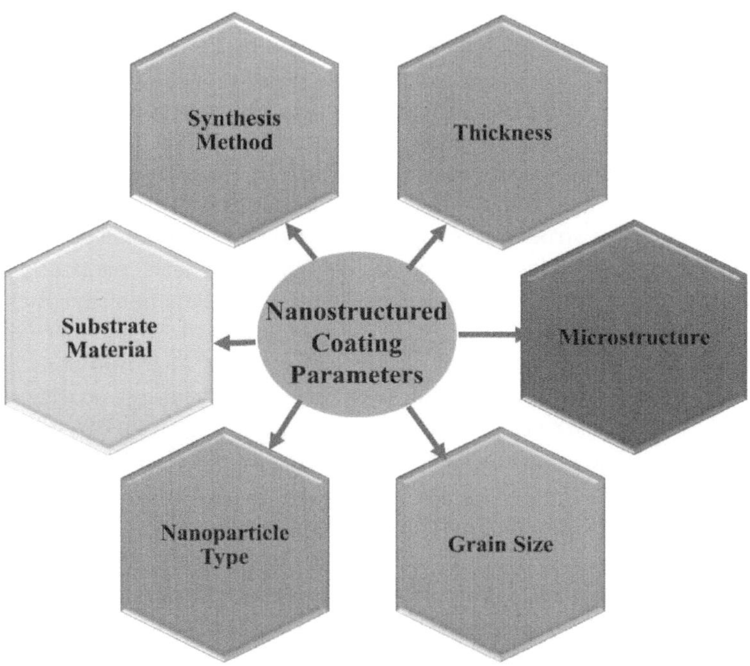

FIGURE 9.1 Various parameters influencing nanostructured coating formation [5]. (CC By 4.0 Attribution.)

phase distribution. These improvements are mainly attributable to the extensive surface area of nanosized particles, which leads to stronger interparticle bonding and densification of the coating during plasma spraying [6].

One of the important benefits of nanostructured zirconia coatings is their increased wear resistance against wear mechanisms such as abrasive wear, erosive wear, and fatigue wear. The availability of a well-adhered, dense ceramic layer on stainless steel surfaces can be considered a protective layer that minimizes metal-to-metal contact and reduces material loss under tribological loading. The increased grain fineness of nanostructured coatings also leads to enhanced fracture toughness by favoring energy dissipation processes such as crack deflection, grain boundary sliding, and stress-induced phase transformations. These properties render nanostructured zirconia coatings highly attractive for applications with high contact stresses, recurrent sliding motion, and hostile environmental conditions [7].

Besides their wear resistance, plasma-sprayed nanostructured zirconia coatings also show very good thermal insulation because of the low thermal conductivity of zirconia (~2.2 W/mK for YSZ). This is an important property for high-temperature applications such as thermal barrier coatings (TBCs) in gas turbines and aero-engine components, where substrate degradation has to be avoided with thermal protection. In addition, the chemical inertness of zirconia guarantees long-term corrosion resistance in aggressive environments, making it an appropriate material for marine and chemical processing applications where seawater, acid solutions, and oxidizing chemicals are common. Despite numerous benefits, the performance of plasma-sprayed nanostructured zirconia coatings depends on several factors such as coating porosity, residual stresses, adhesion strength, and phase stability. Porosity above the threshold can undermine the mechanical integrity of the coating and accentuate susceptibility toward wear-induced failure, while residual tensile stresses can provoke delamination or spallation upon cyclically loaded conditions. Thus, heavy emphasis has been placed on how to optimize plasma spray processing parameters and secondary post-processing techniques such as laser remelting, hot isostatic pressing, and surface sealing to improve the structural integrity and longevity of zirconia coatings [8,9].

Plasma-sprayed nanostructured zirconia coatings offer an exceptionally effective solution for enhancing the wear resistance of stainless steel components for biomedical, marine, and industrial use. Capitalizing on the specific benefits of nanostructured materials, including increased mechanical strength, improved phase stability, and more precise microstructural characteristics, these coatings provide dramatic increases in surface hardness, friction reduction, and long-term performance. Continued research activities focus on new coating architectures, hybrid composite coatings, and new deposition methods to further extend the use of nanostructured zirconia coatings in various engineering applications [10].

9.2 PLASMA SPRAYING PROCESS FOR NANOSTRUCTURED ZIRCONIA COATINGS

Plasma spraying is a common thermal spray method for coating stainless steel substrates with zirconia-based coatings, resulting in improved wear resistance, corrosion protection, and thermal stability. In this process, molten or semi-molten particles of zirconia are fed into a hot plasma jet, which accelerates and projects them onto the substrate. The particles on impact instantly solidify, developing a protective coating. The final microstructure and performance of the coating largely rely on a range of process parameters, such as plasma arc power, feedstock properties, substrate temperature, and spray distance. Plasma spraying, an established thermal spray method, consists of the injection of ceramic feedstock powders into a hot plasma jet, where the particles are melted or partially melted and then accelerated onto the substrate surface to create a dense coating. The most important parameters controlling the deposition quality are plasma power, carrier gas flow rate, spray distance, and powder feed rate. The application of nanostructured zirconia powders in plasma spraying adds more complexity, with nanosized particles having increased surface energy and reactivity, which can affect their melting behavior and deposition properties. Research has identified that tuning process parameters such as plasma arc temperature, particle velocity, and substrate cooling significantly affect the microstructure and properties of the deposited nanostructured coatings [11]. Figure 9.2 shows the plasma spraying setup.

The selection of feedstock has a significant impact on coating properties. Nanostructured zirconia feedstock is generally prepared using spray drying or sol-gel synthesis techniques, resulting in a uniform distribution of nano-sized grains. The addition of YSZ is especially useful since yttria stabilizes the phase and suppresses unwanted phase changes that might degrade the mechanical strength of the coating. YSZ coatings are very tough and resistant to wear over long periods and therefore find applications in extreme conditions such as biomedical implants and TBCs [13].

In plasma spraying, the coating develops a unique lamellar microstructure of partially melted and fully melted zones. The presence of nanograins in the coating leads to improved mechanical performance, especially toughness and wear resistance.

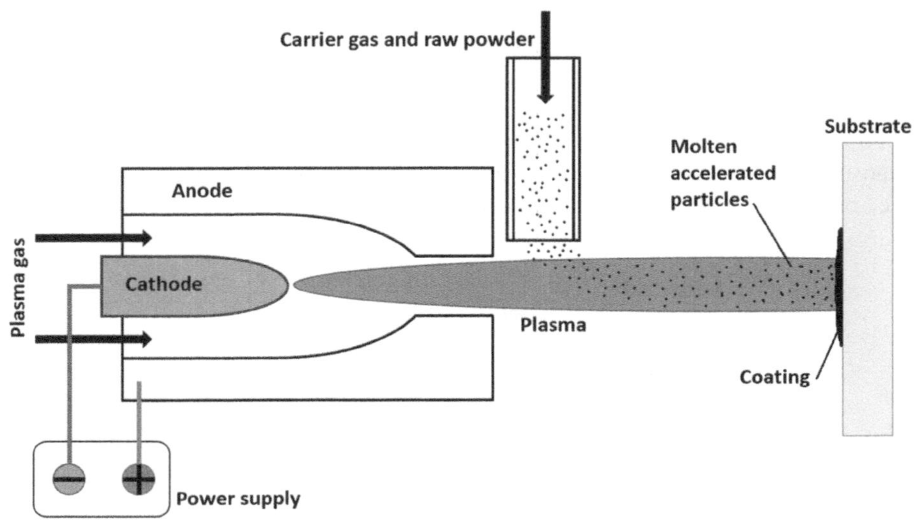

FIGURE 9.2 Schematic illustration of the plasma spraying setup [12]. (CC By 4.0 Attribution.)

However, this process is bound to produce porosity and microcracking, which can affect the behavior of the coating under certain service conditions. In water environments, microstructural characteristics can affect corrosion resistance and long-term durability, requiring post-treatment methods such as sealing and heat treatment to counteract possible degradation. The optimization of plasma spraying parameters is necessary to retain coatings with the optimum performance features. The control of plasma arc power, particle velocity, and substrate preheating can have a substantial impact on coating density, adhesion strength, and porosity. In addition, the careful adjustment of feedstock characteristics to particle size distribution and phase composition provides uniform deposition and enhanced mechanical properties [14].

9.3 WEAR MECHANISMS UNDER DISTILLED WATER CONDITIONS

Plasma-sprayed nanostructured zirconia coatings have been identified as potential wear-resistant materials, especially in sliding contact environments in distilled water. The coatings are characterized by superior mechanical properties, such as high hardness, toughness, and thermal stability, which are ideal for harsh tribological applications. Nevertheless, the wear behavior of the coatings is intricate, with several wear mechanisms being simultaneously active. The prevailing wear mechanisms are a function of applied load, sliding velocity, temperature, and microstructural properties of the coating. Knowledge of these mechanisms is important for maximizing the coating performance and the service life of the components being protected [15]. Figure 9.3 shows that the wear performance of two-body wet sliding is affected by several factors such as material properties, surface properties, lubrication, and contact conditions. Hardness plays an important role, with materials that are harder tending to have improved wear resistance, whereas toughness and resistance to fracture contribute to preventing crack initiation and propagation. Microstructure, such as grain size, phase composition, and porosity, also significantly affects wear behavior. Surface properties such as roughness, coating adhesion, and porosity affect friction and material loss, with increased roughness increasing wear and reduced roughness decreasing it.

Abrasive wear is a prevalent mechanism that has been found in plasma-sprayed nanostructured zirconia coatings subjected to sliding contact. It is caused by the presence of hard particles, like zirconia grains or wear debris, between the two surfaces

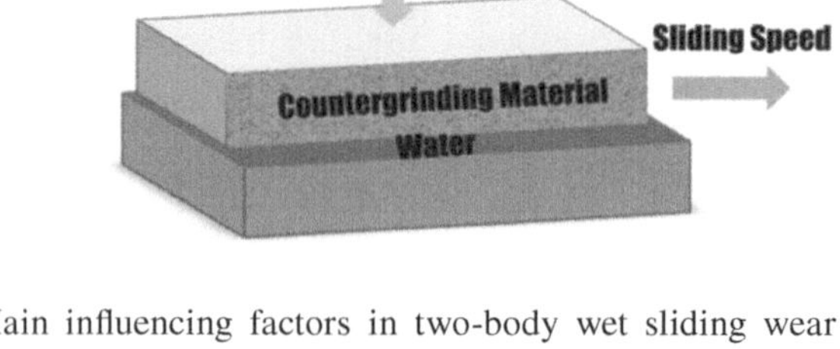

FIGURE 9.3 Main influencing factors in two-body wet sliding wear [16]. (CC By 4.0 Attribution.)

in contact. These particles function as third bodies that wear down the coating and counterface through microcutting or plowing, causing material removal. The amount of abrasive wear is determined by parameters such as the hardness of the abrasive particles, applied load, and sliding velocity. In certain instances, the wear debris produced during sliding may agglomerate and add to material loss, accelerating the wear rate. Optimal control of the microstructure of the coating, including grain size and porosity, can prevent abrasive wear and increase the durability of the coating [17].

Adhesive wear is also a major wear mechanism that can affect the performance of plasma-sprayed nanostructured zirconia coatings. This wear mode occurs when adhesive bonds between the surface of the coating and the counterface material are very strong during sliding contact. With continuing relative motion, the bonds will likely break, resulting in material transfer or removal from the coating. The degree of adhesive wear is influenced by the material compatibility of contact materials, surface roughness, and the availability of lubricating media. Under distilled water conditions, adhesive wear can be affected by the interaction between water molecules and the surface, which can change adhesion properties. Surface engineering methods, including coating composition modification and surface treatment, can be used to minimize adhesive wear and enhance the overall wear resistance of the coatings [18]. Figure 9.4 shows the setup of a pin on disk wear test.

Tribochemical wear becomes especially important under distilled water conditions since it is brought about by chemical reactions between the coating material, counterface, and ambient environment. Tribochemical wear in zirconia coatings, for instance, results from hydrothermal degradation of zirconia in the presence of water. This is commonly referred to as low-temperature degradation or aging and

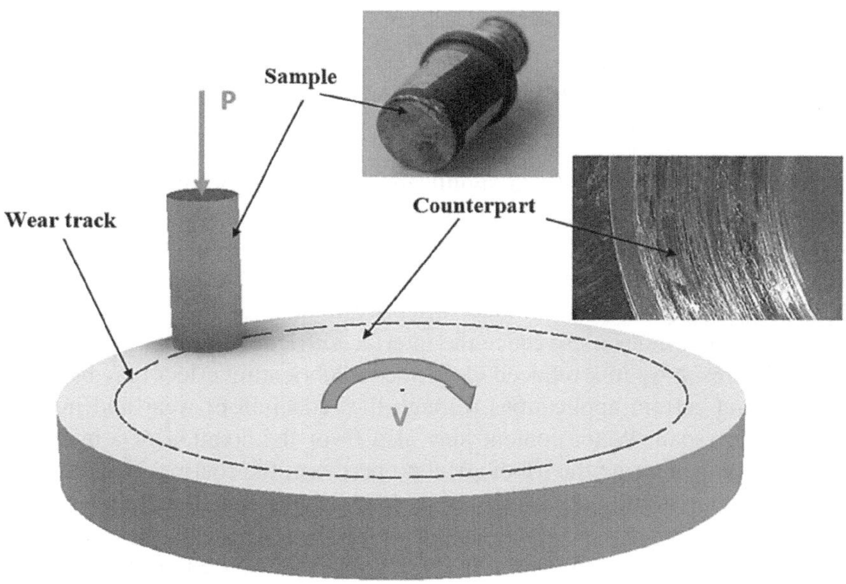

FIGURE 9.4 Illustration of a tribological pin on disk test. [19]. (CC By 4.0 Attribution.)

is responsible for changing the tetragonal zirconia phase into the monoclinic phase through microcracking and enhanced wear rates. In addition, there can be a chemical interaction between the surface oxide layer and water molecules to enhance wear product formation, which can be wear-enhancing or wear-reducing based on their character. Techniques such as doping of zirconia with stabilizing elements (e.g., yttria) and optimization of spraying parameters can improve the tribochemical wear resistance of the coatings. The interactions between these mechanisms are complicated, and their relative importance is determined by the prevailing operating conditions in designing coatings with specific wear resistance characteristics [20,21].

The use of plasma-sprayed nanostructured zirconia coatings in different industries is motivated by their potential to improve wear resistance, decrease the coefficient of friction (CoF), and increase the life of important parts. In casing and tool joint applications, they offer a wear and corrosion barrier, thus enhancing equipment reliability and lowering maintenance expenses. Through optimization of the microstructure and composition of the coatings, their performance under operating conditions can be improved further by researchers and engineers. The evolution of future coating technology, such as the production of hybrid coatings and multilayered systems, promises to yield even better wear resistance and durability [22].

Plasma-sprayed nanostructured zirconia coatings demonstrate a multiplicity of wear mechanisms, viz., abrasive, adhesive, and tribochemical wear, upon sliding contact in distilled water. The prevailing wear mechanism is a function of various parameters, such as applied load, sliding velocity, temperature, and microstructure. Identifying these mechanisms and incorporating measures to counteract them will allow the performance and lifespan of such coatings to be enhanced in diverse industrial applications. Future research and development in coating materials and process methodologies will continue to advance their wear resistance and extend their application areas in harsh environments [23].

9.4 INFLUENCE OF PROCESS PARAMETERS ON WEAR PERFORMANCE

Environmental conditions play a significant role in determining the wear performance of zirconia coatings, and interaction with distilled water is one of the factors considered. Although it may seem to be a neutral medium chemically, distilled water interacts with zirconia coatings in several ways, influencing their tribological behavior accordingly. Distilled water affects zirconia coatings primarily through lubrication. Water molecules reduce the interfacial friction between the counterface and coating, resulting in a reduced CoF. Such a lubricating effect may have beneficial effects on certain applications through the reduction of wear and increase in the coating lifespan. Water content may also favor the creation of a hydrated top layer on zirconia, altering the tribological properties of the coating. Nonetheless, this lubricating action is strongly influenced by load, sliding speed, and the existence of surface flaws [24]. Figure 9.5 shows an inclined plane testing apparatus employed for friction analysis. Figure 9.5a is the free-body diagram of the forces on the Haber–Bosch sample, and Figure 9.5b is the experimental apparatus with a sled resting on an inclined plane to determine the CoF.

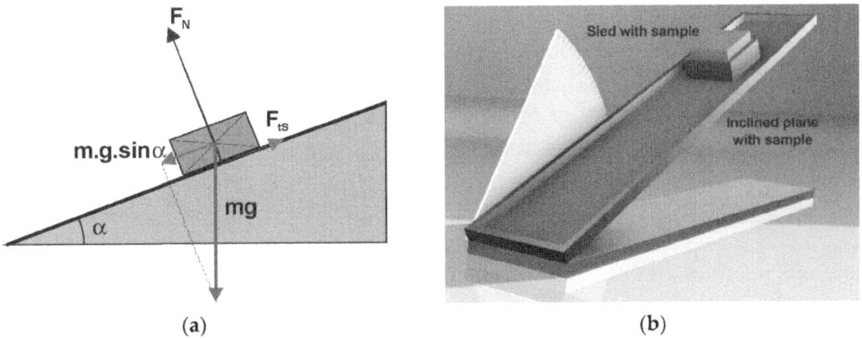

FIGURE 9.5 (a) Method of inclined plane for the measurement of the static coefficient of friction; (b) arrangement of measurement device for the measurement of friction using the inclined plane [25]. (CC By 4.0 Attribution.)

In addition to its lubricating action, distilled water can also cause negative effects on zirconia coatings. As a polar solvent, distilled water easily picks up ions and minerals from the environment. If the coating has microstructural flaws such as pores, cracks, and grain boundary defects, water penetration can promote degradation mechanisms. This is of greater concern where stresses that amplify the propagation of cracks and delamination of the coating are present. With time, this degrades the mechanical integrity of the zirconia layer and impairs its resistance to wear while making it easier to fail during cyclic loading. In addition, one should not disregard the corrosion potential of distilled water. The lack of dissolved minerals makes distilled water more aggressive in leaching ions from the zirconia coating and the underlying metallic substrate. This ion leaching can be especially troublesome in biomedical applications where zirconia coatings are employed to improve the biocompatibility and wear resistance of surgical-grade stainless steel or titanium implants. The stripping of essential metallic constituents from the substrate can cause local weak points, which may result in accelerated corrosion and even failure of the implant. Temperature is also another significant parameter affecting the role of distilled water in zirconia coatings [26,27].

High temperatures tend to accelerate chemical reactions, such as the hydrothermal degradation of zirconia. This phenomenon, or low-temperature degradation (LTD), is the process by which the metastable tetragonal phase converts to the monoclinic phase with an increase in volume and roughening of the surface. Distilled water, added to the above process, aggravates LTD through the medium it creates for hydration reactions, causing microcracking and a drastic reduction in wear performance. In real-world applications, the synergistic effects of lubrication, ion leaching, and hydrothermal degradation should be taken into account when employing zirconia coatings in conditions where distilled water is present. Methods of alleviating such effects involve minimizing the porosity of the coating through the optimization of the microstructure, protective surface treatments, and environmental control, including temperature and pH. Hybrid coating systems based on secondary phases or reinforcement elements can also be employed to improve resistance to corrosion and wear [28].

Distilled water can be used as a lubricant to reduce the friction, and its capacity to cause corrosion and degradation of zirconia coatings should not be discounted. The synergy between tribological and chemical effects emphasizes the necessity of thorough testing and optimization of zirconia coatings to ascertain their performance and longevity in practical applications [29].

9.5 FUTURE DIRECTION

There is a need for future studies to focus on maximizing the process parameters to enhance the hydrothermal stability of thermally sprayed hydroxyapatite (HA) coatings on surgical-grade stainless steel. Hydrothermal degradation continues to be a key challenge in biomedical applications, especially for load-bearing implants subjected to body fluids for long durations. Spray temperature, feedstock composition, particle velocity, and post-deposition treatments are significant factors affecting coating crystallinity, porosity, adhesion strength, and dissolution behavior. Systematic research should focus on determining the optimum sets of these parameters to improve the durability of the coating and reduce phase transformations, especially the development of amorphous phases and tricalcium phosphate, which destabilize the structure. Furthermore, new characterization methods such as *in situ* high-temperature X-ray diffraction and Raman spectroscopy should be used to track phase development under mimicked physiological conditions [30].

Another promising direction is the creation of hybrid HA-based coatings with self-healing capability to counteract microcracks and local delamination, which may result in implant failure. The addition of bioactive glass, zirconia, or polymeric phases into the HA matrix may bring in crack-healing mechanisms that are activated by physiological conditions. Self-healing mechanisms may be derived from biomimetic strategies and the release of calcium and phosphate ions from embedded reservoirs to heal damaged areas. Furthermore, nanostructured coatings with *in situ* drug delivery nanoparticles might also impart antibacterial properties and osteointegration while ensuring mechanical stability. The long-term wear behavior is also a crucial issue that needs extensive study to determine the clinical reliability of HA-coated implants. *In vivo* and *in vitro* research should evaluate the wear resistance of the coatings under dynamic loading conditions simulating realistic physiological stresses. Accelerated aging tests, such as tribocorrosion testing in simulated body fluids, should be performed to assess coating degradation mechanisms. The interaction of HA coatings with proteins, inflammatory reactions, and tissue integration should also be investigated over long-term periods [31].

Computational modeling strategies, such as finite element analysis, should also be addressed in future work to predict coating behavior under physiological loading. Such models can provide a simulation of stress distribution, cracking growth, and dissolution rates and help develop coatings with greater long-term efficiency. In addition, collaborations between researchers involving materials science, bioengineering, and computation models will prove valuable in achieving the next step for HA coating technologies in the area of biomedical implants. With solutions to these challenges, thermally sprayed HA coatings can be designed to provide enhanced performance, improving the durability and biocompatibility of orthopedic and dental implants [32].

9.6 CONCLUSION

Plasma-sprayed nanostructured zirconia coatings have demonstrated significant potential in enhancing the wear resistance of stainless steel substrates under distilled water conditions. The unique nanostructured morphology, characterized by reduced porosity, refined grain boundaries, and enhanced adhesion, contributes to improved mechanical integrity and tribological performance. The superior hardness and fracture toughness of these coatings provide effective resistance against material degradation, particularly in aqueous environments where conventional coatings may fail due to delamination, spallation, or accelerated wear. This review highlights that wear mechanisms in distilled water conditions are predominantly governed by a combination of microabrasion, tribochemical reactions, and fatigue wear. The presence of water can lead to surface hydration effects, influencing frictional behavior and material transfer. Despite these challenges, nanostructured zirconia coatings exhibit lower friction coefficients and reduced wear rates compared with conventional microstructured coatings, making them suitable for applications requiring prolonged service life in wet environments. However, several challenges remain, including optimizing processing parameters to achieve uniform coatings with minimal defects, improving the long-term stability of the coatings in aqueous conditions, and understanding the interplay between phase transformations and tribological performance. Future research should focus on tailoring the microstructure through advanced deposition techniques and incorporating reinforcement phases to further enhance wear resistance. In addition, systematic studies on the combined effects of thermal cycling, mechanical loading, and electrochemical interactions will be crucial for predicting the real-world performance of these coatings.

REFERENCES

1. Shahed CA, Ahmad F, Günister E, Altaf K, Ali S, Raza A, Malik K, Haider W. Microstructure and mechanical performance of low-cost biomedical-grade Titanium-316L alloy. *Journal of Materials Research and Technology*. 2023 Nov 1;27:8008–22.
2. Zhang Z, Ji G, Shi Z. Tribological properties of ZrO_2 nanofilms coated on stainless steel in a 5% NaCl solution, distilled water and a dry environment. *Surface and Coatings Technology*. 2018 Sep 25;350:128–35.
3. Kumar N, Choubey VK. Recent trends in coating processes on various AISI steel substrates: A review. *Journal of Materials Science*. 2024 Jan;59(2):395–422.
4. Brar SS, Brar GS, Chawla V. High temperature corrosion and role of plasma spray coatings-a review. *Technology*. 2017;4:5.
5. Singh VP, Sil A, Jayaganthan R. Wear of plasma sprayed conventional and nanostructured Al_2O_3 and Cr_2O_3, based coatings. *Transactions of the Indian Institute of Metals*. 2012 Feb;65:1–2.
6. Farooq SA, Raina A, Mohan S, Arvind Singh R, Jayalakshmi S, Irfan Ul Haq M. Nanostructured coatings: Review on processing techniques, corrosion behaviour and tribological performance. *Nanomaterials*. 2022 Apr 12;12(8):1323.
7. Zhang C, Xu J, Sun G, Wei X, Xiao J, Zhang G, Yin S. Wear behaviors of 5 wt% SiO_2–Ni60 coatings deposited by atmospheric plasma spraying under dry and water-lubrication sliding conditions. *Wear*. 2021 Apr 15;470:203621.

8. Awang M, Khalili AA, Pedapati SR. A review: Thin protective coating for wear protection in high-temperature application. *Metals*. 2019 Dec 25;10(1):42.
9. Ahangari A, Mokhtarzade A, Mahmoodi M. In vitro tribological behavior and corrosion resistance of $SiO_2/TiO_2/ZrO_2$ coatings formed on the tantalum substrate using plasma electrolytic oxidation for orthopedic implant applications. *Materials Chemistry and Physics*. 2023 Oct 15;308:128210.
10. Thakur A, Kumar A, Kaya S, Marzouki R, Zhang F, Guo L. Recent advancements in surface modification, characterization and functionalization for enhancing the biocompatibility and corrosion resistance of biomedical implants. *Coatings*. 2022 Oct 3;12(10):1459.
11. Gan JA, Berndt CC. Nanocomposite coatings: Thermal spray processing, microstructure and performance. *International Materials Reviews*. 2015 May;60(4):195–244.
12. Ramezani M, Mohd Ripin Z, Pasang T, Jiang CP. Surface engineering of metals: Techniques, characterizations and applications. *Metals*. 2023 Jul 20;13(7):1299.
13. Ahmad Z, Khan AU, Farooq R. Plasma-sprayed nanostructured coatings. In *High Temperature Corrosion*. 2016 Sep 7:123. In Tech, Croatia. https://doi.org/10.5772/64316
14. Nowakowska M, Łatka L, Sokołowski P, Szala M, Toma FL, Walczak M. Investigation into microstructure and mechanical properties effects on sliding wear and cavitation erosion of Al_2O_3–TiO_2 coatings sprayed by APS, SPS and S-HVOF. *Wear*. 2022 Nov 15;508:204462.
15. Yugeswaran S, Kobayashi A, Ucisik AH, Subramanian B. Characterization of gas tunnel type plasma sprayed hydroxyapatite–nanostructure titania composite coatings. *Applied Surface Science*. 2015 Aug 30;347:48–56.
16. Song S, Zhu Z, Du S, Li Y, Liu C. Research on polymer wear under water conditions: A review. *Lubricants (2075–4442)*. 2024 Sep 1;12(9).
17. Kim GE, Brochu M. Thermal spray nanostructured ceramic and metal-matrix composite coatings. In M Aliofkhazraei (Eds.), *Anti-Abrasive Nanocoatings*. 2015 Jan 1:481–511. Cambridge: Woodhead Publishing.
18. AlYafei, Huda Fadol S. *Tribo-Mechanical, Biocompatibility, and Corrosion Properties Investigation of Zirconium and SST 304 by Application of Hydroxyapatite/Reduced Graphene Oxide/Palladium-Platinum Nanocomposite Coatings*. Ph.D. Thesis, Qatar University, Doha, Qatar; January 2020. http://hdl.handle.net/10576/16312
19. Pashechko M, Dziedzic K, Jozwik J. Analysis of wear resistance of borided steel C45. *Materials*. 2020 Dec 3;13(23):5529.
20. Sathish M, Radhika N, Saleh B. Current status, challenges, and future prospects of thin film coating techniques and coating structures. *Journal of Bio-and Tribo-Corrosion*. 2023 Jun;9(2):35.
21. Lima RS, Marple BR. Thermal spray coatings engineered from nanostructured ceramic agglomerated powders for structural, thermal barrier and biomedical applications: A review. *Journal of Thermal Spray Technology*. 2007 Mar;16:40–63.
22. Boulos MI, Fauchais PL, Heberlein JV, Boulos MI, Fauchais PL, Heberlein JV. Nanocrystalline and nanostructured coatings. *Thermal Spray Fundamentals: From Powder to Part*. 2021:749–828.
23. Loghman-Estarki MR, Nejati M, Edris H, Razavi RS, Jamali H, Pakseresht AH. Evaluation of hot corrosion behavior of plasma sprayed scandia and yttria co-stabilized nanostructured thermal barrier coatings in the presence of molten sulfate and vanadate salt. *Journal of the European Ceramic Society*. 2015 Feb 1;35(2):693–702.
24. Fauchais PL, Heberlein JV, Boulos MI, Fauchais PL, Heberlein JV, Boulos MI. Nanostructured or finely structured coatings. *Thermal Spray Fundamentals: From Powder to Part*. 2014:981–1111.

25. Kelemenová T, Dovica M, Božek P, Koláriková I, Benedik O, Virgala I, Prada E, Miková Ľ, Kot T, Kelemen M. Specific problems in measurement of coefficient of friction using variable incidence tribometer. *Symmetry*. 2020 Jul 27;12(8):1235.
26. Hu H, Mao L, Yin S, Liao H, Zhang C. Wear-resistant ceramic coatings deposited by liquid thermal spraying. *Ceramics International*. 2022 Nov 15;48(22):33245–55.
27. Xu J, Zhang C, Sun G, Xiao J, Zhang L, Zhang G. Role of SiC nanoparticles on tribological properties of atmospheric plasma sprayed 5 wt% SiC–Ni60 coatings. *Tribology International*. 2020 Jun 1;146:106220.
28. Bolleddu V, Racherla V, Bandyopadhyay PP. Characterization of air plasma-sprayed yttria-stabilized zirconia coatings deposited with nitrogen. *The International Journal of Advanced Manufacturing Technology*. 2017 Jun;90:3437–49.
29. Yang J, Jia J, Li X, Lu C, Feng X. Synergistic lubrication of Ag and Ag_2MoO_4 nanoparticles anchored in plasma-sprayed YSZ coatings: Remarkably-durable lubricating performance at 800° C. *Tribology International*. 2021 Jan 1;153:106670.
30. Li S, Zhao X, An Y, Liu D, Zhou H, Chen J. YSZ/MoS_2 self-lubricating coating fabricated by thermal spraying and hydrothermal reaction. *Ceramics International*. 2018 Oct 15;44(15):17864–72.
31. Wu Z, Li S, Zhang P, Wang C, Deng C, Mao J, Li W, Tu X. Controllable in-situ synthesis of MoS_2/C in plasma-sprayed YSZ coatings: Microstructure, mechanical and tribological properties. *Surface and Coatings Technology*. 2022 Oct 25;448:128895.
32. Liu L, Wang S, Zhang B, Jiang G, Liu H, Yang J, Wang J, Liu W. Present status and prospects of nanostructured thermal barrier coatings and their performance improvement strategies: A review. *Journal of Manufacturing Processes*. 2023 Jul 7;97:12–34.

10 Case Study on the Erosion Wear Performance of ZrO_2-Coated SS-410 Using RSM and TOPSIS

Gurmeet Singh, Pulkit Jain, and Perumal Kumar

10.1 INTRODUCTION

Erosion wear poses a critical challenge in slurry transportation systems, commonly observed in industries such as coal mining, mineral processing, and thermal power generation. When components are exposed to abrasive particles suspended in the slurry, they undergo material degradation, compromising operational efficiency and increasing maintenance requirements (Sedai et al. 2021, Singh et al. 2019a & Desale et al. 2008). Erosion occurs when minute abrasive particles, known as erodents, are transported by water through pipelines, valves, or bends and strike the surface of the target material. This repeated interaction leads to surface degradation through mechanisms such as small craters, microcutting, and plowing. Specific wear mechanisms depend on the nature of the material, with ductile materials typically undergoing plowing and microcutting and brittle materials experiencing crack formation and surface fragmentation (Singh et al. 2024, Gee et al. 2007, Tan et al. 2012 & Aiming et al. 1995).

Furthermore, numerous factors significantly influence the erosion wear rate, including slurry rotational speed, particle concentration, particle shape and size, slurry pH value, exposure duration, and the microstructure, density, and hardness of the substrate material. Numerous researchers have designed various test rigs to not only evaluate the erosion rate performance under diverse conditions but also rank the significance of these influencing parameters on materials (Chandel et al. 2012, More et al. 2017 & Grewal et al. 2013).

Apart from these, many popular methods, including laser cladding, cement-based coatings, and high-velocity oxyfuel (HVOF) thermal spray coatings, are used to protect equipment used in slurry transportation systems from erosion. Literature (Prasanna et al. 2018, Singh et al. 2019b & Ramesh et al. 2010) demonstrates that HVOF coating is particularly effective in mitigating erosion wear as it forms a dense, smooth, adherent, and homogeneous protective layer, outperforming other

DOI: 10.1201/9781003519300-12

coating techniques in terms of durability and performance. Moreover, Hussain et al. (2022) and Haque et al. (2020) have investigated the erosion wear performance of various grades of stainless steel using different test rigs. The results of their detailed research indicate that tungsten carbide (WC) coatings are highly effective in resisting erosion wear, particularly in components or parts exposed to erosive conditions for prolonged durations. Recent research studies also suggest that incorporating rare elements like zirconium oxide (ZrO_2) significantly enhances the erosion and corrosion resistance of stainless steel. This improvement is attributed to its superior hardness, toughness, and enhanced bond strength. Zirconium, being a transition metal, exhibits remarkable thermal stability, catalytic activity, mechanical strength, and biodegradability. These properties make it highly effective in resisting erosion and corrosion wear, enabling its application across various fields, including mechanical and biomedical industries. Moreover, zirconium oxide (ZrO_2) exists in three crystallographic forms: monoclinic, tetragonal, and cubic. The monoclinic phase remains stable at room temperature, transitioning to the tetragonal phase at temperatures above 1170°C and further converting into the cubic phase at 2370°C. Among these phases, the tetragonal phase exhibits the highest hardness and strength (Krishnamurthy et al. 2012, Wang et al. 2024, & Sliem et al. 2023), making it particularly advantageous for applications requiring superior mechanical properties.

Apart from above techniques, many researchers, including Sharma et al. (2019), Purohit et al. (2020), Lakshmi et al. (2021), and Behzadian et al. (2012), have applied different design of experiments techniques such as Taguchi, response surface methodology (RSM), and ANOVA to not only obtain effective solutions but also minimize the number of trials and hence the cost of experimentation. These techniques also help identify the impact of each significant parameter on erosion wear rate in a more accurate and cost-effective manner.

In this chapter, a novel approach incorporating the integration of RSM and technique for order preference by similarity to ideal solution (TOPSIS), followed by Pearson's correlation analysis, has been used to examine the wear rate performance of ZrO_2-coated and uncoated SS-410 stainless steel under various operating parameters. RSM, a mathematical and statistical tool, is used to optimize trials through modeling and problem analysis, while TOPSIS, in combination with Pearson's correlation approach, is used to obtain the closeness score of each experimental trial and rank them accordingly for further identification of the most dominant parameter resulting in maximum erosion rate using correlation analysis.

10.2 METHODOLOGY

10.2.1 Material Selection

The main aim of this chapter is to evaluate the slurry erosion of equipment used in slurry transportation systems such as pumps, pipelines, impellers, and bends. SS-410 stainless steel was selected as the substrate material due to its favorable mechanical properties and inherent resistance to erosion. To further enhance its erosion resistance and hardness, the substrate was coated with zirconium. The material was prepared

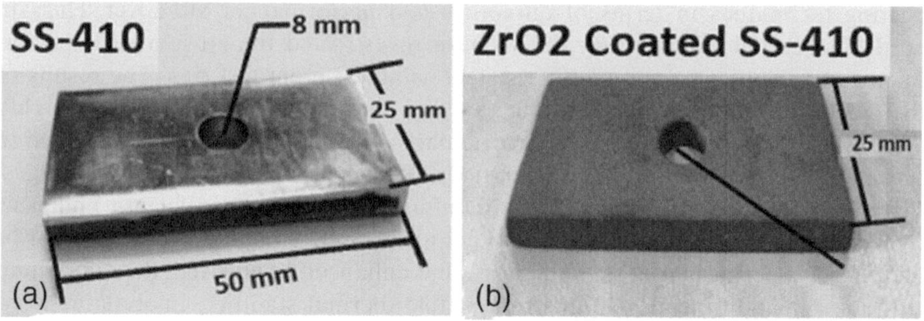

FIGURE 10.1 Specifications of (a) uncoated SS-410 and (b) ZrO_2-coated SS-410.

in rectangular specimens with the dimensions of 50×25×5 mm, featuring an 8-mm centrally drilled hole, as illustrated in Figure 10.1. These dimensions were chosen for the steady and uniform rotation of the specimen inside the slurry pot during testing.

10.2.2 Microhardness Testing

Vickers–Rockwell hardness tester was used to measure the microhardness (HV) of uncoated and ZrO_2-coated SS-410. ZrO_2-coated SS-410 showed significantly higher hardness compared with uncoated SS-410, as evident from the results in Table 10.1. This improvement is attributable to the HVOF coating process, which forms a dense, smooth, and homogeneous layer with minimal porosity on the substrate material, thereby enhancing its durability and hardness. Furthermore, ZrO_2 coating establishes a strong bond with the substrate, ensuring superior hardness and mechanical integrity.

TABLE 10.1
Microhardness of ZrO_2-Coated and Uncoated SS-410

Materials	Indentation Mark	Roughness (μm)	Average Hardness (HV)
Uncoated stainless steel (SS-410)	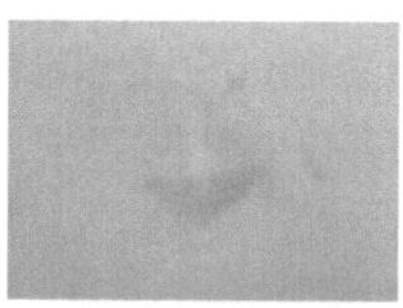	1.53	344
ZrO_2-coated SS-410	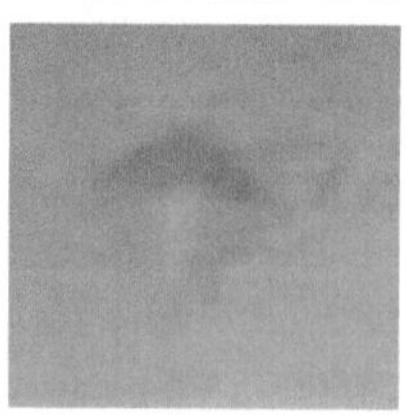	2.24	689

10.2.3 Coating Powders and Techniques

The coating powder used in this study was thoroughly blended using a mixer to obtain a smooth and uniform coating while maintaining a particle size of 15–30 μm. The coating powder, consisting of tungsten (WC), cobalt (Co), chromium (Cr), and 3% zirconium (ZrO_2), was procured from Metalizing Equipments, Rajasthan. The final composition was prepared as WC-10Co-4Cr+3% ZrO_2, resulting in a unique and homogeneous coating mixture. Coating experiments were conducted as illustrated in Figure 10.2 at Harsha Welders, Panchkula, using an HVOF system of spraying with an HP-2700 torch. The flow of oxygen, fuel, and air was precisely controlled using an HVOF control panel to achieve optimal coating conditions, as shown in Table 10.2. Figure 10.3 presents both an SEM image of ZrO_2 powder and energy-dispersive spectroscopy (EDS) analysis.

10.2.4 Erodent Classification

The fly ash erodent material used in this study was procured from Guru Gobind Singh Super Thermal Plant, Ropar. The slurry mixture (fly ash and water) at varying concentrations was then used for erosion wear testing. A British sieve shaker

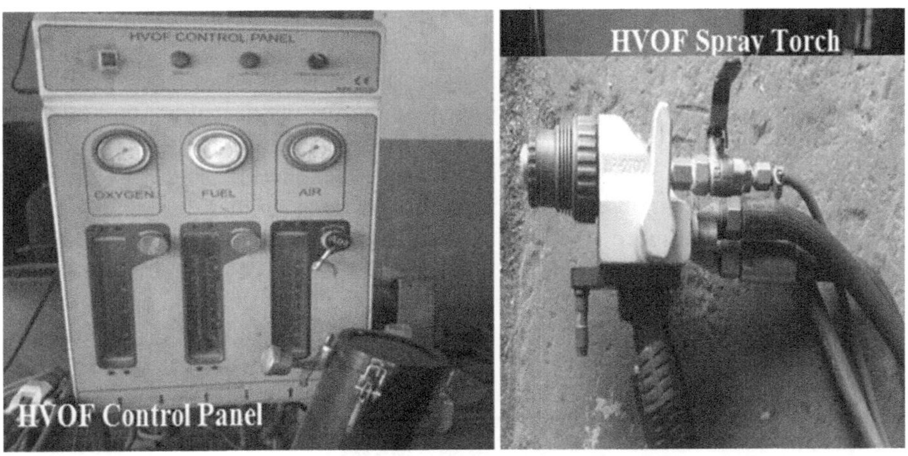

FIGURE 10.2 HVOF thermal spray control panel and spray torch.

TABLE 10.2
Optimal Process Parameters for HVOF Thermal Spray Coating

Interface	Spray Injection Gap (mm)	Powder Deposition Rate (g/min)	Flow Rate (L/min)	Fuel Pressure (kPa)
Air	145	35	550	12
Oxygen			250	7
Fuel			60	5

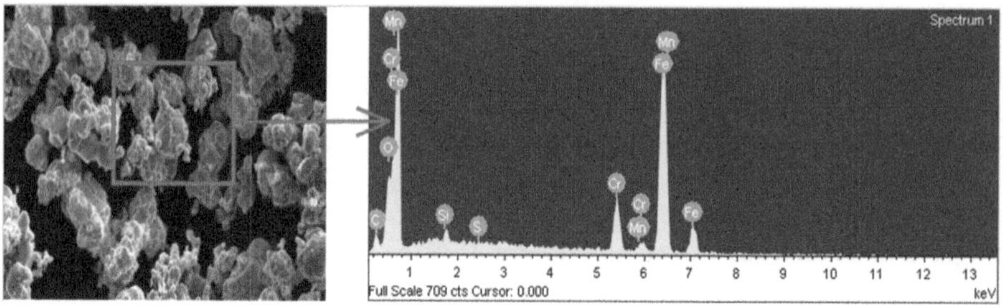

FIGURE 10.3 Microstructural and compositional analysis of WC-10Co-4Cr+3% ZrO_2 powder using SEM and EDS.

was used to classify the particles based on their size to carry out particle size distribution analysis of the erodent.

Furthermore, several researchers, including Walker and Hambe (2015), Bahadur and Badruddin (1990), and Singh et al. (2020), have investigated the significance of erodent shape, particularly the circularity factor (CF), in the erosion wear of slurry transportation systems, which is calculated as follows:

$$C.F = \frac{4\pi A}{P^2} \tag{10.1}$$

where P and A represent the perimeter and area, respectively, measured in meters and square meters. Since the obtained CF value was 0.94 (close to 1), it indicates that fly ash particles exhibit a nearly cenospherical shape, as shown in Figure 10.4a. Based on observations from the particle size distribution analysis of fly ash, as shown in Figure 10.4b, the particle size ranges between 53 and 250 μm. The percentage of particles retained in each mesh size is calculated and depicted in Figure 10.4b.

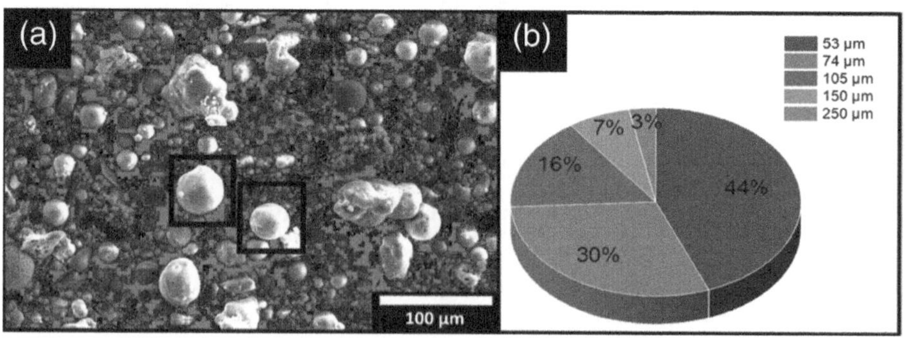

FIGURE 10.4 (a) Mixed SEM sample of fly ash particles and (b) particle size distribution of fly ash particles.

10.2.5 Response Surface Methodology

A RSM (Sharma et al. 2012 & Singh Sidhu et al. 2024) analysis using a face-centered design with two levels for each factor was applied in this chapter. The factors considered include particle size, rotational speed, time duration, and particle concentration. These factors are represented by R (rotational speed, rpm), C (particle concentration, %), T (time duration, min), and S (particle size, µm). For each factor, the upper and lower levels are coded as +1 and 0, respectively.

10.2.6 Technique for Order Preference by Similarity to Ideal Solution

TOPSIS is a decision-making method used to determine the best alternative from a set of options based on multiple criteria (Behzadian et al. 2012). It is widely used in decision-making processes such as supplier selection, product evaluation, and project selection. In order to determine the major dominating factor affecting mass loss, TOPSIS method has been applied in this chapter. It includes the major steps as depicted in the flowchart shown in Figure 10.5.

The description of all the steps involved in applying a combination of TOPSIS and correlation approach is explained below:

i. **Decision Matrix Formulation**: The major parameters considered, namely concentration, rpm, size, and time duration, are arranged to form the decision matrix.
ii. **Decision Matrix Normalization (N)**: The following relation is used to normalize each element of the decision matrix formed in step 1:

$$N_{ij} = \frac{Y_{ij}}{\sqrt{\sum_{i=1}^{n} Y_{ij}^2}} \tag{10.2}$$

where *N* and *X* represent normalized and original value for the *i*th row and *j*th column and '*n*' refers to total samples.

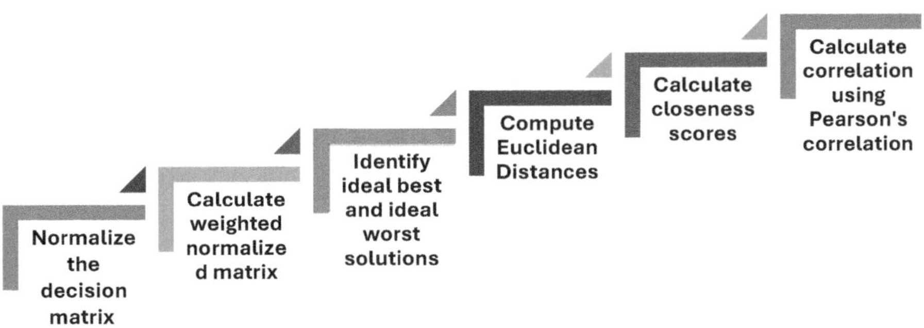

FIGURE 10.5 Steps involved in computing the major dominating factor affecting output parameters using a combination of TOPSIS and correlation.

iii. **Assigning Weights to Normalized Matrix (*W*)**: Multiplying each normalized value with a specific weight value gives higher significance to some parameters over others. Here, weights are assigned equally in this chapter:

$$W_{ij} = N_{ij} * w_j \tag{10.3}$$

where w_j represents the weight assigned to each individual parameter (here $w = 0.25$ each).

iv. Find the ideal best and ideal worst solutions by considering minima for cost parameters and maxima for beneficial parameters for the best case and vice versa for the worst case.

v. **Euclidean Distance (E.D)**: Compute this distance from ideal best ($E.D_{i+}$) and ideal worst solutions ($E.D_{i-}$).

vi. Find the relative closeness score (*C*) of each sample with respect to the ideal solution:

$$C = \frac{E.D_{i-}}{E.D_{i+} + E.D_{i-}} \tag{10.4}$$

vii. Rank the samples on the basis of their relative closeness score to identify top-ranked samples (here five samples) having the maximum impact on output (mass loss). The sample with the highest closeness score is found to be nearest to the ideal solution and is given the first rank.

viii. Compute correlation between factors (input parameters) and output (mass loss) using Pearson's correlation (C.S. Lai et al. 2019) using the following relation:

$$p_r = \frac{(P_i - \bar{P})(Q_i - \bar{Q})}{\sqrt{(P_i - \bar{P})^2 (Q_i - \bar{Q})^2}} \tag{10.5}$$

where

p_r = coefficient of Pearson's correlation
P_i and Q_i = individual data points for two variables being compared
$\bar{P}$ and $\bar{Q}$ = mean values of *P* and *Q*, respectively.

The value of p_r ranges from −1 to 1 where '1' indicates perfect positive correlation, '−1' indicates perfect negative correlation, and '0' value indicates no correlation.

10.3 RESULTS AND DISCUSSION

10.3.1 Erosion Wear Rate

The erosion wear behavior of coated and uncoated SS-410 stainless steel, using fly ash as the erodent, was analyzed using RSM, as presented in Table 10.3. Experiments were conducted based on the RSM matrix, varying the influencing

TABLE 10.3
Predicted Mass Loss for Different Influencing Parameters Using RSM Approach and TOPSIS Closeness Score Results

Alternative	R (rpm)	C (%)	S (μm)	T (min)	Predicted Mass Loss	Closeness Score
Case 1	600	65	150	60	0.00528	0.5036
Case 2	600	65	150	40	0.00179	0.407
Case 3	800	55	150	50	0.00528	0.466
Case 4	700	45	53	30	0.0017875	0.0658
Case 5	500	35	53	60	0.00179	0.2643
Case 6	900	65	225	200	0.00528	0.7954
Case 7	800	65	150	80	0.00528	0.578
Case 8	1100	65	250	200	0.00528	0.7842
Case 9	800	65	190	200	0.00179	0.7519
Case 10	1100	65	250	200	0.00179	0.8563
Case 11	800	65	200	50	0.00179	0.5228
Case 12	600	65	150	50	0.00179	0.3837
Case 13	1000	65	250	100	0.00179	0.6838
Case 14	800	65	200	100	0.00528	0.6722
Case 15	900	65	225	200	0.00179	0.8389
Case 16	600	65	125	50	0.00528	0.3787
Case 17	1000	65	225	100	0.00528	0.7243
Case 18	700	55	125	60	0.00528	0.4437
Case 19	600	55	90	50	0.00528	0.315
Case 20	700	65	90	50	0.00528	0.2487
Case 21	700	45	150	60	0.00179	0.4545
Case 22	700	45	90	40	0.00528	0.2255
Case 23	500	65	90	40	0.00528	0.3046
Case 24	700	55	53	30	0.00528	0.175
Case 25	600	65	100	40	0.00179	0.3093
Case 26	1000	65	250	200	0.00179	0.8114
Case 27	700	55	53	40	0.00528	0.178
Case 28	700	65	53	40	0.00528	0.2164
Case 29	1000	65	200	60	0.00528	0.6519
Case 30	900	65	200	60	0.00179	0.6332
Case 31	800	65	200	80	0.00528	0.63

parameters accordingly. The wear rate was quantified in terms of mass loss, which results from the interaction between the surface of the erodent and the suspended slurry particles.

10.3.2 TOPSIS Results

The closeness score for each of the cases considered in the RSM matrix are obtained after applying the TOPSIS method as shown in Table 10.3.

TABLE 10.4
Top Five Results by TOPSIS Based on Closeness Score

Rank	Parameters (Rotational Speed, Concentration, Size, and Time)	Predicted Mass Loss	Closeness Score	Alternative
1	1100, 65, 250, 200	0.00179	0.8563	Case 10
2	900, 65, 225, 200	0.00179	0.8389	Case 15
3	1000, 65, 250, 200	0.00179	0.8114	Case 26
4	900, 65, 225, 200	0.00528	0.7954	Case 6
5	1100, 65, 250, 200	0.00528	0.7842	Case 8

The ranking of the cases (top five) based on their relative closeness to the ideal solution is tabulated in Table 10.4, where case 10 (1100 rpm, 65% concentration, 250 μm size, and 200 min time duration) is found to have the highest closeness score.

10.3.3 Correlation Results

Results of the Pearson's correlation are given in Table 10.5, indicating the individual impact of all major parameters considered on the predicted mass loss.

This table shows that particle size has a correlation of −0.2199 (moderate negative correlation), followed by rotational speed with a weak negative correlation of −0.1666, particle concentration with a further weaker negative correlation of −0.0879, and time duration with a very weak negative correlation of −0.0865. The results indicate that erosion wear increases nonlinearly with increasing influencing parameters. It can be concluded that erodent size was the most influential parameter resulting in the highest erosion rate, followed by rpm, slurry concentration, and time period, as shown in Table 10.5 and Figure 10.6. The hierarchy of the impact is as follows: erodent size > slurry concentration > rpm > time period.

The experiments were conducted with an error bar range of ±10%. Furthermore, Figure 10.6 clearly demonstrates that the erosion wear of ZrO_2-coated SS-410 is consistently lower than that of uncoated SS-410 in all test cases, which is due to the high resistance offered by ZrO_2 coating against the suspended particles. In addition,

TABLE 10.5
Pearson's Correlation Results to Identify the Most Influential Parameter

Parameter Correlation Matrix	R	C	S	T	Predicted Mass Loss
Rotational speed	1	0.52389	0.98618	0.87095	−0.16663
Concentration	0.52389	1	0.54968	0.39561	−0.08786
Particle size	0.98618	0.54968	1	0.85678	−0.21991
Time duration	0.87092	0.39563	0.85670	1	−0.08650
Predicted mass loss	−0.1666	−0.0879	−0.2199	−0.0865	1

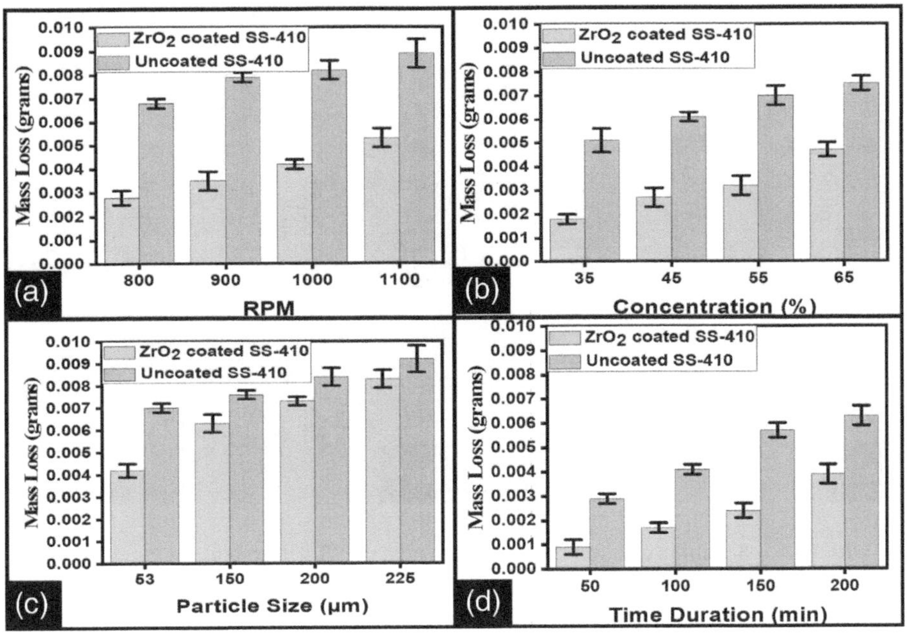

FIGURE 10.6 Impact of influencing parameters on the erosion wear rate of ZrO_2-coated and uncoated steel.

ZrO_2-coated SS-410 (689 HV) exhibits greater hardness than uncoated SS-410 (344 HV), leading to its superior wear resistance. Lindgren et al. (2015), Patel et al. (2020), and Adam et al. (2023) also reported similar observations. It can be easily concluded that ZrO_2 coating is far superior in reducing erosion wear compared with uncoated steel. Tandel et al. (2023), Llewellyn et al. (2004), and Peng et al. (2021) also studied the similar impact force phenomena, whose results are consistent with those obtained in this chapter. Regarding the most dominant parameter, particle size showed the most pronounced effect on erosion wear rate for both coated and uncoated steel. Particles with larger size showed a greater impact than smaller particles due to their higher velocity and momentum during whirling, resulting in stronger impingement on the target surface.

10.3.4 Erosion Wear Mechanism

In order to examine the erosion wear behavior of both coated and uncoated SS-410 under maximum impact conditions, advanced SEM was utilized, as depicted in Figure 10.7a and b. It can be easily observed that both exhibited erosion at different locations, but the material loss obtained in uncoated SS-410 is significantly higher than that of its ZrO_2-coated counterpart. The dominant wear mechanism identified was microcutting in the case of uncoated SS-410, resulting in its substantial surface degradation. In addition, localized deformations caused by high-velocity suspended particles resulted in the creation of craters at various spots due to their impact and rebound.

FIGURE 10.7 Scanning electron microscopy of (a) bare SS-410 and (b) ZrO_2-coated SS-410 at maximum influencing parameters.

On the other hand, a dominant ductile wear mechanism was observed in the case of ZrO_2-coated SS-410. Due to continuous exposure to fly ash particles, the surface exhibited signs of plowing, followed by microcutting and crack formation in specific regions. However, the protective coating effectively minimized material loss compared with the uncoated substrate. Researchers Singh et al. (2024) and Wang et al. (2024) also reported similar findings.

10.4 CONCLUSIONS

This chapter investigated and compared the erosion wear performance of ZrO_2-coated and uncoated SS-410 using fly ash as the erodent. Analysis was carried out using the integration of approaches, including RSM, TOPSIS, and Pearson's correlation. The following important conclusions are drawn:

- ZrO_2-coated SS-410 showed significantly better erosion resistance than its uncoated counterpart (344 HV), primarily due to its higher hardness (689 HV).
- The hierarchy of the impact of parameters was as follows: erodent size > slurry concentration > rpm > time period.
- Erosion wear showed a nonlinear increase with an increase in influencing parameters for both coated and uncoated SS-410.
- SEM examination confirmed that the erosion wear mechanism was predominantly ductile in both uncoated and coated specimens.
- ZrO_2 coating was effectively applied to SS-410 using the HVOF thermal spray technique.

REFERENCES

Adam, I., Singh, G., & Gill, H. S. (2023). CFD erosion wear simulation in three. In *Mechatronics and Automation Technology: Proceedings of ICMAT 2022* (Vol. 33, p. 357). International Conference on Mechatronics and Automation Technology, Wuhan, China, on October 29–30.

Aiming, F., Jinming, L., & Ziyun, T. (1995). Failure analysis of the impeller of a slurry pump subjected to corrosive wear. *Wear*, 181, 876–882.

Bahadur, S., & Badruddin, R. (1990). Erodent particle characterization and the effect of particle size and shape on erosion. *Wear*, 138(1–2), 189–208.

Behzadian, M., Otaghsara, S. K., Yazdani, M., & Ignatius, J. A. (2012). A state-of the-art survey of TOPSIS applications. *Expert Systems with Applications*, 39(17), 13051–13069.

Chandel, S., Singh, S. N., & Seshadri, V. (2012). Experimental study of erosion wear in a centrifugal slurry pump using coriolis wear test rig. *Particulate Science and Technology*, 30(2), 179–195.

Desale, G. R., Gandhi, B. K., & Jain, S. C. (2008). Slurry erosion of ductile materials under normal impact condition. *Wear*, 264(3–4), 322–330.

Gee, M. G., Gant, A., & Roebuck, B. (2007). Wear mechanisms in abrasion and erosion of WC/Co and related hardmetals. *Wear*, 263(1–6), 137–148.

Grewal, H. S., Agrawal, A., & Singh, H. (2013). Design and development of high-velocity slurry erosion test rig using CFD. *Journal of Materials Engineering and Performance*, 22, 152–161.

Haque, U., Singh, G., Gill, H. S., & Sehgal, S. S. (2020). Tribo-performance of 80% WC+ 20% TIO_2 HVOF thermal spray coating on SS-404. *Materials Today: Proceedings*, 33, 1680–1689.

Hussain, A., Singh, G., & Gill, H. S. (2022). Solid particle erosion behaviour of industrial epoxy resin composite against different parameters. *Materials Today: Proceedings*, 48, 1492–1496.

Krishnamurthy, N., Murali, M. S., Venkataraman, B., & Mukunda, P. G. (2012). Characterization and solid particle erosion behavior of plasma sprayed alumina and calcia-stabilized zirconia coatings on Al-6061 substrate. *Wear*, 274, 15–27.

Lai, C. S., Tao, Y., Xu, F., Ng, W. W., Jia, Y., Yuan, H., … & Locatelli, G. (2019). A robust correlation analysis framework for imbalanced and dichotomous data with uncertainty. *Information Sciences*, 470, 58–77.

Lakshmi, K. J., Sogalad, I., Basavarajappa, S., & Raghavendra, C. R. (2021). Optimization of erosive wear parameters on NiCrAlY based nano composite coating by RSM. *Materials Today: Proceedings*, 46, 763–766.

Lindgren, M., Suihkonen, R., & Vuorinen, J. (2015). Erosive wear of various stainless steel grades used as impeller blade materials in high temperature aqueous slurry. *Wear*, 328, 391–400.

Llewellyn, R. J., Yick, S. K., & Dolman, K. F. (2004). Scouring erosion resistance of metallic materials used in slurry pump service. *Wear*, 256(6), 592–599.

More, S. R., Bhatt, D. V., & Menghani, J. V. (2017). Study of the parametric performance of solid particle erosion wear under the slurry pot test rig. *Tribology in Industry*, 39(4), 471.

Patel, M., Kumar, A., Pardhi, B., & Pal, M. (2020). Abrasive, erosive and corrosive wear in slurry pumps–A review. *International Research Journal of Engineering and Technology*, 7(3), 2188–2195.

Peng, G., Fan, F., Zhou, L., Huang, X., & Ma, J. (2021). Optimal hydraulic design to minimize erosive wear in a centrifugal slurry pump impeller. *Engineering Failure Analysis*, 120, 105105.

Prasanna, N. D., Siddaraju, C., Shetty, G., Ramesh, M. R., & Reddy, M. (2018). Studies on the role of HVOF coatings to combat erosion in turbine alloys. *Materials Today: Proceedings*, 5(1), 3130–3136.

Purohit, A., Satapathy, A., & Swain, P. T. R. (2020). Analysis of erosion wear behavior of LD sludge filled polypropylene composites using RSM. In *Materials Science Forum* (Vol. 978, pp. 222–228). Trans Tech Publications Ltd, Switzerland.

Ramesh, M. R., Prakash, S., Nath, S. K., Sapra, P. K., & Venkataraman, B. (2010). Solid particle erosion of HVOF sprayed WC-Co/NiCrFeSiB coatings. *Wear*, 269(3–4), 197–205.

Sharma, R., Singh, G., & Kumar, S. (2019). Optimization and erosion wear response of HVOF coated pipeline material at different impingement angles. *Indian Journal of Science and Technology*, 12(8), 1–8.

Sharma, S., Dwivedi, D. K., & Jain, P. K. (2012). Study and model development of erosive wear by RSM. *Proceedings of the Institution of Mechanical Engineers, Part J: Journal of Engineering Tribology*, 226(1), 57–70.

Sliem, M. H., Fayyaz, O., Shakoor, R. A., Bagherifard, S., Mansoor, B., Abdullah, A., & Mohamed, A. M. A. (2023). The influence of different preparation methods on the erosion behavior of NiP-ZrO_2 nanocomposite coating. *Tribology International*, 178, 108014.

Singh, G. (2024). Effect of suspended particles on erosion characteristics of impeller steels utilizing combined CFD and DOE approach. *Industrial Lubrication and Tribology*, 77(2), 202–210.

Singh, G., Jain, P., & Gill, H. S. (2024). Tribo-erosion analysis of HVOF colmonoy-88-coated pump impeller steel in coal ash particle slurries. *Journal of Tribology*, 146(8), 081702.

Singh, J., Kumar, S., & Mohapatra, S. K. (2019a). Erosion wear performance of Ni-Cr-O and NiCrBSiFe-WC (Co) composite coatings deposited by HVOF technique. *Industrial Lubrication and Tribology*, 71(4), 610–619.

Singh, J., Kumar, S., & Mohapatra, S. K. (2019b). Tribological performance of yttrium (III) and zirconium (IV) ceramics reinforced WC–10Co4Cr cermet powder HVOF thermally sprayed on X2CrNiMo-17-12-2 steel. *Ceramics International*, 45(17), 23126–23142.

Singh, J., Kumar, S., & Mohapatra, S. K. (2020). Erosion tribo-performance of HVOF deposited stellite-6 and colmonoy-88 micron layers on SS-316L. *Tribology International*, 147, 105262.

Singh Sidhu, R., Singh, G., & Gill, H. S. (2024). Evaluation of the anti-erosion performance of PLA and ABS materials with two different micro textures. *Pigment & Resin Technology*, 53(5), 587–594.

Sedai, A., Singh, G., Dhakal, R., Khatiwada, A., Khanal, K., Kumal, B., … & Mishra, A. K. (2021, November). Technical and economic feasibility of a fully solar-powered airport in Nepal. In *2021 IEEE International Conference on Intelligent Systems, Smart and Green Technologies (ICISSGT)* (pp. 122–127). IEEE, Visakhapatnam, India.

Tan, Y., Zhang, H., Yang, D., Jiang, S., Song, J., & Sheng, Y. (2012). Numerical simulation of concrete pumping process and investigation of wear mechanism of the piping wall. *Tribology International*, 46(1), 137–144.

Tandel, R. R., Patel, R. N., & Jain, S. V. (2023). Correlation development of erosive wear and silt erosion failure mechanisms for pump as turbine. *Engineering Failure Analysis*, 153, 107610.

Wang, S., Li, Z., Zhang, H., Liang, X., He, W., Chen, Y., … & Li, B. (2024). Study on the sand erosion resistance of ZrN and ZrAlSiN coatings. *Surface and Coatings Technology*, 488, 131081.

Walker, C. I., & Hambe, M. (2015). Influence of particle shape on slurry wear of white iron. *Wear*, 332, 1021–1027.

Section III

Energy and Catalytic Applications

11 Catalytic Wonders
The Power of Zirconium in Modern Chemistry

Hardeep Kaur, Akshita Singh, Milind Shrivastava, Shinar Athwal, Kashish Garg, and Noushi Zaidi

11.1 INTRODUCTION

Zirconium (Zr) is an exotic element that can be extracted from zircon ($ZrSiO_4$) and baddeleyite (ZrO_2). Since zirconium is an abundant metal compared to other metals like zinc, copper, lead, or nickel, it can be widely used for practical applications [1]. In nature, zirconium is formed by two ores, free oxides and silicate oxides. Zirconium has properties such as a large surface area, high thermal stability, high charge, large size, high mechanical stability, acid-base characteristics, and tetravalent characteristics [2]. These properties of zirconium can be used for various purposes as an adsorbent, in dental and bone implants, PET scan agents, chelators, radiotracers, and catalysts [3–8]. Look into the catalytic properties of zirconium-based materials and find the wonders they can do. For ages, scientists have been exploring zirconium for catalytic applications because of certain characteristics, oxidizing–reducing ability, short ionic radius, ability to accept lone pairs of electrons, and spherical geometry of electron orbital [9]. Zirconium can be used as a catalyst in various forms like Zr Metal-Organic Framework, Zr phosphates, Zr nitride, and Zr oxide [10–12]. For instance, Zr nitride is a better catalyst compared to platinum as it is cost-effective, and enhances the applicability of electrochemical energy platforms [13,14]. Zr-based materials have active sites to execute chemical reactions and these sites are responsible for their catalytic activity. The sites allow complex reactions like hydrogenation and oxygenation to take place, catalyzing several compounds' conversion processes [15,16].

In this chapter, we have discussed the surface properties of Zr which contribute toward catalysis and which are tunable acidic and basic sites. The Lewis acidic and basic sites of Zr play an important role in the catalytic reactions [17–20]. Meerwein-Ponndorf-Verley (MPV)-reduction reaction is a catalytic reaction that involves the conversion of aldehydes into ethanol to recover and synthesize various organic products [21,22]. Also, zirconium is modified to enhance the unit's overall catalytic performance. There Zr has different types of modification and the most common modification forms are mesoporous Zr, titanium dioxide-augmented Zr, iron-precipitated Zr, sulfated Zr, and ceria-zirconia unit (Figure 11.1) [23–25]. All these Zr modifications have shown positive results. Further, the integration of

DOI: 10.1201/9781003519300-14

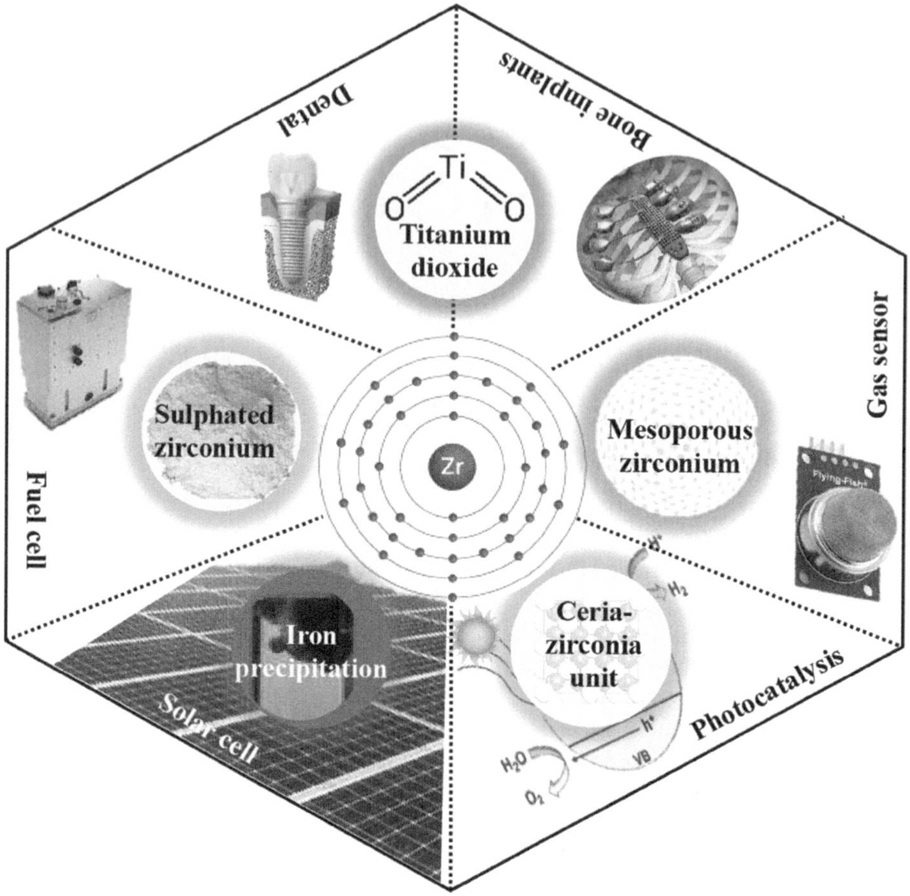

FIGURE 11.1 Commonly used modification products are titanium dioxide, mesoporous zirconium, ceria-zirconia unit, iron precipitation, and sulfated zirconium. (Created by Authors.)

nanotechnology has increased the practical utility of zirconium-based materials as their incorporation increases the surface area and decreases the band gap of the composite [14]. Widely used nanoparticles are gold, iron, platinum, molybdenum, and hafnium [26–28]. Considering the application part of zirconium catalysts, it can be used for various purposes such as the synthesis of organic compounds, green production of biofuel, ketonization of industrial waste, free fatty acid esterification of oils, and development of microbial fuel cells (Figure 11.1) [29–33]. Thus zirconium-based catalyst has a significant practical utility and can be used to carry out the recovery, synthesis, and adsorption of various compounds only contributing toward sustainable development.

Zr can be used for various purposes such as the synthesis of organic compounds, green production of biofuel, ketonization of industrial waste, free fatty acid esterification of oils, and development of microbial fuel cells.

11.2 SURFACE PROPERTIES OF ZIRCONIUM THAT CONTRIBUTE TO THE CATALYTIC BEHAVIOR

The ratio of acidic-basic sites on the surface of zirconium greatly impacts the catalytic nature of the Zr-based unit (Figure 11.2). The significance of these sites has been discussed below.

11.2.1 Acidic Sites

Daniel et al. stated that yield and selectivity of desired propene can be achieved when acidity tuning of the ZSM-5 catalyst is achieved for the acidic properties of zirconium. The acidity assessment identified both Lewis with Brønsted acids present across weak to medium acidic sites on the surface of the catalyst. Due to the elevation of Zr loading, accumulation of Brønsted acidic sites was reduced increasing Lewis acidity. This in turn resulted in obstruction in propene over-oxidation and an increase in propene selectivity. The best yield of propene was shown by Zr/ZSM-5 within the catalysts as they converted 57.19% of propane and had 75.54% propene selectivity with high stability [34]. El-Aal et al. reflected in their study that different physicochemical techniques were employed for the synthesis and characterization of catalysts of two types of series. To determine the catalysts' acidity, isopropyl alcohol dehydration was carried out along with active adsorption of 2,6-dimethyl pyridine with pyridine. The proficiency of the catalyst was checked in a reactor called a fixed bed to convert methanol to DME. At a temperature of 400°C, 95% yield of DME was maximized as compared to $SO_4^{2-}/ZrO_2/SCBFA$ catalyst and maintained the same regulation for 90 hours showing amazing stability. Acidity along with surface area

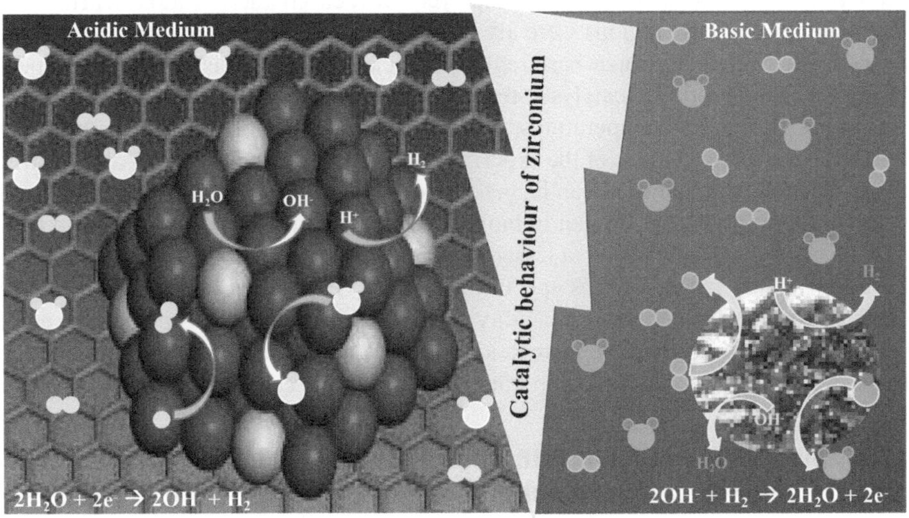

FIGURE 11.2 The ratio of acidic–basic sites on the surface of zirconium greatly impacts the catalytic nature of the Zr-based unit. (Created by Authors.)

were important factors that determined the performance of the catalyst. Therefore, SCBFA-Zr altered along with SO_4^{2-} are cheap, sustainable, potent, and reusable catalysts suited for DME production [35]. Similarly, Wu et al. reported that sulfated zirconia is an effective super acid catalyst for the formation of n-butane from i-butane, and can be used as a substitute for Pt-catalysts. A method called one-step was used to prepare Cu-SO_4^{2-}/ZrO_2 and its performance was compared in contrast to commercial Pt-Cl/Al_2O_3, with mean selectivity of n-butane as 86.1% and mean conversion of i-butane as 43.8% and a steady n-butane yield for 120 hours. So, Lewis active regions act as a descriptor for correlating the yield of n-butane and give an account for the volcano-type relation of n-butane and Cu content. This work illustrated the practicality of Cu tempering and gave an overlook of the promotion procedure which might be expanded over other super acid catalysts other than sulfated zirconia [36]. Thus, acidic sites play a critical role in the catalytic performance of Zr.

11.2.2 Basic Sites

For the basic properties of zirconium, Hatta et al. described a successful method of production for FSZr using a micro-emulsion method usually used for the CO Methanation procedure. The results of the performance depict the unique performance of the catalyst FSZr in contrast to industrially available ZrO_2. The extraordinary actions are due to the FSZr's different fibrous form, a wide area of surface, and unique basic properties. The increased surface area provides a better approach to active regions along with strong base characteristics for easy reactant adsorption. The efficiency of CO methanation is increased due to these factors when combined [37]. Further, Song et al. in their work developed new basic catalysts for converting glycerol to produce GC and GD. KF was loaded over La-Zr catalyst to formulate these catalysts and various methods were employed for their characterization. During this interaction, weak base regions were allocated to surface OH groups when LaOF was formed, whereas strong-super base regions associated with Lewis base were formed. Using the latest produced catalysts, the generation of GD by decarboxylation of GC was impractical at low temperatures. A credible mechanism of reaction has been put forward based on the results of the experiment and characterization [38]. Additionally, Ma et al. explained that BZC can be generated in huge amounts at the industrial level to study the transfer of hydrogen in biomass-obtained carboxides and displayed brilliant characteristics of catalyst at lesser temperatures mainly during the production of FFA from FUR. When ATR-IR spectrometry and XPS were performed for different catalysts, results showed that the acidity of BZC was strongest as compared to other catalysts. Also, the kinetics of the reaction were taken into account [39]. Therefore, the basic sites of the Zr surface play an important part in the catalytic activity.

11.3 SIGNIFICANCE OF THE ZIRCONIUM CATALYST FOR MEERWEIN-PONNDORF-VERLEY-REDUCTION REACTION

Meerwein, Schmidt, and Verley in the year 1925 demonstrated that aluminum ethoxide can catalyze the reduction reaction of aldehydes in ethanol presence following its stoichiometric oxidation to aldehyde which is its analog (Figure 11.3).

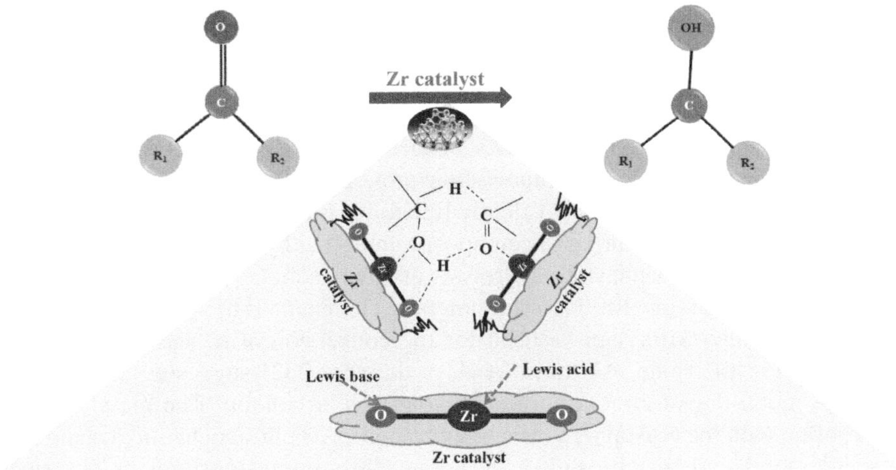

FIGURE 11.3 Illustration depicting the role of Zr in MPV reduction reaction. (Created by Authors.)

Ponndorf concluded the same in the upcoming years. In the 1950s, metal alkaloids and zirconium showed successful tests in this reduction reaction. Thereby, the MPV reaction has been examined using several catalysts ranging from zirconium and aluminum alkaloids to exotic rare-earth metals [40]. Monoclinic zirconium has more intrinsic activity relative to other zirconium phases owing to an ample and adequate reactant fragment distribution on the surface of the catalyst revealing a better structure–activity relationship for MPV reaction along with Zr-beta zeolite. Optimizing the crystallography of Zr material, a simple and cost-effective catalyst can be obtained for MPV reaction [41]. Temperature-programmed-desorption (TPD) and Fourier transform infrared spectroscopy estimations indicated that ZrO_2 has an ample proportion of base sites that are responsible for the activation of methyl groups present at 2-propanol which is bonded to Lewis acid sites. Isotopic-2-propanol suggested that the complementary effect of Lewis acid and base sites along with the density and strength of ZrO_2 enhanced the MPV reduction reaction over zirconium-based materials [42]. To achieve an efficient cascade of MPV reaction, certain factors are required namely, the hydrophobic catalyst having large microspores which combine dehydration activities and hydrogen transfer. To get such conditions, Zr-zeolite has proven to be a better zirconium substrate as it has a higher ratio of open/closed zirconium sites and hydrophobicity [43]. To catalyze the MPV reduction of cyclohexanone, Li et al. in the year 2015 prepared a partially crystalline zircono-silicate derived from amorphous zircono-silicate. This crystalline form demonstrated high reactivity in the reduction reaction of cyclohexanone in the presence of 2-propanol. The critical catalytic factors of this material are mesopore volume, amount of Lewis acid, and surface hydrophobicity. Also, zircono-silicate exhibited the ability of water tolerance [44]. Likewise, Li et al. in the year 2020 fabricated a heterogeneous catalyst comprising Zr, Lanthanum, and beta zeolite to reduce cyclohexane via MPV reaction.

Lanthanum doping did not alter the concentration of acid sites on the surface of zirconium; rather, it increased the synergistic effect between basic and acidic sites ratio of the solid. 1 wt% of La-Zr-Beta depicted the best catalytic reaction leading to 95% conversion of cyclohexanone and 995 of cyclohexanone selectivity [45]. To enhance the catalytic activity of zirconium, Koreniuk et al. in the year 2015 proposed the use of a continuous-flow monolithic silica-based micro-reactor. It was indicated that Zr^{4+} ions covalently linked to silica while ending with ligands of isopropoxy/hydroxo making the composite catalytically more active than simply supported zirconium. Undesired hydrolysis of reactive metals was controlled in the micro-reactor achieving 100% conversion of cyclohexane [46]. Not only for cyclohexanone but also MPV can be used for the conversion of α, and β-unsaturated aldehydes. For the same, Akinnow et al. in the year 2021 suggested the incorporation of metal-doped zirconium and mesoporous zirconium. The metal dopants incorporate with the crystal structure of Zr changing its physicochemical characteristics like total acid-basicity and surface area. Chromium deposition on Zr formed the most sustainable and stable catalytic unit for aldehyde transformation [47]. Further, Zhang et al. demonstrated an enhanced method to incorporate Zr with beta zeolite via the Mechano chemistry-based liquid synthesis (MCLA) process. Through this route, the authors were able to develop high-tolerance material which consequently enhanced the catalytic activity of the Zr unit [48]. Focusing on a renewable approach, Leng et al. in the year 2019 fabricated a Zr-tannin catalytic unit. The characterization technique showed that the unit contained robust Lewis acid–base sites and a strong organic–inorganic framework. This improved catalytic version of Zr can be used to reduce a wide range of carbonyl compounds into alcohols in the presence of 2-propanol [49]. Thus, undoubtedly, the Zr catalyst has a significant role in MPV reduction reaction as it is responsible for recovering and synthesizing many organic compounds which hold industrial relevance. Zr can catalyze the reduction reaction of aldehydes in ethanol presence following its stoichiometric oxidation to aldehyde which is its analog.

11.4 MODIFICATIONS OF ZIRCONIUM TO ENHANCE THE CATALYSIS

Zirconium is reinforced with various compounds and elements to improve the overall activity of the developed unit. Commonly used modification products are titanium dioxide, mesoporous zirconium, ceria-zirconia unit, iron precipitation, and sulfated zirconium [50] which have been briefly discussed below and diagrammatically represented in Figure 11.3. The chemical properties and catalytic activity achieved due to the modification of the Zr surface have been discussed in Table 11.1.

11.4.1 Titanium Dioxide

Titanium dioxide has emerged as an efficient modification material (Figure 11.4). Qi et al. described a method of cathode catalyst preparation of microbial fuel cell (MFC) using a hydrothermal reaction by using S-TiO_2 combined with Zr based on MOF-808. The size of the particle was around 5 μm and the particle of S-TiO_2 was

TABLE 11.1
An Insight into the Zr Modifications and Their Properties

Modification Element	Synthesis	Chemical Properties	Zirconium Type	Catalytic Activity	References
Titanium dioxide	Hydrothermal reaction	The size of the particle was around 5 µm. Bonds like O-H, Zr-O, and C-O exist in S-TiO_2@MOF-808.	Zr based on MOF-808	The catalytic stability of S-TiO_2@MOF-808 was increased due to the distortion in the framework of MOF-808 by S-TiO_2.	[51]
Mesoporous zirconium	Addition of Zr^+ sulfation	3-D interrelated mesopores are present along with strong Brønsted acid regions.	Zr-KIT-6	During hydrolysis of cellulose, catalyst-sulfated Zr-KIT-6 shows high catalytic activity due to strong Brønsted acid regions, which immediately provide photons in cellulose for cleaving β-1,4 glycosidic bonds.	[52]
Ceria-zirconia unit	EISA	Low-crystalline nature of the catalysts along with more surface area.	Zr/CeO_2-Al_2O_3	The soot oxidation was significantly increased by the catalyst.	[53]
Iron precipitation	FTS	A unique solvent-independent solid-state.	Fe/ZrO_2	The performance of the FTS was increased with a high rate of CO conversion and a low selectivity rate of CH_4. Formation of χ-Fe_5C_2 occurs which ultimately leads to the production of precious products of C_{5+} hydrocarbon.	[54]
Sulfated zirconium	Wet impregnation	Bigger density of acid and higher acidic strength.	Zr-S-400	80% conversion of glycerol in 1 hour at 40°C.	[55]

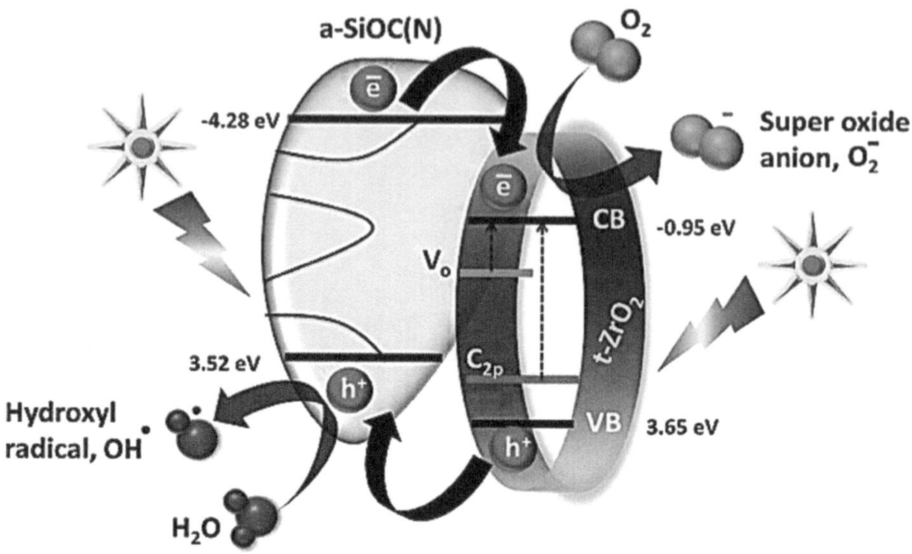

FIGURE 11.4 Catalytic activity of titanium oxide-doped zirconium catalyst [58]. (Permissions under Creative Commons Attribution 4.0 International License.)

joined to MOF-88 surface as a solid non-uniform block. Bonds like O-H, Zr-O, and C-O exist in S-TiO_2@MOF-808. When S and TiO_2 were combined, it decreased the resistance of charge transfer, increased the rate of electron transfer as well as enhanced the oxidation reduction Reaction (ORR). The density of power of S-TiO_2@MOF-808-MFC was almost twice that of S-TiO_2-MFC. 205 mV was the max vol. in S-TiO_2@MOF-808-MFC and solidity was kept constant for 6 days [51]. Al-Fatesh et al. addressed a method of production of hydrogen via the decomposition of methane with the goal of production of clean energy. A rigorous process of characterization was involved in exploring the usage of 10 TiZr in Fe and catalysts based on doped Fe resulting in the decomposition of methane. A surge in the catalytic activity was observed for longer periods due to the in situ production of catalytic active regions. Moreover, decreased crystallinity was observed when Ni was introduced to the 30Fe/10TiZr catalyst, demonstrating higher dispersion. In the case of Co-induced 30Fe/10TiZr, 80% of the methane was converted with hydrogen yield within a period of 3 hours. Whereas in Ni-induced 30Fe/10TiZr, 87% of methane was converted along with 82% hydrogen yield within 3 hours [56]. Likewise, Li et al. explained that H_2-ICEs are developed as an alternative to traditional fuel engines to decrease emissions of CO_2 but NO_x has to be catalytically reduced with H_2 to remove NO_x from H_2-ICEs. A high quantity of Pt oxide over Pt/TiO_2 causes a decrease in H_2-SCR activity as compared to Zr-altered Pt/TiO_2 catalysts which show large NO_x conversion along with nitrogen selectivity at different temperatures. The strong interconnection of Pt with TiO_2 gets weakened due to Zr which in turn activates more species of Pt^0 which act as active regions for induction of hydrogen and NO_x within H_2-SCR reaction. This work illustrates an in-depth view of the advancement of H_2-SCR catalysts along with the regulation of the temperature of conversion [57]. Thus titanium dioxide along with Zr brings efficient catalytic results.

11.4.2 Mesoporous Zirconium

Mesoporous Zr has been widely explored as an augmentation material (Figure 11.5) [59]. Guo et al. explained that due to the increasing demand for clean as well as sustainable management, a requirement for strong acid catalysts is present for systematic cellulose hydrolysis. In KIT-6 silica, 3-D interrelated mesopores are present which are considered above mesoporous forms with 1- or 2-D channels as reactants as well as products diffuse faster when reaction occurs. Functioning catalyst KIT-6 is produced when Zr is added along with subsequent sulfation to examine microcrystalline cellulose conversion by catalysis. During hydrolysis of cellulose, catalyst-sulfated Zr-KIT-6 shows high catalytic activity due to strong Brønsted acid regions, which immediately provide photons in cellulose for cleaving *β*-1,4 glycosidic bonds. A credible pathway during the initial state of hydrolysis of cellulose on catalyst was introduced [52,60]. Wang et al. described various methods of preparation of mesoporous ZrO_2 and Co_xO_y with VAN as models to traverse the execution of the catalysts during HDO reaction. The morphology, chemical configuration and structure of the catalyst were tested via many methods of characterization. During ideal conditions, catalysis of VAN to MMP resulted in 99% yield, along with the production of ZrO_2, i.e., the catalyst via precipitation method, the solvent was isopropanol, the

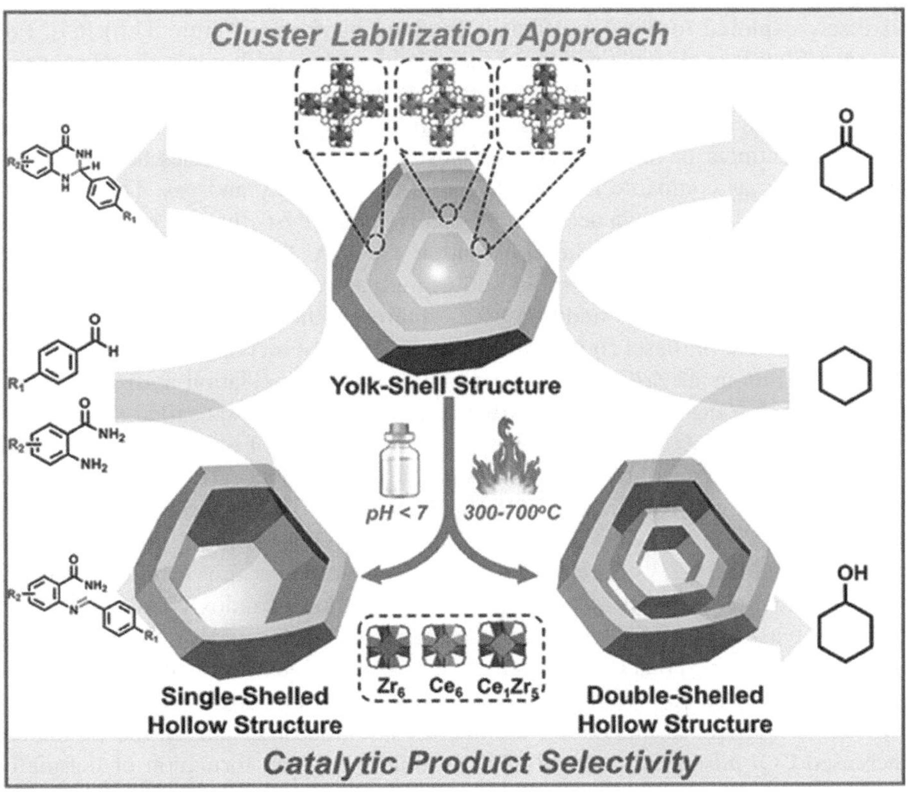

FIGURE 11.5 Hollow Zr/Ce-UiO-66 structure for selective catalysis [63]. (Permissions under Creative Commons Attribution 4.0 International License.)

pressure of hydrogen as 5 MPa, 24 hours as the time of reaction, and 180°C as the temperature of reaction. When HDO of VAN was carried out, 99% yield of MCYL was obtained, along with Co_xO_y as catalyst produced via precipitation method, the solvent as isopropanol, 4 MPa pressure of hydrogen, 24 hours as the time of reaction, and 170°C as the temperature of reaction [61]. Moreover, Xu et al. described that the important raw materials required for the manufacture of excellent pharmaceuticals and chemicals are carbonyl compounds. A cheap and eco-friendly way of producing carbonyl compounds is the catalytic oxidation of hydrocarbons as well as alcohols via heterogeneous catalysts. The results of the characterization confirmed the embedment of species of vanadium in the M-ZrPO support framework. Dissipation of vanadium regions resulted in the formation of catalysts with increased activity to selectively oxidize hydrocarbons and alcohols to carbonyl compounds. In comparison to well-known catalysts, 20 V/M-ZrPO particularly has shown superior performance with key standards like reusability, time of reaction, and product yield [62]. Therefore, mesoporous Zr can significantly impact the performance of zirconium catalysts.

11.4.3 Ceria-Zirconia Unit

Ceria-zirconia unit synergistically enhances the catalytic performance. It has been adversely explored to develop efficient catalytic platforms (Figure 11.6) [64]. For instance, Shukla et al. explained that a major source of air pollution is diesel soot and it has adverse effects on both human health as well as the environment. The usage of noble metals as catalysts has resulted in an increase in the cost of the systems for the after-treatment of diesel. Thus, the feasibility of diesel vehicles has decreased commercially as compared to vehicles based on electricity and gas. Their work is based on a catalyst that can act as an alternative to catalysts that are based on noble metals to reduce the cost of diesel oxidation. CeO_2-Al_2O_3 are the catalysts known for their capacity to store oxygen as well as high activity of redox and are good for the oxidation of soot. The addition of Zr improved the activity of the catalyst in case of oxidation of diesel soot. Drastic changes were observed in T_{50} soot at different concentrations of Zr/CeO_2-Al_2O_3, which resulted in oxidation at different temperatures [53]. Yang et al. described that a sequence of catalysts called mesoporous nanoparticle P-CexZr1-xOy (PCZ) having high vacancies of oxygen along with the abundant specific area of the surface was produced. The data calculated showed that the size of the average particle of PCZ was 10 nm with 108 m^2/g as a specific area of surface. When effective Zr doping along with P123 templating was done, vacancies of oxygen within PCZ were regulated. This changed the cerium's chemical state as well as reduction properties resulting in the formation of redox couple Ce^{3+}/Ce^{4+}, which increased the ability of catalytic oxidation of PCZ in indoor conditions [65]. Singh et al. explained that Ce-Zr metal oxide was produced along with raw Ce and Zr by a method called co-precipitation. The calculated results of the experiment along with the theoretical results confirmed that the addition of Zr ions in the Ce matrix increased CO_2 adsorption as well as activation due to the formation of thermally constant surface species, further detected through DRIFTS. In addition, calculations of DFT disclosed that vacancy of oxygen resulted in the alteration of electronic

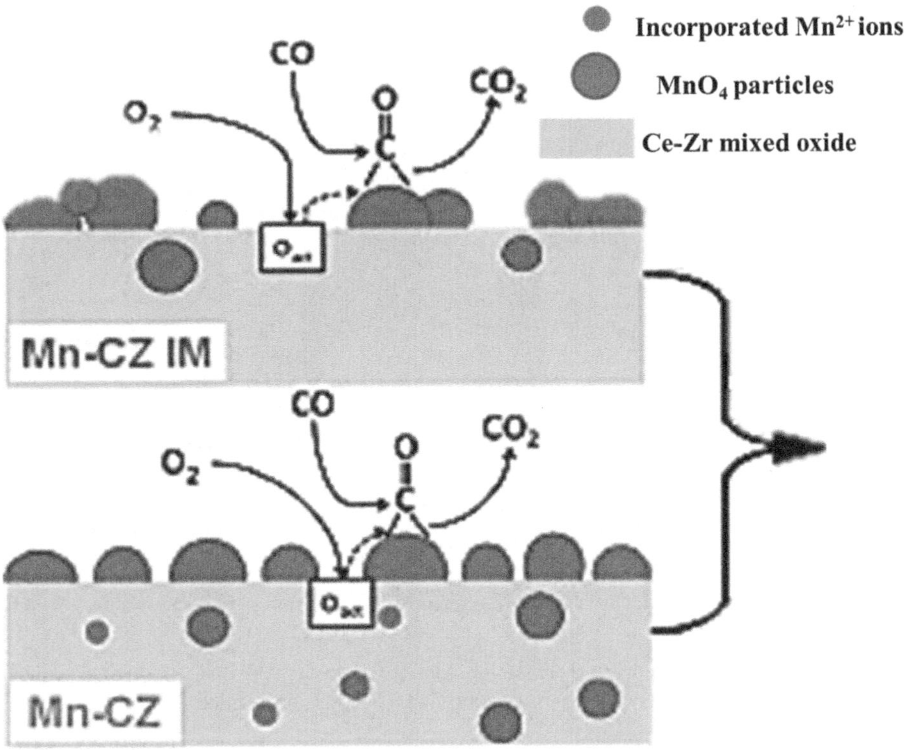

FIGURE 11.6 Demonstration of increase in catalytic activity of CO oxidation by using ceria-zirconia catalysts [67]. (Permissions under Creative Commons Attribution 4.0 International License.)

structure in Ce-Zr mixed oxide, believed to improve the activity of the catalyst [66]. Therefore, the ceria-zirconia platform is one of the best ways to enhance the catalytic performance of the device.

11.4.4 Iron Precipitation

Many scientists have worked on developing zirconium catalytic units precipitated with iron (Figure 11.7) [68]. For example, Ge et al. [69] explained that efforts for the utilization of syngas for direct production into higher alcohols have been unsuccessful. The disruption in the process is due to a lack of understanding of the catalyst structure which usually functions by a combination of different metals, supports, and promoters. When Fe is promoted by Zr, it acts as an excellent catalyst. Formation of vacancies for oxygen occurs along with healing in clusters of amorphous ZrO_2 which increases the activation of CO and H_2 with the production of Fe_5C_2 due to Fe_3O_4 contact. According to correlations of activity composition, the active phase is the Fe_3O_4-Fe_5C_2 interface. Thus, monometallic iron has emerged as a promising catalyst for the future. Zhang et al. described that the development of extensive and

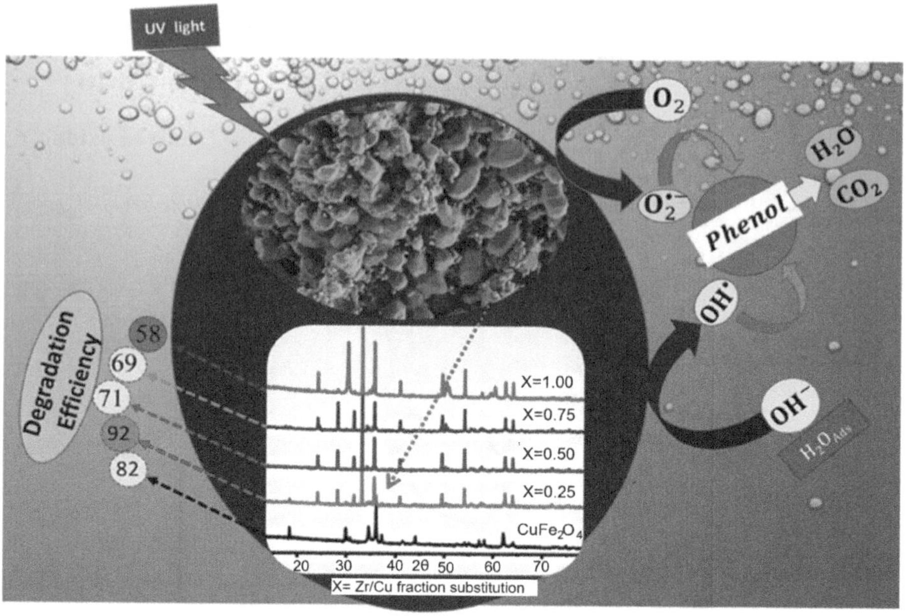

FIGURE 11.7 Illustration of increase in catalytic activity of Zr-Cu-Fe nanocomposites in removing pollutants from waste water [71]. (Permissions under Creative Commons Attribution 4.0 International License.)

ecological solid-state production methods is extremely important for the production of mainly Fe-based precipitated catalysts used in FTS. The interrelation between activity as well as structure of the catalyst was calculated by elaborate characterization techniques. A rigorous examination of the influence of Na promoter along with the role of Cu and Mn in increasing the catalytic performance in FTS was carried out. Therefore, the formation of χ-Fe_5C_2 occurs which further encourages chain propagation. This ultimately leads to the production of precious products of C_{5+} hydrocarbon which have the potential to be the ideal candidates for industrial implementation in the future for Fe-based FTS [54]. Likewise, Shi et al. studied the activity of the catalyst as well as cyclic catalysis in various methyl benzoates via a sequence of solid Lewis acidic catalysts. The reactivity between 31 types of benzoic acids and various substituents as well as methanol by the catalyst was analyzed via the activity of the catalyst as well as cyclic catalysis. Additionally, the mechanism of the catalyst was verified by the acid strength, microstructure, specific area of surface, and methyl benzoate production via GC-MS. Zr ions significantly impacted the catalytic activity which was further enhanced by a certain amount of iron and titanium ions with the ratio Fe:Zr: Ti=2:1:1 illustrating ideal catalytic activity. The surface of the catalyst as well as the quantity of acid regions were enhanced due to the addition of Fe ion leading to elevation in the activity of the catalyst [70]. Thus, the precipitation of iron over Zr brings forth better results concerning catalysis.

11.4.5 Sulfated Zirconium

A considerable amount of research has been done to develop sulfated zirconium (Figure 11.8). Vannucci et al. described the production of a sequence of catalysts of acidic nature from industrial zirconium oxide and 0.5M solution of H_2SO_4 via wet impregnation. The results of the performance showed an interrelation between the temperature of calcination and acid regions of materials. Additionally, reaction kinetics, as well as thermodynamics, were calculated along with the impact of the molar proportion of reactants as well as the water content of the mixture of the reaction. Taking into consideration that the water adsorption by the catalyst did not influence the available active regions, an easy pseudo-homogeneous expression of the kinetics was produced and was further added to the undergoing study of the experimental data [55]. On the same line, Yan et al. explained the synthesis of mesoporous SBA-15 via the incorporation of Mg and grafting of sulfated Zr. The addition of Mg in mesoporous SBA-15 had quite an impact on the base strength along with a significant

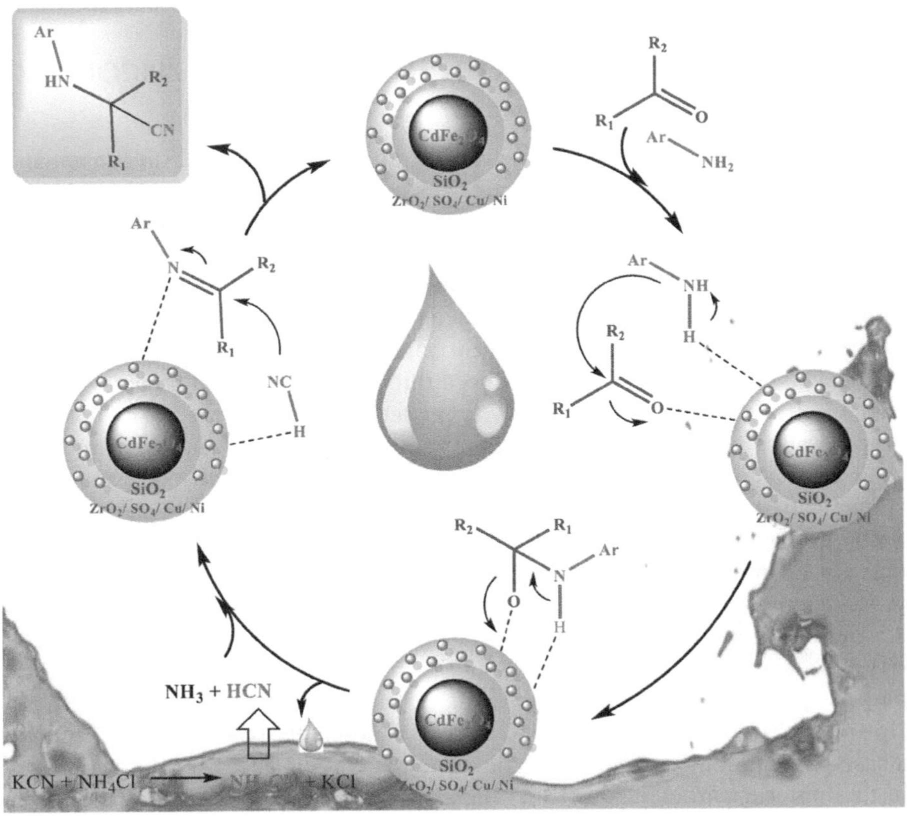

FIGURE 11.8 Representation of $CdFe_2O_4@SiO_2@ZrO_2/SO_4^{2-}/Cu/Ni$ nano-catalyst used for the production of α-amionitriles [74]. (Permissions under Creative Commons Attribution 4.0 International License.)

impact on the Brønsted/Lewis acidity due to the SZ group. The conversion of fructose into 5-HMF and iPrOH/DMSO system was carried out in the presence of SZ@Mg-SBA-15 catalyst. 98.6% 5-HMF was produced under ideal reaction conditions. The result depicted that the dehydration of fructose to form 5-HMF occurred due to the amalgamation of the Brønsted acid region and basic region [72]. Furthermore, Abedin et al. described a straightforward approach called MDHA for the conversion of CH4 into aromatics along with the formation of H_2. The most constant and active catalysts for the production of MDHA are considered to be Mo-doped with HZSM-5/HMCM-22. There is significant precipitation of carbon. So, MDHA is collated with promoters like W, Pt, Cr, and Pd for the catalyst Mo/SZ to reduce carbon (C) precipitation and enhance aromatics selectivity. Various promoters showed various activities. The deactivation was slower and C precipitation was lower with Cr, Pd, and Pt catalysts, whereas with W, C precipitation was higher with faster deactivation. Cr as well as Pt enhanced the Mo/SZ aromatics selectivity in comparison to W as well as Pd [73]. Thereby, sulfate grafting over Zr improves the chemical properties of the unit and consequently enhances the catalysis.

11.5 INTEGRATION OF NANOTECHNOLOGY WITH ZIRCONIUM AND ITS EFFECT ON THE CATALYTIC PROPERTY

Nanotechnology has become an integral part of material science as it can enhance the overall activity of the composite in which it's being used [75,76]. Concerning zirconium then nanoparticles combined with it only brings promising results. This preferred combination has a crucial role in enhancing the catalytic property of the zirconium platform. Scientists have been exploring this integration extensively to achieve advanced catalytic units with efficient performance [77]. Reddy et al. in the year 2019 synthesized a copper-doped zirconium oxide nanoparticle with a tetragonal structure. The particles were produced through hydrothermal process and their photoelectrocatalytic and photocatalytic properties were analyzed. The doping of copper was responsible for increasing the nanostructure's final surface area and the band gap value was also decreased. The unit could catalytically degrade methyl orange with an efficiency of 93.7% [78]. Likewise, Reddy et al. in the year 2020 doped iron with zirconium oxide nanoparticles via hydrothermal method. The doping or iron ions developed a reasonable band gap which harvested more photons and consequently stimulated the production of more electrons. The increased presence of electrons in the valence band positively impacted the unit's photocatalytic property [79]. Chen et al. in the year 2018 fabricated a nanoparticle consisting of zirconium and platinum to catalytically oxidize 5-hydroxymethylfurfural (5-HMF) to 2,5-furandicarboxylic acid (2,5-FDCA). The nano-catalysts are best performed in a base-free environment. The catalytic oxidation was enhanced due to the increase in the uniform particle size and its high dispersion. The presence of the ketone group improved the adsorption between reactants and catalysts [80]. Reinforcement of molybdenum disulfide (MoS) onto zirconium can form an efficient nanosheet that can carry out catalytic applications. The doped MoS was in a hexagonal phase and showed the presence of an electron-hole pair with better migration dynamics which can be a reason for the increased catalytic performance of the sheets [81].

Zirconium nanoparticles can be used to catalyze polyolefins via hydrogenolysis. This catalytic activity can be achieved by placing the amorphous nano-sized Zr nanoparticles between the plates of mesoporous silica. The macromolecules are transferred from bulk to active Zr nanoparticles through radical mesopores leading to hydrogenolytic cleavage forming a C-18-centered distribution [82]. Modification of SiO_2/Co with Zr can form a nano-oxidative unit to catalyze the dehydrogenation of propane. The active catalyst in this combination was tetrahedral Co^{2+} ions. With the addition of Zr, the bond strength between Co-O and oxygen-donating ability both changed which was the primary factor for enhanced catalysis [83]. Qui et al. in the year 2019 developed a photocatalytic unit by functionalizing zirconium metal-organic framework (MOF) with a noble metal. The authors suggested that the catalytic oxidation of aromatic alcohols into aldehydes increased in the presence of Au/UiO-66 while both Au/UiO-66 and Pt increased Cr(VI) reduction [84]. Also, zirconium can be used to directly convert carbon dioxide to lower olefins. For the same, Dang et al. in the year 2018 fabricated a combined unit of zirconium and indium to form a bi-functional catalyst. Here, zirconium particles played an important part in the catalytic activity and physiochemical performance of the catalytic unit. The incorporation of zirconium increased the oxygen-vacant sites, preventing the sintering of nanoparticles, and stabilizing the intermediated formed after the hydrogenation of carbon [85]. Therefore nano-science has a positive impact on zirconium-based catalytic units. Both the components have a synergistic effect which ultimately enhances the catalytic performance of the platform. More research can be focused on how to increase the practical applications of the zirconium nano units to form desirable, cost-effective, and sustainable catalytic systems.

11.6 APPLICATIONS OF ZIRCONIUM CATALYST

Zirconium has many practical applications that can be used for the betterment of the human being and environment. These applications and their practical usage have been mentioned in Table 11.2 as well as discussed briefly below.

11.6.1 Biofuel

Biomass, which can be any organic material originating from plants, animals, or microorganisms, is the source of liquid or gaseous fuels known as biofuels. Since they can be replenished more quickly than fossil fuels, which take millions of years to develop, they are regarded as renewable energy sources. As the world nears the exhaustion of fossil fuels, it becomes paramount to shift efforts to improve the biofuel production industry and Zr catalysts can assist in this endeavor. One of the methods is using Zr catalysts for improving biomass processing, biomasses like waste cooking oil (WCO), and Algal biomass (Figure 11.9). For WCO, we have many Zr-based catalysts like Fe_2O_3-MnO-SO_4^{2-}/ZrO_2 nanoparticle solid catalyst, produced by the impregnation reaction and 3 hours of calcination at 600°C. These catalysts have been shown to produce up to 96.5% of biodiesel under optimal conditions and stability of catalyst after multiple cycles of reaction without significant deterioration [94]. Other catalysts that can be used for waste corn oil are the ZrO_2–SrO_2

TABLE 11.2
Various Practical Applications of Zirconium-Based Materials

Zirconium Form	Synthesis	Working Principle	Substrate	Application	References
Solid solution	Calcination	The accumulation of m-HCOO on the surface of the catalyst.	Copper and cerium.	CO_2 hydrogenation into methanol	[16]
Zirconium oxide	Dry impregnation method	The formation of oxygen vacancies reduced coke formation.	Nickel and ceria	Production of syngas	[86]
Zirconium β-diketonate complexes	Thermal treatment	Zirconium form in the presence of PU acts as an improved carbamate exchanger.	Zr acetylacetonate	Polyurethane synthesis	[87]
Zirconium oxide	Sol-gel process	Zr doping increased active oxygen molecules and catalytic performance.	Co_3O_4	Propane oxidation	[88]
Zirconium oxide	Co-precipitation method	Vanadium doping increases hydrogen pretreatment and structural transformation.	Vanadium	Propane dehydrogenation	[89]
Zr MOF	Chemical synthesis	Porphyrin rings topological structure enhanced size-selective catalytic activity.	Porphyrins	Target-oriented functions	[90]
Magnetic Zr	Solvent method	Zr doping caused the adjustment of acidic-basic sites.	Boron	Hydrogenation of furfural	[91]
Zr^{4+} ions	Hydrothermal process	Increased basicity and crystallization enhanced the trans-esterification process.	CaO	Biodiesel	[92]
Zirconia	Ammonia evaporation	The lattice bond of oxygen plays a crucial role in the dissociation of the C-H bond.	Ceria	Methane activation	[93]

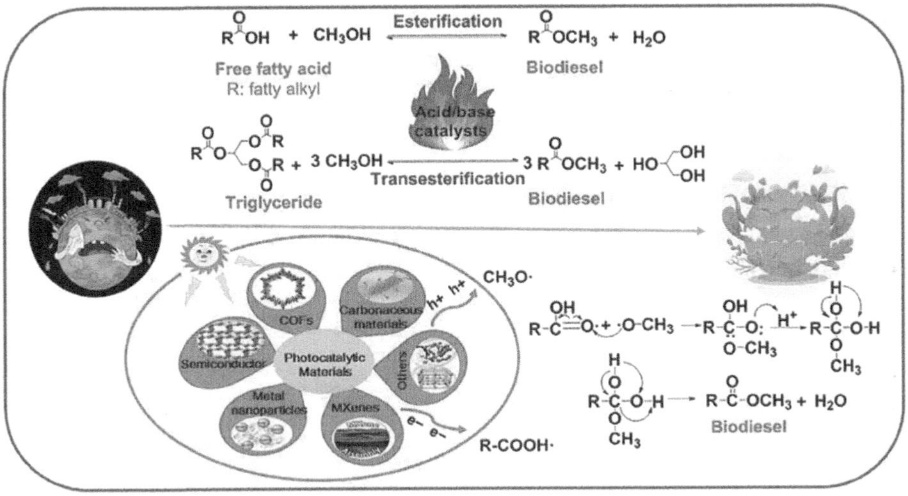

FIGURE 11.9 Production of biodiesel from fatty acids [98]. (Permissions under Creative Commons Attribution 4.0 International License.)

and ZrO_2–CuO mixed oxides produced by co-precipitation and the Pechini sol-gel method. These catalysts have been shown to produce high yields of biodiesel from cheap feed stocks like WCO specially catalyzing the trans-esterification of WCO when the molar ratio of catalysts used was 2:1 (Ze:Sr) for 2 cycles giving biodiesel output of 83%–88% [95]. Microalgal (*Scenedesmus obliquus*) lipids trans-esterification can also be catalyzed into biodiesel using WO_3/ZrO_2 (tungstated zirconia). They are prepared by initially Zirconium oxychloride hydrate ($ZrOCl_2\_8H_2O$) which was precipitated with a 25% ammonia solution at a pH of 8.0–8.5 to create $Zr(OH)_4$. After completely washing Zr $(OH)_4$ in water to get rid of any chloride salts, it was dried for 24 hours at 100°C. In a mortar, the dried $Zr(OH)_4$ was ground into powder. Ammonium meta-tungstate hydrate $[(NH_4)_6(H_2W_{12}O_{40})\ nH_2O]$ (50 wt%) was then used to impregnate $Zr(OH)_4$. Tungstated zirconia was created by calcining the impregnated powder for 4 hours at 800°C. This catalyst has been proven to produce a biodiesel yield of 94.58% up to 3 cycles of reuse without any catalyst activity degradation with a low requirement of methanol [96]. Aside from biomass-specific catalysts, there are general-purpose catalysts like mesoporous zirconium oxophosphates which are created by the evaporation-induced self-assembly (EISA) technique in severely acidic circumstances with pluronic F127, an amphiphilic block copolymer, as a starting point. They have strong Lewis and Brønsted acid sites present and the wide surface area and high acid density could be used to create an effective biodiesel catalyst. Moreover, a high pore size that facilitates the diffusion of big organic molecules helps to improve catalytic capacity. These catalysts effectively converted various alcohols to their corresponding ester with glacial acetic acid and long-chain fatty acids to their corresponding methyl ester with methanol. Under ideal reaction conditions, the observed biodiesel yield in a typical reaction was greater than 91%. Additionally, this mesoporous solid acid catalyst has been recycled five times with only a minor loss of activity, indicating that the mesoporous zirconium oxophosphate framework has a high degree of chemical stability and reusability [97]. There zirconium-based materials can be widely used to develop biofuel and can help to meet the fuel shortage around the globe.

11.6.2 Ketonization of Industrial Waste

The process of converting organic waste materials into ketones through chemical reactions is known as "ketonization of industrial waste." This strategy is part of a broader effort to manage industrial waste more sustainably while exploring its potential for beneficial use. This process can be catalyzed using zirconium catalysts to make those processes faster as well as cheaper. For various types of organic industrial waste, different types of catalysts are employed. Initially, Zr-Mg-Y-O catalysts that were previously used for primary alcohol transformations were modified to allow for the processing of waste esters or alcohol fractions. Tests have been conducted on the ketonization of alcohol fractions obtained from industrial installations and methyl ester fractions originating from fat. Mostly ketone fractions were produced in both situations. This process makes it possible to manage trash intelligently, and the proportion that is produced—which is primarily made up of ketones—opens new possibilities for ketone applications. From a technological perspective, the suggested approach is extremely straightforward and converts primary alcohol and aldehydes found in waste alcohol fractions or esters in methyl esters fractions of waste fat origin. Conversion uses a specific material and doesn't involve the addition of any other substances; it happens continuously. A diluent is advised in the long-chain methyl esters fraction situation. It is in line with current chemical industry trends that aim to maximize the number of atoms that go from the raw material to the finished product. This approach satisfies the requirements of contemporary processes as well as the presumptions of eco-development strategy from a technological and financial standpoint [99]. Another industrial waste type is carboxylic acids for which separated Zr centers supported on high surface area silica can be utilized for aldol condensation of ketones to dimeric enones and ketonic decarboxylation (ketonization) of carboxylic acids to internal ketones. Cp_2ZrCl_2 was grafted onto the surface of amorphous silica to create catalysts. However, increasing the Zr amount after a certain amount in the catalyst decreases its catalytic abilities as the ketonization of carboxylic acid and aldols require the presence of strong Lewis acid adsorption sites for catalysis [30]. Zr catalysts can also be employed for the ketonization of acetic acids for these we use sulfated Zr catalysts. This sulfation of Zr catalysts shows increased Brønsted and Lewis acid strength but decreased Lewis's basicity. These changes improve selectivity toward the ketonization process, but conversion rates decrease significantly whereas the productivity of conversion of acetone in the vapor phase increases due to a decrease in basic sites in the sulfated Zr catalysts [100]. Lastly for the ketonization of free fatty acid industrial wastes like palm fatty acid distillate (PFAD), and palm kernel oil (PKO) which is a very common industrial waste from oil industries. For this, we use La/ZrO_2 catalysts. These catalysts are further improved by ultrasonication techniques creating large cavitation on the catalyst's surface. Catalysts like La/ZrO_2-DU performed better after this treatment with 80% conversion FFA to 51% ketone. However, increasing the La compositions in the catalysts decreased this improvement. Overall, the use of La/ZrO_2-DU for the Ketonization of FFA industrial waste made the process cheaper and efficient [101]. Thus, Zr can be successfully used to convert toxic industrial waste into less toxic substances via a ketonization reaction.

11.6.3 Microbial Fuel Cells

Microbial fuel cells are electrochemical contraptions that use microbes to generate electrons which when passed through an anode and then to a cathode generate electricity. These fuel cells are eco-friendly and sustainable sources of electricity, but the use of expensive metals like platinum in their creation increases their cost of use higher. So, by using inexpensive zirconium in its creation, we can lower the cost of MFCs and make them industrially more workable. There are many ways to incorporate zirconium catalysts in MFCs; one of them is using a mix of platinum cathode with some amount of zirconium catalysts like sulfonated zirconium ($SZrO_2$) which have shown to catalyze ORR at the same rate or even higher compared to all normal platinum cathodes [102]. Zirconium can also be used with other cathode catalysts; zirconium palladium oxide composite cathode has been shown to provide greater current conduction and capacitance compared to manganese palladium oxide composite cathode but slightly less cathode performance comparatively [103]. To further improve upon the application of zirconium, the other way to utilize zirconium catalysts is to use standalone Zr catalysts as cathodes like a Zr-MOF (zirconium-metal-organic framework) cathode instead of platinum has shown to produce high ORR catalytic activity with lesser resistance to charge transfer increasing electricity generation by

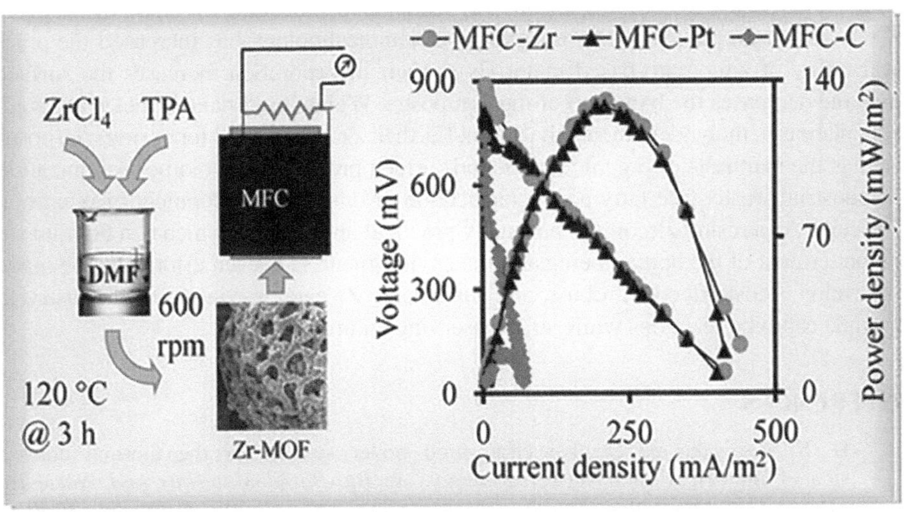

FIGURE 11.10 Demonstration of Zr-based microbial fuel cell for the oxygen reduction reactions (ORRs) [32]. (Permissions under Creative Commons Attribution 4.0 International License.)

the MFCs (Figure 11.10). In comparison with other non-platinum-based cathodes, iron phthalocyanine–MnOx, MOF-derived Co–N–C composites, graphene oxide/SnO_2 nanocomposites, and graphene oxide-supported magnesium oxide deliver volumetric power densities of 4.72, 9.99, 4.8, and 3.5 W/m³, respectively. Compared to these Zr-MOF produced 9.84, which is proof of its higher catalytic activity [32]. Non-platinum-based cathode can also be made out of other zirconium catalysts, one of which is ZrCNO which has been shown to contain sufficient catalytic activity. The catalytic ability can be further improved by modifying the ZrCNO production method leading to many fold increases in its catalytic activity but still, the activity is one-fourth of Pt cathode. So, it is only used in MFCs that use synthetic wastewater treatment [104]. As a result, the Zr catalyst can develop environmentally friendly microbial cells as they enhance the power density and catalytic efficiency.

11.7 CONCLUSION AND PROSPECTS

In a nutshell, it can be concluded that zirconium-based materials have emerged as a catalytic wonder with practical usage. The physical and chemical characteristics contribute toward this catalytic property. For instance, the Lewis acidic and basic sites contribute to the catalytic activity of Zr. Various research have been conducted to alter these sites to achieve enhanced performance of the catalytic unit. The Zr has a considerable part in MPV reduction reaction as it can catalyze the reduction reaction of aldehydes in ethanol presence, following its stoichiometric oxidation to aldehyde which is its analog. Zirconium is reinforced with various compounds and elements to improve the overall activity of the developed unit. Commonly used modification products are titanium dioxide, mesoporous zirconium, ceria-zirconia unit, iron precipitation and sulfated zirconium. Concerning zirconium then nanoparticles combined with it only bring promising results. This preferred combination has a crucial role in enhancing the catalytic property of the zirconium platform. The integration of nanotechnology has increased the practical utility of zirconium-based materials as their incorporation increases the surface area and decreases the band gap of the composite. Widely used nanoparticles are gold, iron, platinum, molybdenum, and hafnium. Further, Zr can be used for various purposes such as the synthesis of organic compounds, green production of biofuel, ketonization of industrial waste, free fatty acid esterification of oils, and development of microbial fuel cells. As a result, zirconium has many practical applications which can be used for the betterment of the human being and the environment. However, efforts can be made to develop a cost-effective, robust, and sustainable Zr catalyst via green chemistry to upgrade catalytic reactions while still conserving nature.

REFERENCES

1. H. S. Vishwakarma et al., "Grate-fired boiler system for the thermal conversion of municipal solid waste to energy," in *Bio-Based Materials and Waste for Energy Generation and Resource Management*, pp. 183–205, 2023, doi: 10.1016/B978-0-323-91149-8.00015-6.

2. S. Sonal and B. K. Mishra, "A comprehensive review on the synthesis and performance of different zirconium-based adsorbents for the removal of various water contaminants," *Chem. Eng. J.*, vol. 424, p. 130509, 2021, doi: 10.1016/j.cej.2021.130509.
3. L. Peng et al., "Zirconium-based catalysts in organic synthesis," *Top. Curr. Chem.*, vol. 380, no. 5, p. 41, 2022, doi: 10.1007/s41061-022-00396-4.
4. J. R. Dilworth and S. I. Pascu, "The chemistry of PET imaging with zirconium-89," *Chem. Soc. Rev.*, vol. 47, no. 8, pp. 2554–2571, 2018, doi: 10.1039/C7CS00014F.
5. C. Hu, J. Sun, C. Long, L. Wu, C. Zhou, and X. Zhang, "Synthesis of nano zirconium oxide and its application in dentistry," *Nanotechnol.* Rev., vol. 8, no. 1, pp. 396–404, 2019, doi: 10.1515/ntrev-2019-0035.
6. J. P. Holland, "Predicting the thermodynamic stability of zirconium radiotracers," *Inorg. Chem.*, vol. 59, no. 3, pp. 2070–2082, Feb. 2020, doi: 10.1021/acs.inorgchem.9b03515.
7. H. K. Sneh Lata, "Diet of silver cyprinid, *Rastrineobola argentea* in Lake Victoria, Kenya. E.," *Int. J. Adv. Res.*, vol. 4, no. 6, pp. 625–634, 2016, doi: 10.21474/IJAR01.
8. C. C. Liu, "Electrochemical sensors," *Med. Devices Syst.*, vol. 07, no. 07, pp. 48-1-48–6, 2006, doi: 10.5796/kogyobutsurikagaku.58.1087.
9. Z. Lu et al., "Node-accessible zirconium MOFs," *J. Am. Chem.* Soc., vol. 142, no. 50, pp. 21110–21121, 2020, doi: 10.1021/jacs.0c09782.
10. H. Lundberg, F. Tinnis, J. Zhang, A. G. Algarra, F. Himo, and H. Adolfsson, "Mechanistic elucidation of zirconium-catalyzed direct amidation," *J. Am. Chem. Soc.*, vol. 139, no. 6, pp. 2286–2295, Feb. 2017, doi: 10.1021/jacs.6b10973.
11. F. Li et al., "Catalytic transfer hydrogenation of butyl levulinate to γ-valerolactone over zirconium phosphates with adjustable Lewis and Brønsted acid sites," *Appl. Catal. B Environ.*, vol. 214, pp. 67–77, 2017, doi: 10.1016/j.apcatb.2017.05.013.
12. H. Kaur, S. Sharma, S. Athwal, M. Thakur, M. Verma, and V. Mutreja, "The potential applications of 2D nanomaterials for water purification," in *2D Nanomaterials*, John Wiley & Sons, Ltd, 2024, pp. 361–395. doi: 10.1002/9781394167883.ch16.
13. Y. Yuan et al., "Zirconium nitride catalysts surpass platinum for oxygen reduction," *Nat. Mater.*, vol. 19, no. 3, pp. 282–286, 2020, doi: 10.1038/s41563-019-0535-9.
14. H. Kaur et al., "Assessment of diabetes biomarker monitoring via novel biosensor activity," *Results Chem.*, vol. 5, no. January, p. 100777, 2023, doi: 10.1016/j.rechem.2023.100777.
15. G. X. Yan, A. Wang, I. E. Wachs, and J. Baltrusaitis, "Critical review on the active site structure of sulfated zirconia catalysts and prospects in fuel production," *Appl. Catal. A Gen.*, vol. 572, pp. 210–225, 2019, doi: 10.1016/j.apcata.2018.12.012.
16. T. Chen et al., "Unraveling highly tunable selectivity in CO_2 hydrogenation over bimetallic In-Zr oxide catalysts," *ACS Catal.*, vol. 9, no. 9, pp. 8785–8797, Sep. 2019, doi: 10.1021/acscatal.9b01869.
17. G. Devi et al., "Toxicity assessment of fluoride-contaminated soil and wastewater in *Solanum tuberosum*," *Water. Air. Soil Pollut.*, vol. 233, no. 7, pp. 1–23, 2022, doi: 10.1007/s11270-022-05694-7.
18. W. Zhang, Z. Wang, J. Huang, and Y. Jiang, "Zirconia-based solid acid catalysts for biomass conversion," *Energy & Fuels*, vol. 35, no. 11, pp. 9209–9227, Jun. 2021, doi: 10.1021/acs.energyfuels.1c00709.
19. Y. Zhou et al., "Solid acid catalyst based on single-layer α-zirconium phosphate nanosheets for biodiesel production via esterification," *Catalysts*, vol. 8, no. 1, 2018, doi: 10.3390/catal8010017.
20. Y. Zhang, Y. Chen, J. Pan, M. Liu, P. Jin, and Y. Yan, "Synthesis and evaluation of acid-base bi-functionalized SBA-15 catalyst for biomass energy conversation," *Chem. Eng. J.*, vol. 313, pp. 1593–1606, 2017, doi: 10.1016/j.cej.2016.11.033.

21. J. Hidalgo-Carrillo, A. Parejas, M. J. Cuesta-Rioboo, A. Marinas, and F. J. Urbano, "MPV reduction of furfural to furfuryl alcohol on Mg, Zr, Ti, Zr–Ti, and Mg–Ti solids: Influence of acid–base properties," *Catalysts*, vol. 8, no. 11, 2018, doi: 10.3390/catal8110539.
22. L. Ye, Y. Han, M. Zhang, Z. Yu, and X. Lu, "Enhanced sorption of carbonyl groups by zirconium hydroxide modified with polydopamine for highly selective production of alcohols via MPV reduction under mild conditions," *Fuel*, vol. 331, p. 125786, 2023, doi: 10.1016/j.fuel.2022.125786.
23. M. Rimoldi et al., "Catalytic zirconium/hafnium-based metal–organic frameworks," *ACS Catal.*, vol. 7, no. 2, pp. 997–1014, Feb. 2017, doi: 10.1021/acscatal.6b02923.
24. X. Yang, X. Ma, X. Yu, and M. Ge, "Exploration of strong metal-support interaction in zirconia supported catalysts for toluene oxidation," *Appl. Catal. B Environ.*, vol. 263, p. 118355, 2020, doi: 10.1016/j.apcatb.2019.118355.
25. D. Li, W. Ni, and Z. Hou, "Conversion of biomass to chemicals over zirconium phosphate-based catalysts," *Chinese J. Catal.*, vol. 38, no. 11, pp. 1784–1793, 2017, doi: 10.1016/S1872-2067(17)62908-5.
26. N. Kumari et al., "Zirconia-based nanomaterials: Recent developments in synthesis and applications," *Nanoscale Adv.*, vol. 4, no. 20, pp. 4210–4236, 2022, doi: 10.1039/d2na00367h.
27. A. Masudi and O. Muraza, "Zirconia-based nanocatalysts in heavy oil upgrading: A mini review," *Energy & Fuels*, vol. 32, no. 3, pp. 2840–2854, Mar. 2018, doi: 10.1021/acs.energyfuels.7b03264.
28. H. Kaur et al., "Mechanistic investigation of formation of highly-dispersed silver nanoparticles using sea buckthorn extract," *Nanotechnology*, vol. 34, no. 8, p. 85703, 2023, doi: 10.1088/1361-6528/aca20d.
29. Y. Wang et al., "Phosphorus-modified zirconium metal organic frameworks for catalytic transfer hydrogenation of furfural," *New J. Chem.*, vol. 44, no. 46, pp. 20308–20315, 2020, doi: 10.1039/D0NJ04285D.
30. S. Shylesh, L. A. Bettinson, A. Aljahri, M. Head-Gordon, and A. T. Bell, "Experimental and computational studies of carbon–carbon bond formation via ketonization and aldol condensation over site-isolated zirconium catalysts," *ACS Catal.*, vol. 10, no. 8, pp. 4566–4579, Apr. 2020, doi: 10.1021/acscatal.9b05176.
31. A. Jrad, M. Hmadeh, B. J. Abu Tarboush, G. Awada, and M. Ahmad, "Structural engineering of Zr-based metal-organic framework catalysts for optimized biofuel additives production," *Chem. Eng. J.*, vol. 382, p. 122793, 2020, doi: 10.1016/j.cej.2019.122793.
32. I. Das, M. T. Noori, M. Shaikh, M. M. Ghangrekar, and R. Ananthakrishnan, "Synthesis and application of zirconium metal–organic framework in microbial fuel cells as a cost-effective oxygen reduction catalyst with competitive performance," *ACS Appl. Energy Mater.*, vol. 3, no. 4, pp. 3512–3520, Apr. 2020, doi: 10.1021/acsaem.0c00054.
33. S. Vipasha, K. Hardeep, K. Tarun, and M. Tullika, "Traditional Indian herb *Cathranthus roseus* used as cancer treatment: A review," *Int. J. Pharmacogn. Phytochem. Res.*, vol. 8, no. 12, pp. 1926–1928, 2016.
34. S. Daniel, C. K. Fonzeu Monguen, L. Wu, and Z. Tian, "Flame synthesis of Zr/ZSM-5 catalysts with tunable acidity for the oxidative dehydrogenation of propane to propene," *J. Therm. Sci.*, vol. 33, no. 1, pp. 268–283, 2024, doi: 10.1007/s11630-023-1873-0.
35. M. Abd El-Aal, M. N. Goda, M. M. M. Abd El-Wahab, N. O. El-Gamal, and A. E.-A. A. Said, "Zirconia–sugarcane bagasse fly ash as a novel solid acid nanocatalyst for selective dehydration of methanol to dimethyl ether," *J. Chinese Chem. Soc.*, vol. 71, no. 3, pp. 333–344, 2024, doi: 10.1002/jccs.202300434.
36. J. Wu et al., "Copper-induced formation of Lewis acid sites enhancing sulfated zirconia catalyzed i-butane normalization," *J. Catal.*, vol. 432, p. 115400, Apr. 2024, doi: 10.1016/J.JCAT.2024.115400.

37. A. H. Hatta, N. Norazahar, M. A. Aziz, and M. Alhassan, "Unravelling the potential of fibrous silica zirconia catalyst for CO methanation in energy production," *E3S Web Conf.*, vol. 516, 2024, doi: 10.1051/e3sconf/202451601002.
38. X. Song et al., "Synthesis of glycerol carbonate over porous La-Zr based catalysts: The role of strong and super basic sites," *J. Alloys Compd.*, vol. 750, pp. 828–837, Jun. 2018, doi: 10.1016/J.JALLCOM.2018.03.392.
39. M. Ma et al., "Simple basic zirconium carbonate: Low temperature catalysis for hydrogen transfer of biomass-derived carboxides," *Green Chem.*, vol. 21, no. 21, pp. 5969–5979, 2019, doi: 10.1039/C9GC03033F.
40. J. F. Miñambres and J. Čejka, "Meerwein-Ponndorf-Verley reduction in current heterogeneous catalysis research: a review," *Catal. Rev.*, pp. 1–42, Apr. 2023, doi: 10.1080/01614940.2023.2197716.
41. F. Gonell, M. Boronat, and A. Corma, "Structure–reactivity relationship in isolated Zr sites present in Zr-zeolite and ZrO_2 for the Meerwein–Ponndorf–Verley reaction," *Catal. Sci. Technol.*, vol. 7, no. 13, pp. 2865–2873, 2017, doi: 10.1039/C7CY00567A.
42. T. Komanoya, K. Nakajima, M. Kitano, and M. Hara, "Synergistic catalysis by lewis acid and base sites on ZrO_2 for Meerwein–Ponndorf–Verley reduction," *J. Phys. Chem. C*, vol. 119, no. 47, pp. 26540–26546, Nov. 2015, doi: 10.1021/acs.jpcc.5b08355.
43. H. Zhang, Z. J. Quek, S. Jaenicke, and G.-K. Chuah, "Hydrophobicity and co-solvent effects on Meerwein-Ponndorf-Verley reduction/dehydration cascade reactions over Zr-zeolite catalysts," *J. Catal.*, vol. 400, pp. 50–61, 2021, doi: 10.1016/j.jcat.2021.05.011.
44. G. Li, W. H. Fu, and Y. M. Wang, "Meerwein–Ponndorf–Verley reduction of cyclohexanone catalyzed by partially crystalline zirconosilicate," *Catal. Commun.*, vol. 62, pp. 10–13, 2015, doi: 10.1016/j.catcom.2014.12.016.
45. W. Li, S. Liu, H. Wang, B. Gao, C. Tu, and Y. Luo, "La-doped Zr-beta zeolite as efficient catalyst for reduction of cyclohexanone to cyclohexanol via the MPV process," *Catal. Commun.*, vol. 133, p. 105845, 2020, doi: 10.1016/j.catcom.2019.105845.
46. A. Koreniuk, K. Maresz, and J. Mrowiec-Białoń, "Supported zirconium-based continuous-flow microreactor for effective Meerwein–Ponndorf–Verley reduction of cyclohexanone," *Catal. Commun.*, vol. 64, pp. 48–51, 2015, doi: 10.1016/j.catcom.2015.01.021.
47. C. A. Akinnawo, N. Bingwa, and R. Meijboom, "Metal-doped mesoporous ZrO_2 catalyzed chemoselective synthesis of allylic alcohols from Meerwein-Ponndorf-Verley reduction of α,β-unsaturated aldehydes," *New J. Chem.*, vol. 45, no. 17, pp. 7878–7892, 2021, doi: 10.1039/d1nj00936b.
48. H. Zhang, S. Jaenicke, K. Okumura, H.-R. Tan, and G.-K. Chuah, "Mechanochemistry-based liquid-assisted synthesis of hydrophobic Zr-Beta with high metal loading for Meerwein-Ponndorf-Verley reduction," *J. Catal.*, vol. 431, p. 115398, 2024, doi: 10.1016/j.jcat.2024.115398.
49. Y. Leng, L. Shi, S. Du, J. Jiang, and P. Jiang, "A tannin-derived zirconium-containing porous hybrid for efficient Meerwein–Ponndorf–Verley reduction under mild conditions," *Green Chem.*, vol. 22, no. 1, pp. 180–186, 2020, doi: 10.1039/C9GC03393A.
50. H. Kaur, N. Kumari, A. Sharma, M. Sachdeva, and V. Mutreja, "Optical and electrochemical microfluidic sensors for water contaminants: A short review," *Mater. Today Proc.*, vol. 48, pp. 1673–1679, 2022, doi: 10.1016/j.matpr.2021.09.534.
51. Q. Qi et al., "Improving bioelectrochemical performance by sulfur-doped titanium dioxide cooperated with Zirconium based metal–organic framework (S-TiO_2@MOF-808) as cathode in microbial fuel cells," *Bioresour. Technol.*, vol. 394, p. 130288, Feb. 2024, doi: 10.1016/J.BIORTECH.2023.130288.
52. L. Guo, W. Xie, and X. Xu, "A robust mesoporous sulfated zirconium-doped KIT-6 solid acid catalyst for the hydrolysis of microcrystalline cellulose," *J. Chem. Technol.* Biotechnol., vol. 99, no. 2, pp. 551–561, 2024, doi: 10.1002/jctb.7559.

53. M. K. Shukla, V. Bangwal, A. Dhar, T. Bhaskar, and A. Kumar, "Catalytic activity of Zr/CeO_2-Al_2O_3 catalyst for diesel soot oxidation: Synthesis, characterization, and performance evaluation," *Environ. Sci. Pollut. Res.*, 2024, doi: 10.1007/s11356-024-34052-9.
54. W. Zhang et al., "Eco-friendly solid-state synthesis of Na-promoted Mn-Fe/ZrO_2 catalyst for Fischer-Tropsch synthesis," *Fuel*, vol. 363, p. 131013, 2024, doi: 10.1016/j.fuel.2024.131013.
55. J. A. Vannucci, N. N. Nichio, and F. Pompeo, "Solketal synthesis from ketalization of glycerol with acetone: A kinetic study over a sulfated zirconia catalyst," *Catal. Today*, vol. 372, pp. 238–245, 2021, doi: 10.1016/j.cattod.2020.10.005.
56. A. S. Al-Fatesh et al., "Optimizing hydrogen production: Influence of promoters in methane decomposition on titania-modified-zirconia supported Fe catalyst," *ACS Omega*, vol. 9, no. 18, pp. 20322–20330, 2024, doi: 10.1021/acsomega.4c00729.
57. Y. Li et al., "Revealing the roles of Zr on enhanced H_2-SCR performances on Pt/TiO_2 catalyst," *Chem. Eng. J.*, vol. 490, p. 151714, Jun. 2024, doi: 10.1016/J.CEJ.2024.151714.
58. S. Bhaskar, E. W. Awin, K. C. H. Kumar, A. Lale, S. Bernard, and R. Kumar, "Design of nanoscaled heterojunctions in precursor-derived t-ZrO_2/SiOC(N) nanocomposites: Transgressing the boundaries of catalytic activity from UV to visible light," *Sci. Rep.*, vol. 10, no. 1, pp. 1–13, 2020, doi: 10.1038/s41598-019-57394-8.
59. H. Kaur, N. Gurjar, and R. Gill, "The noni fruit (*Morinda citrifolia* L.): A systematic review on anticancer potential and other health beneficial pharmacological activities," *J. Med. Plants Stud.*, vol. 86, no. 2, pp. 86–93, 2018.
60. S. Sharma, H. Kaur, M. Thakur, and S. Athwal, "Engineering 2D nanomaterials for biomedical applications," in *2D Nanomaterials*, John Wiley & Sons, Ltd, 2024, pp. 333–360. doi: 10.1002/9781394167883.ch15.
61. X. Wang, Y. Zhang, J. Chen, and Y. Xu, "Mesoporous zirconium oxides and cobalt oxides for selective hydrodeoxygenation of vanillin," *Catal. Le*tt., vol. 154, no. 7, pp. 3195–3211, 2024, doi: 10.1007/s10562-023-04516-7.
62. W. Xu, H. Huang, L. Bi, H. Xu, Z. Miao, and M. Wu, "One-pot fabrication of vanadium-doped ordered mesoporous zirconium oxophosphate catalyst for the synthesis of carbonyl compounds," *New J. Chem.*, vol. 48, no. 24, pp. 10820–10829, 2024, doi: 10.1039/D4NJ01661K.
63. H. H. He et al., "Yolk–shell and hollow Zr/Ce-UiO-66 for manipulating selectivity in Tandem reactions and photoreactions," *Journal of the American Chemical Society*, vol. 145, no. 31, pp. 17164–17175, 2023.
64. H. Kaur, A. Bhosale, and S. Shrivastav, "Biosensors: Classification, fundamental characterization and new trends: A review," *Int. J. Heal. Sci. Res.*, vol. 8, no. June, p. 315, 2018, [Online]. Available: www.ijhsr.org
65. Z. Yang et al., "Regulation of oxygen vacancies in ceria-zirconia nanocatalysts by pluronic P123-templated for room temperature formaldehyde total oxidation," *Catal. Le*tt., vol. 154, no. 2, pp. 503–512, 2024, doi: 10.1007/s10562-023-04321-2.
66. P. P. Singh, A. P, S. Kamaliny, K. Ray, and S. Sengupta, "The pivotal role of oxygen vacancy and surface hydroxyl in the adsorption and activation of CO_2 on ceria-zirconia mixed oxide," *Mol. Catal.*, vol. 555, p. 113855, 2024, doi: 10.1016/j.mcat.2024.113855.
67. I. Y. Kaplin et al., "Efficiency of manganese modified CTAB-templated ceria-zirconia catalysts in total CO oxidation," *Appl. Surf. Sci.*, vol. 485, pp. 432–440, 2019, doi: 10.1016/j.apsusc.2019.04.206.
68. H. Kaur and S. Athwal, "Paper-based microfluidic devices for pesticide detection," pp. 126–152, 2023, doi: 10.4018/978-1-6684-6952-1.ch007.
69. Y. Ge, T. Zou, A. J. Martín, T. Block, R. Pöttgen, and J. Pérez-Ramírez, "Defective zirconia promotes monometallic iron catalysts for higher alcohol synthesis," *Chem Catal.*, vol. 4, no. 6, Jun. 2024, doi: 10.1016/j.checat.2024.101010.

70. C. Shi, X. Yu, W. Wang, H. Wu, A. Zhang, and S. Liu, "The activity and cyclic catalysis of synthesized iron-supported Zr/Ti solid acid catalysts in methyl benzoate compounds," *Catalysts*, vol. 13, no. 6. 2023. doi: 10.3390/catal13060971.
71. M. A. Wahba and A. A. Badawy, "Novel Zr–Cu–Fe nanocomposite metal oxides: Structural, magnetic and composition activity effects on photodegradation of phenols," *J. Sol-Gel Sci. Technol.*, vol. 94, no. 3, pp. 637–647, 2020, doi: 10.1007/s10971-019-05190-1.
72. C. Yan et al., "Sulfated zirconia catalysts supported on mesoporous Mg-SBA-15 with different morphologies for highly efficient conversion of fructose to 5-hydroxymethylfurfural," *Microporous Mesoporous Mater.*, vol. 328, p. 111507, 2021, doi: 10.1016/j.micromeso.2021.111507.
73. M. A. Abedin, S. Kanitkar, S. Bhattar, and J. J. Spivey, "Methane dehydroaromatization using Mo supported on sulfated zirconia catalyst: Effect of promoters," *Catal. Today*, vol. 365, pp. 71–79, 2021, doi: 10.1016/j.cattod.2020.06.071.
74. M. A. Nasseri, S. Ramezani-Moghadam, M. Kazemnejadi, and A. Allahresani, "Cu/Ni-doped sulfated zirconium oxide immobilized on $CdFe_2O_4$ NPs: A cheap, sustainable and magnetically recyclable inorgano-catalyst for the efficient preparation of α-aminonitriles in aqueous media," *Res. Chem. Intermed.*, vol. 46, no. 9, pp. 4233–4256, 2020, doi: 10.1007/s11164-020-04203-x.
75. H. Kaur et al., "Effect of synthesis methods and conditions on properties and applications of carbon dots for the detection of potential water contaminants: A review," *Crit. Rev. Anal. Chem.*, vol. 53, no. 4, pp. 751–774, 2023, doi: 10.1080/10408347.2021.1977608.
76. H. Kaur, S. Athwal, and K. Garg, "Futuristic approaches in biofertilizer industry: Challenges, opportunities, and future directions," in *Metabolomics, Proteomics and Gene Editing Approaches in Biofertilizer Industry: Volume* II, S. Kaur, V. Dwibedi, and P. K. Sahu, Eds., Singapore: Springer Nature Singapore, 2024, pp. 15–33. doi: 10.1007/978-981-97-2910-4_2.
77. H. Kaur, S. Sareen, V. Mutreja, and M. Verma, "Spinach-derived carbon dots for the turn-on detection of chromium ions (Cr^{3+})," *Journal of Inorganic and Organometallic Polymers and Materials*, vol. 33, no. 12, pp. 3703–3715, 2023.
78. C. V. Reddy, I. N. Reddy, K. R. Reddy, S. Jaesool, and K. Yoo, "Template-free synthesis of tetragonal Co-doped ZrO_2 nanoparticles for applications in electrochemical energy storage and water treatment," *Electrochim. Acta*, vol. 317, pp. 416–426, 2019, doi: 10.1016/j.electacta.2019.06.010.
79. C. V. Reddy et al., "Efficient removal of toxic organic dyes and photoelectrochemical properties of iron-doped zirconia nanoparticles," *Chemosphere*, vol. 239, p. 124766, 2020, doi: 10.1016/j.chemosphere.2019.124766.
80. H. Chen et al., "Atomic layer deposition of Pt nanoparticles on low surface area zirconium oxide for the efficient base-free oxidation of 5-hydroxymethylfurfural to 2,5-furandicarboxylic acid," *Appl. Catal. A Gen.*, vol. 555, pp. 98–107, 2018, doi: 10.1016/j.apcata.2018.01.023.
81. M. Ikram et al., "Promising performance of chemically exfoliated Zr-doped MoS_2 nanosheets for catalytic and antibacterial applications," *RSC Adv.*, vol. 10, no. 35, pp. 20559–20571, 2020, doi: 10.1039/d0ra02458a.
82. S. Chen et al., "Ultrasmall amorphous zirconia nanoparticles catalyse polyolefin hydrogenolysis," *Nat. Catal.*, vol. 6, no. 2, pp. 161–173, 2023, doi: 10.1038/s41929-023-00910-x.
83. Y. Zhao et al., "Zirconium modification promotes catalytic activity of a single-site cobalt heterogeneous catalyst for propane dehydrogenation," *ACS Omega*, vol. 3, no. 9, pp. 11117–11127, Sep. 2018, doi: 10.1021/acsomega.8b00862.
84. J. Qiu et al., "Noble metal nanoparticle-functionalized Zr-metal organic frameworks with excellent photocatalytic performance," *J. Colloid Interface Sci.*, vol. 538, pp. 569–577, 2019, doi: 10.1016/j.jcis.2018.12.024.

85. S. Dang et al., "Role of zirconium in direct CO_2 hydrogenation to lower olefins on oxide/zeolite bifunctional catalysts," *J. Catal.*, vol. 364, pp. 382–393, 2018, doi: 10.1016/j.jcat.2018.06.010.
86. A. S. Al-Fatesh, Y. Arafat, S. O. Kasim, A. A. Ibrahim, A. E. Abasaeed, and A. H. Fakeeha, "In situ auto-gasification of coke deposits over a novel Ni-Ce/W-Zr catalyst by sequential generation of oxygen vacancies for remarkably stable syngas production via CO_2-reforming of methane," *Appl. Catal. B Environ.*, vol. 280, p. 119445, 2021, doi: 10.1016/j.apcatb.2020.119445.
87. S. Kim et al., "Thermal activation of zirconium(IV) acetylacetonate catalysts to enhance polyurethane synthesis and reprocessing," *Macromolecules*, vol. 57, no. 14, pp. 6759–6768, Jul. 2024, doi: 10.1021/acs.macromol.4c00625.
88. G. Chai, W. Zhang, L. F. Liotta, M. Li, Y. Guo, and A. Giroir-Fendler, "Total oxidation of propane over Co_3O_4-based catalysts: Elucidating the influence of Zr dopant," *Appl. Catal. B Environ.*, vol. 298, p. 120606, 2021, doi: 10.1016/j.apcatb.2021.120606.
89. N. Jeon, H. Choe, B. Jeong, and Y. Yun, "Propane dehydrogenation over vanadium-doped zirconium oxide catalysts," *Catal.* Today, vol. 352, pp. 337–344, 2020, doi: 10.1016/j.cattod.2019.12.012.
90. L. Yang et al., "Ligand-directed conformational control over porphyrinic zirconium metal–organic frameworks for size-selective catalysis," *J. Am. Chem. Soc.*, vol. 143, no. 31, pp. 12129–12137, Aug. 2021, doi: 10.1021/jacs.1c03960.
91. D. Li, J. Zhang, Y. Liu, H. Yuan, and Y. Chen, "Boron doped magnetic catalysts for selective transfer hydrogenation of furfural into furfuryl alcohol," *Chem. Eng. Sci.*, vol. 229, p. 116075, 2021, doi: 10.1016/j.ces.2020.116075.
92. M. Sakti La Ore et al., "The synthesis of SO_4/ZrO_2 and Zr/CaO catalysts via hydrothermal treatment and their application for conversion of low-grade coconut oil into biodiesel," *J. Environ. Chem. Eng.*, vol. 8, no. 5, p. 104205, 2020, doi: 10.1016/j.jece.2020.104205.
93. S. Das et al., "Role of lattice oxygen in methane activation on Ni-phyllosilicate@$Ce_{1-x}Zr_xO_2$ core-shell catalyst for methane dry reforming: Zr doping effect, mechanism, and kinetic study," *Appl. Catal. B Environ.*, vol. 290, p. 119998, 2021, doi: 10.1016/j.apcatb.2021.119998.
94. F. H. Alhassan, U. Rashid, and Y. H. Taufiq-Yap, "Synthesis of waste cooking oil-based biodiesel via effectual recyclable bi-functional $Fe_2O_3 - MnO - SO_4^{2-}/ZrO_2$ nanoparticle solid catalyst," *Fuel*, vol. 142, pp. 38–45, 2015, doi: 10.1016/j.fuel.2014.10.038.
95. M. Afsharizadeh and M. Mohsennia, "Catalytic synthesis of biodiesel from waste cooking oil and corn oil over zirconia-based metal oxide nanocatalysts," *React. Kinet. Mech. Catal.*, vol. 128, no. 1, pp. 443–459, 2019, doi: 10.1007/s11144-019-01622-9.
96. A. Guldhe, P. Singh, F. A. Ansari, B. Singh, and F. Bux, "Biodiesel synthesis from microalgal lipids using tungstated zirconia as a heterogeneous acid catalyst and its comparison with homogeneous acid and enzyme catalysts," *Fuel*, vol. 187, pp. 180–188, 2017, doi: 10.1016/j.fuel.2016.09.053.
97. S. K. Das, M. K. Bhunia, A. K. Sinha, and A. Bhaumik, "Synthesis, characterization, and biofuel application of mesoporous zirconium oxophosphates," *ACS Catal.*, vol. 1, no. 5, pp. 493–501, 2011, doi: 10.1021/cs200005z.
98. J. Huang, Y. Jian, P. Zhu, O. Abdelaziz, and H. Li, "Research progress on the photo-driven catalytic production of biodiesel," *Front. Chem.*, vol. 10, no. April, pp. 1–7, 2022, doi: 10.3389/fchem.2022.904251.
99. R. Klimkiewicz, H. Grabowska, and H. Teterycz, "Application of Zr-Mg-Y-O catalyst for ketonization of ester and alcohol type industrial wastes," *Polish J. Environ. Stud.*, vol. 12, no. 1, pp. 67–71, 2003.

100. M. Delarmelina, G. Deshmukh, A. Goguet, C. R. A. Catlow, and H. Manyar, "Role of sulfation of zirconia catalysts in vapor phase ketonization of acetic acid," *J. Phys. Chem.* C, vol. 125, no. 50, pp. 27578–27595, 2021, doi: 10.1021/acs.jpcc.1c06920.
101. S. Zaidi et al., "Facile synthesis of nanosized La/ZrO_2 catalysts for ketonization of free fatty acid and biomass feedstocks," *J. Taiwan Inst. Chem. Eng.*, vol. 121, pp. 217–228, 2021, doi: 10.1016/j.jtice.2021.04.013.
102. A. Iannaci, T. Pepè Sciarria, B. Mecheri, F. Adani, S. Licoccia, and A. D'Epifanio, "Power generation using a low-cost sulfated zirconium oxide based cathode in single chamber microbial fuel cells," *J. Alloys Compd.*, vol. 693, pp. 170–176, 2017, doi: 10.1016/j.jallcom.2016.09.159.
103. D. A. Jadhav, P. A. Deshpande, and M. M. Ghangrekar, "Enhancing the performance of single-chambered microbial fuel cell using manganese/palladium and zirconium/palladium composite cathode catalysts," *Bioresour. Technol.*, vol. 238, pp. 568–574, 2017, doi: 10.1016/j.biortech.2017.04.085.
104. O. Ichihashi, K. Matsuura, K. Hirooka, and T. Takeguchi, "Application of zirconium-based materials as catalyst of air-cathode in microbial fuel cells," *J. Water Environ. Technol.*, vol. 14, no. 2, pp. 106–113, 2016, doi: 10.2965/jwet.15-081.

12 Zirconia-Based Materials for Electrochemical Energy Storage

Jibrin Muhammad Yelwa

12.1 INTRODUCTION

Due to the rapid developments in the field of electrochemical energy storage devices, the development of zirconia-based materials has become imperative in order to meet the growing global demand for sustainable energy storage (Roseline, 2024). Zirconia, ZrO_2, displays excellent thermal stability along with very high mechanical strength and ionic conductivity (Yasin *et al.*, 2017; Manzoor *et al.*, 2021); thus, it boosts the performance and lifetime of energy storage devices significantly (Matei *et al.*, 2024). The reason for such versatility in zirconia is because of its polymorphic nature and the effect of doping, hence surprisingly allowing various phases to be stabilized in a wide range of electrochemical batteries, supercapacitors, and fuel cells (Akhkozov *et al.*, 2022; Steele, 2000).

Efficient energy storage systems are a necessity, especially in accommodating intermittent renewable energy sources, such as solar and wind (Zhu *et al.*, 2024). In achieving this goal, electrochemical energy storage systems (EESSs) play the central role in enabling the storage of energy and its release in a regulated manner. Above all, cubic phase-stabilized zirconia-based materials doped with typical dopants, such as yttria, namely, Y_2O_3, have demonstrated remarkable ionic conductivity. Owing to this fact, such materials are ideal candidates for serving as a solid electrolyte in both batteries and fuel cells (Ryu *et al.*, 2023). Such materials not only improve effectively ionic transport but also exhibit very high chemical stability—a very critical requirement for ensuring the durability of electrochemical devices, as was reportedly done lately by Goodenough & Park (2013).

Recent developments have been made on nanostructured zirconia-based materials, which could significantly increase the surface area and active sites, thus greatly enhancing electrochemical performance (Naeem Ashiq *et al.*, 2021). For example, nanostructured yttria-stabilized zirconia (YSZ) has demonstrated distinct improvements in ionic conductivity and mechanical strength (Maiti *et al.*, 2022; Aydın *et al.*, 2022), both of which are important features for high-performance solid oxide fuel cells (Beshiwork *et al.*, 2022). Besides, zirconia has been explored in conjunction with other components, such as carbon-based nanomaterials, in developing composite electrodes with enhanced electrochemical performance (Zhao *et al.*, 2020).

DOI: 10.1201/9781003519300-15

Many challenges argue against the development of such materials. Phase stability, sintering behavior, and the price of dopants are critical factors that affect practical applications (Zhang *et al.*, 2015). Continuous developments in material synthesis and processing techniques result in overcoming these challenges so as to take advantage of zirconia on its own to optimize the properties for the particular electrochemical applications (Kumari *et al.*, 2022). In addition, the application does not end with conventional batteries and fuel cells (Akhkozov *et al.*, 2022). Thus, its applicability on new lithium sulfur batteries or supercapacitors is from its versatility and capability of revolutionizing energy storage systems in general (Jia *et al.*, 2020). The work presented in these studies indicates that there is ongoing research and various innovative ways in which efforts have been developed to use all that zirconia-based materials could offer (Matei *et al.*, 2024). Zirconia-based materials are considered to be the most promising methods for the development of high-performance, durable, and efficient energy storage due to their ability to store electrochemical energy (Zakaria & Kamarudin, 2021). Being in this field, this constant movement is characteristic where serious scientific and technological elicits are driven by the global imperative for sustainable energy solutions (Goodenough & Park, 2013). There can be no doubt about the fact that complex properties of zirconia and its incorporation into various electrochemical devices will make the future landscape of energy storage technologies (Ye *et al.*, 2020).

Electrochemical energy storage involves the process of conversion between electrical and chemical energy and the reverse (Temam *et al.*, 2023). It provides the basis for a number of energy storage systems, including but not confined to batteries and supercapacitors for applications varying in scale from portable electronics up to large-scale energy storage systems that help in grid stabilization, as pointed out by Nisar *et al.* (2020). In fact, the importance of EESSs lies in their balancing of energy systems and demand, enhancement of reliability of renewable sources of energy, and reduction of greenhouse gas emissions. Figure 12.1 illustrates the different devices used for electrochemical energy storage and conversion, with zirconia-based materials providing one route to optimizing efficiency and performance. This system plays

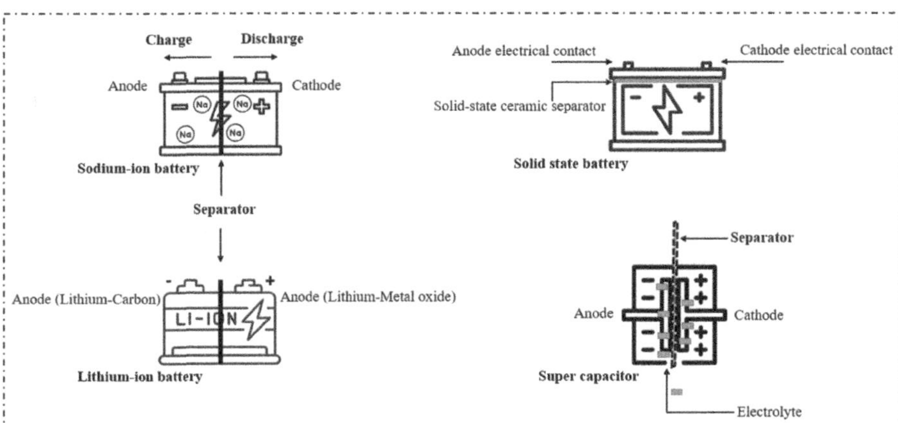

FIGURE 12.1 Electrochemical energy storage and conversion systems.

a vital role in enabling a shift toward a sustainable energy future because it delivers efficient and reliable energy storage solutions (Goodenough & Park, 2013). Recent developments in energy storage materials have focused on approaches for enhancing energy density, cycle life, and safety. New development techniques, such as solid-state electrolytes, advanced cathode materials, and novel nanostructures, point the way toward next-generation batteries and supercapacitors.

12.2 ELECTROCHEMICAL ENERGY STORAGE MATERIALS

12.2.1 Lithium-Ion Batteries

The various applications of lithium-ion batteries (LIBs) are due to their great energy density, cyclability, and relatively low self-discharge rate. Conventionally, the main components of such batteries are graphite anodes, lithium cobalt oxide cathodes, and liquid electrolytes (Joo & Choi, 2008). Over recent years, great interest has been paid to enhancing the safety and energy density of these batteries while making them more economically viable (Whittingham, 2004).

12.2.2 Sodium-Ion Batteries

Sodium-ion batteries (SIBs) have been unfolding as potential variants from LIBs, owing to the abundance and low cost of sodium. Indeed, they operate on the same principles as LIBs, except that active materials will consist of sodium-based compounds. Further research is done to improve their performance and stability (Hwang *et al.*, 2017).

12.2.3 Supercapacitors

Energy is stored in supercapacitors through the electrostatic separation of charges. They are characterized by high power density, rapid charge-discharge cycling, and long cycle life. Activated carbon, graphene, and metal oxides are some of the materials used in supercapacitors (Zhang et al, 2022).

12.2.4 Solid-State Batteries

Working principle: In solid-state batteries (SSBs), a liquid electrolyte is replaced by a solid one, enhancing their safety by allowing higher energy density (Joo & Choi, 2008). However, this technology is under development, and its research is mainly oriented toward finding suitable solid electrolytes and improving the interface between electrolytes and electrodes (Banerjee *et al.*, 2020; Sun & Kamat, 2021). Table 12.1 describes materials used for electrochemical energy storage, with their energy density, power density, cycle life, efficiency, working temperature, and applications.

12.2.5 Zirconia-Based Materials

Zirconia-based materials are gaining attention for their potential in electrochemical energy storage applications. Materials possessing thermal and chemical stability,

TABLE 12.1
Comparison of Materials Used in Electrochemical Energy Storage

Material	Energy Density (Wh/kg)	Power Density (W/kg)	Cycle Life (Cycles)	Efficiency (%)	Operating Temperature (°C)	Application	References
Lithium-Ion Batteries	150–250	200–2000	1000–5000	85–95	−20 to 60	Portable electronics, EVs	Whittingham (2004)
Sodium-Ion Batteries	100–150	100–1500	500–3000	80–90	−10 to 50	Grid storage, portable electronics	Hwang *et al.* (2017)
Supercapacitors	5–50	1000–10,000	>100,000	90–98	−40 to 70	Power backups, hybrid EVs	Zhang and Zhao (2021); Zhang and Huang (2017)
Solid-State Batteries	200–400	300–2500	>5000	90–97	−30 to 80	Advanced battery systems	Banerjee *et al.* (2020)
Zinc-Ion Batteries	50–120	500–2000	2000–4000	75–85	−10 to 55	Grid storage, wearable electronics	Liu *et al.* (2021)
Flow Batteries	30–60	200–800	>10,000	70–85	5 to 45	Large-scale energy storage	Chen *et al.* (2019)

along with chemical resistance and ionic conductivity, could potentially be used as components in batteries or supercapacitors depending on integration with other technologies (Badwal, 1992). The following sections took a material based on zirconia synthesis approach, its properties and applications, in the context of EESSs, i.e., zirconia (Kim *et al.*, 2018).

12.3 PROPERTIES OF ZIRCONIA-BASED MATERIALS FOR ELECTROCHEMICAL ENERGY STORAGE

12.3.1 Thermal Stability and Chemical Resistance

A number of properties make ZrO_2 a very promising material for electrochemical energy storage. Therefore, it is highly desirable to develop a high thermal stability material, such as ZrO_2, offering integrity performance for these applications. In addition, zirconia is quite chemically resistant leading to minimal degradative alteration with time and providing better storage lifetime (Kim *et al.*, 2024).

12.3.2 Ionic Conductivity

Ion ionic conductivity of zirconia is another important property. Among its doped forms (e.g., YSZ), it exhibits very high ionic conductivity and is recognized as one of the best electrolytes for solid oxide fuel cells and batteries (Beshiwork *et al.*, 2022; Joo & Choi, 2008). A fundamental element to the performance of electrochemical cell is that high-conductivity ions can move fast (Ye *et al.*, 2020).

12.4 TYPES AND SYNTHESIS OF ZIRCONIA-BASED MATERIALS

Figure 12.2 shows the structural arrangement of zirconia materials, their nanostructures, and their electrochemical applications' stability.

12.4.1 Types of Zirconia-Based Materials

12.4.1.1 Zirconia Nanotubes

As Aydın *et al.* (2022) mentions, Zirconia nanotubes are cylindrical nanostructures with a large surface area to volume ratio and therefore increase their electrochemical activity. Energy storage devices, such as lithium-ion batteries and supercapacitors, are key for increasing the charge storage capacity and cycling stability (Jia *et al.*, 2020) of such materials that have such a large range of applications.

12.4.1.2 Zirconia Nanoparticles

Zirconia nanoparticles have spheric diameter particles that lie within the nanometer diameter range. Due to their small size and high surface area, nanoparticles might

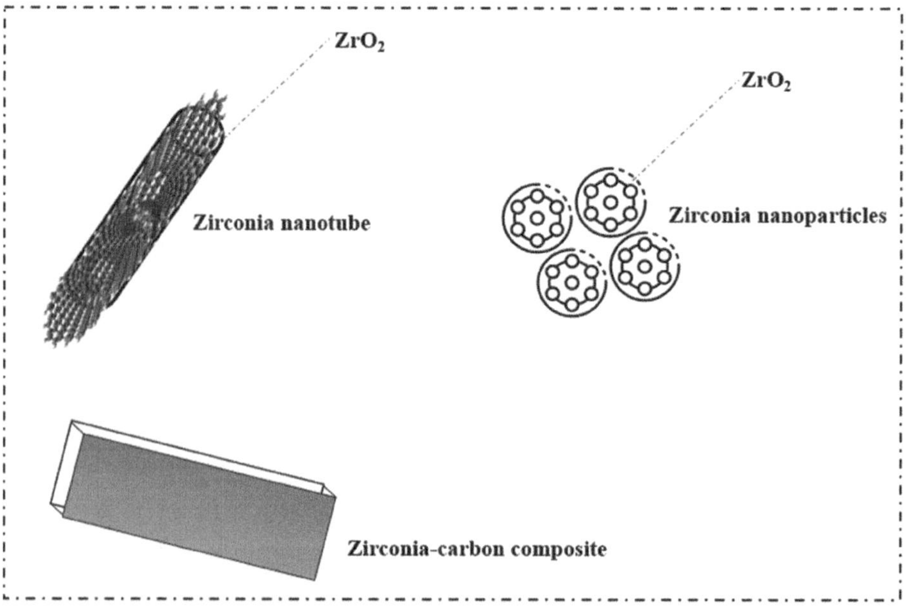

FIGURE 12.2 Structure of zirconia-based materials.

give better electrochemical performance. They can be applied in a few sorts of battery technologies (Bruce *et al.*, 2012). Normally, nanoparticles are added to batteries to enhance the conductivity and stability of electrode materials (Bortamuly *et al.*, 2023).

12.4.1.3 Zirconia-Based Composites

In most instances, materials have been combined to form one with enhanced properties in a zirconia composite. For example, a composite of carbon and zirconia incorporates the high conductivity of carbon into stable zirconia. The composites are specifically tailored to improve the overall performance metrics such as energy density and cycle life (Wang *et al.*, 2025).

12.4.2 Synthesis Methods

12.4.2.1 Sol-Gel Method

In general, sol-gel processing involves the evolution from a solution-phase precursor to a solid gel-phase material (Nayak *et al.*, 2022). The method allows for the appropriate processing of chemical composition and morphology in the case of zirconia-based materials (Kumari *et al.*, 2022). Sol-gel processing is one of the methods for synthesizing high-purity homogeneous nanoparticles and thin films of zirconia (Lv *et al.*, 2019).

12.4.2.2 Hydrothermal Synthesis

Hydrothermal synthesis generally involves the crystallization of materials from aqueous solutions under conditions of high temperature and pressure. This method usually has an advantage in the synthesis of zirconia nanostructures of well-defined shape and size (Heiba *et al.*, 2025). Hydrothermal synthesis most often results in the fabrication of different nanostructures like nanotubes (Yang & Wu, 2022).

12.4.2.3 Co-Precipitation Method

Co-precipitation refers to the simultaneous precipitation of several components from solutions. The technique has gained broad applications for the synthesis of composites based on zirconia due to its homogeneous distribution in their phases. Indeed, the method is very effective in getting improved doped zirconia with high ionic conductivity, such as YSZ (Ye *et al.*, 2020). Figure 12.3 illustrates the different types of synthesis methods for zirconia-based materials, including sol-gel, hydrothermal, precipitation, and solid state.

12.5 CHARACTERIZATION TECHNIQUES FOR ZIRCONIA-BASED MATERIALS

12.5.1 X-ray Diffraction (XRD)

In all zirconia-based studies, X-ray diffraction is one of the key means of determining the crystalline structure of the material. Thus, from XRD studies, information about phase composition, crystallite size, and lattice parameters has been obtained, the necessary base for understanding their electrochemical properties (Wang *et al.*, 2020).

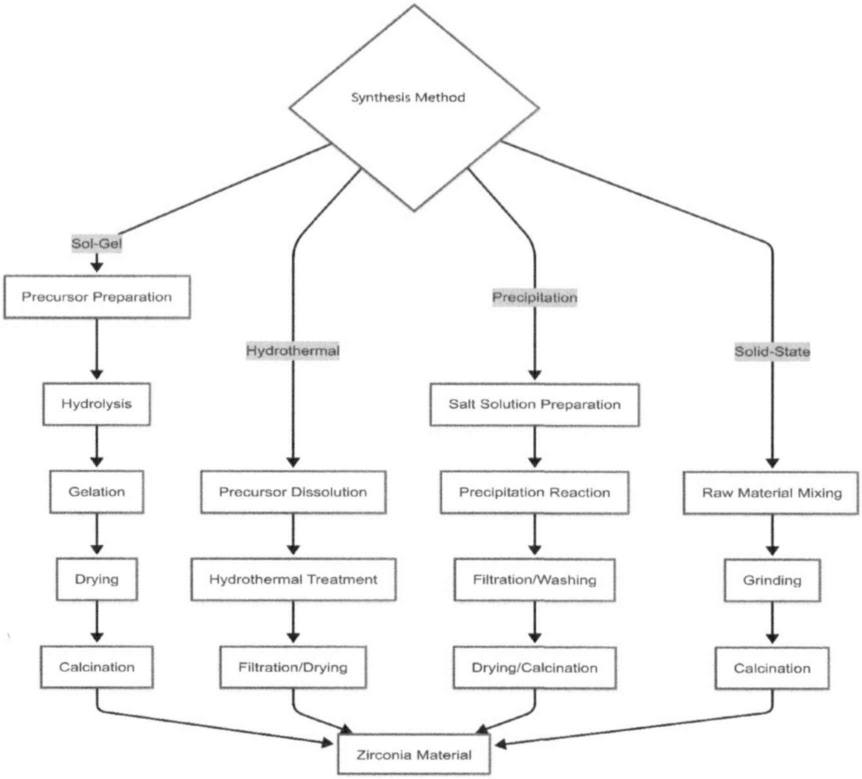

FIGURE 12.3 Flowchart of synthesis methods for zirconia-based materials.

12.5.2 Scanning Electron Microscopy (SEM)

In particular, scanning electron microscopy (SEM) observation has been widely used in order to investigate the surface morphology and microstructure of zirconia-based materials. The result from SEM is able to give high-resolution images, which allows for clear identification of the texture, grain size, and porosity of the material under investigation, which are generally considered to be important factors influencing electrochemical performance (Ali *et al.*, 2022).

12.5.3 Electrochemical Impedance Spectroscopy (EIS)

Electrochemical impedance spectroscopy (EIS) measures the impedance of zirconia-based materials within a running frequency range. It is employed to study ionic conductivity and changes in charge transfer resistance, which describes the material's suitability for electrochemical energy storage (Lv *et al.*, 2019). Table 12.2 outlines some characterization methods for zirconia materials, their function, relevant parameters, sensitivity, and suitable sample types.

TABLE 12.2
Characterization Techniques for Zirconia-Based Materials

Technique	Purpose	Key Metrics	Sensitivity	Sample Type	References
X-ray Diffraction (XRD)	Phase identification and crystallinity	Lattice parameters, crystallite size	High	Crystalline materials	Ali *et al.* (2022)
Scanning Electron Microscopy (SEM)	Surface morphology analysis	Grain size, porosity	High	Powders, thin films	Ali *et al.* (2022)
Transmission Electron Microscopy (TEM)	Atomic-scale imaging	Crystal defects, nanoparticle shape	Very High	Nanomaterials	Naeem Ashiq *et al.* (2021)
Electrochemical Impedance Spectroscopy (EIS)	Ionic conductivity and charge transfer resistance	Impedance, Resistance	Medium	Electrodes, electrolytes	Lv *et al.* (2019); Joo and Choi (2008)
Raman Spectroscopy	Chemical composition analysis	Vibrational modes, phase identification	Medium	Powders, films	Kim *et al.* (2024)

12.6 APPLICATIONS OF ZIRCONIA-BASED MATERIALS IN BATTERIES

12.6.1 Lithium-Ion Batteries

LIBs are considered the most popular rechargeable batteries due to their high-energy densities and very long cycle life. However, challenges such as thermal instability and capacity degradation have arisen, limiting the performance of such batteries. Therefore, several zirconia-based materials have been considered to overcome such challenges by enhancing thermal stability and ionic conductivity in LIBs (Park *et al.*, 2023).

12.6.1.1 Electrolyte Applications

Highly dense, stabilized zirconia exhibits considerable ionic conductivity, enabling YSZ to become well-known as a solid electrolyte in LIBs (Joo & Choio, 2008). Chen *et al.* (2021) show evidence that the introduction of YSZ into the electrolytes increased the ionic conductivity for better battery performance. Lv *et al.* (2019), in their preparation of composite-based electrolyte, were further able to improve its characteristics, viz. the incorporation of YSZ together with other material additions, such as polymer composite materials.

TABLE 12.3
Performance Metrics of Zirconia-Based Anodes vs. Traditional Anodes

Anode Material	Capacity Retention (%)	Cycle Stability (Cycles)	Conductivity (S/cm)	Structural Stability	Charge-Discharge Rate (C-rate)	References
Graphite (traditional)	70	500	10^{-6}	Moderate	0.5–1C	Zhu *et al.* (2024)
Zirconia-coated graphite	90	1000	10^{-4}	High	1–2C	Zhu *et al.* (2024)
Zirconia-carbon composite	95	2000	10^{-3}	Very High	2–5C	Wang *et al.* (2020)
Lithium-titanium oxide (LTO)	85	5000	10^{-3}	Very High	5–10C	Naeem Ashiq *et al.* (2021)
Silicon-based anode	60	300	10^{-6}	Low	0.1–0.5C	Liu *et al.* (2020)

12.6.1.2 Anode and Cathode Materials

Zirconia-based materials have also been explored for both anode and cathode components in LIBs (Zhang & Huang, 2017). The cycling stability of anodes coated with zirconia was significantly improved, along with a decrease in capacity fading, a critical factor with regard to the life of LIBs. For example, Zhu *et al.* (2024) reported capacity retention above 90% after 500 cycles with zirconia-coated graphite anodes, whereas only 70% was observed for uncoated ones. Table 12.3 compares the operation of zirconia-based anodes with traditional anodes, considering capacity retention, cycling stability, and conductivity.

12.6.2 Sodium-Ion Batteries

Recently, SIBs have drawn increasing interest because sodium is an abundant and low-cost element that can overcome the weak points of LIBs. However, the existing SIBs still have several challenges to overcome, such as low-energy density and cycle life. Therefore, further investigations have been performed with zirconia-based materials for improved SIB performance (Hwang *et al.*, 2017).

12.6.2.1 Applications of Electrolyte

YSZ is also applied in SIBs as solid electrolytes. YSZ has a high ion-conducting ability which thus better transports sodium ions inside the system that is so crucial for SIB performances (Ye *et al*, 2020). As Liu *et al.* (2016) state, addition of zirconia to electrolytes increases ionic conduction and stability in the batteries.

12.6.2.2 Anode and Cathode Materials

However, the development of zirconia-based anode materials for SIBs had opened more possibilities to continue to enhance cycling stability and capacity retention as well as the thermal stability of ZrO_2-coated cathode that was better to electrochemical performance that was fit for high-performance SIBs (Hwang *et al.*, 2017; Zhang & Huang, 2017).

12.6.3 Solid-State Batteries

In SSBs, the replacement of the liquid electrolyte is carried out by a solid one. Due to such replacement, its safety is increased and there is the possibility of higher energy density (Joo & Choi, 2008). The YSZ-based material is among a wide group of investigated materials, which is commonly applied as a solid electrolyte in SSBs because of the high ionic conductivity and stability (Ye *et al.*, 2020).

12.6.3.1 Applications of Electrolyte

It has been shown that SSBs employ YSZ as one of the basic elements with high ionic conductivity and stability (Xue *et al.*, 2019). Despite considerable efforts to microstructure and composition of YSZ for superior performance as a solid electrolyte, its performance remains currently poor. In the second paper, we develop a variety of composite electrolytes including YSZ along with some other materials for the purpose of increasing the ionic conductivity and mechanical properties of SSBs (Lv *et al.* 2019).

12.6.3.2 Interface Engineering

The interfacial problems of solid electrolytes with electrodes are considered as one of the major challenges of SSB. This interface was improved with the better contact by means of different types of materials that have the potential such as zirconia materials. Ji *et al.* (2024) alleviate the performance and efficiency issues of SSBs by this improvement.

12.7 APPLICATIONS OF ZIRCONIA-BASED MATERIALS IN SUPERCAPACITORS

12.7.1 Symmetric Supercapacitors

Symmetric supercapacitors are characterized by their pair of electrodes that consist of same material. High power density and fast charge–discharge cycling rate as well as long life cycles are demonstrated in a symmetric supercapacitor. Typically, zirconia-based materials were used for symmetric supercapacitors due to their high surface area and excellent electrochemical performances in chemical stability (Park *et al.*, 2023).

12.7.1.1 Zirconia Nanoparticles

ZrO_2 nanoparticles provide a high surface area, hence allowing raising the symmetric supercapacitor capacitance. One way such supercapacitors based on zirconia nanoparticles could enhance electrochemical performance is by having higher active sites in charge storage and thus fast ion transport movement (Naeem Ashiq *et al.*, 2021).

12.7.1.2 Zirconia-Based Composites

Zirconia-based composite with conductive polymer or carbon-based material provides an improved performance to the symmetric supercapacitors. The high conducting nature of a polymer or a carbon material along with zirconia is employed for improved energy storage capabilities (Lv *et al.*, 2019).

12.7.2 Asymmetric Supercapacitors

Asymmetric supercapacitors have electrodes of different composition in both the positive and negative electrodes and thus offer a wider operating voltage range and better energy density than symmetric supercapacitors. Attempts to optimize asymmetric supercapacitors for both electrodes using zirconia-based material performance were made (Kim *et al.*, 2024).

12.7.2.1 Zirconia-Coated Electrodes

Among all asymmetric supercapacitors, zirconia has been used to coat their electrodes, thus giving even better cycling stability with some very good electrochemical performances. Indeed, the zirconia coating protects the active material from degradation by enhancing the durability of the electrode material (Kumari *et al.*, 2022).

12.7.2.2 Hybrid Electrodes

Hybrid electrodes of zirconia combined with other metal oxides or conductive materials showed better performance in asymmetric supercapacitors (Kreuer, 2003). The hybrid electrodes leverage the synergistic interactions occurring among the various components to enhance capacitance, energy density, and cycling stability (Park *et al.*, 2023). Table 12.4 contrasts zirconia-based supercapacitors with others in terms of their capacitance, energy density, power density, cycle stability, and charging time.

TABLE 12.4
Comparison of Zirconia-Based Supercapacitors with Other Materials

Supercapacitor Material	Capacitance (F/g)	Energy Density (Wh/kg)	Power Density (W/kg)	Cycle Stability (Cycles)	Charge Time (s)	References
Activated carbon	150–250	5–10	1000–3000	>100,000	<10	Naeem Ashiq *et al.* (2021)
Graphene	200–350	8–15	2000–5000	>100,000	<5	Naeem Ashiq *et al.* (2021)
Zirconia-based composite	250–400	12–20	3000–8000	>150,000	<5	Park et al., (2023)
Carbon nanotubes (CNT)	180–300	10–18	4000–9000	>120,000	<3	Liu *et al.* (2021)
MXene-based supercapacitors	300–500	15–25	5000–10,000	>200,000	<2	Chen *et al.* (2021)

12.7.3 Performance Comparison

12.7.3.1 Energy Density and Power Density

Zirconia-based supercapacitors have been proven to exhibit competitive energy density and power density relative to supercapacitors based on other fundamental materials. Because of the high surface area, stability, and efficiency of energy storage and delivery, zirconia is considered an important parameter in supercapacitor research. Manzoor *et al.* (2021) reported that the potentials of zirconia-based supercapacitors may achieve performances comparable to or even surpass those developed using activated carbon, graphene, and other metal oxides.

12.7.3.2 Cycling Stability

Among the most important parameters are those related to the cyclic ability of supercapacitors; a good supercapacitor must ensure a long duration of its device, ensuring its reliability at all times. A very slight capacity loss due to thousands of charge/discharge cycles was noticed during zirconia-based supercapacitor operation. The chemical stability brought about by zirconia from repeated cycling resulted in lesser degradation (Bortamuly *et al.*, 2023).

12.7.3.3 Cost and Scalability

While there exists a lot of performance gain because of the zirconia-based material, there lie cost and scalability as significant influencing factors too. Relatively conventional synthesis methods of obtaining zirconia-based materials are scalable most easily up to mass produce (Wen & Oh, 2014). The general use of zirconium in a combination that also reduces the price of the finally beneficial materials will gradually decrease the supercapacitor's cost (Zhao *et al.*, 2024) and economically validate it toward commercialization at large (Lv *et al.*, 2019).

12.8 APPLICATIONS OF ZIRCONIA-BASED MATERIALS IN ELECTROCHEMICAL ENERGY STORAGE

12.8.1 Lithium-Ion Batteries

There is a wide range of research on zirconia-based materials for their applications in LIBs. The YSZ is commonly used as a solid electrolyte material owing to its excellent ionic conductivity and good thermal stability. Thermal stability and cyclability can be enhanced by zirconia coatings on active material anodes and cathodes (Sun *et al.*, 2023), making them suitable for high-performance LIBs (Park *et al.*, 2023).

12.8.2 Sodium-Ion Batteries

Zirconia-based materials contribute to an enhancement of ionic conductivity and stability of the electrolyte in SIBs. YSZ, with its high sodium-ion conductivity, has been integrated into SIBs to improve their performance. Besides, the improvement in cycling stability and reduction of capacity fading for the zirconia-coated electrodes is very important for long-term operation (Hwang *et al.*, 2017).

12.8.3 Solid-State Batteries

Other very good applications of new zirconia-based materials are SSBs. The use of YSZ as a solid electrolyte in SSBs has many advantages, such as high ionic conductivity, thermal stability, and safety. The majority of research efforts are devoted to the optimization of the interface between the solid electrolyte and electrodes for improving the overall performance of SSBs (Ye *et al.*, 2020).

12.8.4 Supercapacitors

Zirconia-based materials have also been applied in supercapacitors due to their extremely high surface area, chemical stability, and excellent electrochemical properties. Thus, zirconia nanoparticles and their composites have been applied to enhance capacitance and energy density in supercapacitors. They offer robust performance with long cycling stability that is appropriate for applications at high power (Naeem Ashiq *et al.*, 2021).

Figure 12.4 shows zirconia-based materials' applications for supercapacitors, highlighting their superior electrochemical properties compared with common materials. It has been demonstrated that YSZ in SSBs can significantly enhance ionic conductivity and thermal stability, leading to improved battery performance. Table 12.5 illustrates the applications, benefits, ionic conductivity, and stability improvement of zirconia-based materials for electrochemical energy storage.

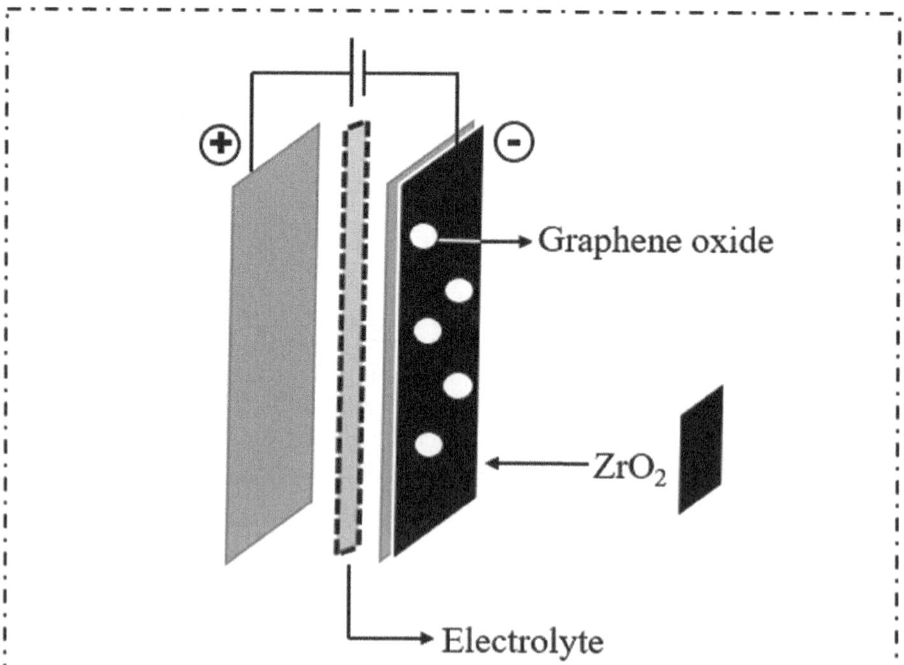

FIGURE 12.4 Zirconia-based supercapacitor.

TABLE 12.5
Applications of Zirconia-Based Materials in Electrochemical Energy Storage

Application	Benefits of Zirconia	Ionic Conductivity (S/cm)	Stability Improvement (%)	References
Lithium-ion batteries	Improved thermal stability, ionic conductivity	10^{-3}	40	Park *et al.*, (2023)
Sodium-ion batteries	Enhanced electrode stability, long cycle life	10^{-3}	35	Hwang *et al.* (2017)
Solid-state batteries	High ionic conductivity, better safety	10^{-2}	50	Ye *et al.* (2020)
Supercapacitors	High surface area, chemical stability	10^{-3}	30	Temam *et al.* (2023)
Zinc-ion batteries	Corrosion resistance, enhanced conductivity	10^{-3}	45	Liu *et al.* (2021)
Fuel cells	Stable ionic transport, high-temperature resilience	10^{-2}	60	Zhang and Huang, *et al.* (2017); Zhao *et al.* (2023); Akhkozov *et al.* (2022)

12.9 CHALLENGES AND FUTURE DIRECTIONS

Most of the challenges associated with the synthesis of zirconia-based materials are scalability for application in electrochemical energy storage. While different preparation methods have so far been engaged, such as sol-gel, hydrothermal, and co-precipitation techniques, there is an urgent need for the production techniques that will be cost-effective with the ability to meet commercial demands (Lv *et al.*, 2019). In this respect, future work has to be directed to the optimization of these procedures for high-value productions on a larger scale. Also, improvement in the interface of zirconia-based materials with other components in electrochemical devices holds the key to better performance due to interfacial resistance and poor adhesion that seriously affects batteries and supercapacitors. These problems will require advanced characterization techniques and new engineering methods to understand these interfaces better and optimize them.

Moreover, despite the many advantages of the application of zirconia-based materials, it also provides great room for performance improvement since the increased ionic conductivity, mechanical strength, and electrochemical stability would result in the better performance of energy devices. Further research should be conducted on new preparation methods and approaches for doping, which could offer improvement in the intrinsic properties of zirconia-based materials. It also plays a vital role in considering the environmental impact of the zirconia-based material in energy storage devices, focusing on synthesizing techniques by environmentally friendly methods and the availability of feedstocks. To make such materials commercially viable, production techniques should be economical and environment-friendly.

In the near future, studies should be oriented toward developing zirconia composites incorporating other advanced materials. This will generate new composites characteristic of improved properties. Advanced characterization methodologies will be used in the deep investigation of material properties and interfaces. Future efforts in the development of synthesis techniques for zirconia-based materials should be scalable and economical to produce quality materials, complemented by in-depth environmental impact studies. Besides, these materials also hold great promise for integration with next-generation energy storage solutions, such as hybrid batteries and flexible supercapacitors, enabling the possibility of advanced applications.

12.10 CONCLUSION

This chapter has defined the huge contribution of zirconia-based materials in developing the technological aspects of electrochemical energy storage devices, including LIBs, SIBs, supercapacitors, and SSBs. Zirconia possesses a unique combination of properties, such as high thermal stability and chemical resistance, while showing excellent ionic conductivity. It is an appropriate material for performance and stability improvement in the above energy storage systems. Accordingly, various forms of zirconia, including nanoparticles, nanotubes, and composites, have been developed and optimized for use in electrodes, electrolytes, and coatings, demonstrating improved energy density, cycling stability, and overall efficiency.

This means that, from this perspective, very promising practical applications are forecasted for zirconia-based materials in future advanced energy storage. However, a number of challenging issues, such as scalable syntheses, interface engineering, and economic viability, have to be met in order for the full practical development of applications to be realized. Actually, answering these challenges will require continuous research and development efforts by unlocking new applications for zirconia-based materials in the rapidly changing field of electrochemical energy storage which is toward more efficient and sustainable energy systems.

REFERENCES

Akhkozov, L., Danilenko, I., Podhurska, V., Shylo, A., Vasyliv, B., Ostash, O., & Lyubchyk, A. (2022). Zirconia-based materials in alternative energy devices-A strategy for improving material properties by optimizing the characteristics of initial powders. *International Journal of Hydrogen Energy*, *47*(97), 41359–41371.

Ali, A., Chiang, Y. W., & Santos, R. M. (2022). X-ray diffraction techniques for mineral characterization: A review for engineers of the fundamentals, applications, and research directions. *Minerals*, *12*(2), 205.

Aydın, H., Kurtan, U., Demir, M., & Karakuş, S. (2022). Synthesis and application of a self-standing zirconia-based carbon nanofiber in a supercapacitor. *Energy & Fuels*, *36*(4), 2212–2219.

Badwal, S. P. (1992). Zirconia-based solid electrolytes: Microstructure, stability and ionic conductivity. *Solid State Ionics*, *52*(1–3), 23–32.

Banerjee, A., Wang, X., Fang, C., Wu, E. A., & Meng, Y. S. (2020). Interfaces and Interphases in all-solid-state batteries with inorganic solid electrolytes. *Chemical Reviews, 120*(14), 6878–6933.

Beshiwork, B. A., Teketel, B. S., Luo, X., Tian, D., Yang, Q., Zhu, S., Chen, Y., Timurkutluk, B. and Lin, B., (2022). Nanoengineering electrode for yttria-stabilized zirconia-based symmetrical solid oxide fuel cells to achieve superior output performance. *Separation and Purification Technology*, *295*, 121174.

Bortamuly, R., Yeasmin, S., Elias, L., Das, M. R., Mahanta, D., & Saikia, P. (2023). Activated carbon derived from oleander seeds supported ceria-zirconia mixed oxides for enhanced supercapacitive behaviour. *Journal of Energy Storage*, *73*, 109029.

Bruce, P. G., Freunberger, S. A., Hardwick, L. J., & Tarascon, J. M. (2012). Li–O_2 and Li–S batteries with high energy storage. *Nature Materials, 11*(1), 19–29.

Chen, W., Zhang, D., Yang, K., Luo, M., Yang, P., & Zhou, X. (2021). Mxene (Ti3C2Tx)/cellulose nanofiber/porous carbon film as free-standing electrode for ultrathin and flexible supercapacitors. *Chemical Engineering Journal*, *413*, 127524.

Chen, X., Zhang, H., Li, Y., Xing, J., Zhang, Z., Ding, X., ... & Wang, S. (2021). Fabrication and performance of anode-supported proton conducting solid oxide fuel cells based on BaZr0. 1Ce0. 7Y0. 1Yb0. 1O3-δ electrolyte by multi-layer aqueous-based co-tape casting. *Journal of Power Sources*, *506*, 229922.

Chen, Y., Zhou, M., Xia, Y., Wang, X., Liu, Y., Yao, Y., ... & Wang, Q. (2019). A stable and high-capacity redox targeting-based electrolyte for aqueous flow batteries. *Joule*, *3*(9), 2255–2267.

Goodenough, J. B., & Park, K. S. (2013). The Li-ion rechargeable battery: A perspective. *Journal of the American Chemical Society, 135*(4), 1167–1176.

Heiba, A. R., Abdel-Salam, M. O., Yoon, T., & El Sawy, E. (2025). Zr-MOF composites with zipped and unzipped carbon nanotubes for high-performance electrochemical supercapacitors. *Nanoscale*, *17*(1), 459–473.

Hwang, J. Y., Myung, S. T., & Sun, Y. K. (2017). Sodium-ion batteries: Present and future. *Chemical Society Reviews, 46*(12), 3529–3614.

Ji, W., Luo, B., Wang, Q., Yu, G., Zhang, Z., Tian, Y., Zhao, Z., Zhao, R., Wang, S., Wang, X. and Zhang, B. (2024). Interface engineering enabling thin lithium metal electrodes down to 0.78 µm for garnet-type solid-state batteries. *Nature Communications*, *15*(1), 9920.

Jia, S., Long, J., Li, J., Yang, S., Huang, K., Yang, N., Liang, Y., & Xiao, J. (2020). Biomineralized zircon-coated PVDF nanofiber separator for enhancing thermo- and electrochemical properties of lithium-ion batteries. *Journal of Materials Science, 55*(31), 14907–14921.

Joo, J. H., & Choi, G. M. (2008). Simple fabrication of micro-solid oxide fuel cell supported on metal substrate. *Journal of Power Sources*, *182*(2), 589–593.

Kim, S. K., Lee, H. J., Moon, J. Y., Jo, Y. R., Lee, J., Park, J. H., ... & Joo, J. H. (2024). Understanding the phase stability of yttria stabilized zirconia electrolyte under solid oxide electrolysis cell operation conditions. *Journal of Materials Chemistry A*, *12*(14), 8319–8330.

Kim, S. H., Park, C. J., Cho, L. R., & Huh, Y. H. (2018). Evaluation of the ceramic liner bonding effect between zirconia and lithium disilicate. *The Journal of Prosthetic Dentistry*, *120*(2), 282–289.

Kreuer, K. D. (2003). Proton-conducting oxides. *Annual Review of Materials Research, 33*, 333–359.

Kumari, N., Sareen, S., Verma, M., Sharma, S., Sharma, A., Sohal, H. S., … & Mutreja, V. (2022). Zirconia-based nanomaterials: Recent developments in synthesis and applications. *Nanoscale Advances*, *4*(20), 4210–4236.

Liu, T., Zhang, X., Wang, X., Yu, J., & Li, L. (2016). A review of zirconia-based solid electrolytes. *Ionics*, *22*, 2249–2262.

Liu, C., Xie, X., Lu, B., Zhou, J., & Liang, S. (2021). Electrolyte strategies toward better zinc-ion batteries. *ACS Energy Letters*, *6*(3), 1015–1033.

Liu, H., Chen, T., Xu, Z., Liu, Z., Yang, J., & Chen, J. (2020). High-safety and long-life silicon-based lithium-ion batteries via a multifunctional binder. *ACS Applied Materials & Interfaces*, *12*(49), 54842–54850.

Lv, H., Bao, J., Qi, S., Jin, Q., & Guo, W. (2019). Optical and mechanical properties of purple zirconia ceramics. *Journal of Asian Ceramic Societies*, *7*(3), 306–311.

Maiti, T. K., Majhi, J., Maiti, S. K., Singh, J., Dixit, P., Rohilla, T., Ghosh, S., Bhushan, S. and Chattopadhyay, S. (2022). Zirconia-and ceria-based electrolytes for fuel cell applications: Critical advancements toward sustainable and clean energy production. *Environmental Science and Pollution Research*, *29*(43), 64489–64512.

Manzoor, S., Ashiq, M., Usman, M., Sadaqat, M., Mahmood, K., Munawar, T., Iqbal, F., Al-Anazy, M., Ashiq, M., & Najam-Ul-Haq, M. (2021). Development of excellent and novel flowery zirconia/cadmium sulfide nanohybrid electrode: For high performance electrochemical supercapacitor application. *Journal of energy storage*, 40, 102718.

Matei, E., Şăulean, A. A., Petriceanu, M., Râpă, M., Piticescu, R. R., Ştefănoiu, R., & Predescu, C. (2024). Environmental sustainability based on zirconium dioxide utilization in non-conventional energy applications. *Environments*, *11*(12), 265.

Naeem Ashiq, M., Aman, A., Alshahrani, T., Faisal Iqbal, M., Razzaq, A., Najam-Ul-Haq, M., Shah, A., Nisar, J., Tyagi, D. and Fahad Ehsan, M. (2021). Enhanced electrochemical properties of silver-coated zirconia nanoparticles for supercapacitor application. *Journal of Taibah University for Science*, *15*(1), 10–16.

Nayak, S., Kittur, A., & Nayak, S. (2022). Green synthesis of Silver-Zirconia composite using chitosan biopolymer binder for fabrication of electrode materials in supercapattery application for sustainable energy storage. *Current Research in Green and Sustainable Chemistry*, *5*, 100292.

Nisar, U., Petla, R., Al-Hail, S. A. J., Quddus, A. A., Monawwar, H., Shakoor, A., … & Amin, R. (2020). Impact of surface coating on electrochemical and thermal behaviors of a Li-rich $Li_{1.2}Ni_{0.16}Mn_{0.56}Co_{0.08}O_2$ cathode. *RSC Advances*, *10*(26), 15274–15281.

Park, S. J., Song, Y. W., Kang, B., Kang, M. J., Kim, M. Y., Choi, Y. J., ... & Hong, Y. (2023). Effect of High-temperature thermal management on degradation of li-ion battery for fast charging. *IEEE Transactions on Transportation Electrification*, *10*(2), 2912–2922.

Roseline, S. (2024). Zirconium's role in advancing electrochemical energy storage devices: A comprehensive review. *Introduction to Functional Nanomaterials*, 180–188.

Ryu, S., Choi, I. W., Kim, Y. J., Lee, S., Jeong, W., Yu, W., … & Cha, S. W. (2023). Nanocrystal engineering of thin-film yttria-stabilized zirconia electrolytes for low-temperature solid-oxide fuel cells. *ACS Applied Materials & Interfaces*, *15*(36), 42659–42666.

Steele, B. C. (2000). Appraisal of $Ce_{1-y}Gd_yO_{2-y/2}$ electrolytes for IT-SOFC operation at 500°C. *Solid State Ionics*, *129*(1–4), 95–110.

Sun, T., Cheng, X., Cao, T., Wang, M., Tian, J., Yan, T., ... & Zhang, Y. (2023). Optimizing LI ion transport in a garnet-type solid electrolyte via a grain boundary design. *Batteries*, *9*(11), 526.

Sun, Y. K., & Kamat, P. V. (2021). Advances in solid-state batteries, a virtual issue. *ACS Energy Lett.*, 6, 2356−2358.

Temam, A. G., Alshoaibi, A., Getaneh, S. A., Awada, C., Nwanya, A. C., Ejikeme, P. M., & Ezema, F. I. (2023). Recent progress on phosphate-based electroactive materials for supercapacitor applications. *Results in Surfaces and Interfaces*, *13*, 100154.

Wang, Z., Zeng, Z., Lu, Z., Liu, C., Dai, W., Hua, N., ... & Wu, H. (2025). Influence of yttrium oxide distribution on microstructure and properties of zirconia based on alternative doping routes. *JOM*, 1–10.

Wang, C., Wang, Y., Zhang, G., Chen, Y., Han, X., Liang, L., ... & Xu, L. (2020). Preparation of a novel transplant material, zirconium oxide (ZrO2) nanotubes, and characterizations research. *Annals of Transplantation*, *25*, e924272-1.

Wen, Y., & Oh, J. K. (2014). Recent strategies to develop polysaccharide-based nanomaterials for biomedical applications. *Macromolecular Rapid Communications*, *35*(21), 1819–1832.

Whittingham, M. S. (2004). Lithium batteries and cathode materials. *Chemical Reviews, 104*(10), 4271–4301. doi: 10.1021/cr020731c

Xue, Q., Huang, X., Zhang, J., Zhang, H., & Feng, Z. (2019). Grain boundary segregation and its influences on ionic conduction properties of Scandia doped zirconia electrolytes. *Journal of Rare Earths*, *37*(6), 645–651.

Yang, H., & Wu, N. (2022). Ionic conductivity and ion transport mechanisms of solid-state lithium-ion battery electrolytes: A review. *Energy Science & Engineering*, *10*(5), 1643–1671.

Yasin, A. S., Obaid, M., Mohamed, I. M., Yousef, A., & Barakat, N. A. (2017). ZrO_2 nanofibers/activated carbon composite as a novel and effective electrode material for the enhancement of capacitive deionization performance. *RSC Advances*, *7*(8), 4616–4626.

Ye, T., Li, L., & Zhang, Y. (2020). Recent progress in solid electrolytes for energy storage devices. *Advanced Functional Materials*, *30*(29), 2000077.

Zakaria, Z., & Kamarudin, S. K. (2021). Advanced modification of Scandia-stabilized zirconia electrolytes for solid oxide fuel cells application—A review. *International Journal of Energy Research*, *45*(4), 4871–4887.

Zhang, H., Zhang, J., Gao, X., Wen, L., Li, W., & Zhao, D. (2022). Advances in materials and structures of supercapacitors. *Ionics*, *28*(2), 515–531.

Zhang, W., Tan, Y., Gao, Y., Wu, J., & Tang, B. (2015). Ultrafine nano zirconia as electrochemical pseudocapacitor material. *Ceramics International*, *41*(2), 2626–2630.

Zhang, C., & Huang, K. (2017). A new composite cathode for intermediate temperature solid oxide fuel cells with zirconia-based electrolytes. *Journal of Power Sources*, *342*, 419–426.

Zhao, F., Irfan, M., Abbas, W., Solre, G. F., Al-Asbahi, B. A., & Qaid, S. M. (2024). Polypyrrole doped zirconium dioxide composite as electrode material for super-capacitor applications. *Indian Journal of Physics*, *99*(4), 1273–1280.

Zhao, H., Liu, B., Li, Y., Li, B., Ma, H., & Komarneni, S. (2020). One-pot green hydrothermal synthesis of bio-derived nitrogen-doped carbon sheets embedded with zirconia nanoparticles for electrochemical sensing of methyl parathion. *Ceramics International*, *46*(12), 19713–19722.

Zhu, Z., Liu, Y., Zhang, H., Xu, Y., Shao, Z., Ge, L., … & Chen, H. (2024). Emerging trends in electrochemical energy storage: A focus on low-temperature pseudocapacitors. *Energy Reviews*, *4*(1), 100118.

13 Zirconia-Based Solid Metal-Organic Framework for Electrochemical Sensors

Supriya Rana, Krishna K. Yadav, and Sunaina

13.1 INTRODUCTION

A huge number of hazardous gases, volatile organic compounds, and other toxic pollutants that are of very low concentrations are directly released into the environment which can cause huge harm to the environment and sustainability, as well as several human health risks [1,2]. Moreover, organic pollutants have high chemical stability and they remain stable as such for longer periods. Therefore, the development of high-performance gas sensors plays an important role in day-to-day life that includes human safety, health diagnosis, environmental monitoring, industrial safety, and measuring food quality [3–5]. A sensing material undergoes changes in physical properties such as conductivity, work function, and dielectric content in the presence of analytes. So, the chemical interference between sensing materials and analytes to be detected is an important factor in evaluating the performance [6]. The various factors associated with the sensing application are dependent on the response time, selectivity, steadiness, durability, cost, and sensitivity of the materials for forming sensing devices [7]. Earlier, various technologies have been developed for sensing, including high-performance liquid chromatography, gas chromatography, and surface-enhanced Raman spectroscopy, which are used for sensing applications. However, these techniques are expensive, non-portable, and complex analysis and sample preparation methodologies are required for sensing [8]. The various limitations associated with earlier techniques enable researchers to develop inexpensive, portable, high-performance sensing materials for practical applications. Developing such novel materials has been carried out for decades, among all metal oxides that have been synthesized which exhibit poor selectivity and stability issues associated with them that hamper their long-term applicability [9]. However, metal-organic frameworks (MOFs) have achieved heightened consideration for exploring the field of sensing applications. MOF materials having high surface area, adjustable pore size/volumes, numerous active sites, high chemical and thermal stability, unique organic linkers, and shape/size characteristics make them a potential candidate for sensing applications [10,11]. Tremendous work is going on to develop MOF-based sensing applications, such as luminescence-based sensors for H_2S detection, NO_x

 DOI: 10.1201/9781003519300-16

detection, non-aromatic compound, chlorofluorocarbon removal, and pesticide detection [7,12]. Therefore, the sensing applicability of MOFs depends on the biodegradability, nontoxicity, and luminescence behavior of the materials [13].

Gas sensors have become very important in detecting harmful gases present in the environment, such as carbon monoxide (CO) and ammonia (NH_3), volatile organic compounds, pesticides, and explosives, and they remain very stable in the environment. Therefore, regular quality check assessment of the environmental hazards is of utmost importance [6,14]. MOF-based materials have tunable pore size, high porosity, high surface area, tunable structural stability, flexible coordination environment, easy synthesis procedure, and high gas accessibility [15]. Therefore, the structural features enable the use of MOF materials for gas-sensing applications [16]. The active sites present in the MOF materials show a significant change in structural physicochemical properties in the presence of analytes during adsorption [10,16]. So, the synthesis of active MOF materials is required to enhance selective adsorption or interaction. The various types of methods used for MOF synthesis include layer-by-layer growth, hydrothermal, solvothermal, microwave, ultrasonic methods, etc., organic linker, metal node, reaction medium such as aqueous, non-aqueous, or combined in an appropriate ratio which plays a significant role in MOFs in synthesis and achieving desired properties [17]. The selection of MOFs is crucial for sensing applications and the activity depends on the electrode area and mass transfer efficiency. Usually, MOFs having a smaller size, porous structure, and unique morphological characteristics such as 2D or hybrid structure provide more accessibility for diffusion and improvement in sensitivity [18,19]. So, synthesizing stable MOFs is a challenging task, and tailoring their structure highly influences sensitivity by providing a significant surface area and high active sites. The sensing ability of MOFs is classified into three different areas which include luminescent MOF sensors, conductive MOF sensors, chemi-resistive gas sensors, and electromechanics sensors [20–22].

Among all MOF materials, zirconia-based MOF (Zr-MOF) is highly studied for gas-sensing purposes due to its larger surface area, unique metal cluster, and unique shape/size characteristics [5,23]. The sensitivity of MOF materials facilitates lower detection limits. High charge density, polarized bond, ability to attain high defect concentration without loss in crystallinity, high oxidation state of Zr^{4+}, and strong Zr and O atoms result in high thermal stability and water stability. Such properties provide Zr-MOFs that have huge potential in the field of adsorbent extraction and sensing applications [12,24]. The major mechanism associated with MOFs is a change in the intermolecular distance among metal ions and ligands, the interaction of the metal with analytes through chemical binding or host–guest chemistry. The high surface area and tunable pore size of the MOF materials suit for sensing a material. Cavka et al. [25] have been the first group that synthesized Zr-MOF having 12 coordination as $Zr_6O_4(OH)_4(CO_2)_{12}$ as in closed-packed metal structures using the solvothermal method in N, N'-dimethylformamide (DMF). By using 1,4-benzene-dicarboxylate (BDC) as an organic ligand to form UIO-66 MOF, 4,4′ biphenyl-dicarboxylate (BPDC) as a ligand to form UiO-67, and terphenyl-dicarboxylate (TPDC) as a ligand to form UiO-68-based Zr-MOFs, a three-dimensional structure is created as shown in Figure 13.1. However, the introduction of the electronic properties of the MOF is done by doping materials, such as ionic species, nanoparticles, and conductive

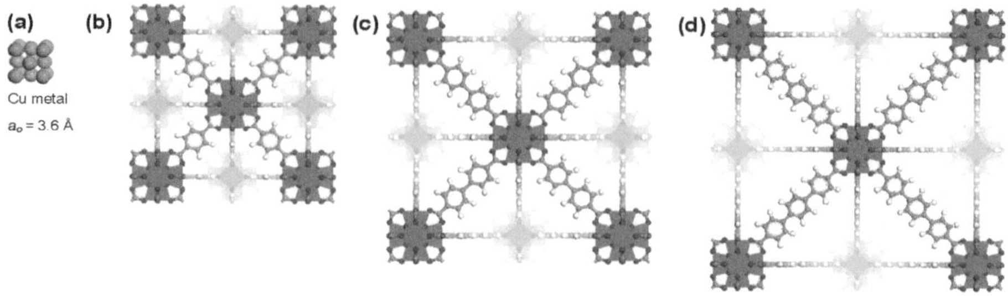

FIGURE 13.1 (a) Single unit cell of copper for scaling, (b) Zr-MOF; UiO-66, (c) Zr-MOF; UiO-67, (d) Zr-MOF; UiO-68. (Reprinted with permission from [25].)

materials-modified organic ligands. These conductive MOFs offer advanced capabilities for sensing and microelectronic devices. Zr-based MOFs have been investigated owing to their low cytotoxicity, colloidal stability, and adaptability to post-synthetic functionalization which improves the physiochemical properties and becomes a potential candidate for the overall modification for sensing applications [2].

So, owing to the solid affinity of Zr with O, it results in poor crystallinity of the Zr-MOF. Therefore, modulating the synthesis method for Zr-MOF to improve crystallinity can be done by using a modulator such as acetic acid which provides rapid metal and organic ligand coordination. This technique also benefits in tuning the physiochemical properties, pore size, and defect sites of Zr-MOFs [26,27]. Zr-based clusters include Zr_6O_8 and Zr_8O_6 having 12 coordination connected as $Zr_6(\mu_3\text{-}O)_4(\mu_3\text{-}OH)_4(CO_2)_{12}$ with Zr_6-octahedron [28,29]. Furthermore, the number of defects present in the Zr-MOFs has drawn a lot of attention which include the presence of missing linkers and missing clusters [29,30]. Furukawa et al. [29] studied in detail the water adsorption ability of Zr-based porous materials and the findings proved that the Zr-MOFs having $Zr_6O_4(OH)_4(CO_2)$ framework, secondary binary units (SBUs), and carboxyl organic linkers form porous frameworks. A detailed list of the Zr-MOFs with other topologies. Defects due to missing linker are due to Zr-containing cluster that is not fully 12-coordinated and the various organic ligands used for Zr-MOF are listed in detail [29]. Zr_6 nodes can be modified to be 4-, 6-, 8-, 10-, or 12-connected by using different organic ligands such as ditopic, tritopic, tetratopic, or hexatopic organic linkers as shown in Figure 13.2a and b [12,31]. Another defect in Zr-MOF is missing clusters which include the Zr cluster removal along with 12 ligands causing vacancy in the coordination number present in the adjacent Zr clusters. Therefore, this defect engineering results in a high surface area, large pore size, and a more accessible framework [32].

Zr-based MOF or nanocomposites have excellent water resistance and stability toward acids, organic solvents, and boiling water tests [25]. However, Zr^{4+} metal ion shows a strong affinity for hydroxide ion in the alkaline medium which causes the replacement with the linker framework and hence decreases the stability of the MOF. The high surface area and larger pore size of Zr-MOF endow with the high adsorption ability of the material [33]. Therefore, designing the structure of MOF, introducing functional groups such as amino, carboxyl, or hydroxy group, and introducing defect engineering provide high interaction with the incoming target molecule. These

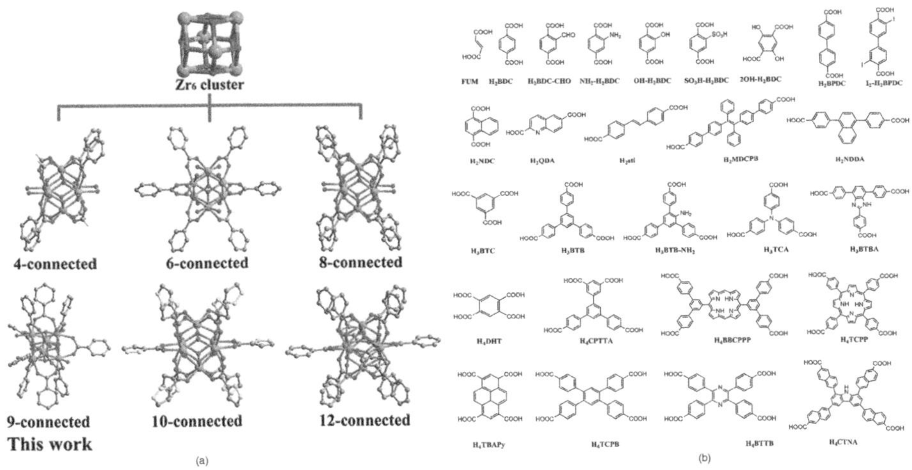

FIGURE 13.2 (a) Schematic showing $Zr_6O_4(OH)_4(CO_2)$ having SBUs with various coordination numbers (Reprinted with permission from [31]) and (b) different ligands used for the building of Zr-MOFs. (Reprinted with permission from [12].)

factors will enhance and improve the overall catalytic performance along with enormous applications [34,35]. In the following chapter, a detailed analysis of Zr-MOF as an electrochemical sensor and electrochemical biosensors has been done.

13.2 ZR-MOF AS ELECTROCHEMICAL SENSOR

Electrochemical sensing is a technique where a change in current is used for sensing analytes. Therefore, the faster the rate of electrochemical response, the faster the sensitivity phenomenon. The sensing process depends on the type of electrocatalyst and its quick response rate for sensing. However, the major limitation associated with MOFs is that they become unstable in water for prolonged periods and have stability issues [12]. The surface of such materials plays a significant role in electrochemical sensing. As per Pearson's hard/soft acid/base principle, high-valent metal ions such as Zr can coordinate with carboxylate-based ligands [2]. Therefore, Zr-MOFs have excellent potential to be efficiently used for sensing applications. UIO-66 and UIO-67 are very popular Zr cluster-based MOFs, where UIO-66 comprises $Zr_6O_4(OH)_4$ cubic framework, and 1,4-bezene-dicarboxylate as a ligand. Whereas, UIO-67 comprises biphenyl-4, 4′-dicarboxylate as a ligand with 2 different cages corresponding to octahedral and tetrahedral cages [8]. The use of solar energy to reduce organic pollutants present in the environment by using photocatalysis is a major process for wastewater treatment. Zheng et al. [36] synthesized NH_2-MIL-125 using the hydrothermal method and the structure consists of TiO_2/ZrO_2 composite self-doped with nitrogen in the MOF framework. They studied in depth the photodegradation efficiency of methylene blue, tetracyclic, and phenols. The findings of the research proved that the catalytic efficiency for degradation for methylene blue is 93% for 80 minutes UV radiation, 86% efficiency for tetracyclic, and 72%

degradation efficiency after 12 minutes UV radiation by using UV visible adsorption photocatalytic measurements.

Water oxidation using solar energy to convert water into H_2 and O_2 is a promising technique for developing materials sustainable and green solutions for energy consumption [37]. Wang et al. [38] have designed low-cost, novel, and efficient materials using Zr-MOF such as UIO-66-NH_2 octahedrons for H_2 production via photocatalytic reaction. The data confirms that graphene-wrapped UIO-66-NH_2 octahedrons possess ultra-high H_2 generation electrocatalysts because of the presence of graphene which prevents the electrons and holes from recombination. Liu et al. [39] synthesized successfully 3D Zr (IV) MOFs consisting of 2,3,5,6-tetrakis(4-carboxyphenyl) pyrazine framework and named as HBU-18 using the solvothermal method. The materials are used as fluorescent probes and serve as efficient materials for selective tyrosine (Tyr) detection at the low-level limit. Similarly, the materials exhibit an extremely low detection limit of 6.2 nM for 2,4,6-trinitrophenol. Furthermore, the materials also serve as efficient selectivity for CO_2 gas adsorption in comparison to N_2 with an uptake of 30.58 cm^3/g at room temperature conditions. Nandi et al. [40] synthesized dinitro-functionalized MOF having a framework similar to UIO-66 with Zr(IV) metal center using a solvothermal technique. The materials show fluorescence behavior toward H_2S sensing with extremely low detection limits corresponding to 14.14 μM and the value matches with the limits of H_2S present in biological samples, i.e., human blood plasma (Figure 13.3). Therefore, the materials exhibit advanced H_2S sensing for both extracellular and intracellular levels.

Wei et al. [41] synthesized low-cost Zr-based UiO-66-NH_2 MOF materials by using Zr source as a metal center and 2-amino terephthalic acid, acetic acid, and tetrabutylammonium bromide as an organic linker in DMF as a solvent via utilizing the more economical electrochemical method. Interestingly, the materials exhibit fluorescence behavior for detecting Fe^{3+} ions present in water at low concentrations of 10^{-8} mol/L and function as promising economical materials for the fluorescence detection of various metal ions. Xu et al. [42] synthesized MOFs-808 which is Zr-based MOF using trimesic acid as a linker via the solvothermal method. The findings proved

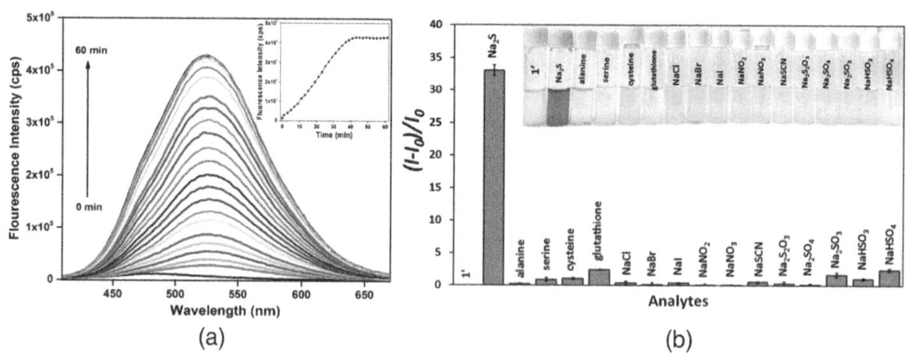

FIGURE 13.3 (a) UiO-66-$(NO_2)_2$ fluorescence increment behavior vs time in HEPES suspension with the addition of Na_2S, (b) corresponding comparative fluorescence response toward different analytes after 40 minutes. (Reprinted with permission from [40].)

that the presence of polar and electron-deficient groups such as polycarboxylic boost the H_2 adsorption. The materials show high stability in acidic medium, high temperatures up to 300°C, high surface area, and crystallinity which provide excellent storage capacity for hydrogen of 7.31 wt% at 77 K under 4 MPa. Karadeniz et al. [43] synthesized UiO-66 derivatives such as UiO-66, UiO-66-NH_2, MOF-801, and MOF-804 having Zr_6-cluster nodes and comprising $Zr_{12}O_8(OH)_8(CH_3COO)_{24}$ zirconium-based dodecanuclear acetate cluster via water-assisted milling mechanochemical synthesis. The materials have high crystallinity and porosity along with a green synthesis approach for MOFs. Catalytic hydrolysis profiles show that the nerve agent simulates dimethyl 4-nitrophenyl phosphate degradation in the presence of UiO-66 and UiO-66-NH_2, and the performance confirmed that the UiO-66-NH_2 prepared by TSE exhibits higher catalytic activity with $t_{1/2} \sim 6$ minutes in comparison to UiO-66 in water (Figure 13.4a). Cheng et al. [44] developed Cd/Zr-MOF where metal ions are Zr^{4+} and Cd^{2+} and terephthalic acid (H_2BDC) as organic linker for making framework via hydrothermal reaction 120°C. Furthermore, the photocatalytic degradation for Rhodamine B was explored using sunlight and the catalytic performance exhibited by molar ratio 5:1 of Cd/Zr-MOF shows 95.82% efficiency after 105 minutes visible light irradiation. Mott–Schottky experiment confirms the presence of n–n heterojunction among Cd-MOF and Zr-MOF. The overall schematic for the photodegradation of Rhodamine B is shown in Figure 13.4b. The study confirmed that the active species involved in the degradation process are superoxide ion (O^{2-}) and electron-hole (h^+). They appeared as effective materials for removing organic pollutants. Dmello et al. [45] developed amine functionalized Zr-based MOF (NH_2-UiO-66) using 1,4-benzene dicarboxylate as an organic linker via the solvothermal method. The study shows that the functionalization results in different interaction pathways with different acidic gases (SO_2, NO_2, CO_2). The highly acidic behavior of SO_2 readily leads to the formation of a complex and follows the reactivity order of $SO_2 > NO_2 > CO_2$.

Liu et al. [46] synthesized UiO-67 MOF-based having two-dimensional nanosheets structures via microwave-assisted method and further, the material is modified by in

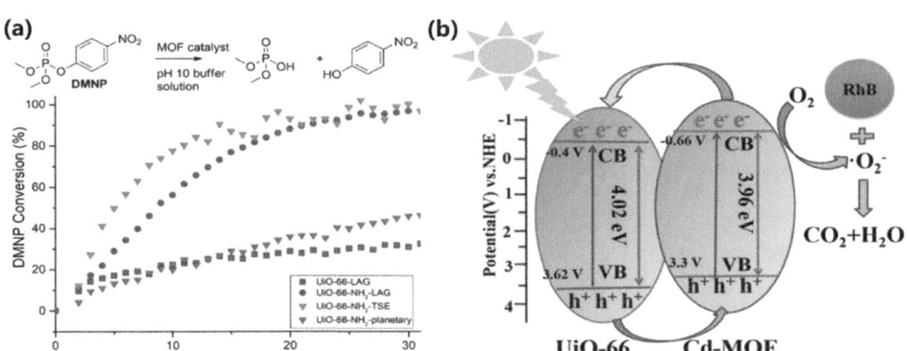

FIGURE 13.4 (a) Hydrolysis profiles showing the degradation of nerve simulant DMNP using UiO-66 and UiO-66-NH_2. (Reprinted with permission from [43].) (b) Photodegradation mechanism of Rhodamine B using Cd/Zr-MOF. (Reprinted with permission from [44].)

situ reduction process by using $HAuCl_4$ in the porous UiO-67 MOF referred to as Au/UiO-67 nanosheet nanocomposite. Au/UiO-67 material shows efficient catalytic performance toward the reduction of 4-nitrophenol to 4-aminophenol. Liang et al. [47] developed Zr-based MOF including UiO-66 and MOF808 containing terephthalic acid as a linker using a microwave-assisted method followed by acid treatment. The synthesized defective UiO-66 exhibits more defects with higher surface area and stability and these factors contribute to the enhancement of catalytic performance, i.e., cyclization of citronella to form 3-methyl-1,3-butanediol. The defective MOF structure of UiO-66 material is activated at 200°C for 12 hours and before microwave irradiation, the material shows a low conversion rate of 5.4% after 4 hours. However, the microwave irradiation of UiO-$66_{MW\text{-}0.5M1h}$ leads to a 68.1% conversion rate under the same conditions. Similarly, Tu et al. [48] synthesized Ti (IV) incorporated $Zr_6O_4(OH)_4$ framework of Zr-based MOF material using a microwave-assisted method, and UiO-66(Zr/Ti)-M demonstrated as an efficient material for photocatalysis through cation exchange. The materials show low detection limits of 0.01 ppb for photocatalytic Se detection along with competent stability in water and acid (HCOOH).

Yan et al. [49] synthesized novel bimetallic MOF consisting of Cu and Zr as metal centers as MOF-818 and 1H-pyrazole-4-carboxylic acid as an organic linker, adding trifluoroacetic acid by using solvothermal method. Further, MOF-818 composite with reduced graphene oxide/multiwalled carbon nanotubes (RGO/MWCNTs) is developed via in situ growth during the solvothermal method as MOF 818@RGO/$MWCNT_{s\text{-}x}$, where x varies from 1 to 4. The synthesized MOF-818@RGO/MWCNTs were loaded onto a glassy carbon electrode (GCE) and used as a working electrode for phenolic acid compound electrocatalytic analysis such as caffeic acid (CA), chlorogenic acid (CGA), and gallic acid (GA) present in biological samples. The MOF-818@RGO/MWCNTs/GCE exhibits a 5.7 nM of sensitivity and a 12.89 μA/μM detection limit in two different linear windows of 0.2–7 μM and 7–50 μM for CA detection. Similarly, the material shows good sensitivity in two linear ranges from 0.1 to 3 μM and 3 to 20 μM with 12.50 μA/μM detection value for CGA and comparable performance for GA. The sensor displays excellent stability, reproducibility,

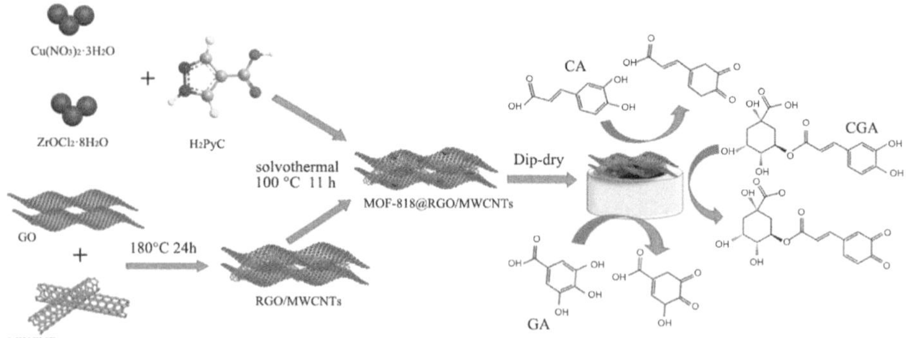

FIGURE 13.5 Schematic showing the overall synthesis of modified MOF-818@RGO/MWCNTs for sensing applications. (Reprinted with permission from [49].)

and performance especially for practical applicability such as human serum and urine-based samples as shown in Figure 13.5. The developed material exhibits a recovery percentage of 97.08%–102.67% for CA, 98.20%–101.50% for CGA, and 97.86.7%–101.52% for GA. The coupling effect of duel metals Cu and Zr and composite with RGO/MWCNTs provides an enhancement in the active sites which helps to improve the conductivity of modified electrode materials.

13.3 ZR-MOF AS ELECTROCHEMICAL BIOSENSORS

The excessive use of antibiotics and other pharmaceutical drugs has created severe environmental concerns, and their over-dependency leads to enormous health issues for mankind. So, many regulations and legislations have been created to limit excessive antibiotic industrialization such as nitrofurazone, ciprofloxacin, etc. [50]. Therefore, quick sensing of such pharmaceutical materials and drugs is of high importance. The electrochemical sensors while used as a biosensor require additional surface modification for enhancing the conductivity of the electrode surface to provide a higher number of electroactive sites [51]. The sensitivity toward biomolecules can be enhanced by introducing additional binding sites. Zr-based MOF fulfills all the required characteristics which include surface area, and adaptable structure, and enables designing for sensing for an efficient material as a biosensor. It provides strong interaction with the biomolecules either by electrostatic forces or π–π stacking [52]. The pharmaceutical industry has a huge responsibility for tackling health issues. However, these drugs cause numerous environmental issues and health concerns. So, the development of MOFs-based materials shows great potential for detecting various drugs and pharmaceutical waste such as cocaine detection. Fang et al. [53] designed composite materials consisting of amine-functionalized Zr(IV)-MOF as NH_2–UiO-66 consisting of reduced graphene oxide (RGO) and used as working electrodes for electrochemical sensing of ciprofloxacin (Cip). Further, the electrode was modified by Cu deposition on the surface of the material by using an anodic stripping voltammetry method and this modification facilitates the formation of a stable complex of Cip with Cu^{2+} (Figure 13.6). Cu^{2+} modified NH_2– UiO-66/RGO electrode shows sensing activity toward Cip by forming Cu^{2+}–Cip complex chemically in laboratory scale and contaminants in real water samples. The detection trace value is down to 6.67 nM and 10.86 μA/μM sensitivity [54]. Moreover, the material exhibits high sensitivity, selectivity, and stability, along with recoveries.

Luteolin has numerous beneficial merits in human beings such as antiallergy, antibacterial, antiviral, and anticancer properties but an excessive amount of luteolin results in oxidative damage to DNA [55]. Wang et al. [56] studied the detection of low concentrations of luteolin using an electrochemical sensing technique. They synthesized Zr-based UIO-66 MOFs composite with electrochemically reduced graphene oxide (ErGO) and the material was loaded onto a GCE as a working electrode for luteolin sensing. The modified UiO-66/ErGO/GCE electrode showed an instant response with a high oxidation peak current and a low 0.75 nM detection limit for luteolin with good recovery. Li et al. [57] designed Zr-porphyrin-based MOF by using *meso*-tetra(4-carboxyphenyl)porphin as a linker composite with graphene oxide and a working electrode is prepared by immersing in an aptasensor-based

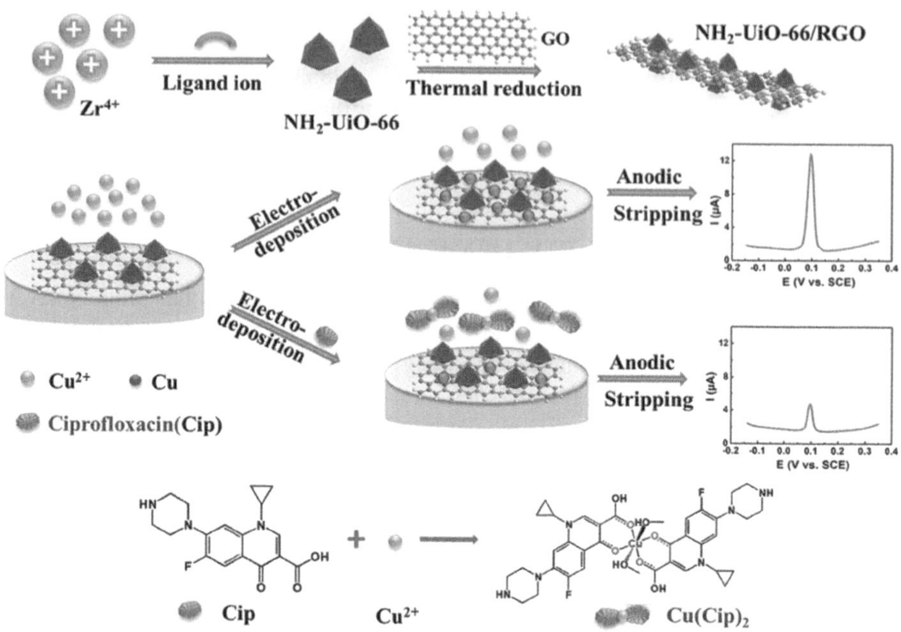

FIGURE 13.6 Schematics showing the overall synthesis technique and the mechanism for electrochemical Cip detection using Cu^{2+} modified NH_2-UiO-66/RGO. (Reprinted with permission from [53].)

electrochemical sensor which modified further by an Au electrode (AE). The resultant electrode apt/PCN-222/GO/AE is used as an electrochemical sensor for the chloramphenicol sensor having a 7.04 pg/mL detection limit in the range of 0.01–50 ng/mL. The modified electrode shows good repeatability and outstanding selectivity as a sensor material and improved performance because of Zr(IV) presence as a metal center which shows elevated affinity with the phosphate group present in aptamer and the presence of GO enhances the overall electrical conductivity of the material.

Malekzadeh et al. [58] synthesized a nanoporous material consisting of zirconium MOF with molecularity imprinted polymer (Zr-MOF/MIP) using the electropolymerization method. The electrode material shows good stability, selectivity, reproducibility, and good sensing ability toward diclofenac (DFC) determination in real and pharmaceutical samples. The CV studies in the potential window of −0.6 to 1.0 V at a scan rate of 50 mV/s demonstrate an electrochemical response with a sensing limit of 0.1 μM and 1.8% concentration. Nirasimappa et al. [59] fabricated an electrochemical sensor using Zr-MOF having three-dimensional rhombohedral microcrystals with cage structures via a solvothermal method. The perfect size of the cages acts as an active center to analyze tetracycle (TNT) which binds to the Zr center via hydrogen bonding. Therefore, the materials act with excellent stability and selectivity to bind TNT in wastewater and tap water samples. Yuan et al. [60] synthesized Zr-MOF UiO-66 assembled with carbon nanotubes (CNT) to form modified composite materials as CNT@UiO-66 and the material acts as high-performance electrochemical aptasensor to detect oxytetracycline (OTC) at low detection limit

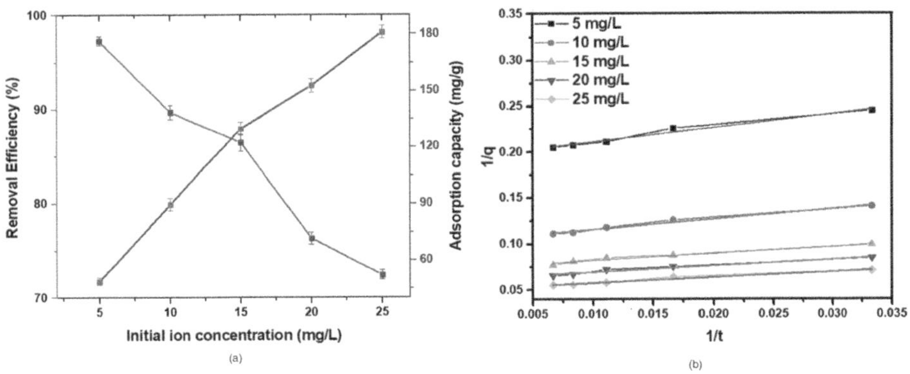

FIGURE 13.7 (a) Effect of adsorption efficiency of As(V) ion with its concentration, (b) adsorption kinetics for As(V) of UiO-66-NDC/GO. (Reprinted with permission from [61].)

corresponding to 1.48 fg/mL with good stability and high conductivity. Simarjeet et al. [61] investigated the As(V) sensing by synthesizing Zr-based MOF via the solvothermal method using 1,4-naphthalene dicarboxylic acid as an organic linker and results in the formation of UiO-66-NDC. Further, the material was decorated onto graphene oxide named UiO-66-NDC/GO nanocomposite. UiO66 NDC nanocomposite with GO shows excellent adsorption of As(V) with q_{max} 147.06 mg/g at ambient conditions by using Langmuir isotherms with high reusability. The outcome shows that the removal percentage for As(V) decreases significantly to 25 mg/L when the As concentration increases. However, the adsorption capacity for the UiO-66-NDC/GO nanocomposite increases from 48.6 to 181.1 mg/g (Figure 13.7a) and follows type 2 pseudo-second-order kinetics (Figure 13.7b).

The electrochemical aptasensors are developed for DNA or RNA oligonucleotide detection. However, the various nanomaterials such as quantum dots, microgel-based composites, and graphene oxide-based composites show low binding affinity and detection limits with low active sites and surface area [62–64]. Therefore, advancement in materials development to fabricate efficient electrochemical sensors is highly required. Among all MOFs, Zr-MOF with proper functionalization shows effective π-π stacking or hydrogen bonding with nucleic acid while sensing miDNA or miRNA [65,66]. Wei et al. [67] fabricated Zr-based MOF as an electrochemical sensor named Apt-MB-UiO-66 and the framework consists of a self-assembled electroactive structure with functionalized magnetic nanoparticles (MNPs) having magnetic dispersible aptasensor for microcrystal-LR (MC-LR) sensing in a lab scale and Eurosceptic water resources were done successfully. Gorgani et al. [68] developed CuZr-based bimetallic MOF for aptamer immobilization. CuZrMOF is used as an electrode material for miR-21 detection and function as a biomarker for cancer diagnosis. The aptasensor displays outstanding sensing performance with a 0.45 zM low detection limit in a CV detection range of 1 zM to 1 pM. The material exhibits a high surface area, strong affinity, and large pore size which support sensing performance.

The addition of pesticides during farming reduces crop damage which leads to severe human disease and excessive concentration of pesticides in food items and their consumption causes human health issues due to their carcinogenicity and nephrotoxicity [69,70]. Qiu et al. [71] fabricated a new aptasensor consisting of a UiO-66 having Zr^{4+} metal center coordinated with a phosphate group-terminated DNA (Zr–O–P) and used for Ochratoxin A (OTA) sensing. OTA binding aptamer (OBA) was immobilized on the surface of the electrode and this unique coordination resulted in enhanced sensitivity with a 0.079 fM detection limit within a range of 0.1 fM–2.0 μM. The modified aptasensor was effectively used for OTA detection present in the red wine samples and it demonstrated a favorable sensing performance for monitoring food safety. Khosropour et al. [72] synthesized Zr-MOF as MOF-808 using trimesic acid via the solvothermal method. Graphene nanoribbons (GNRs) were synthesized separately from oxidative unzipping of multi-walled carbon nanotubes (MWCNTs). Finally, the MOF-808 is modified by loading the Au nanoparticles onto the GNR/MOF-808 nanocomposite. The modified electrode as AuNP@GNR/MOF-808 is used as an electrochemical aptasensor for fungicide carbendazim (CBZ) sensing evaluated through electrochemical impedance spectroscopy (EIS) and differential pulse voltammetry (DPV) and the schematic showing the overall sensing behavior for the materials is elaborated in Figure 13.8.

The synthetic food dyes industry uses various inexpensive dyes such as azo dyes, quinolone, tartrazine, etc., which give color to various food items such as beverages. These synthetic dyes are unstable at high heat, light, and pH, and result in negative consequences for human health [73]. Lv et al. [74] synthesized composite material containing CTAB-functionalized graphene with Cu^{2+} modified Zr-MOF. The modified CTAB-Gr-Cu/Zr-MOF composite electrode exhibits sensing application

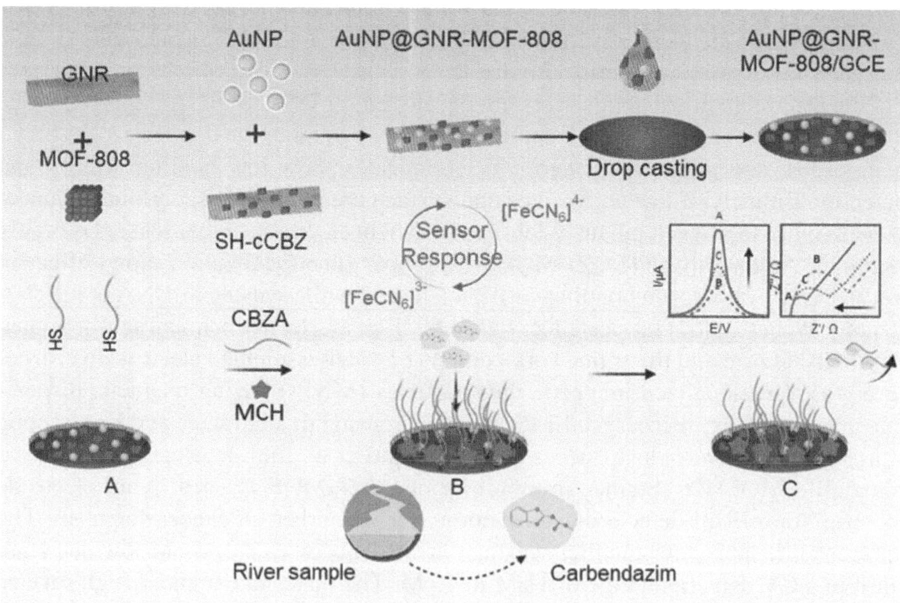

FIGURE 13.8 Schematic showing the overall sensing of CBZ using AuNP@GNR/MOF-808-GCE. (Reprinted with permission from [72].)

toward sunset yellow (SY). The composite is used as an electrochemical sensor and shows a 6.68 nM detection limit which is neatly 6.53-fold high compared to bare electrodes. The outcome proved that the synergetic effect of Cu/Zr-MOF and graphene enhances the sensitivity of the electrode along with high stability and reproducibility. Hydrogen peroxide is a widely used product in pharmaceutical industries, fuel cell applications, organic synthesis, and food-based industries, and one of the side products of enzymatic biosensor activities. So, the sensing of hydrogen peroxide using electrochemical techniques is highly critical, and the development of an effective electrochemical sensor is required [75]. Yin et al. [76] synthesized a novel porphyrinic-based Zr-MOF composite (Zr-PorMOF) via solvothermal method. The composite materials are grown onto macroporous carbon (MPC) and the modified Zr-PorMOF/MPC composites are used for electrochemical reduction of H_2O_2 as an electrode material via an electrochemical sensing technique. This enzyme-free H_2O_2 sensor exhibits 66 μA/mM sensitivity and a low value of 0.18 μM detection limit. Zr-PorMOF/MPC composites display multifunctional performance as a sensor for detecting substances such as uric acid (UA), xanthine (XA), and hypoxanthine (HX). The findings proved that the synergy effect of Zr-PorMOF and MPC provides electrochemical stability and enhances the conductivity due to the facile charger transfer process. Chang et al. [77] developed Zr-based MOF (MOF-808) crystals via the solvothermal method and further, the pore of MOF-808 was modified by confining active cobalt. The cobalt is electrodeposited on the surface of MOF-808 crystals. The thin films of Co@MOF-808 deposited on the electrode surface are used for electrochemical H_2O_2 sensors in aqueous electrolytes with 1.3 μM detection limits and 382.27 μA/mM/cm^2 sensitivity. The overall schematic showing the electrochemical sensing ability of Co@MOF-808 toward H_2O_2 is shown in Figure 13.9.

Heavy metals such as lead show extreme toxicity and non-biodegradability which cause huge environmental concerns when present in water reservoirs and food items. The excessive amount results in severe human health concerns such as lead poisoning in the body [78]. Chang et al. [79] developed a facile electrochemical sensing technique to detect selectivity in the presence of dopamine (DA) in the human body and the presence of ascorbic acid (AA) and uric acid (UA) as interferents. They developed Zr-based MOF as UiO-66 having a missing linker that results in defects in the MOF structure. Defective UiO-66 thin film electrodes act as an efficient sensor for DA examination. Further, the defective UiO-66 material is modified by being coated onto

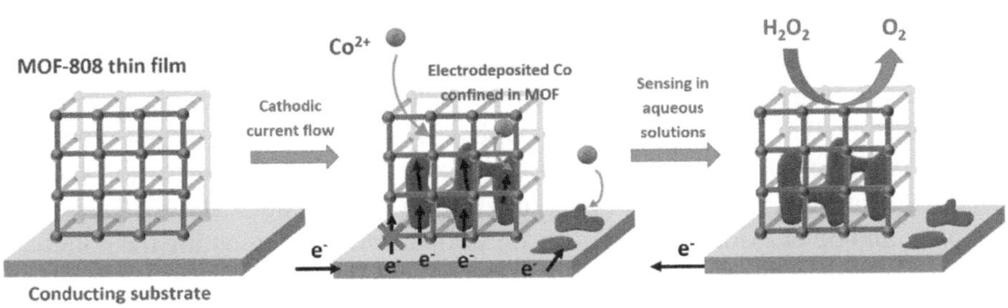

FIGURE 13.9 The overall schematic showing electrochemical H_2O_2 sensing of Cu/MOF-808. (Reprinted with permission from [77].)

graphene oxide (GO) and the resultant modified electrode exhibits nearly 6.4-fold higher sensitivity and effective selectivity for DA versus AA and UA. Similarly, Radha et al. [80] fabricated a Zr-NDI-based MOF composite with MWCNT for dopamine sensing via fluorescence and electrochemical techniques. The sensitivity shows a low 2 μM detection limit and 0.6 μM with fluorescence microscopy and DPV methods in an aqueous phosphate buffer solution as electrolyte. The material shows 95% retention during recyclability tests with a great recovery rate.

13.4 LIMITATIONS

The practical applicability of such Zr-MOFs or composite materials still has some limitations such as stability issues in harsh conditions, water, and other solvents, and limited detection of analytes. Another major limitation associated with such materials is the requirement of toxic raw materials for synthesis and the synthesis requires a very high energy during processing. Most of the research materials are yet in the experimental and analysis stages to advance the practical application. Therefore, these challenges can be overcome by discovering Zr-based MOF or composites with advancements in methodologies and synthesis parameters. Another major problem belonging to Zr-MOFs is the insulting behavior of the organic ligand present in MOF, which leads to the inadequate electrical conductivity of the materials. Furthermore, the stability of Zr-based MOF or composites is an important factor in the presence of analytes as they undergo structural changes depending on the environment during analysis. So, the enhancement in the stability of the structural properties of MOFs is crucial for real-world application by advancing synthetic protocols, strengthening the bond of metal ions with organic linkers, regulating the coordination bond with incoming analytes, and creating the advanced structural architecture of MOFs. Further advancement is required to evaluate the luminescence characteristic associated with Zr-MOFs to understand the host–guest chemistry during analysis, which is crucial to increase the sensitivity and selectivity for sensing applications.

13.5 CONCLUSION AND FUTURE PERSPECTIVE

The development of Zr-based MOF or composites shows exceptional performances in electrochemical sensing and biosensor applications such as biomarkers, reducing pesticides, food additives, drug detection, etc. Therefore, the advancement in structure, design, stability, and easy production is of utmost importance. The sensing ability of Zr-MOF can be enhanced by rational design synthesis which provides large surface area, structural stability, sensitivity, selectivity, and high response in the presence of analytes. The chapter deals with a detailed analysis of earlier reports on Zr-MOF-based electrochemical sensors and biosensors for analyst detection and pollutant removal. The various processes, such as adsorption, photocatalysis, or electrocatalysis, and other oxidation processes are well known for eliminating the pollutants in the environment. However, Zr-based MOF or related nanocomposite materials are attractive alternatives because of their structural features, large surface, and pore size, and tunable composition. The future perspectives involve the advancement of Zr-MOF to enhance the robustness during practical application. Gaining insight into binding or detecting the mechanism will be highly beneficial to evaluate analyses and help to improve the sensing performance. The development of easier and more convenient instrumentation for sensing applications, such as Zr-MOF films, strips, etc., is highly encouraging for large-scale applications in routine lifestyles. Moreover, the main aim is to create or develop manually handled, safe, wearable, and inexpensive sensor materials for faster pollutant detection without involving complex systems. The portable designs for sensors-based devices are another major concern for the upcoming electrochemical sensing applications. Finally, the researchers need to develop human and environmentally friendly materials with scalable and easy synthesis processes for increasing practical applications.

REFERENCES

1. S. Głowniak, B. Szczęśniak, J. Choma, and M. Jaroniec, "Mechanochemistry: Toward green synthesis of metal–organic frameworks," *Mater. Today*, vol. 46, pp. 109–124, Jun. 2021, doi: 10.1016/J.MATTOD.2021.01.008.
2. H. Zhang, P. Xiong, G. Li, C. Liao, and G. Jiang, "Applications of multifunctional zirconium-based metal-organic frameworks in analytical chemistry: Overview and perspectives," *TrAC – Trends Anal. Chem.*, vol. 131, p. 116015, 2020, doi: 10.1016/j.trac.2020.116015.
3. S. Rana, S. K. Guchhait, K. K. Yadav, S. Devi, S. K. Mehta, and M. Jha, "Hierarchical nanostructure engineering of 3D nickel cobaltite ultrafine nanoparticles assemblies for synergistic electrocatalytic water and urea oxidation," *J. Electroanal. Chem.*, vol. 966, p. 118378, Aug. 2024, doi: 10.1016/J.JELECHEM.2024.118378.
4. K. K. Yadav, S. K. Guchhait, S. Rana, and M. Jha, "Excellent hydrogen generation from ultrathin nanosheets of cobalt cyclotetraphosphate," *Mater. Sci. Eng. B Solid-State Mater. Adv. Technol.*, vol. 265, p. 114983, Mar. 2021, doi: 10.1016/j.mseb.2020.114983.
5. N. Miura, T. Sato, S. A. Anggraini, H. Ikeda, and S. Zhuiykov, "A review of mixed-potential type zirconia-based gas sensors," *Ionics*, vol. 20, no. 7, pp. 901–925, 2014, doi: 10.1007/s11581-014-1140-1.
6. T. Rasheed and K. Rizwan, "Metal-organic frameworks based hybrid nanocomposites as state-of–the-art analytical tools for electrochemical sensing applications," *Biosens. Bioelectron.*, vol. 199, no. June 2021, 2022, doi: 10.1016/j.bios.2021.113867.
7. G. L. Yang, X. L. Jiang, H. Xu, and B. Zhao, "Applications of MOFs as luminescent sensors for environmental pollutants," *Small*, vol. 17, no. 22, pp. 1–19, 2021, doi: 10.1002/smll.202005327.
8. H. Khosropour, M. Keramat, F. Tasca, and W. Laiwattanapaisal, "A comprehensive review of the application of Zr-based metal–organic frameworks for electrochemical sensors and biosensors," *Microchim. Acta*, vol. 191, no. 8, 2024, doi: 10.1007/s00604-024-06515-w.
9. S. A. Younis, N. Bhardwaj, S. K. Bhardwaj, K. H. Kim, and A. Deep, "Rare earth metal–organic frameworks (RE-MOFs): Synthesis, properties, and biomedical applications," *Coord. Chem. Rev.*, vol. 429, p. 213620, Feb. 2021, doi: 10.1016/J.CCR.2020.213620.
10. S. M. Majhi et al., "Metal-organic frameworks for advanced transducer based gas sensors: Review and perspectives," *Nanoscale Adv.*, vol. 4, no. 3, pp. 697–732, 2022, doi: 10.1039/d1na00798j.
11. S. Rana, K. K. Yadav, S. Devi, S. K. Mehta, and M. Jha, "Oxalate-mediated synthesis of hybrid nickel cobalt-based nanostructures for boosting water and urea electrooxidation efficiency," *J. Alloys Compd.*, p. 174241, Mar. 2024, doi: 10.1016/J.JALLCOM.2024.174241.

12. C. Jia, T. He, and G. M. Wang, "Zirconium-based metal-organic frameworks for fluorescent sensing," *Coord. Chem. Rev.*, vol. 476, p. 214930, 2023, doi: 10.1016/j.ccr.2022.214930.
13. S. Qiu, M. Xue, and G. Zhu, "Metal–organic framework membranes: From synthesis to separation application," *Chem. Soc. Rev.*, vol. 43, no. 16, pp. 6116–6140, Jul. 2014, doi: 10.1039/C4CS00159A.
14. B. Wang et al., "Highly stable Zr(IV)-based metal-organic frameworks for the detection and removal of antibiotics and organic explosives in water," *J. Am. Chem. Soc.*, vol. 138, no. 19, pp. 6204–6216, May 2016, doi: 10.1021/jacs.6b01663.
15. S. Rana et al., "Tailoring crystalline Mn_5O_8 nanostructures: Kinetic control and stabilization for enhanced electrocatalytic O_2 evolution," *Fuel*, vol. 367, p. 131485, Jul. 2024, doi: 10.1016/J.FUEL.2024.131485.
16. S. Tajik et al., "Performance of metal–organic frameworks in the electrochemical sensing of environmental pollutants," *J. Mater. Chem.* A, vol. 9, no. 13, pp. 8195–8220, Apr. 2021, doi: 10.1039/D0TA08344E.
17. K. Chattopadhyay, M. Mandal, and D. K. Maiti, "A review on zirconium-based metal-organic frameworks: Synthetic approaches and biomedical applications," *Mater. Adv.*, vol. 5, no. 1, pp. 51–67, 2023, doi: 10.1039/d3ma00735a.
18. H. L. Jiang, D. Feng, T. F. Liu, J. R. Li, and H. C. Zhou, "Pore surface engineering with controlled loadings of functional groups via click chemistry in highly stable metal-organic frameworks," *J. Am. Chem. Soc.*, vol. 134, no. 36, pp. 14690–14693, Sep. 2012, doi: 10.1021/ja3063919.
19. S. Goel and S. Rana, "Metallosurfactants as catalysts in organic reactions and energy-based applications," *Met. Fundam. Catal. Biomed. Appl.*, pp. 83–102, Jan. 2022, doi: 10.1002/9783527831289.CH5.
20. W. P. Lustig, S. Mukherjee, N. D. Rudd, A. V. Desai, J. Li, and S. K. Ghosh, "Metal-organic frameworks: Functional luminescent and photonic materials for sensing applications," *Chem. Soc. Rev.*, vol. 46, no. 11, pp. 3242–3285, 2017, doi: 10.1039/c6cs00930a.
21. A. Zuliani, N. Khiar, and C. Carrillo-Carrión, "Recent progress of metal–organic frameworks as sensors in (bio)analytical fields: Towards real-world applications," *Anal. Bioanal. Chem.*, vol. 415, no. 11, pp. 2005–2023, 2023, doi: 10.1007/s00216-022-04493-7.
22. K. K. Yadav et al., "Utilization of waste coir fibre architecture to synthesize porous graphene oxide and their derivatives: An efficient energy storage material," *J. Clean. Prod.*, vol. 276, p. 124240, 2020, doi: 10.1016/j.jclepro.2020.124240.
23. K. K. Yadav et al., "Synthesis of zirconium diboride and its application in the protection of stainless steel surface in harsh environment," *J. Solid State Electrochem.*, 2019, doi: 10.1007/s10008-019-04408-0.
24. H. Singh, Sunaina, K. K. Yadav, V. K. Bajpai, and M. Jha, "Tuning the bandgap of m-ZrO_2 by incorporation of copper nanoparticles into visible region for the treatment of organic pollutants," *Mater. Res. Bull.*, vol. 123, no. July 2019, p. 110698, 2020, doi: 10.1016/j.materresbull.2019.110698.
25. J. H. Cavka et al., "A new zirconium inorganic building brick forming metal organic frameworks with exceptional stability," *J. Am. Chem. Soc.*, vol. 130, no. 42, pp. 13850–13851, Oct. 2008, doi: 10.1021/ja8057953.
26. M. Taddei et al., "Band gap modulation in zirconium-based metal–organic frameworks by defect engineering," *J. Mater. Chem.* A, vol. 7, no. 41, pp. 23781–23786, Oct. 2019, doi: 10.1039/C9TA05216J.
27. V. Bon, I. Senkovska, I. A. Baburin, and S. Kaskel, "Zr- and Hf-based metal-organic frameworks: Tracking down the polymorphism," *Cryst. Growth Des.*, vol. 13, no. 3, pp. 1231–1237, Mar. 2013, doi:10.1021/cg301691d.

28. D. Feng, H. L. Jiang, Y. P. Chen, Z. Y. Gu, Z. Wei, and H. C. Zhou, "Metal-organic frameworks based on previously unknown Zr_8/Hf_8 cubic clusters," *Inorg. Chem.*, vol. 52, no. 21, pp. 12661–12667, Nov. 2013, doi: 10.1021/ic4018536.
29. H. Furukawa et al., "Water adsorption in porous metal-organic frameworks and related materials," *J. Am. Chem. Soc.*, vol. 136, no. 11, pp. 4369–4381, Mar. 2014, doi: 10.1021/ja500330a.
30. H. Wu et al., "Unusual and highly tunable missing-linker defects in zirconium metal-organic framework UiO-66 and their important effects on gas adsorption," *J. Am. Chem. Soc.*, vol. 135, no. 28, pp. 10525–10532, Jul. 2013, doi: 10.1021/ja404514r.
31. Y. Zhao et al., "Robust corrole-based metal-organic frameworks with rare 9-connected Zr/Hf-oxo clusters," *J. Am. Chem. Soc.*, vol. 141, no. 36, pp. 14443–14450, Sep. 2019, doi: 10.1021/jacs.9b07700.
32. B. Shan, S. M. McIntyre, M. R. Armstrong, Y. Shen, and B. Mu, "Investigation of missing-cluster defects in UiO-66 and ferrocene deposition into defect-induced cavities," *Ind. Eng. Chem. Res.*, vol. 57, no. 42, pp. 14233–14241, Oct. 2018, doi: 10.1021/acs.iecr.8b03516.
33. J. N. Hao and B. Yan, "A water-stable lanthanide-functionalized MOF as a highly selective and sensitive fluorescent probe for Cd^{2+}," *Chem. Commun.*, vol. 51, no. 36, pp. 7737–7740, Apr. 2015, doi: 10.1039/C5CC01430A.
34. H. T. Zhang, J. W. Zhang, G. Huang, Z. Y. Du, and H. L. Jiang, "An amine-functionalized metal–organic framework as a sensing platform for DNA detection," *Chem. Commun.*, vol. 50, no. 81, pp. 12069–12072, Sep. 2014, doi: 10.1039/C4CC05571C.
35. M. Chen et al., "A novel aptamer- metal ions- nanoscale MOF based electrochemical biocodes for multiple antibiotics detection and signal amplification," *Sensors Actuators B Chem.*, vol. 242, pp. 1201–1209, Apr. 2017, doi: 10.1016/J.SNB.2016.08.185.
36. J. Zheng et al., "Hydrothermally synthesized Ti/Zr bimetallic MOFs derived N self-doped TiO_2/ZrO_2 composite catalysts with enhanced photocatalytic degradation of methylene blue," *Colloids Surfaces A Physicochem. Eng. Asp.*, vol. 623, p. 126629, Aug. 2021, doi: 10.1016/J.COLSURFA.2021.126629.
37. A. Belal et al., "Electrochemical oxygen generation from VO_2 nanoflakes decorated onto graphite sheet," *J. Alloys Compd.*, vol. 976, no. November 2023, p. 173058, 2024, doi: 10.1016/j.jallcom.2023.173058.
38. Y. Wang et al., "Hydrogen production with ultrahigh efficiency under visible light by graphene well-wrapped UiO-66-NH_2 octahedrons," *J. Mater. Chem.* A, vol. 5, no. 38, pp. 20136–20140, Oct. 2017, doi: 10.1039/C7TA06341E.
39. J. M. Liu, J. X. Hou, J. Liu, X. Jing, L. J. Li, and J. L. Du, "Pyrazinyl-functionalized Zr(IV)-MOF for ultrasensitive detection of tyrosine/TNP and efficient CO_2/N_2 separation," *J. Mater. Chem.* C, vol. 7, no. 38, pp. 11851–11857, Oct. 2019, doi: 10.1039/C9TC03096D.
40. S. Nandi, S. Banesh, V. Trivedi, and S. Biswas, "A dinitro-functionalized metal–organic framework featuring visual and fluorogenic sensing of H_2S in living cells, human blood plasma and environmental samples," *Analyst*, vol. 143, no. 6, pp. 1482–1491, Mar. 2018, doi: 10.1039/C7AN01964E.
41. J.Z.Weietal.,"Rapidandlow-costelectrochemicalsynthesisofUiO-66-NH_2withenhanced fluorescence detection performance," *Inorg. Chem.*, vol. 58, no. 10, pp. 6742–6747, May 2019, doi: 10.1021/acs.inorgchem.9b00157.
42. J. Xu et al., "Optimized synthesis of Zr(IV) metal organic frameworks (MOFs-808) for efficient hydrogen storage," *New J. Chem.*, vol. 43, no. 10, pp. 4092–4099, Mar. 2019, doi: 10.1039/C8NJ06362A.
43. B. Karadeniz et al., "Benign by design: Green and scalable synthesis of zirconium UiO-metal-organic frameworks by water-assisted mechanochemistry," *ACS Sustain. Chem. Eng.*, vol. 6, no. 11, pp. 15841–15849, Nov. 2018, doi: 10.1021/acssuschemeng.8b04458.

44. W. Cheng, Y. Wang, S. Ge, X. Ding, Z. Cui, and Q. Shao, "One-step microwave hydrothermal preparation of Cd/Zr-bimetallic metal–organic frameworks for enhanced photochemical properties," *Adv. Compos. Hybrid Mater.*, vol. 4, no. 1, pp. 150–161, Mar. 2021, doi: 10.1007/s42114-020-00199-5.
45. M. E. DMello, N. G. Sundaram, A. Singh, A. K. Singh, and S. B. Kalidindi, "An amine functionalized zirconium metal-organic framework as an effective chemiresistive sensor for acidic gases," *Chem. Commun.*, vol. 55, no. 3, pp. 349–352, 2019, doi: 10.1039/C8CC06875E.
46. X. Liu et al., "Microwave-assisted synthesis of 2D Zr-MOF nanosheets supported gold nanocomposites as efficient catalysts for the reduction of 4-nitrophenol," *J. Alloys Compd.*, vol. 922, p. 165939, Nov. 2022, doi: 10.1016/J.JALLCOM.2022.165939.
47. Y. Liang et al., "Microwave-assisted acid-induced formation of linker vacancies within Zr-based metal organic frameworks with enhanced heterogeneous catalysis," *Chinese Chem. Lett.*, vol. 32, no. 2, pp. 787–790, Feb. 2021, doi: 10.1016/J.CCLET.2020.06.042.
48. J. Tu et al., "Microwave-induced fast incorporation of titanium into UiO-66 metal–organic frameworks for enhanced photocatalytic properties," *Chem. Commun.*, vol. 53, no. 23, pp. 3361–3364, Mar. 2017, doi: 10.1039/C7CC00076F.
49. Y. Yan, X. Bo, and L. Guo, "MOF-818 metal-organic framework-reduced graphene oxide/multiwalled carbon nanotubes composite for electrochemical sensitive detection of phenolic acids," *Talanta*, vol. 218, p. 121123, Oct. 2020, doi: 10.1016/J.TALANTA.2020.121123.
50. R. Rani, A. Deep, B. Mizaikoff, and S. Singh, "Zirconium metal organic framework based opto-electrochemical sensor for nitrofurazone detection," *J. Electroanal. Chem.*, vol. 909, p. 116124, Mar. 2022, doi: 10.1016/J.JELECHEM.2022.116124.
51. D. Grieshaber, R. MacKenzie, J. Vörös, and E. Reimhult, "Electrochemical biosensors - sensor principles and architectures," *Sensors 2008, Vol.* 8, Pages 1400–1458, vol. 8, no. 3, pp. 1400–1458, Mar. 2008, doi: 10.3390/S80314000.
52. F. Su et al., "Two-dimensional zirconium-based metal-organic framework nanosheet composites embedded with Au nanoclusters: A highly sensitive electrochemical aptasensor toward detecting cocaine," *ACS Sensors*, vol. 2, no. 7, pp. 998–1005, Jul. 2017, doi: 10.1021/acssensors.7b00268.
53. X. Fang et al., "Nanocomposites of Zr(IV)-based metal-organic frameworks and reduced graphene oxide for electrochemically sensing ciprofloxacin in water," *ACS Appl. Nano Mater.*, vol. 2, no. 4, pp. 2367–2376, Apr. 2019, doi: 10.1021/acsanm.9b00243.
54. I. Turel, N. Bukovec, and E. Farkas, "Complex formation between some metals and a quinolone family member (ciprofloxacin)," *Polyhedron*, vol. 15, no. 2, pp. 269–275, Jan. 1996, doi: 10.1016/0277-5387(95)00231-G.
55. M. Cao, X. Yin, X. Bo, and L. Guo, "High-performance electrocatalyst based on metal-organic framework/macroporous carbon composite for efficient detection of luteolin," *J. Electroanal. Chem.*, vol. 824, pp. 153–160, Sep. 2018, doi: 10.1016/J.JELECHEM.2018.07.049.
56. Q. Wang, C. Gu, Y. Fu, L. Liu, and Y. Xie, "Ultrasensitive electrochemical sensor for luteolin based on zirconium metal-organic framework UiO-66/reduced graphene oxide composite modified glass carbon electrode," *Molecules*, vol. 25, no. 19, p. 4557, Oct. 2020, doi: 10.3390/MOLECULES25194557.
57. H. K. Li et al., "Artful union of a zirconium-porphyrin MOF/GO composite for fabricating an aptamer-based electrochemical sensor with superb detecting performance," *Chinese Chem. Lett.*, vol. 32, no. 9, pp. 2851–2855, Sep. 2021, doi: 10.1016/J.CCLET.2021.02.042.
58. M. Malekzadeh, A. Mohadesi, M. A. Karimi, and M. Ranjbar, "Development of a new electrochemical sensor based on Zr-MOF/MIP for sensitive diclofenac determination," *Anal. Bioanal. Electrochem*, vol. 12, no. 3, pp. 402–414, 2020.

59. P. Narasimhappa, R. Varshney, S. Singh, T. S. K. Naik, and P. C. Ramamurthy, "3D rhombohedral microcrystals metal-organic frameworks for electrochemical and fluorescence sensing of tetracycline," *Chemosphere*, vol. 333, p. 138977, Aug. 2023, doi: 10.1016/J.CHEMOSPHERE.2023.138977.
60. R. Yuan, H. Chen, J. Liu, R. Li, and H. He, "An electrochemical impedimetric platform formed by a CNT@UiO-66 nanocomposite for quantitative analysis of oxytetracycline," *Dalt.* Trans., vol. 52, no. 33, pp. 11552–11557, Aug. 2023, doi: 10.1039/D3DT01980B.
61. S. Singh et al., "A systematic study of arsenic adsorption and removal from aqueous environments using novel graphene oxide functionalized UiO-66-NDC nanocomposites," *Sci. Rep.*, vol. 12, no. 1, pp. 1–15, Sep. 2022, doi: 10.1038/s41598-022-18959-2.
62. C. Zhu, L. Li, Z. Wang, M. Irfan, and F. Qu, "Recent advances of aptasensors for exosomes detection," *Biosens. Bioelectron.*, vol. 160, p. 112213, Jul. 2020, doi: 10.1016/J.BIOS.2020.112213.
63. X. Cong et al., "A sensitive photoelectrochemical aptasensor for miRNA-21 based on the sensitization effect of CdSe quantum dots," *Electroanalysis*, vol. 30, no. 6, pp. 1140–1146, Jun. 2018, doi: 10.1002/ELAN.201800079.
64. H. S. Dinani et al., "Fabrication of Au/Fe3O4/RGO based aptasensor for measurement of miRNA-128, a biomarker for acute lymphoblastic leukemia (ALL)," *Eng. Life Sci.*, vol. 22, no. 8, pp. 519–534, Aug. 2022, doi: 10.1002/ELSC.202100170.
65. N. Zhou et al., "Two-dimensional oriented growth of Zn-MOF-on-Zr-MOF architecture: A highly sensitive and selective platform for detecting cancer markers," *Biosens. Bioelectron.*, vol. 123, pp. 51–58, Jan. 2019, doi: 10.1016/J.BIOS.2018.09.079.
66. W. Xue, Y. Wang, L. Guo, and H. Zhang, "Zr-MOF functionalized nanochannels: Application to regenerative and sensitive electrochemical aptasensing platform," *Sensors Actuators B Chem.*, vol. 381, p. 133455, Apr. 2023, doi: 10.1016/J.SNB.2023.133455.
67. T. Wei et al., "Self-assembled electroactive MOF–magnetic dispersible aptasensor enables ultrasensitive microcystin-LR detection in eutrophic water," *Chem. Eng. J.*, vol. 466, p. 142809, Jun. 2023, doi: 10.1016/J.CEJ.2023.142809.
68. L. Gorgani, M. Mohammadi, G. N. Darzi, and J. B. Raoof, "Electrochemical aptasensor based on bimetallic CuZr-MOF for ultrasensitive detection of miR-21," *Sensors Actuators B Chem.*, vol. 378, p. 133194, Mar. 2023, doi: 10.1016/J.SNB.2022.133194.
69. X. Dong, S. Qi, M. Qin, N. Ding, Y. Zhang, and Z. Wang, "A novel ternary Y-DNA walker amplification strategy designed fluorescence aptasensor based on Au@SiO_2@Fe_3O_4 nanomaterials for ochratoxin A detection," *Microchim. Acta*, vol. 190, no. 11, pp. 1–11, Nov. 2023, doi: 10.1007/s00604-023-06018-0.
70. H. Khosropour et al., "A comprehensive review on electrochemical and optical aptasensors for organophosphorus pesticides," *Microchim. Acta*, vol. 189, no. 9, pp. 1–23, Aug. 2022, doi: 10.1007/S00604-022-05399-Y.
71. W. Qiu et al., "Specific coordination between Zr-MOF and phosphate-terminated DNA coupled with strand displacement for the construction of reusable and ultrasensitive aptasensor," *Anal. Chem.*, vol. 92, no. 16, pp. 11332–11340, Aug. 2020, doi: 10.1021/acs.analchem.0c02018.
72. H. Khosropour, N. Maeboonruan, C. Sriprachuabwong, A. Tuantranont, and W. Laiwattanapaisal, "A new double signal on electrochemical aptasensor based on gold nanoparticles/graphene nanoribbons/MOF-808 as enhancing nanocomposite for ultrasensitive and selective detection of carbendazim," *OpenNano*, vol. 8, p. 100086, Nov. 2022, doi: 10.1016/J.ONANO.2022.100086.
73. R. Sun et al., "Simple and sensitive electrochemical detection of sunset yellow and Sudan I in food based on AuNPs/Zr-MOF-Graphene," *Food Control*, vol. 145, p. 109491, Mar. 2023, doi: 10.1016/J.FOODCONT.2022.109491.
74. R. Lv et al., "Cu^{2+} modified Zr-based metal organic framework-CTAB-graphene for sensitive electrochemical detection of sunset yellow," *Food Chem. Toxicol.*, vol. 166, p. 113250, Aug. 2022, doi: 10.1016/J.FCT.2022.113250.

75. K. Yu et al., "MOF-818 nanozyme-based colorimetric and electrochemical dual-mode smartphone sensing platform for in situ detection of H_2O_2 and H_2S released from living cells," *Chem. Eng. J.*, vol. 451, p. 138321, Jan. 2023, doi: 10.1016/J.CEJ.2022.138321.
76. D. Yin, J. Liu, X. Bo, M. Li, and L. Guo, "Porphyrinic metal-organic framework/macroporous carbon composites for electrocatalytic applications," *Electrochim. Acta*, vol. 247, pp. 41–49, Sep. 2017, doi: 10.1016/J.ELECTACTA.2017.06.176.
77. Y. S. Chang, J. H. Li, Y. C. Chen, W. H. Ho, Y. Da Song, and C. W. Kung, "Electrodeposition of pore-confined cobalt in metal–organic framework thin films toward electrochemical H_2O_2 detection," *Electrochim. Acta*, vol. 347, p. 136276, Jul. 2020, doi: 10.1016/J.ELECTACTA.2020.136276.
78. L. Wang, X. Peng, H. Fu, C. Huang, Y. Li, and Z. Liu, "Recent advances in the development of electrochemical aptasensors for detection of heavy metals in food," *Biosens. Bioelectron.*, vol. 147, p. 111777, Jan. 2020, doi: 10.1016/J.BIOS.2019.111777.
79. Y. N. Chang, C. H. Shen, C. W. Huang, M. D. Tsai, and C. W. Kung, "Defective metal-organic framework nanocrystals as signal amplifiers for electrochemical dopamine sensing," *ACS Appl. Nano Mater.*, vol. 6, no. 5, pp. 3675–3684, Mar. 2023, doi: 10.1021/acsanm.2c05402.
80. G. Radha, K. S. Kumar, K. N. Chappanda, and H. Aggarwal, "Zr-NDI MOF based optical and electrochemical dual sensor for selective detection of dopamine," *J. Phys. Chem.* C, vol. 127, no. 18, pp. 8864–8872, May 2023, doi: 10.1021/acs.jpcc.3c01377.

14 Zirconium-Based MOFs
Advances in Electrochemical Sensing and Biosensing

Amrinder Mehta, Hitesh Vasudev, and Sule Emmanuel Oshioriameh

14.1 INTRODUCTION

Organic frameworks of zirconium or more precisely zirconium-based metal-organic frameworks (Zr-MOFs) have attracted great attention in the chemistry and materials science communities in recent years. These three-dimensional crystals are made from metal clusters of zirconium being bonded through organic linkers in a network structure [1]. Figure 14.1 highlights the major synthesis approaches normally employed in the synthesis of Zr-MOFs. It is a flowchart starting with a title at the top of the page "Zr-MOFs Synthesis Methods," followed by four separate tracks. These synthesis approaches include the solvothermal method, which is a solution-based synthesis method, the microwave-assisted method, which uses microwave radiation in the synthesis process, the electrochemical-assisted method, which uses electricity in the synthesis of the framework, and mechanochemical synthesis, which uses mechanical force during the synthesis process. Each method is presented in a "wave" box and below each box, there is an arrow pointing downward toward the main banner. The layout is clean and professional with primary directional arrows guiding the viewer in a hierarchical manner through the information they provide [2].

Figure 14.2 illustrates the versatile applications and properties of Zr-MOFs in biomedical applications through a circular diagram. At the center, there is a molecular structure of Zr-MOFs, surrounded by various MOF types, including UiO-66, UiO-67, PCN-222, PCN-224, MOF-525, NU-1000, and MOF-808. The diagram is organized in concentric circles, with the outermost ring highlighting the key characteristics of these materials: "Tunable Porous Structure, High Surface Area, Biocompatible Nature, and Functionalization-friendly Properties." The middle ring focuses on "Biomedical Applications," which branches into several key areas: biosensing, bioimaging, drug/cargo delivery, targeted therapy/treatment, and antibacterial applications. The porous nature and high surface area of these MOFs make them highly effective for medical applications. The systematic arrangement of the diagram effectively communicates how the inherent properties of Zr-MOFs directly contribute to their various biomedical applications, making them valuable materials in modern medical science and therapeutic approaches [3].

DOI: 10.1201/9781003519300-17

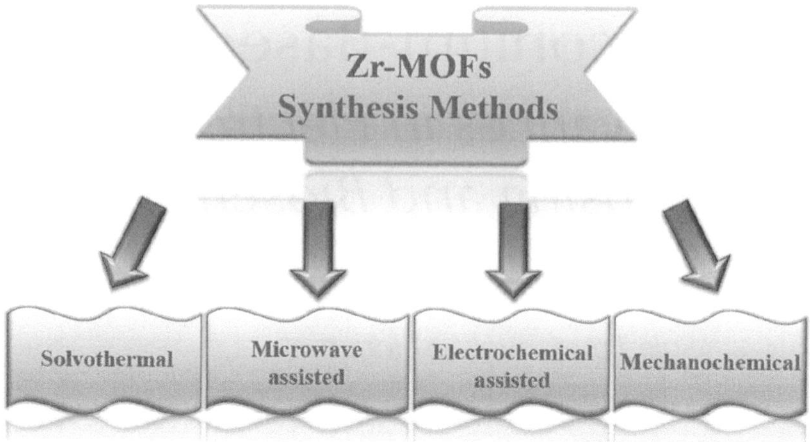

FIGURE 14.1 Synthetic methods to prepare Zr-MOFs [2]. (Reproduction permission from Springer Nature.)

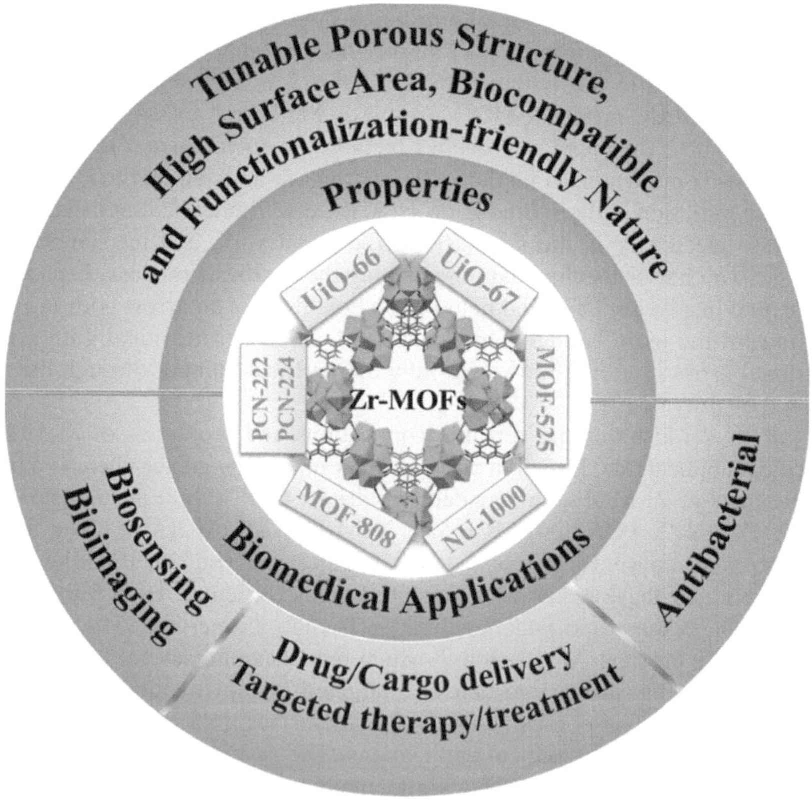

FIGURE 14.2 Various properties and biomedical applications of Zr-MOFs [3]. (CC By 3.0 Attribution.)

The structure of Zr-MOFs is demonstrated by the thermal and chemical stability of such materials, which differentiates Zr-MOFs from other existing types of MOFs. Among all the synthesized Zr-MOFs, UiO-66 is one of the most studied materials capable of a reversible transition between its dehydroxylated and hydroxylated states. This structural flexibility is due to the repetitive processes of hydration and dehydration, which prove the kinetic characteristics of these materials. The weak acid sites found within the framework of Zr-MOFs have been shown to augment the chemistry of these materials. Based on infrared measurements, these sites participate in the material's ability to interact with guest molecules. Recent studies have indicated that the adsorption characteristics of UiO-66/MOF exhibit natural variation, as demonstrated through both Monte Carlo simulations and microcalorimetry. These studies revealed that the adsorption interactions showed moderate affinity for various molecules, including methane, CO, and CO_2. This property, coupled with the impressive adsorption capacity in MOFs of Zr, pushes them toward gas-based applications where the recovery of carbon dioxide for industrial usage is required [4]. Figure 14.3 presents reactions of a molecular system with different adsorption mechanisms. The left side features an X-shaped large structure where an octahedral or diamond is made of metallic frame with spherical particles outside. This structure suggests both the physisorption and multilayer adsorption characteristics. On the right side, there is a much more detailed molecular interaction in which a secondary building unit (SBU) is shown to be involved in chemisorption processes. The SBU has a complex organic molecule (referred to as AO7) through coordination bonding by means of Lewis acid–base interactions. Skeletal formula can be described by dashed and solid lines; the organic molecule includes aromatic rings, S groups, as well as nitrogen-containing functional groups. The illustration effectively demonstrates the occurrence of physisorption, chemisorption, and multilayer adsorption within a molecular system, which adds complexity to the understanding of molecular interactions at the surface [5].

Another interesting property of Zr-MOFs obtained using the current method is their mechanical characteristics. An analysis of the elastic constants of these materials using density functional theory (DFT) has shown that the behavior is rich and

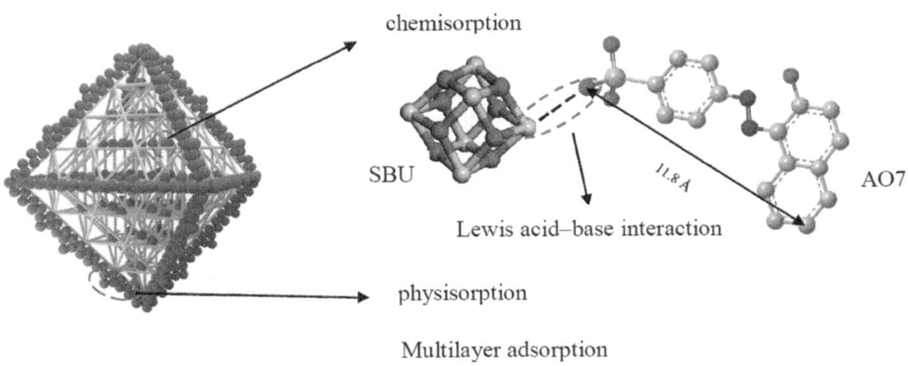

FIGURE 14.3 The model of zirconium-based metal-organic frameworks (Zr-MOF) adsorption of acid orange 7 (AO7) [5]. (Open Access (CC by 4.0 Attribution.)

anisotropic. The mechanical behaviors of Zr-MOFs can be either anisotropic-positive linear compressibility and Poisson's ratio or anisotropic-negative. This flexibility is explained by the exceptional properties of the molecules within the described framework, which provides stability and sufficient elasticity at the same time [6].

In addition to the adsorption of gases and liquids, Zr-MOFs have been synthesized for use in energy and environmental applications. Research has centered on the intentional incorporation of defects that are introduced within the Zr-MOF structure as a way of increasing their functionality. These defective Zr-MOFs are proving to be diverse in many applications such as environmental pollutant sequestration, chemisorption of chemical warfare agents, solar photovoltaic conversion, and electrochemical usage [7].

Defect engineering in Zr-MOFs is essential for improving the performance of MOFs. These defects are thus useful in generating open sites within the structure that allow for efficient adsorption and elimination of pollutants. They also enhance the catalytic ability of Zr-MOFs for various energy conversion processes such as hydrogen generation and CO_2 conversion. This control over the distribution and type of defects provides researchers with opportunities to fine-tune the properties of the developed Zr-MOFs for targeted applications with improved selectivity and efficiency [8]. In addition, tuning the pore size of the Zr-MOFs was found to affect the adsorbent capacity and transport characteristics of the materials. Such high level of control also broadens their applicability toward environmental cleanup and conversion of energy systems. The high chemical and thermal stability, as well as the synthetic tunability of defective Zr-MOFs, allow their practical use under various environmental conditions and in various applications [9]. The general understanding of Zr-MOFs provides insight into a new class of materials with unique structures, functions, and applications. The key features of thermal stability, structural reversibility, and structure defect engineering have placed Zr-MOFs on the map of materials science and technology for solving major problems in energy conversion and environmental cleanup. Further advancements in this research field would make Zr-MOFs increasingly significant for the development of novel strategies for sustainable technologies and industrial processes.

14.2 RECENT ADVANCES IN ELECTROCHEMICAL SENSORS UTILIZING ZIRCONIUM MOFs

The latest achievements in the development of electrochemical sensors based on zirconium-containing metal-organic frameworks (Zr-MOFs) indicate the direction of modern analytical chemistry development. These novel materials have garnered interest because of their outstanding characteristics, such as high thermal and chemical stability, tunable porosity, and diverse functionalization possibilities. Zr-MOFs have been named as potential candidates for the electrochemical sensing of a range of analytes. However, one of the exceptional advances is the electrotechnological management of Zr-MOFs for electrochemical identification of 4-nitrobenzaldehyde [10]. A new conductive hybrid composite of P(3MT) and a porphyrin-based Zr-MOF, PCN-222(Fe), was successfully prepared via in situ polymerization. This composite combines the large specific surface of MOFs and low electric huddle of conductive polymer that enable it to sensitive and selective sensor for 4-nitrobenzaldehyde with 0.233 μM. Another improved use of Zr-MOFs is chiral identification. The UiO-88-LP crystalline material with non-clustered and very high ordering was obtained during the synthesis using the L-proline-tuned Zr-MOF. When coated on the glass carbon electrode, this material was proven to give the configuration of tryptophan enantiomers [11,12]. The electrochemical sensing interface for L-tryptophan or D-tryptophan demonstrated redox peaks with a peak-to-current ratio of 2.47, suggesting possible use in chiral sensing systems. Figure 14.4 depicts a two-step chess and biosystems synthesis which are used in the construction of chemical and biosensors. Organic frameworks (Zr-MOFs) have demonstrated significant progress in the field of analytical chemistry. These innovative materials have garnered attention due to their exceptional properties, including high thermal and chemical stability, tunable porosity, and versatile functionalization capabilities. Zr-MOFs have emerged as promising candidates for electrochemical sensing applications, particularly in the detection of various analytes. One notable development is the use of Zr-MOFs for the efficient electrochemical detection of 4-nitrobenzaldehyde. A novel hybrid composite integrating conductive poly-3-methoxythiophene and PCN-222(Fe), a porphyrin-based Zr-MOF, was synthesized using an in situ polymerization strategy. This composite leverages the large specific area of MOFs and the low electrical resistance of conductive polymers, resulting in a sensitive and selective sensor for 4-nitrobenzaldehyde with a detection limit of 0.233 μM. Another significant advancement involves the application of Zr-MOFs in chiral recognition. A highly ordered and non-clustered crystalline material, UiO-88-LP, was synthesized using L-proline-tuned Zr-MOF. When modified on a glassy carbon electrode, this material demonstrated the ability to recognize the configuration of tryptophan enantiomers. The electrochemical sensing interface showed

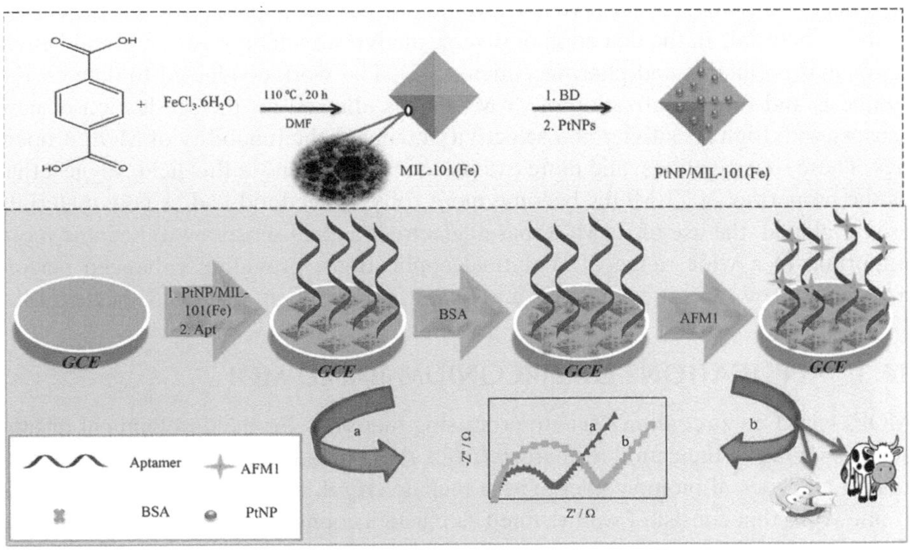

FIGURE 14.4 Schematic diagram for the preparation schematic diagram and electrochemical test [13]. (Reproduction permission from Elsevier.)

clear redox peaks for L-tryptophan and D-tryptophan, with a peak-to-current ratio of 2.47, indicating its potential for chiral sensing applications [13].

Other researchers have also utilized hydrophobic Zr-MOFs as innovative electrode chasers for detecting specific compounds. One study evaluated the use of hydrophobic Zr-MOFs in the electrochemical analysis of flibanserin, highlighting the potential of these materials in the pharmaceutical industry. The development of bimetallic MOFs incorporating zirconium has further expanded the possibilities for electrochemical sensors. The paper describes an ISE design created from a zirconium-copper bimetallic MOF hybrid material modified with an ionic liquid, aimed at quantifying nitrite in food matrices [14]. This approach takes advantage of the benefits of Zr-MOFs, bimetallic systems, and ionic liquids, which can improve the sensitivity and selectivity. The effect of mixed linkers in Zr-MOFs was investigated using density functional theory calculations on the bandgap modulation of the material systems. This present study showed that the bandgaps of mixed-linker systems are not midway between the bandgaps of single-linker systems, which contravenes Vegard's law for semiconductors [15]. The build-up of mid-gap states in these mixed-linker MOFs results in bandgap narrowing, and the extent of such effects differs in some specific structures because of the influence of π-π stacked interactions. This discovery provides a basis for designing the electronic characteristics of Zr-MOFs for specific sensing applications. Zr-MOFs have also been incorporated with other nanomaterials for electrochemical sensing application as well. For instance, the carbon dot–UiO-66-NH2 hybrid was employed as a surface modifier for glassy carbon electrodes for the electrochemical determination of Cd(II) and Pb(II) in water samples [16]. This sensor was highly sensitive with good linearity for both metal ions, and we concluded that the hybrid Zr-MOF materials have high application potential in environmental sensing. The latest developments in electrochemical sensors based on Zr-MOFs have established the applicability and efficiency of these materials in the detection of diverse analytes, including gas and liquid environmental pollutants and pharmaceuticals [17]. The work conducted to date on the synthesis and functionalization of Zr-MOFs has allowed for the production of new sensors with high sensitivity and selectivity owing to the tunability of MOF properties. These investigations, and more extensive investigations in this field, suggest that as the properties of Zr-MOFs become more fully understood and as new materials are developed, the use of Zr-MOF-based electrochemical sensors will become more important in a wide range of analytical applications, providing enhanced performance and new capacities to detect and quantify a variety of chemical species [18].

14.3 APPLICATIONS OF ZIRCONIUM-BASED MOF

MOFs based on zirconium (Zr) are promising materials for the development of biosensors owing to their unique characteristics that make them suitable for the detection of all types of biomolecules. These include chiral, crystalline, and mesoporous frameworks that consist of well-defined Zr clusters connected by a variety of organic linkers giving them characteristics such as high surface area, structural flexibility, chemical and thermal stability, and tunability. Another benefit of Zr-MOFs biosensing is that they can be post-synthetically modified to perform various tasks [19].

Scholars have designed electrochemical biosensors with reference to analogous Zr-MOFs for protein identification, and have identified that the effectiveness of these sensors is tied to the pore dimension and associated atmosphere of the MOFs. Such tunability affords variations in the binding interactions between aptamers and targeted proteins, and hence the high selectivity and sensitivity of biosensors. For example, an optimized Zr-MOF-based sensor offers a high selectivity to lysozyme with a lowest detection limit of 3.6 pg mL-1, and the repeatable and stable performance for practical sample tests. Zr-MOFs have also been used in constructing electrochemiluminescence (ECL) biosensors [20].

Figure 14.5 depicts the synthesis of routes and modifications in cases of UiO-68-An MOFs. The process starts with two organic linkers, 1 and 2, which undergo metalation to ZrCl4 DMF, followed by acetic acid at 90°C to the first UiO-68-An/Ma framework structure. This framework then passes through two distinct modification routes. This pathway involves a reversible Diels-Alder (D-A) reaction with a furan alcohol derivative, affording UiO-68-An/DA. The second route involves using the thiol-ene reaction of a thiol alcohol species to introduce UiO-68-An/TE. In both modifications, the basic MOF scaffold of Zr nodes and organic linkers is preserved, and new functional groups are added. The balls correspond to the zirconium metal centers and the organic connectors have anthracene

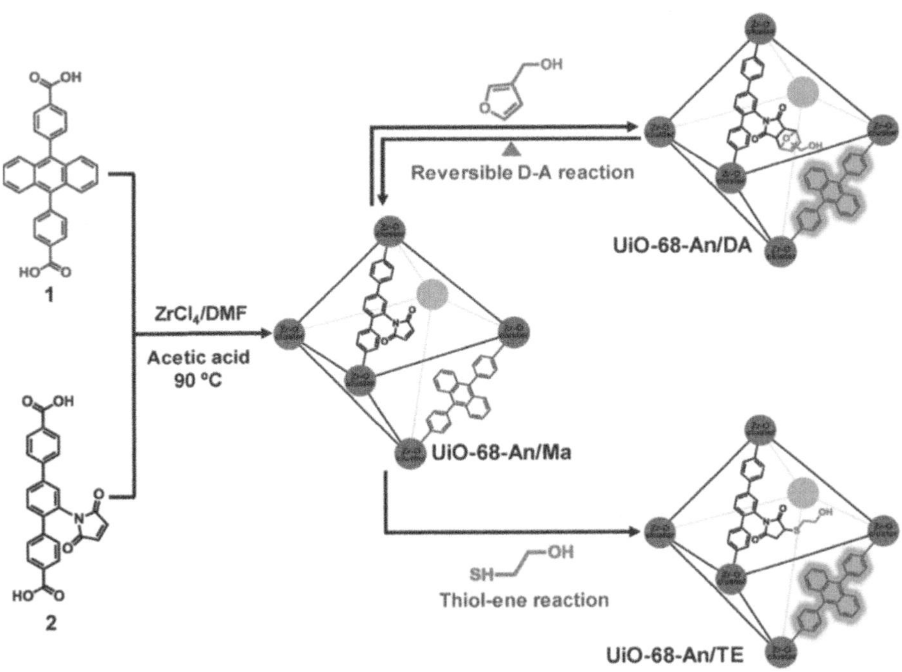

FIGURE 14.5 Synthesis of UiO-68-An/Ma and tuning the fluorescent PET in the MOF through the reversible D–A reaction or thiol-ene reaction [21]. (Reproduction permission from John Wiley and Sons.)

units, highlighted in the altered structures. These changes also show how post-synthetic MOF functionalization is an effective strategy [21].

One of these is intramolecular coordination-induced ECL, where Zr-MOFs act as emitters for the efficient detection of the clinical drug bleomycin at low concentrations. Using 4,4′,4‴-nitrilotribenzoic acid ligands, the materials combine the Zr-MOFs coordinating with Zr(IV) ions and the effective co-reactants for promotion of ECL reaction. This has made bleomycin detection sensitive with detection limit at 0.50 nm, through this innovative design. Another potentially fascinating branch of Zr-MOFs in the field of biosensing is the creation of ratiometric ECL biosensors. Some Zr-MOFs have been designed to show both anodic and cathodic bi-ECL signals including Zr-DPA@TCPP [22]. This feature has been utilized to develop radiometric ECL biosensors for application in the detection of individual DNA sequences including, HPV-16 DNA. Owing to the dual-emission properties of Zr-MOFs and DNA recycling amplification reactions, the sensitivities of these biosensors are high, and the detection limits are in the attomolar range. Zr-MOFs demonstrate flexibility in the field of impedimetric apt sensors for cancer biomarker detection. For instance, a zirconium-cobalt MOF has been used in an impedimetric aptamer sensor for carcinoembryonic antigen detection, demonstrating that Zr-MOFs have applications in cancer diagnosis [23]. There have also been some recent improvements in the synthesis of core-actant-free Zr-MOFs for radiometric ECL detection. These new materials, including Zr-TBAPY, exhibit both cathodic and anodic ECL without the need for additional core actants. This property has been exploited to develop ultrasensitive biosensors for the detection of HIV-DNA, which indicates the future implications of Zr-MOFs in the diagnosis of infectious diseases. Zr-MOFs afford enhanced biosensing properties, and integration with other nanomaterials and biomolecules has advanced their functions. Other studies have investigated the incorporation of Zr-MOF with aptamers, enzymes, and other functional moieties to improve the sensitivity, selectivity, and efficiency of biosensors [24]. Biosensing is one of the most promising areas to which Zr-MOFs have shown great promise being used for the detection of proteins, nucleic acids, and small molecules. Their tunable nature, high stability, and easy compatibility with diverse detection modes make them potential candidates for next-generation biosensors. As research in this field grows, Zr-MOF-based biosensors are likely to become progressively more relevant in medical diagnostics, environmental monitoring, and various other analytical uses, due to the enhanced selectivity, sensitivity, and reliability of the substance when detecting numerous biological analytes [25,26].

Figure 14.6 depicts an advanced nanomedicine delivery approach to cancer therapy. In Figure 14.6a, the fabrication process of a nanoparticle is demonstrated as individual components, including PFA, TPZ, UiO66, and PDA, come together to form a spherical morphology having a spiked surface. Figure 14.6b demonstrates the therapeutic application in two sections: On the left, it presents the treatment delivery through 808 nm laser to a tumor site in the mouse model where nanoparticles are taken up at the tumor site. To the right, a closer view of the structure of a cell shows what takes place in a cancer cell including a sequence of molecular events. This diagram depicts collage of the nanoparticle uptake into the cell followed by the activation of different pathways such as oxygen-dependent process, altered gene expression, and cell death. Several molecular markers and DNA structures within

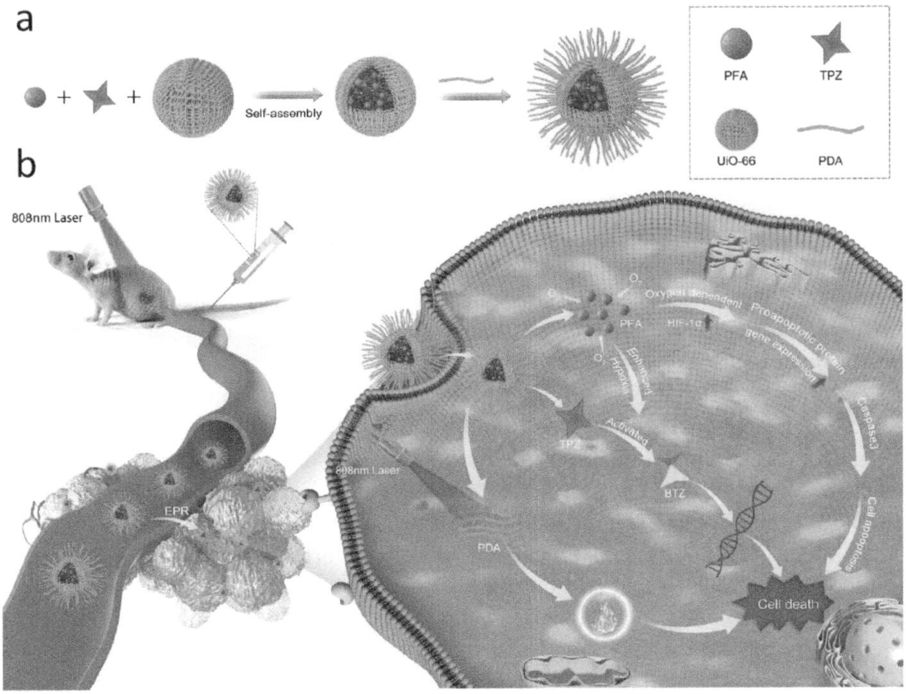

FIGURE 14.6 (a) Schematic illustration of the preparation of TPZ/PFA@UiO-66@PDA NPs and (b) tail vein-injected TPZ/PFA@UiO-66@PDA NPs combined with laser radiation for hypoxia-activated osteosarcoma therapy [27]. (Under the Creative Commons Attribution 4.0.)

the cellular environment signify complex biological processes of this chosen type of cancer treatment [27].

In recent years, MOFs based on zirconium have demonstrated superior performance when used as sensing materials compared to other MOFs. Owing to their characteristics, such as high thermal and chemical stability, tunable pore size, and accessibility to the surface for functionalization, MOFs are suitable for several sensing applications. Another advantage of Zr-MOFs in sensing applications is their highly stable structure. Compared to some other MOFs that can supply their structure under severe conditions, Zr-MOFs retain their structural integrity, making them more effective sensors. Such stability is important in applications where the sensor must be used for a long duration or is exposed to difficult conditions [28]. Owing to the tunability of Zr-MOFs, they can be clearly distinguished from other materials in the field of sensing. The function of Zr-MOF-based sensors can be enhanced by successfully controlling the pore dimensions, in addition to variations in the chemical environment. Such flexibility enables the development of remarkably selective and sensitive biosensors, as demonstrated by the development of a Zr-MOF biosensor for the detection of lysozyme with a detection limit of 3.6 pg mL – 1. Zr-MOFs have been found to be most appropriate for use in electrochemical sensing. Owing to their functionalizability and ability to incorporate other materials, new sensing platforms have been developed [29]. The hybrid material compositing of Zr-MOF and conductive polymers to fabricate a highly sensitive electrochemical sensing platform for 4-nitrobenzaldehyde clearly indicates the advantage of integrating Zr-MOFs with other functional materials. In the field of optical sensing, the authors reported that Zr-MOFs have several valuable characteristics. The development of ratiometric ECL biosensors through the use of MOFs, particularly Zr-MOFs like Zr-DPA@TCPP, has enabled the production of innovative electrochemiluminescent ratiometric biosensors with enhanced sensitivity [30]. These sensors have demonstrated the ability to identify as few as several atoms per liter of solution containing the target DNA sequences, thereby surpassing many other MOF-based sensors. Additionally, Zr-MOFs have been effectively utilized in vapor sensing applications, yielding positive outcomes.

Zr-MOFs have also been used in vapor sensing applications with good results achieved. Being guest-responsive, their fluorescence permits the identification of distinct solvent vapors depending on the fluorescence emission peaks. This capability makes Zr-MOFs highly useful for environmental analysis and industrial process monitoring and control. The hybrid property of Zr-MOFs with other nanomaterials has also boosted up the sensing aspects of this material [31]. Composite materials formed from Zr-MOFs and carbon dots were employed to create a sensing platform that can detect heavy metal ions in water. Generally, the sensing performance of MOFs is improved by Zr-MOFs because of their high stability and multifunctional characteristics. Although other MOFs called copper-MOFs or aluminum-MOFs have been used for sensing, they suffer from poor chemical stability or structural flexibility compared to Zr-MOFs. The zirconium nodes in Zr-MOFs are solid and can endure stressful environments while the incorporation of functional groups that are essential in sensing can still be achieved [32]. Moreover, ligand design and post-synthetic modification provide more flexibility in the sensor design than many other Zr-MOF systems. Such flexibility enables researchers to fine-tune Zr-MOF sensors for specific analyte species and sensing conditions. Zr-MOFs were less prone to problems in sensitivity and reusability compared to other MOFs used in sensing applications. Their high robustness, often programmable characteristics and easy over-functionalization render them ideal for the design of future sensor systems. With the development of this field in the future, based on Zr-MOF materials, sensors are destined to win a more important position in many analysis processes, including high sensitivity, selectivity, and reliability for the detection of diverse analytes and different modes [33].

14.4 DEVELOPMENT OF ZIRCONIUM-BASED ELECTROCHEMICAL SENSORS

Zirconium-based electrochemical sensors have challenges and limitations, which are listed in Table 14.1. One of the key issues is the improvement of the sensor response in the presence of interfering compounds. However, when used in real-life samples such as food and biological fluids, the stability and specificity of these sensors are affected by interferences substances [34].

This issue is especially important for Zr-MOF-based sensors because they possess a high surface area, and porosity can cause the non-specific adsorption of

TABLE 14.1
Challenges and Limitations in the Development of Zirconium-Based Electrochemical Sensors

Challenges	Limitations	Key Findings
Optimization in complex matrices	Reproducibility	MOFs show promising potential for electrochemical sensing applications [35].
Long-term stability	Selectivity	MOF composites offer large surface area, high porosity, and strong catalytic efficiency [36].
Integration with other components	Cost considerations	Functionalization with polymers and nanoscale building blocks enhances sensing capabilities [37].
Electrode fouling	Standardization of fabrication processes	MOFs can be used as electrode materials or signal tags in sensor design [38].
Environmental and health concerns	Detection limits in complex samples	UiO-66(Zr) functionalized with SO3H group shows rapid detection of Pb2+ ions [39].
Scalability of synthesis	Long-term stability in aqueous media	MOF composites demonstrate good stability and electrical conductivity [40].
Sample preparation for complex matrices	Interference from similar analytes	Integration with optical fibers allows for remote, rapid detection of heavy metals [41].

molecules to the sensing interface, thereby reducing the selectivity of the sensor. The last concern with Zr-MOF-based sensors is their stability over long periods of time. Nonetheless, Zr-MOFs have demonstrated good thermal and chemical stability, but after some extended time of exposure to highly abusive conditions or cycling, the MOFs' sensor performance may degrade [42]. This limitation is significant especially for all the applications that should always be running or operating in tough environments. Another limitation of Zr-MOF synthesis in sensor application is the reproducibility and scalability of the fabrication process. In order to obtain comparable sensor performance, it is necessary to control the synthesis of MOFs which can be challenging in large-scale processes. This problem is aggravated by the increased variability of the Zr-MOF structure and composition, meaning that even minor changes can affect the sensitivity and selectivity of the chosen sensor [43]. Electrode fouling is another factor that must be addressed when developing an electrochemical sensor based on Zr-MOF. Some concentrations may occlude the electrode surface; therefore, the sensitivity and accuracy of the electrode may decline over time. However, this problem can, to some extent, be solved using the advantages of Zr-MOFs; nevertheless, the problem of fouling remains one of the most significant challenges in the development of efficient strategies for electrode coverage. However, the ability to integrate Zr-MOFs with other components of the sensor system, including transducers and signal processing units, is another challenge [44,45]. One of the critical aspects of the Zr-MOF sensor is making sure that there is good contact between the electrical current being carried and the electrode material. However, as with other MOFs, much research is still being conducted to improve the immobilization techniques of biomolecules or other recognition elements on the surface of Zr-MOFs without considerably affecting the structure and function of MOFs. Cost factors are also relevant to the fabrication of Zr-MOF-based sensors. Although Zr-MOFs have many interesting characteristics, the fabrication of high-performance materials and integration of MOFs into sensor devices may be costly, hampering their application in real-world applications. Whereas the sensitivity and measurement range of Zr-MOF-based sensors can be impressive in many cases, they may still require enhancement for specific tasks [46].

Figure 14.7 shows electrochemical nanobiosensors and their utility in medical diagnostics and regenerative medicine. At the top, it showcases three main categories of biomolecule sensing: molecular probe biosensors that employ hybridization, enzymes biosensors with MMP-cleaved systems, and hormone biosensors that operate under the antibody-antigen reactions. In close proximity to these are the whole cell-based biosensors using 2D and 3D cell culture with ECM and Matrigel coatings for cancer and stem cell applications. The middle section explains the different electrochemical detection techniques used in these biosensors. These methods, such as cyclic voltammetry, differential pulse voltammetry, and electrochemical impedance spectroscopy, are recognized by the distinct signal output graphs they produce. Electrochemical detection and molecular biomarkers are illustrated as detection mechanisms used by detection mechanisms. The top panel of this figure describes the principles of these nanobiosensor technologies, while the lower part shows their practical use in medicine. The left side demonstrates diagnostic medicine applications in which biosensors identify certain molecular targets. In the central panel, biosensing is presented with the option of using computers for the online control of the entire setup. On the right side, there is an example of regenerative medicine using technology: from patient-specific stem cells to drug screening. In the entire image, aspects of merging different sensing platforms with medical aspects are highlighted;

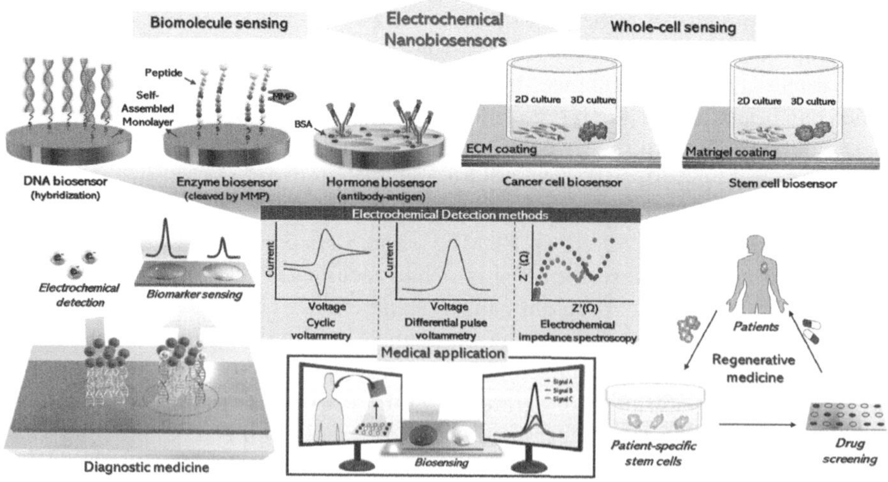

FIGURE 14.7 Advanced electrochemical sensors for various targets [47]. (CC By 4.0 Attribution.)

good examples are shown to illustrate how these technologies act as a link between detection at molecular level and putting it into practical and applicable healthcare needs. The diagram serves the purpose of explaining the role and complexity of nanobiosensor technology and its importance in developing the next-generation diagnostic and therapeutic tools [47].

The challenge is to achieve low detection limits while simultaneously achieving a broad dynamic range for the analytical system, particularly when analyzing complex molecules or in the presence of potentially interfering agents. Environmental and health risks associated with nanomaterials also need to be addressed. Because Zr-MOFs are usually employed in the nanoscale, the potential hazards associated with the release of such materials into the environment or with their interaction with living tissues and organs should be investigated thoroughly and controlled appropriately [48]. Furthermore, the need to standardize the fabrication process of fabric sensors, as well as the calibration and evaluation of their performance, can greatly benefit from the application of Zr-MOF-based electrochemical sensors. The absence of such standards can complicate comparison of the results of the diverse studies and the subsequent application of these sensors beyond laboratory environments. Thus, the development of Zr-MOF-based electrochemical sensors represents a highly promising approach, although further investigations are needed to overcome the indicated problems and limitations. The elimination of these challenges will provide a path toward building enhanced sensing platforms based on the intrinsic traits of Zr-MOFs [49].

14.5 FUTURE TRENDS IN ZIRCONIUM MOF

Promising developments in the chemistry of zirconium MOFs, with the potential to improve sensor effectiveness, may redefine the existing prospects for analytical chemistry and materials science. Further research on the characteristics of Zr-MOF will open up more opportunities to enhance the performance of sensors. Among them, a key focus is made on the synthesis of composites based on Zr-MOFs and other functional components. The incorporation of silver nanoparticles into Zr-MOFs has been considered to improve fluorescence-based detection techniques. This approach, as used for the UiO-66-NH2@AgNPs system, enhances the sensitivity and reduces the detection limits of analytes such as chlorpyrifos. Further research could examine similar systems in which Zr-MOFs are integrated with diverse nanoparticles, quantum dots, or other materials to form synergistic multilayered sensing platforms [50].

Figure 14.8 shows a structural MOF analysis of Zr-DPT. DPA with an 8% content of the material was labeled as Zr-based. Figure 14.8 contains four important elements. Considering Figure 14.8a, there is an SEM picture showing that the synthesized product is crystalline particles with approximately 185 ± 20 nm in average size according to the histogram in the inset. Figure 14.8b is a nitrogen adsorption isotherm whose shape implies a highly microporous material based on the Type I classification of the International Union of Pure and Applied Chemistry (IUPAC), and a pore size distribution plot is provided in the insert. In Figure 14.8c, the authors show a highly enlarged representation of the MOF's internal framework, where tetrahedral and octahedral capacious pockets between the rods can also be observed.

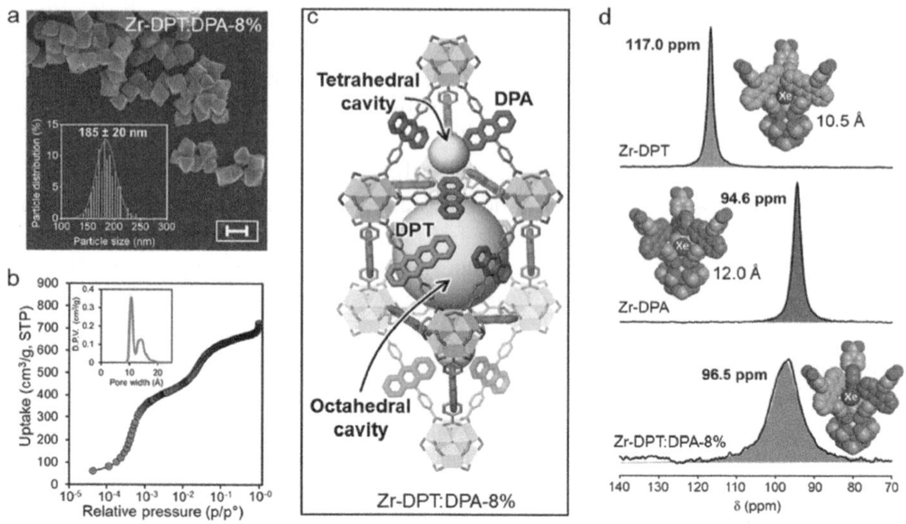

FIGURE 14.8 Structural characteristics of hetero-ligand zirconium-based metal-organic framework (MOF) nanocrystals. (a) A picture of MOF nanocrystals captured by scanning electron microscopy. (b) The 77 K N2 adsorption isotherm was obtained. (c) Structure of Zr-DPT:DPA crystal. (d) Hetero-ligand nanocrystals' hyperpolarized laser-assisted 129Xe NMR spectra in comparison to homo-ligand MOFs based on DPA and DPT [51]. (CC By 4.0 Attribution.)

The structure reveals a rather complicated arrangement of diphenylthiophene (DPT) and diphenylamine (DPA) ligands surrounding zirconium clusters, forming a solid three-dimensional structure. Figure 14.8d presents Xe-129 NMR spectra comparing three different variants: The chemical shift values analyzed in the current work include Zr-DPT at 117.0 ppm, Zr-DPA at 94.6 ppm, and Zr-DPT.DPA-8% at 96.5 ppm. The molecular structures depicted close to each spectrum also display different cavity dimensions of approximately 10.5 Å and 12.0 Å that reveal how different ligands affect the framework pore chemistry and environment [51].

A promising feature on this front is the possibility of designing specific Zr-MOF structures for particular sensor uses. To enable the selective sensing of a broad range of analytes, researchers have endeavored to control the pore size, functionality, and surface chemistry of Zr-MOFs. This approach could pave the way for the design of Zr-MOF-based sensors that can identify and selectively sense various structural descriptors in complex matrices, such as food products and physiological fluids. In addition, the connectivity of Zr-MOFs with modern analytical tools is envisaged to open new avenues in the future [52]. The incorporation of Zr-MOFs with surface-enhanced Raman spectroscopy (SERS) might lead to highly sensitive and non-reporter-based detection systems. Despite the recent interest in the use of other 2D materials such as MXenes for SERS, Zr-MOFs are likely to exhibit comparable improvements in Raman-based sensing. Future developments in the synthesis and fabrication of Zr-MOF-based sensors should address aspects such as duplicability, ease of fabrication, and affordability. Thus, researchers can continue to investigate novel approaches for the controlled growth and deposition of Zr-MOFs on different substrate materials, including flexible and wearable materials, to broaden the scope of sensing applications [53]. Possible future developments of Zr-MOFs include their integration in the construction of multimodal sensing platforms. By integrating EChem, optical, and gravimetric sensing mechanisms into one Zr-MOF platform, researchers hope to build more complex and accurate sensors that can supply supplementary data about analytes. Machine learning and artificial intelligence will continue to become significant in the formulation of Zr-MOF sensor [54]. These computational tools can be used to predict and design Zr-MOF structures for certain sensing functionalities and to improve data processing and evaluation from a myriad of sensor systems. Zr-MOF sensor may find more significant applications in environmental and biomedical fields, and it can be expected that these fields will define the future of Zr-MOF sensor development. The creation of new, efficient, and selective Zr-MOF-based sensors for the fast and precise detection of pollutants, toxins, and biomarkers may open new avenues in environmental and healthcare diagnostics [55]. Finally, it is essential to manage the difficulties associated with selecting the appropriate quantity of functional materials in Zr-MOFs to achieve long-term stability and performance stability under real conditions for the practical application of Zr-MOF-based sensors. Future work will thus involve research on enhancing the applicability of these materials in real systems and difficult conditions, in addition to researching methods for minimizing fouling and sensor degradation within the initial periods of sensor utilization [56]. By applying the properties of the above materials and integrating them with other technologies and methods of fabrication, researchers will be able to design a new generation of efficient, selective, and universal sensing platforms that could change many fields, from environmental monitoring to healthcare.

14.6 CONCLUSION

MOFs are a rapidly developing class of materials that can be applied in the creation of new-generation sensors owing to their unique structural and property features. This study describes the use of Zr-MOFs. The use of MOFs in the creation of electrochemical sensors has stimulated much interest due to their organometallic nature and the presence of metal nodes and organic linkers. The stability of Zr-MOFs at high and low temperatures, high chemical resistance, ability to modulate pore size, and ease of surface modification have made them one of the most researched materials in analytical chemistry and materials science. Recent improvements in Zr-MOF-based electrochemical sensors support the findings of this study, as numerous analytes, including environmental pollutants, pharmaceuticals, and biological markers, can be detected using this framework. Recent enhancements, such as functionalization, the use of mixed linkers, and the incorporation of companion materials, have enabled the synthesis of highly sensitive and selective sensors with low detection limits and high dynamic ranges.

Nevertheless, the development of Zr-MOF-based sensors is not without difficulties, both conceptually and in terms of real-life applications. These encompass enhancing performance in complicated matrices, guaranteeing lasting stability, enhancing the repeatability and ampleness of the synthesis, and contemplating ecological and well-being factors connected to nanomaterials. There are several potential directions for the future studies of Zr-MOFs to increase the sensor performance. These include the enhancement of Zr-MOFs by incorporating momentous segments attached to the framework, modulation of Zr-MOF structures for one purpose or another, incorporation of sophisticated analysis protocols, and use of artificial intelligence and machine learning for sensor tweaking and data interpretation. With further development in this area of study, Zr-MOF-based electrochemical sensors are likely to have wider uses in a variety of applications in the analysis field. It has the potential to overhaul applications such as environmental, medical, and industrial process monitoring. Thus, it can be foreseen that future generations of Zr-MOF-based sensors will provide outstanding performance in terms of the sensitivity, selectivity, and reliability of the detection and quantification of various chemical species, with regard to the current limitations and challenges that have been discussed in this review. The continuous advancements in Zr-MOF-based sensors can be categorized as a compelling trend in material and analytical chemistry and are beneficial for solving the significant problems arising in the energy, environment, and healthcare industries.

REFERENCES

1. Jia C, He T, Wang G-M. Zirconium-based metal-organic frameworks for fluorescent sensing. *Coordination Chemistry Reviews*. 2023;476:214930.
2. Khosropour H, Keramat M, Tasca F, Laiwattanapaisal W. A comprehensive review of the application of Zr-based metal–organic frameworks for electrochemical sensors and biosensors. *Microchimica Acta*. 2024;191:449.
3. Chattopadhyay K, Mandal M, Maiti DK. A review on zirconium-based metal–organic frameworks: Synthetic approaches and biomedical applications. *Materials Advances*. 2024;5:51–67.
4. Kung C-W, Goswami S, Hod I, Wang TC, Duan J, Farha OK, Hupp JT. Charge transport in zirconium-based metal–organic frameworks. *Accounts of Chemical Research*. 2020;53:1187–95.
5. Zhan X-Q, Tsai F-C, Xie L, Zhang K-D, Liu H-L, Ma N, Shi D, Jiang T. Ligands-coordinated Zr-based MOF for wastewater treatment. *Nanomaterials*. 2018;8:655.
6. Du Q, Rao R, Bi F, Yang Y, Zhang W, Yang Y, Liu N, Zhang X. Preparation of modified zirconium-based metal-organic frameworks (Zr-MOFs) supported metals and recent application in environment: A review and perspectives. *Surfaces Interfaces*. 2022;28:101647.
7. Taddei M. When defects turn into virtues: The curious case of zirconium-based metal-organic frameworks. *Coordination Chemistry Reviews*. 2017;343:1–24.
8. Su H, Lv S, Song H, Shi K, Zhu J, Zhang Y. Recent advances in continuous zirconium-based metal–organic framework membranes for high-precision separation. *Separation Purification Technology*. 2024:127318.
9. Wang B, Wang P, Xie L-H, Lin R-B, Lv J, Li J-R, Chen B. A stable zirconium based metal-organic framework for specific recognition of representative polychlorinated dibenzo-p-dioxin molecules. *Nature Communications*. 2019;10:3861.
10. Malekzadeh M, Mohadesi A, Karimi MA, Ranjbar M. Development of a new electrochemical sensor based on Zr-MOF/MIP for sensitive diclofenac determination. *Analytical Bioanalytical Electrochemistry*. 2020;12:402–14.
11. Rani R, Deep A, Mizaikoff B, Singh S. Zirconium metal organic framework based opto-electrochemical sensor for nitrofurazone detection. *Journal of Electroanalytical Chemistry*. 2022;909:116124.

12. Liu L, Wang S, Zhang B, Jiang G, Liu H, Yang J, Wang J, Liu W. Present status and prospects of nanostructured thermal barrier coatings and their performance improvement strategies: A review. *Journal of Manufacturing Processes*. 2023;97:12–34.
13. Jahangiri–Dehaghani F, Zare HR, Shekari Z. Measurement of aflatoxin M1 in powder and pasteurized milk samples by using a label–free electrochemical aptasensor based on platinum nanoparticles loaded on Fe–based metal–organic frameworks. *Food Chemistry*. 2020;310:125820.
14. Gao N, Tan R, Cai Z, Zhao H, Chang G, He Y. A novel electrochemical sensor via Zr-based metal organic framework–graphene for pesticide detection. *Journal of Materials Science*. 2021;56:19060–74.
15. Zhang H, Xiong P, Li G, Liao C, Jiang G. Applications of multifunctional zirconium-based metal-organic frameworks in analytical chemistry: Overview and perspectives. *TrAC Trends in Analytical Chemistry*. 2020;131:116015.
16. Xue W, Wang Y, Guo L, Zhang H. Zr-MOF functionalized nanochannels: Application to regenerative and sensitive electrochemical aptasensing platform. *Sensors Actuators B: Chemical*. 2023;381:133455.
17. Chuang CH, Kung CW. Metal– organic frameworks toward electrochemical sensors: Challenges and opportunities. *Electroanalysis*. 2020;32:1885–95.
18. Li H-K, Ye H-L, Zhao X-X, Sun X-L, Zhu Q-Q, Han Z-Y, Yuan R, He H. Artful union of a zirconium-porphyrin MOF/GO composite for fabricating an aptamer-based electrochemical sensor with superb detecting performance. *Chinese Chemical Letters*. 2021;32:2851–5.
19. Su F, Zhang S, Ji H, Zhao H, Tian J-Y, Liu C-S, Zhang Z, Fang S, Zhu X, Du M. Two-dimensional zirconium-based metal–organic framework nanosheet composites embedded with Au nanoclusters: A highly sensitive electrochemical aptasensor toward detecting cocaine. *ACS Sensors*. 2017;2:998–1005.
20. Fang X, Chen X, Liu Y, Li Q, Zeng Z, Maiyalagan T, Mao S. Nanocomposites of Zr (IV)-based metal–organic frameworks and reduced graphene oxide for electrochemically sensing ciprofloxacin in water. *ACS Applied Nano Materials*. 2019;2:2367–76.
21. Gui B, Meng Y, Xie Y, Tian J, Yu G, Zeng W, Zhang G, Gong S, Yang C, Zhang D. Tuning the photoinduced electron transfer in a Zr-MOF: Toward solid-state fluorescent molecular switch and turn-on sensor. *Advanced Materials*. 2018;30:1802329.
22. Sun R, Lv R, Li Y, Du T, Chen L, Zhang Y, Zhang X, Zhang L, Ma H, Sun H. Simple and sensitive electrochemical detection of sunset yellow and SUDAN I in food based on AuNPs/Zr-MOF-graphene. *Food Control*. 2023;145:109491.
23. Zhang L, Han Y, Sun M, Li S. Non-enzymatic electrochemical sensor based on ionic liquid [BMIM][PF6] functionalized zirconium-copper bimetallic MOF composite for the detection of nitrite in food samples. *Food Chemistry*. 2024:140023.
24. Han J, Li F, Zhao M, Guo M, Liu Y, Guo X, Ran Q, Wang Z, Zhao H. Ultrasensitive electrochemical sensing platform based on Zr-based MOF embellished single-wall carbon nanotube networks for trace analysis of methyl parathion. *Microchemical Journal*. 2024;200:110323.
25. Zhou N, Su F, Guo C, He L, Jia Z, Wang M, Jia Q, Zhang Z, Lu S. Two-dimensional oriented growth of Zn-MOF-on-Zr-MOF architecture: A highly sensitive and selective platform for detecting cancer markers. *Biosensors Bioelectronics*. 2019;123:51–8.
26. Mehta A, Singh G. Consequences of hydroxyapatite doping using plasma spray to implant biomaterials. *Journal of Electrochemical Science Engineering Failure Analysis*. 2023;13:5–23.
27. Chen H, Fu Y, Feng K, Zhou Y, Wang X, Huang H, Chen Y, Wang W, Xu Y, Tian H. Polydopamine-coated UiO-66 nanoparticles loaded with perfluorotributylamine/tirapazamine for hypoxia-activated osteosarcoma therapy. *Journal of Nanobiotechnology*. 2021;19:1–18.

28. Li P, Luo L, Cheng D, Sun Y, Zhang Y, Liu M, Yao S. Regulation of the structure of zirconium-based porphyrinic metal–organic framework as highly electrochemiluminescence sensing platform for thrombin. *Analytical Chemistry*. 2022;94:5707–14.
29. Rageh AH, Said MI, Abdel-Aal FA. Zirconium-based hydrophobic-MOFs as innovative electrode modifiers for flibanserin determination: Exploring the electrooxidation mechanism using a comprehensive spectroelectrochemical study. *Microchimica Acta*. 2024;191:236.
30. Gupta R, Rahi Alhachami F, Khalid I, Majdi HS, Nisar N, Mohamed Hasan Y, Sivaraman R, Romero Parra RM, Al Mashhadani ZI, Fakri Mustafa Y. Recent progress in aptamer-functionalized metal-organic frameworks-based optical and electrochemical sensors for detection of mycotoxins. *Critical Reviews in Analytical Chemistry*. 2024;54:1707–28.
31. Bu L, Jiang D, Song Q, Shan X, Wang W, Chen Z. A novel three-dimensional molecularly imprinted polypyrrole electrochemical sensor based on MOF derived porous carbon and nitrogen doped graphene for ultrasensitive determination of dopamine. *Analyst*. 2022;147:5194–202.
32. Li Y, Hu M, Huang X, Wang M, He L, Song Y, Jia Q, Zhou N, Zhang Z, Du M. Multicomponent zirconium-based metal-organic frameworks for impedimetric aptasensing of living cancer cells. *Sensors Actuators B: Chemical*. 2020;306:127608.
33. Sun Z, Wang L, Wu S, Pan Y, Dong Y, Zhu S, Yang J, Yin Y, Li G. An electrochemical biosensor designed by using Zr-based metal–organic frameworks for the detection of glioblastoma-derived exosomes with practical application. *Analytical Chemistry*. 2020;92:3819–26.
34. Zhang Z-H, Duan F-H, Tian J-Y, He J-Y, Yang L-Y, Zhao H, Zhang S, Liu C-S, He L-H, Chen M. Aptamer-embedded zirconium-based metal–organic framework composites prepared by de novo bio-inspired approach with enhanced biosensing for detecting trace analytes. *ACS Sensors*. 2017;2:982–9.
35. Li X, Meng F, Li Z, Li R, Zhang Y, Zhang M. Dual-signal aptasensor based on Zr-MOF for ultrasensitive detection of AFB1 in corn. *Sensors Actuators B: Chemical*. 2023;394:134372.
36. Ngwem MCN, Kemmegne-Mbouguen JC, Langmi HW, Musyoka NM, Mokaya R. Electrochemical sensor for ascorbic acid, acetaminophen and nitrite based on organoclay/Zr-MOF film modified glassy carbon electrode. *ChemistrySelect*. 2022;7:e202202308.
37. Biswas S, Chen Y, Xie Y, Sun X, Wang Y. Polypyrrole merged zirconium-based metal-organic framework NU-1000 for detection of levodopa. *Microchimica Acta*. 2020;187:1–7.
38. Zhang G-Y, Zhuang Y-H, Shan D, Su G-F, Cosnier S, Zhang X-J. Zirconium-based porphyrinic metal–organic framework (PCN-222): Enhanced photoelectrochemical response and its application for label-free phosphoprotein detection. *Analytical Chemistry*. 2016;88:11207–12.
39. Nazari M, Amini A, Eden NT, Duke MC, Cheng C, Hill MR. Highly-efficient sulfonated UiO-66 (Zr) optical fiber for rapid detection of trace levels of Pb^{2+}. *International Journal of Molecular Sciences*. 2021;22:6053.
40. Xu G, Zhu C, Gao G. Recent progress of advanced conductive metal–organic frameworks: Precise synthesis, electrochemical energy storage applications, and future challenges. *Small*. 2022;18:2203140.
41. Ding M, Cai X, Jiang H-L. Improving MOF stability: Approaches and applications. *Chemical Science*. 2019;10:10209–30.
42. Mathew AE, Jose S, Babu AM, Varghese A. State of the art MOF-composites and MXene-composites: Synthesis, fabrication and diverse applications. *Materials Today Chemistry*. 2024;36:101927.

43. Liu K-G, Sharifzadeh Z, Rouhani F, Ghorbanloo M, Morsali A. Metal-organic framework composites as green/sustainable catalysts. *Coordination Chemistry Reviews*. 2021;436:213827.
44. Musyoka NM, Ren J, Langmi HW, North BC, Mathe M, Bessarabov D. Synthesis of rGO/Zr-MOF composite for hydrogen storage application. *Journal of Alloys Compounds*. 2017;724:450–5.
45. Mehta A, Vasudev H, Singh S. Sustainable manufacturing approach with novel thermal barrier coatings in lowering CO_2 emissions: Performance analysis with probable solutions. *International Journal on Interactive Design Manufacturing*. 2023:1–13.
46. Heiba AR, Abdel-Salam M, Yoon T, El Sawy E. Zr-MOF composites with zipped and unzipped carbon nanotubes for high-performance electrochemical supercapacitors. *Nanoscale*. 2025;17:459–73.
47. Suhito IR, Koo K-M, Kim T-H. Recent advances in electrochemical sensors for the detection of biomolecules and whole cells. *Biomedicines*. 2020;9:15.
48. Heu R, Ateia M, Awfa D, Punyapalakul P, Yoshimura C. Photocatalytic degradation of organic micropollutants in water by Zr-MOF/GO composites. *Journal of Composites Science*. 2020;4:54.
49. Lan Y-C, Kamal S, Lin C-C, Liu Y-H, Lu K-L. Ultra-thin Zr-MOF/PVA/Melamine composites with remarkable sound attenuation effects. *Microporous Mesoporous Materials*. 2023;360:112668.
50. Guo L, Chen S, Yu Y-L, Wang J-H. A smartphone optical device for point-of-care testing of glucose and cholesterol using Ag NPs/UiO-66-NH_2-based ratiometric fluorescent probe. *Analytical Chemistry*. 2021;93:16240–7.
51. Perego J, Bezuidenhout CX, Villa I, Cova F, Crapanzano R, Frank I, Pagano F, Kratochwill N, Auffray E, Bracco S. Highly luminescent scintillating hetero-ligand MOF nanocrystals with engineered Stokes shift for photonic applications. *Nature Communications*. 2022;13:3504.
52. Jiang H-L, Li N, Wang X, Wei X-Y, Zhao R-S, Lin J-M. A zirconium-based metal-organic framework material for solid-phase microextraction of trace polybrominated diphenyl ethers from milk. *Food Chemistry*. 2020;317:126436.
53. Sunil J, Narayana C, Kumari G, Jayaramulu K. Raman spectroscopy, an ideal tool for studying the physical properties and applications of metal–organic frameworks (MOFs). *Chemical Society Reviews*. 2023;52:3397–437.
54. Wu W, Li Y, Song P, Xu Q, Long N, Li P, Zhou L, Fu B, Wang J, Kong W. Metal-organic framework (MOF)-based sensors for exogenous contaminants in food: Mechanisms, advances, and prospects. *Trends in Food Science Technology*. 2023;138:238–71.
55. Li R, Bu T, Zhao Y, Sun X, Wang Q, Tian Y, Bai F, Wang L. Polydopamine coated zirconium metal-organic frameworks-based immunochromatographic assay for highly sensitive detection of deoxynivalenol. *Analytica Chimica Acta*. 2020;1131:109–17.
56. Li Y, Wang C, Li Z, Wang M, He L, Zhang Z. Zirconium-porphyrin complex as novel nanocarrier for label-free impedimetric biosensing neuron-specific enolase. *Sensors Actuators B: Chemical*. 2020;314:128090.

Section IV

Medical and Dental Applications

15 Pharmaceutical and Device Applications of Zirconia

Nigar K. Mujawar, Vikas R. Dhole, Sanganna Burli, and Sawanta S. Mali

15.1 INTRODUCTION TO ZIRCONIA

Zirconia (ZrO_2), commonly known as zirconium dioxide, is a versatile ceramic material with exceptional biocompatibility, thermal stability, and mechanical properties. Three main structural phases—cubic, tetragonal, and monoclinic—of zirconia are determined by temperature and dopant additions. Its great wear resistance, excellent fracture toughness, and chemical inertness have attracted a lot of interest. It finds numerous applications, mainly in the biological and pharmacological domains (Chevalier & Gremillard, 2009).

Zirconia is often stabilized with yttria (YO_3), magnesia (MgO), or ceria (CeO_2) to preserve its favorable properties across a wide temperature range. One of the most often used varieties, yttria-stabilized zirconia (YSZ), offers improved hardness and durability at high temperatures. Because of its great Young's modulus, low heat conductivity, and corrosion resistance, zirconia is a popular material for structural and functional uses, including dental implants, orthopedic prostheses, and drug administration systems (Lughi & Sergo, 2010).

Zirconia is widely used in biomedical and pharmaceutical domains due to its excellent biocompatibility and mechanical performance. Medical applications include zirconia-based biomaterials, such as orthopedic and dental implants, bone scaffolds, and tissue engineering. Unlike other metals, zirconia does not trigger allergic or biological reactions, thus being accepted as a long-term medical implant (Piconi and Maccauro, 1999).

An example of zirconia in the dental field can be seen when used for crowns, bridges, and soft and hard tissue dental implants because of its durability, aesthetic appearance, and strength (similar to natural teeth). Research has demonstrated that zirconia implants have better osseointegration characteristics than classic titanium implants, decreasing the chances of inflammation and implant failure (Zhang et al., 2014).

Zirconia is even used in orthopedic applications as implant devices. It is an excellent option for hip and knee joint replacements thanks to its high wear resistance and low friction coefficient. Moreover, non-magnetic and radiopaque find applications in medical imaging and diagnostic tools. Zirconia is used in medication delivery systems in pharmacology to serve as a carrier allowing regulated drug release. This chemical

DOI: 10.1201/9781003519300-19

inertness and stability feature enhances the efficacy of numerous pharmacological devices through targeted and persistent therapeutic effects. Further adding to their importance in the field of biomedical sciences are the uses of nanoparticles based on zirconia in cancer therapy and bioimaging (Lee et al., 2016).

Although zirconia has been around for millennia, its early uses were mostly restricted to ceramic and refractory materials. Zirconia's resistance to high temperatures and beautiful look have led to its historical usage in pottery and ornamental ceramics. Improvements in materials science during the twentieth century made it possible to employ high-performance zirconia ceramics in both industry and medicine (Chevalier et al., 2007).

One of zirconia's first notable uses in medicine was as a dental material in the 1970s. It is widely employed in modern orthopedics and dentistry because of better stabilization and material manufacturing procedures that have been developed throughout time and have increased its mechanical qualities. Applications for zirconia have changed since October 2023; currently, zirconia-based ceramics are extensively utilized and have promise in the electrical, automotive, and aerospace sectors (Garvie et al., 1975).

Together with nanotechnology, zirconia is enabling new uses: antimicrobial surfaces and individualized treatment approaches (Koch et al., 2010).

All things considered, zirconia has evolved into a vital component with distinctive qualities like great strength, biocompatibility, and outstanding chemical stability in many spheres of science and technology, particularly in biomedicine and pharmaceuticals.

15.2 SYNTHESIS AND CHARACTERIZATION OF ZIRCONIA

15.2.1 Methods of Synthesizing Zirconia Nanoparticles and Bulk Materials

The synthetic approaches used to synthesize zirconia nanoparticles and bulk materials include sol-gel processing, hydrothermal or solvothermal synthesis, co-precipitation, and solid-state reactions. Different approaches have unique benefits, allowing for control over particle size, phase composition, and surface properties (Ramesh et al., 2017). The sol-gel method is a popular strategy for preparing high-purity zirconia with outstanding phase stability, whereas hydrothermal synthesis delivers nanostructures with improved crystallinity (Mahmoudi et al., 2018).

15.2.2 Structural and Surface Characterization (XRD, SEM, TEM, FTIR, Etc.)

The structural and surface properties of zirconia are analyzed using various characterization techniques:

15.2.2.1 X-ray Diffraction (XRD)

X-ray diffraction (XRD) greatly simplifies the zirconia (ZrO_2) crystalline structure analysis. This method identifies which phases occur via diffraction patterns among monoclinic, tetragonal, and cubic structures. Phases can show many different kinds.

Two crystalline solids or glass materials with the same substance can display distinct phases by phase transition. These materials mainly rely on phase stability to ensure suitable performance: steel components and zirconia-based dental ceramics are often coupled with thermal barrier coatings that function as phase-change indicators. The XRD-based study provides exact descriptions of material phases and structural features, which help academics and industry experts improve material attributes like durability, strength, and temperature tolerance in systems based on zirconia.

15.2.2.2 Scanning Electron Microscopy (SEM)

Scanning electron microscope is the essential analytical instrument for investigating zirconia (ZrO_2) shape and particle dimensions. To assist assess material-based performance in ceramics and coatings applications, this method produces exact pictures demonstrating three fundamental characteristics: surface texture, grain structure, and porosity. SEM-based minute morphological studies find homogeneity levels and agglomeration properties in zirconia that change mechanical properties and sintering behavior. Analyzing particle size distribution through SEM enables analysts to find process conditions that deliver optimal strength and stability results. SEM's elemental analysis through energy-dispersive X-ray spectroscopy (EDS) allows researchers and industries to obtain important chemical composition data about zirconia.

15.2.2.3 Transmission Electron Microscopy (TEM)

Zirconia (ZrO_2) nanostructures can be studied at atomic resolutions with transmission electron microscopy (TEM) because this instrument allows precise observation of grain boundaries and defects and phase distributions within ZrO_2 nanostructures. The crystal structure of zirconia, its phase stability, and nanostructured morphology can be thoroughly studied by TEM to aid the development of advanced materials used in catalysis bioceramics and thermal barrier coatings. The TEM examination provides information about nanoparticle dimensions while displaying their forms and distribution arrangements, directly impacting mechanical characteristics and thermal response. Selected area electron diffraction (SAED) and high-resolution TEM (HRTEM) advanced techniques enable scientists to investigate zirconia nanoscale characteristics better, thus leading to improved materials for various high-tech applications.

15.2.2.4 Fourier Transform Infrared Spectroscopy (FTIR)

Using Fourier transform infrared spectroscopy (FTIR) as an analytical method, one may qualitatively assess functional groups and chemical bonds in ZrO_2. The method detects unique zirconia vibrational modes that expose phase-related composition information and show surface hydroxylation states and interactions with bound molecules. Analyzing zirconia with surface modifications, organic and inorganic functional groups define important characteristics, including biocompatibility and catalytic activity, which FTIR provides with the required precision. The work helps to identify structural modifications, adsorbed species, and manufacturing-related pollutants. Optimizing zirconia-based materials for their use in ceramics alongside biomaterials and catalytic uses depends mainly on the exact molecular data offered by FTIR (Lamas et al., 2020).

15.3 SURFACE MODIFICATIONS FOR ENHANCED BIOCOMPATIBILITY

Zirconia surfaces undergo various modifications to improve biocompatibility, including plasma treatments, surface coatings, and functionalization with bioactive molecules. These modifications enhance osseointegration and reduce bacterial adhesion, making zirconia more suitable for medical implants (Almeida et al., 2019).

15.4 Properties of Zirconia Relevant to Biomedical and Pharmaceutical Applications

Table 15.1 summarizes the properties of zirconia relevant to biomedical and pharmaceutical applications.

15.5 USE OF ZIRCONIA IN DENTAL CROWNS, BRIDGES, AND IMPLANTS

Zirconia stands out for dental crown implants because it combines high aesthetic appeal, superior strength, and outstanding biocompatibility. Zirconia crowns serve as dental restorations for both front and back teeth because they show natural tooth appearance through their ability to match tooth color and translucency quality. Zirconia material demonstrates strong durability under chewing pressures while resisting damage from daily wear. Patients who need tooth replacement should consider using zirconia bridges because their strong structure provides better durability against breakage than regular ceramic bridges. Patients who need treatment because of poor bone quality find zirconia implants particularly helpful due to their muscular bone-to-implant junction, as they blend well with biological tissues. Zirconia stands out among other benefits because it provides excellent mechanical properties, including strength and durability and reduced plaque build-up, for dental implant practitioners who need long-lasting solutions (Zhu et al., 2020).

15.6 Advantages of Zirconia in Dental Applications

Because of its better qualities than conventional materials like metals and other ceramics, zirconia—a kind of ceramic material—has grown in favor of dental uses. Its favored selection in restorative dentistry is based on its improved strength, cosmetic appeal, and biocompatibility.

15.6.1 Strength and Durability

Zirconia's great strength qualities are the key advantage over conventional dental materials like metals and ceramics. Zirconia's great flexural strength beats all other dental ceramics, which causes resistance against both fractures and cracks (Zhang & Lawn, 2018a). Because zirconia resists intense chewing pressures, implant crowns and bridges gain the most from it.

TABLE 15.1
Properties of Zirconia Relevant to Biomedical and Pharmaceutical Applications

Sr. No.	Property	Description
1	Biocompatibility and bioactivity	Thanks in great part to its low toxicity and robust integration with biological tissues, modern medical research identifies zirconia (ZrO_2) as a highly biocompatible substance. Its limited ion release and chemical stability assist in avoiding negative medical responses, therefore guaranteeing its safety in clinical settings. Zirconia is a great alternative for dental and orthopedic implants as its porous surface supports better cell adhesion, therefore aiding cell development and specialization. The ability of zirconia to connect straight with bone via osseointegration is one of its main benefits as it improves implant durability and lifetime. Zirconia reduces patient risk of adverse responses by means of better resistance to corrosion than metal implants. Modern biomedical equipment, especially dental implants and joint reconstruction operations, favor its mechanical strength, longevity, and cosmetic appeal (Hickok et al., 2019).
2	Chemical stability and inertness	Because of its great chemical stability and corrosion resistance, zirconia (ZrO_2) is clearly a top choice for long-term biomedical implants. In body fluids, it stays inert, greatly lowering the possibility of ion leakage or material breakdown. Its great biocompatibility guarantees that it won't corrode or set off inflammatory responses, enabling implants to operate safely and consistently throughout time. Excellent wear resistance makes zirconia especially appropriate for the demanding settings of dental and orthopedic uses. Under physiological conditions, it maintains excellent structural integrity, minimizing the risk of fractures and extending the lifespan of implants. These properties make zirconia a highly effective implant material, offering both improved durability and better patient outcomes (Wang et al., 2020).
3	Mechanical strength and wear resistance	For biomedical uses such joint prosthesis, which must endure great stress, zirconia (ZrO_2) is the finest choice as it exhibits extraordinary mechanical strength, great fracture toughness, and excellent wear resistance. Its hardness and resistance against the spread of cracks help boost product durability, therefore reducing the possibility of implant failure under several loads. Stress causes zirconia to change phase, enhancing its toughness by avoiding destructive fractures. The low friction coefficient of zirconia allows it to minimize wear, increasing the implants' lifetime. Because of its biocompatible qualities, corrosion resistance, and exceptional physical traits that result in longer implants, which promote patient comfort and mobility, zirconia is a preferred implant material (Rogers et al., 2018).

(Continued)

TABLE 15.1 (*Continued*)
Properties of Zirconia Relevant to Biomedical and Pharmaceutical Applications

Sr. No.	Property	Description
4	Thermal and electrical properties	Zirconia (ZrO_2) proves outstanding as an insulator because it maintains its structure when exposed to extreme heat. Zirconia is extensively utilized in dental prosthesis and orthopedic implants as its low heat conductivity gives stability in surroundings with temperature variations. Long-term medicinal uses depend on this quality as it helps shield surrounding tissues from significant temperature fluctuations. Zirconia's great resistivity helps it to efficiently isolate electrical currents, thereby ensuring its relevance in biomedical devices requiring low electrical conductivity. This function has proven helpful for sensors, pacemakers, and several electrical implants. Zirconia helps the medical sector as it shows compatibility with biological systems and offers better thermal and electrical insulation qualities, which also stay robust mechanically (Zhao et al., 2021).

15.6.2 Aesthetic Appeal

Compared with metal-based materials, the results achieved with zirconia have a superior aesthetic appeal. The translucency of zirconia makes it possible to create naturally appearing crowns and bridges without using extra porcelain as the metal counterpart does. This natural quality occurs because zirconia allows light transmission similar to human tooth enamel (Sailer et al., 2007). The attractive character of zirconia stands out as its key advantage for displaying in the mouth's visible regions.

15.6.3 Biocompatibility

Zirconia is a highly compatible material with gums because it does not trigger adverse reactions to bodily tissues. Zirconia stands apart from other metals since it demonstrates safety and non-reactivity features that create an excellent fit for patients with metal sensitivities (Göllner et al., 2017a).

15.6.4 Longevity

The extended lifespan is a significant benefit that makes zirconia popular in dental restorative applications. Due to its wear and tear resistance, zirconia surpasses traditional ceramics, while its high corrosion resistance maintains its integrity over many years (Götz et al., 2016).

15.6.5 Ease of Fabrication

CAD/CAM technology is used by medical professionals to effectively process zirconia to produce extremely exact dental prosthesis. The manufacturing process through this technology creates precise products that avoid production mistakes while ensuring proper fit, so extra placement adjustments become unnecessary (Hausch et al., 2017).

15.7 Advantages of Zirconia over Traditional Materials

- Standard ceramics and metal materials have inferior strength aspects and resistance to fracturing.
- The material shows excellent aesthetic properties through its natural translucency for tooth-like visual effects.
- The material demonstrates excellent compatibility with human tissue structures, thus lowering the potential for allergic response and skin discomfort.
- The materials provide enduring strength and resist fracturing, corrosion, and wear damage.
- The use of CAD/CAM technology for fabrication allows precise tooth construction, minimizing manufacturing discrepancies.

15.8 ZIRCONIA LONG-TERM DURABILITY AND PERFORMANCE

Zirconia earns high recognition in dentistry because of its outstanding resistance to long-term wear and use in dental appliances. Both dentists and patients choose zirconia frequently for dental restorations since it resists wear, fracture, and corrosion due to its leading position as a material for implants, bridges, and crowns. Patients and dental practitioners widely select this material because of its strong performance and enduring structural maintenance properties.

15.8.1 Resistance to Wear and Tear

The main reason zirconia remains popular for dental restoration applications over time is that it withstands mechanical stress very well. Zirconia resists abrasion better than porcelain because the material shows improved longevity when opposing teeth come into contact (Zhang & Lawn, 2018a). Zirconia demonstrates superior resistance to wear when used for posterior restoration because chewing forces reach their maximum there.

15.8.2 Fracture Toughness

The fracture toughness rating for zirconia surpasses the values of standard ceramic materials and standard metals. Zirconia crowns, bridges, and implants demonstrate exceptional durability because of their resistance to stresses encountered during everyday biting functions (Götz et al., 2016). Zirconia demonstrates strong resistance to cracking throughout its lifespan because this strength helps it achieve long-lasting performance within high-stress areas of the mouth.

15.8.3 Corrosion Resistance

Corrosion resistance is a significant factor in extending zirconia's lifespan because of its high resistance capability. Maintaining its stability throughout the oral environment, zirconia is completely resistant to oxidation and corrosion, preserving the repair's structural integrity and functional characteristics (Göllner et al., 2017b). Patients with metal sensitivity or corrosion worries might consider zirconia as it resists this harm.

15.8.4 Stable Aesthetic Properties

One major benefit of zirconia is retaining aesthetic qualities as it keeps it looking good throughout its lifetime. Zirconia distinguishes itself by maintaining its inherent translucence and original look even at obvious oral places and resisting material degradation (Sailer et al., 2007). Even after long-term use, the zirconia material has consistent aesthetic qualities until restorations spontaneously mix with tooth enamel.

15.8.5 Longevity of Implants

Zirconia implant research has provided rather remarkable long-term clinical findings. Because of their biocompatible character, structural strength, and corrosion-resistant properties, zirconia implants provide patients seeking tooth replacements with lasting alternatives (Hausch et al., 2017). Many clinical investigations reveal that zirconia implants maintain patient function and appearance over a lengthy time as a long-lasting alternative to titanium systems.

15.8.6 Key Points on Zirconia's Long-Term Durability and Performance

Key points on zirconia's long-term durability and performance are shown in Figure 15.1.

15.9 ZIRCONIA IN ORTHOPEDIC AND MEDICAL IMPLANTS

Medical science utilizes zirconia as an essential material for manufacturing orthopedic implants, including hip and knee replacement components. Zirconia has gained popularity among medical and orthopedic communities because it provides excellent mechanical performance, tissue compatibility, and sufficient strength for load-bearing applications. Zirconia-based joint replacements show exceptional performance regarding mechanical strength and osseointegration, together with biomechanical effectiveness, which improves patient recovery results.

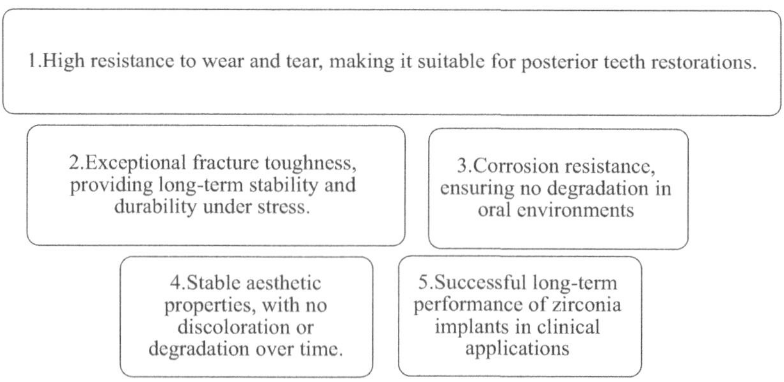

FIGURE 15.1 Key points on zirconia's long-term durability and performance.

15.9.1 Applications in Hip and Knee Replacements

The application of zirconia-based materials continues to rise for hip and knee joint replacements, mainly for prostheses bearing surfaces. Zirconia material exceeds polyethylene and metal alloy wear resistance characteristics leading to improved implant durability (He et al., 2017). Medical studies confirm that zirconia works well as femoral heads and tibial components due to its proven ability to decrease wear and provide long-lasting motion surfaces (Li et al., 2016).

15.9.2 Mechanical Strength for Load-Bearing Applications

The excellent mechanical resistance of zirconia, combined with its high fracture toughness and wear resistance, make it a suitable choice for orthopedic implant load-bearing applications. Zirconia demonstrates superior flexural strength compared to other ceramics to support the high stresses which occur when walking or running (Zhang and Lawn, 2018b). The mechanical strength of zirconia supports implant durability by minimizing frame distortions while facing intense pressure.

15.9.3 Biomechanical Performance and Osseointegration

Zirconia shows superior mechanical strength and is ideally suited for bone integration into the body. The surface of an implant integrates through osseointegration, which allows bone cells to bond with it. Zirconia possesses biological compatibility properties, enabling bone integration that leads to stable implant placement (Göllner et al., 2017b). Zirconia enables improved natural bone regeneration by enhancing cell affinity and cellular differentiation since its surface remains smooth.

15.10 ZIRCONIA IN DRUG DELIVERY SYSTEMS

The pharmaceutical world has recently shown strong interest in zirconia as a drug delivery material because of its essential properties: biocompatibility, stability, and surface adaptability. Medical research involving pharmaceuticals expanded by incorporating zirconia for developing controlled-release devices, targeted drug systems, and surface treatment methods that boost drug absorption rates. The applications are depicted in Figure 15.2, which we discuss in detail hereafter.

15.10.1 Role in Controlled Release Formulations

The pharmaceutical substance zirconia (ZrO_2) is a key material for controlled drug delivery systems, providing a durable drug-carrying container that gradually releases active pharmaceutical ingredients (APIs). By means of its strong chemical and biological environment resistance, the substance stabilizes and guards API pharmaceuticals from degradation, thus maintaining therapeutic performance. Without regular dosage modifications, zirconia-based carriers allow patients to get medication with fewer doses and increase medication adherence. The great biocompatibility of zirconia makes it a perfect material for long-term medication delivery uses without

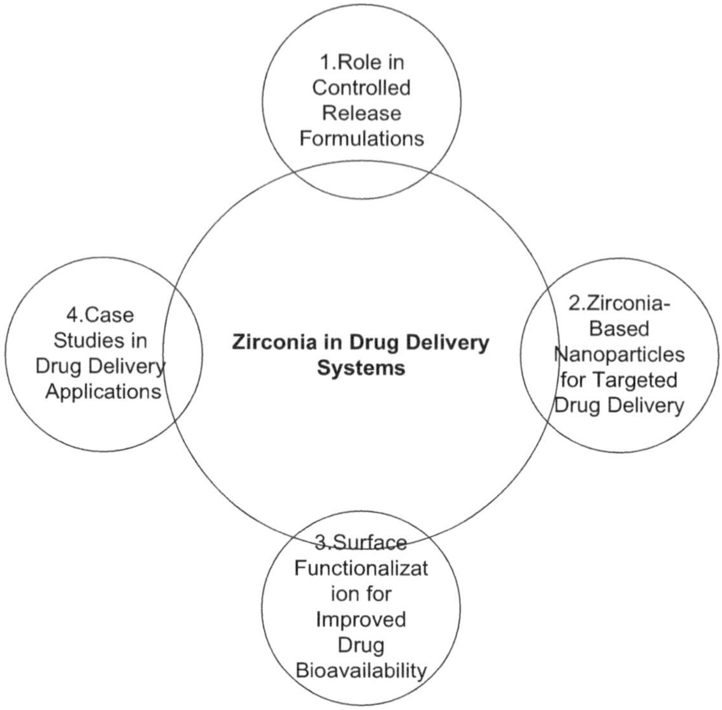

FIGURE 15.2 Zirconia in drug delivery systems.

inducing negative effects inside the body. Zirconia supports effective treatment through its ability to provide steady drug amounts that show specific benefits during the management of chronic diseases, implantable drug systems, and targeted therapeutic approaches. The application of zirconia-based matrices enabled ibuprofen delivery through sustained durations, which minimized the requirement for regular dosing and reduced adverse reactions (Miller et al., 2018).

15.10.2 Zirconia-Based Nanoparticles for Targeted Drug Delivery

Zirconia-based nanoparticles serve as emerging drug delivery vehicles that bring precisely targeted therapy and decrease patient side effects. Zirconia nanoparticles maintain a large surface area combined with adjustable surface properties, enabling effective drug-loading capabilities that serve as ideal carriers for chemotherapy, antibiotics, and additional therapeutic agents. Engineered zirconia nanoparticles can recognize particular tissue types and cell locations, enabling them to deliver drugs effectively and avoid adverse side effects in the human body. The human body can safely interact with biocompatible zirconia materials because these materials minimize the immune system's responses. The drug delivery revolution in precise medical care depends on stable zirconia nanoparticles, which discharge drugs systematically while offering multiple functional properties. Zirconia nanoparticles received folic acid modifications to allow directed treatment of cancer cells. Facilitated by zirconia nanoparticles, these delivery vehicles use doxorubicin as a chemotherapeutic agent, which can specifically target tumors for improved therapy outcomes and decrease side effects for the patient (Zhao et al., 2017).

15.10.3 Surface Functionalization for Improved Drug Bioavailability

Zirconia nanoparticles with surface functionalization show greater utilization in drug delivery processes because they improve the transport of poorly water-soluble drugs. The addition of polymers or the combination with targeting ligands on nanoparticles leads to enhanced cellular and tissue binding, increasing drug delivery availability. These strategic alterations add to drug dissolvability while ensuring stable drug entry into cells and safeguarding the drugs from destructive processes in bloodstream circulation. The functional aspect of these nanoparticles enhances drug stability with controlled release for sensitive drugs, enabling better treatment of cancer infections and chronic diseases. The approach improves both therapeutic results while reducing the side effects and extending drug shelf life. Zirconia nanoparticles receive PEG coating to enhance the solubility and bioavailability of curcumin and related poorly water-soluble drugs, which improves their gastrointestinal absorption potential (Cheng et al., 2019).

15.10.4 Case Studies in Drug Delivery Applications

Different studies reveal how zirconia has potential as a component of a medication delivery system. According to Xie et al. (2019), researchers administer anticancer medications using zirconia nanoparticles, improving their therapeutic success and reducing possible adverse effects. Studies have revealed that zirconia nanoparticles effectively carry antibiotics straight to affected locations, therefore enhancing treatment outcomes even with lowered bacterial resistance risks. Published studies show that zirconia has great use in stabilizing drugs and site-specific delivery while enhancing drug absorption, therefore providing the appropriate basis for exact, individualized medicine in treating diseases. For instance, studies showing zirconia nanoparticles mixed with vancomycin antibiotic delivery to sick human tissues revealed improved therapeutic outcomes, more significant antibiotic efficacy, and greater drug durability (Gao et al., 2018).

15.11 ZIRCONIA-BASED COATINGS IN PHARMACEUTICALS AND MEDICAL DEVICES

Zirconia-based coatings in pharmaceuticals and medical devices are discussed in Table 15.2.

15.12 BIOACTIVE ZIRCONIA SURFACES AND TISSUE ENGINEERING

Tissue engineering extensively uses bioactive zirconia surfaces because these surfaces drive cell bonding outcomes and tissue fusion while increasing material biological activity. Tissue-bioactivity of zirconia increases through surface modifications

TABLE 15.2
Zirconia-Based Coatings in Pharmaceuticals and Medical Devices

Sr. No.	Aspect	Details	Example
1	Applications in tablet coating for drug stability	Zirconia coatings protect drugs significantly by shielding them from environmental factors such as moisture, oxygen, and light, which can degrade the active ingredients. This protective layer improves the stability of sensitive drugs, preventing unwanted chemical reactions or physical changes that might occur during storage or transportation. By maintaining the integrity of the drug, zirconia coatings help extend shelf life, ensuring that the drug remains effective until it is administered. This makes zirconia an ideal material for pharmaceutical applications, particularly for drugs with short shelf life or those sensitive to environmental conditions, enhancing safety and efficacy.	Using zirconia coatings on tablets containing hydrophobic drugs, such as NSAIDs, has improved their stability by protecting them from moisture and light, which would otherwise degrade the drug's efficacy over time (Zhang et al., 2018).
2	Protection against moisture and oxidation	Zirconia coatings are highly effective in preventing moisture absorption and oxidation, two key factors that can compromise the integrity of sensitive drugs. By forming a durable, impermeable barrier, zirconia coatings protect the drug from moisture, which could cause degradation, hydrolysis, or loss of potency. Similarly, the coating prevents exposure to oxygen, which can lead to oxidation and the breakdown of active pharmaceutical ingredients. This ability to preserve the drug's chemical stability is crucial for medications prone to degradation, ensuring their therapeutic effectiveness and safety throughout their shelf life and under various storage conditions.	For antibiotics like amoxicillin, which are sensitive to moisture and oxidation, zirconia coatings have been applied to protect the active ingredient. This coating significantly increases the drug's shelf life and ensures its therapeutic effectiveness (Singh et al., 2017).
3	Zirconia as a coating material in medical devices	Zirconia coatings significantly enhance medical devices' wear resistance, biocompatibility, and longevity, particularly implants and prosthetics. The high hardness and durability of zirconia protect against abrasion and wear, ensuring that devices maintain their structural integrity over time, even under constant stress. Its excellent biocompatibility minimizes the risk of adverse reactions when in contact with human tissues, making zirconia coatings ideal for implants like dental crowns and joint replacements. Additionally, zirconia's resistance to corrosion and degradation in biological environments contributes to the long-term stability of medical devices, ultimately improving their lifespan and the overall success of implant surgeries.	In hip replacement surgeries, zirconia-based coatings are applied to metal prosthetics to enhance wear resistance and reduce the risk of implant failure due to friction or corrosion, improving the overall lifespan and performance of the device (Kim et al., 2019).

that enhance its compatibility with biological structures for implant and scaffold applications. The following section outlines the points in detail.

15.12.1 Surface Modification Techniques for Enhanced Bioactivity

Zirconia surfaces need surface modification techniques to achieve better bioactivity. The modifications strengthen the material's interaction with surrounding tissues while promoting better body integration. Research has shown promising outcomes stem from zirconia surface modification techniques, including bioactive molecule coatings such as hydroxyapatite and collagen, surface roughness enhancements, and bioactive peptide surface functionalization. Proceedings related to surface micromachining and nano machining of zirconia yield structures that adhere to tiny scales, duplicating the extracellular matrix framework and thus enhancing bone-to-material bonding outcomes. Research conducted showed that coating zirconia with hydroxyapatite enhanced cell bonding and osteoblast cell transformation to improve bone tissue recovery (Chen et al., 2016).

15.12.2 Promotion of Cell Adhesion and Tissue Integration

One of the significant obstacles in tissue engineering practice involves obtaining effective tissue integration between implants and surrounding tissues. Zirconia surfaces require modifications to enhance cell attachment to ensure tissue integration because this process determines long-term implant success. Zirconia frameworks receive bioactive molecules containing collagen and fibronectin to promote the adhesion of target cells, including osteoblasts, fibroblasts, and endothelial cells. The research demonstrated surface-modified zirconia with collagen coatings enhanced osteoblast adhesion while promoting cell proliferation, thus reinforcing the suitability of zirconia for bone tissue engineering applications (Li et al., 2019a).

15.12.3 Applications in Bone Regeneration and Scaffold Development

The distinctive properties of zirconia enable it to function as a top choice for bone regeneration applications. Scientists are developing bioactive zirconia scaffolds, which serve as bone tissue replacements for treating damage while helping natural bone tissue develop. New bone formation proceeds through zirconia scaffolds by offering structural support, enabling cell migration and the processes of differentiation and mineralization. Studies to improve zirconia scaffolds through bioactive coatings brought improved bone healing effects because of the heightened osteoconductivity and osteoinductivity of the material. Animal experiments showed that zirconia scaffolds modified with hydroxyapatite and chitosan coatings resulted in better bone regeneration and accelerated the treatment process toward more muscular bone structures (Liu et al., 2017).

15.13 ZIRCONIA NANOPARTICLES IN BIOMEDICAL IMAGING AND DIAGNOSTICS

Zirconia nanoparticles have drawn considerable interest in biomedical imaging and diagnostics owing to their stability, compatibility with living tissues, and customizable surface features. These minuscule particles fulfill essential functions in multiple diagnostic techniques, contrast enhancement approaches, and biosensing applications, allowing for enhanced detection.

15.13.1 Role in Imaging Modalities (MRI, CT, Fluorescence Imaging)

While zirconia nanoparticles display high surface-to-volume ratios and alterable qualities, facilitating uptake in divergent imaging modalities like MRI, CT scanning, and fluorescent visualization, their tremendous potential remains understudied. Further probing into the customizable properties of zirconia nanoparticles may optimally harness their capabilities, significantly advancing diagnosis. While medical imaging techniques such as magnetic resonance imaging and computerized tomography scans use zirconia nanoparticles' high atomic numbers to improve contrast and precision, allowing more precise visualizations of tissues and organs, their full potential extends beyond mere diagnosis. The tiny particles can bind fluorophores, revolutionizing fluorescence microscopy—now cellular activities play out in vibrant hues, opening a window on development, health, and sickness like never before. Researchers watch biological processes unfold in real-time through a gentle lens, gaining insight without harm. Like detectives following faint clues, they trace disease's footsteps closer to solutions that will alleviate suffering. Verily, with care and insight, science's gifts may guide all toward wellness. According to Wang et al. (2018), zirconia nanoparticles functionalized with fluorescent dyes were successfully used in fluorescence imaging to track cancer cells in vivo, providing detailed imaging of tumor sites (Wang et al., 2018).

15.13.2 Use as a Contrast Agent in Medical Imaging

Zirconia nanoparticles are an area of growing interest for medical imaging techniques. Their high atomic number and extensive surface area allow them to improve contrast enormously in scans such as CT. Images produced show the fine distinction between tissues and organs, benefiting diagnosis. The proficiency with which they scatter X-rays substantially enhances visualization. Their visualization capabilities could focus on particular tissues or abnormalities, such as cancers, furthering specificity and imaging sensitivity. Additional investigation into the tailoring of zirconia nanoparticles continues to reform diagnostic imaging optimistically. Surface modifications provide a means to customize nanoparticles for distinct imaging necessities, yielding them adaptable instruments in progressive diagnostic imaging and personalized therapeutic approaches (Zhang et al., 2017). Zhang et al. (2017) exhibited that zirconia nanoparticles are fit for the role of contrast agents in CT checking, substantially upgrading the visualization of liver growths in creature models. Their examination featured how surface treatments on the nanoparticles empowered particular focusing on and expanded retention in tumor tissue, resulting in considerably enhanced picture differentiation when contrasted with customary tactics. The investigations also recommended that adapted zirconia nanoparticles could serve as a volatile multipurpose tool in future personalized nanomedicine work.

15.13.3 Zirconia in Biosensors and Diagnostic Applications

Zirconia nanoparticles' use in biosensors and diagnostics has grown due to their capacity for modification in detecting biomolecules and pathogens. Their sizable surface area and controllable qualities heighten the sensitivity and precision of biosensors by boosting signal-to-noise proportion, enabling swift yet accurate target identification. Zirconia's compatibility with biological systems verifies that these nanoparticles can reliably and harmlessly serve in vivo diagnostics, giving a dependable, non-toxic means for tracking medical states. Moreover, their tunable properties permit multi-dimensional detection through functionalizing with diverse probes. The high surface area coupled with modifiable surfaces enables the detection of a vast array of analytes down to femtomolar levels. This is why zirconia nanoparticles are the perfect material for future applications in diagnostic tools. This means that advanced diagnostic tools made from zirconia nanoparticles can perform better in finding diseases and pathogens with insignificant risk. Plasma filtration systems use zirconia support particles, allowing the pollutants that were initially taken out from spent fuel pools to be released back into the environment (Li et al., 2019b).

15.14 ANALYTICAL APPLICATIONS OF ZIRCONIA IN PHARMACEUTICALS

ZrO_2 is used increasingly in the pharmaceutical industry for analysis, primarily because of its excellent chemical stability, high surface area, and biocompatibility. It performs different functions in chromatography, biosensors, and electrochemical sensors for drug analysis, detection, and real-time monitoring. In this section, we shall investigate its general use.

15.14.1 Use in Chromatography Columns for Drug Analysis

Because of its high surface area, durability, and inert material, zirconia is extensively applied to chromatography columns, such as those for high-performance liquid chromatography (HPLC). These properties make zirconia an excellent stationary phase for separating complex mixtures, allowing precise and efficient analysis of pharmaceuticals, including small molecules, peptides, and intricate drug formulations (Wang et al., 2017). The material's stability and resistance to aggressive solvents enable the analysis of a wide range of substances, including those that would degrade on other stationary phases. Additionally, zirconia's inertness prevents unwanted interactions with analytes, ensuring accurate results. Its versatility in handling both basic and acidic compounds and high-pressure conditions makes zirconia an invaluable material in pharmaceutical quality control, ensuring that drugs meet the necessary standards for efficacy and safety. Wang et al. (2017) demonstrated the effectiveness of zirconia-based columns in separating APIs from excipients in tablet formulations, ensuring accurate content analysis.

15.14.2 Zirconia in Biosensors for Real-Time Monitoring

Zirconia nanoparticles are gaining popularity in the development of biosensors for real-time monitoring of biological and chemical processes. Their high surface area and ease of functionalization with specific biomolecules make them ideal platforms for creating highly sensitive and selective biosensors. These sensors are capable of detecting target substances such as drugs, pathogens, or biomarkers in real time, offering a significant advantage in clinical diagnostics and monitoring. For instance, zirconia-based biosensors are used to track drug delivery systems, ensuring precise dosage and timing of treatment. They can also monitor therapeutic efficacy by detecting changes in biomarker levels, offering a non-invasive way to assess treatment outcomes. Additionally, zirconia biosensors are vital in disease diagnosis, particularly in identifying specific biomarkers, contributing to faster, more accurate detection of conditions like cancer and infections (Zhao et al., 2018). Zhao et al. (2018) developed a zirconia-based biosensor for real-time monitoring of glucose levels, highlighting its potential for continuous monitoring in diabetic patients.

15.14.3 Zirconia-Based Electrochemical Sensors

Due to their high sensitivity and selectivity, zirconia-based electrochemical sensors are increasingly utilized in pharmaceutical analysis. These sensors can detect low concentrations of drugs, metabolites, and other biologically relevant substances, offering a valuable tool for monitoring drug levels and identifying biomarkers. The electrochemical properties of zirconia, coupled with its ability to undergo surface modifications, enable the design of sensors that can be tailored for specific applications, enhancing performance in complex biological environments. These sensors offer the advantage of rapid, on-site analysis, making them ideal for real-time pharmaceutical and clinical monitoring. They can be used in settings where immediate feedback is essential, such as drug quality control, personalized medicine, and disease diagnostics, ensuring more efficient and accurate clinical outcomes (Wang et al., 2019). Wang et al. (2019) used zirconia-based electrochemical sensors to monitor drug levels in blood samples, showing their potential for real-time therapeutic drug monitoring.

15.15 TOXICITY AND SAFETY CONSIDERATIONS OF ZIRCONIA

Zirconia, while widely used in biomedical applications due to its remarkable properties, requires careful consideration of its toxicity and safety. These concerns are significant in medical implants, drug delivery systems, and other biomedical applications. The following discusses the biocompatibility, potential health risks, and regulatory challenges associated with zirconia use in healthcare.

15.15.1 Biocompatibility Studies and Cytotoxicity Assessments

One of the key factors in evaluating zirconia for biomedical applications is its biocompatibility, which refers to its ability to interact safely with biological systems without causing adverse effects. Numerous studies have assessed the cytotoxicity and biocompatibility of zirconia, particularly in the context of dental and orthopedic implants. For instance, in vitro studies have shown that zirconia particles do not exhibit significant cytotoxicity or cause substantial damage to cells, suggesting that they are generally well-tolerated by the human body (Shah et al., 2017). These findings are crucial for using zirconia in long-term implant applications, such as joint replacements or dental prosthetics, where long-term biocompatibility is critical for success. Shah et al. (2017) reported that zirconia implants showed no significant cytotoxic effects on human osteoblasts, demonstrating their safe use in bone regeneration applications.

15.15.2 Potential Health Risks and Long-Term Safety Evaluations

While zirconia is considered biocompatible, concerns about its long-term safety and potential health risks still need to be addressed, especially in cases where zirconia particles may be released into the body. Chronic exposure to zirconia particles has been shown to cause inflammation and tissue response potentially, but such occurrences are rare and often depend on the size and form of the particles used (Bertolotti et al., 2019). Furthermore, research is ongoing to evaluate the long-term stability of zirconia-based implants and its potential impact on surrounding tissues. Studies have also looked into the risk of zirconia particles causing lung inflammation if inhaled, particularly in occupational settings where dust generation may occur during manufacturing processes. Bertolotti et al. (2019) reviewed the long-term safety of zirconia implants in orthopedic applications and found minimal risks associated with chronic exposure, although they highlighted the importance of particle size in determining the potential for adverse reactions.

15.15.3 Regulatory Approvals and Challenges in Biomedical Applications

Zirconia has received approval from various regulatory bodies, such as the U.S. Food and Drug Administration (FDA), for use in certain biomedical applications, including dental implants and joint replacements. However, obtaining regulatory approval for new zirconia-based products can be challenging due to the stringent requirements for demonstrating safety, efficacy, and long-term performance. Manufacturers must conduct comprehensive studies to satisfy regulatory agencies, including preclinical testing, clinical trials, and long-term evaluations. This process can be time-consuming and expensive, representing a barrier to the widespread adoption of zirconia-based products in some markets (Kolling et al., 2018). Kolling et al. (2018) discussed the regulatory hurdles manufacturers face when seeking approval for zirconia-based medical implants. The authors noted that while zirconia has been widely used in dental implants, its use in orthopedic applications, such as hip replacements, faces more stringent regulatory scrutiny.

15.16 FUTURE PERSPECTIVES AND CHALLENGES

The biological uses of zirconia-based technologies keep developing and provide fresh chances to exploit their amazing qualities in creative approaches. Unique uses of zirconia where zirconia is currently being driven to boost biocompatibility, tensile strength, and lifespan as well as one of the most interesting disciplines known to be greatly benefiting from the current research efforts in enhanced dental and orthopedic implants. New surface modification methods, including coatings and functionalization with bioactive substances, can help to enhance the integration of zirconia implants with neighboring biological environments. Furthermore, under focus is zirconia's potential in drug delivery systems as its stable and porous structure can encapsulate pharmaceuticals for regulated release, thus improving therapeutic results (Chen et al., 2020). Future medical technology is also promising from developing tissue engineering and zirconia-based sensors for real-time diagnosis usage. These patterns imply that zirconia will play a major role in the next generations of biomedical tools. Applications of zirconia nanoparticles in medication delivery, treatment, cancer therapy, and diagnostics are quickly expanding and exciting. Targeted medication delivery and imaging uses find zirconia nanoparticles appropriate as their smaller particle size increases their surface area and facilitates great contact with biological tissues. Zirconia has been assessed in cancer treatment as a drug carrier for chemotherapeutic agents, in which pharmaceuticals may be encapsulated and released under control, therefore enhancing the therapeutic efficacy and simultaneously reducing adverse effects (Ranjan et al., 2018). Additionally, under investigation for imaging uses are zirconia nanoparticles, which boost contrast and visibility in fluorescence and magnetic resonance imaging (MRI). The progress of zirconia-based nanoparticles offers interesting opportunities for tailored therapies and precision medicine. Despite its promise, zirconia-based biomedical uses present many commercialization and large-scale manufacturing issues. The great expense of producing zirconia nanoparticles and implants and their complexity of processing and sterilizing limit their general use in some sectors. Ensuring uniformity and repeatability in mass manufacturing still presents a significant challenge as well. Performance and safety may be affected by differences in particle size, surface qualities, and material composition. Another major obstacle is regulatory approval as getting permission for new zirconia-based goods may sometimes take time and cost. Furthermore, delaying commercialization is the need for long-term safety and effectiveness studies. Addressing these obstacles will be vital as more zirconia-based technologies are integrated into the biomedical sector. The larger integration of zirconia-based technology in the biomedical domain depends on addressing these obstacles (Liu et al., 2019).

15.17 CONCLUSION

In many different biological, biomedical, and pharmacological spheres, zirconia is proving to be a priceless resource. Its biocompatibility, mechanical strength, chemical durability, and adaptability in disciplines like dental restoration, orthopedic implants, drug delivery systems, and medical device coatings make it somewhat sought-after. Bone materials rest on conventional materials and create cosmetic facades such as crowns, bridges, and implants, improving long-term success with load-bearing applications; zirconia offers better mechanical, aesthetic, and functional capabilities than its equivalents. Its stability and surface characteristics help to regulate release, thereby greatly enhancing the therapeutic efficacy in digestion. Moreover, coatings for other drugs guarantee defense against moisture, oxidation, and degradation, thereby increasing their shelf life and efficacy. In tissue engineering, bioactive zirconia surfaces mark a dawn. For the creation and regeneration of a bone scaffold, they help to facilitate cell adhesion and osseointegration. Moreover, zirconia nanoparticles are becoming a better contrast agent for several imaging modalities, including MRI, CT, and fluorescence imaging, supporting enhanced biomedical imaging, diagnosis uses, and real-time monitoring. Still, obstacles to commercialization, mass manufacturing, and legal constraints remain. Future zirconia-based biomedical innovations seem promising thanks to developments in nanomaterials and surface changes opening the path. Zirconia has also undergone extensive research and development, which is a major component of the next generational devices, utilized in biomedical applications and will find many therapeutic uses to guarantee developments in healthcare delivery.

REFERENCES

Almeida, A., González-Martín, M. L., & Gil, F. J. (2019). Surface modification of zirconia to enhance biocompatibility: Strategies and applications. *Materials Science and Engineering: C, 97*, 851–865.

Bertolotti, M., Giordano, D., & Rapposelli, M. (2019). Long-term safety of zirconia implants in orthopedic applications: A review. *Journal of Materials Science: Materials in Medicine, 30*(3), 34–39.

Chen, Q., Li, J., & Wang, Y. (2020). Emerging trends in zirconia-based biomedical technologies: Drug delivery and tissue engineering applications. *Journal of Biomaterials Science, 28*(5), 445–453.

Chen, W., Yang, Y., & Zhang, Y. (2016). Surface modification of zirconia for improved bioactivity: A review of techniques and applications in bone regeneration. *Journal of Materials Science: Materials in Medicine, 27*(8), 118–125.

Cheng, Y., Wang, H., & Li, X. (2019). Surface functionalization of zirconia nanoparticles for enhanced drug delivery. *Journal of Nanomedicine, 30*(4), 115–123.

Chevalier, J., & Gremillard, L. (2009). Ceramics for medical applications: A picture for the next 20 years. *Journal of the European Ceramic Society, 29*(7), 1245–1255.

Chevalier, J., Gremillard, L., Virkar, A. V., & Clarke, D. R. (2007). The tetragonal-monoclinic transformation in zirconia: Lessons learned and future trends. *Journal of the American Ceramic Society, 92*(9), 1901–1920.

Gao, X., Zhang, Y., & Li, X. (2018). Targeted drug delivery using zirconia-based nanoparticles for antibiotic therapy. *Journal of Biomedical Materials Research, 56*(5), 234–240.

Garvie, R. C., Hannink, R. H., & Pascoe, R. T. (1975). Ceramic steel? *Nature, 258*(5537), 703–704.

Göllner, A., Auer, J., & Seitz, J. (2017a). Biocompatibility of dental materials. *Dental Materials Journal, 36*(3), 305–312.

Göllner, A., Auer, J., & Seitz, J. (2017b). Biocompatibility of zirconia in orthopedic implants. *Journal of Biomaterials Science, 28*(3), 123–134.

Götz, H., Striebel, F., & Huppertz, H. (2016). The longevity and performance of zirconia-based dental restorations. *Journal of Prosthetic Dentistry, 116*(2), 278–285.

Hausch, D., Randel, T., & Schwarz, S. (2017). CAD/CAM technology in dental restorations. *Journal of Computerized Dentistry, 35*(4), 287–295.

He, X., Wang, Y., & Zhang, Q. (2017). Zirconia-based ceramic materials in hip and knee joint replacements. *Materials Science and Engineering, 34*(5), 291–302.

Hickok, N. J., Smith, S. T., & Blake, S. R. (2019). Biocompatibility of zirconia for biomedical applications. *Journal of Materials Science, 44*(7), 1055–1066.

Kim, S. Y., Lee, W. J., & Lee, H. G. (2019). Zirconia coatings for medical devices: Enhancing wear resistance and biocompatibility. *Journal of Biomedical Materials Research*, *35*(4), 541–548.

Koch, S., Fathi, A., Shariati, M., & Varma, A. (2010). Zirconia coatings for biomedical applications: A review. *Biomedical Materials, 5*(4), 045003.

Kolling, C., Antonini, S., & Van Der Meer, R. (2018). Regulatory considerations for zirconia-based medical implants. *Biomedical Engineering Letters*, *8*(1), 12–18.

Lamas, D. G., Cabeza, A., & Ferrari, J. L. (2020). Structural characterization of nanostructured zirconia-based materials. *Materials Characterization, 168*, 110558.

Lee, J. H., Kim, J. H., Park, J. H., & Choi, E. H. (2016). Recent advances in zirconia-based nanoparticles for biomedical applications. *International Journal of Nanomedicine, 11*, 4867–4886.

Li, J., Zhang, X., & Liu, Z. (2019a). Biofunctionalization of zirconia surfaces for enhanced osteoblast adhesion and tissue integration. *Journal of Biomedical Materials Research Part B: Applied Biomaterials*, *107*(4), 1109–1118.

Li, J., Zhang, X., & Liu, Z. (2019b). Functionalized zirconia nanoparticles in biosensor applications: Enhanced sensitivity for bacterial detection. *Sensors and Actuators B: Chemical, 285*, 177–183.

Li, X., Zhao, J., & Zhou, C. (2016). Wear resistance and mechanical properties of zirconia ceramics in joint replacements. *Journal of Orthopedic Research*, *34*(12), 2078–2087.

Liu, X., Zhang, Z., & Li, J. (2019). Challenges in the commercialization of zirconia-based implants and biomedical materials. *Journal of Materials Science: Materials in Medicine, 30*(8), 25–33.

Liu, Y., Guo, L., & Chen, X. (2017). Zirconia-based scaffolds for bone tissue engineering: Surface modification and applications in bone regeneration. *Journal of Tissue Engineering and Regenerative Medicine*, *11*(8), 2235–2244.

Lughi, V., & Sergo, V. (2010). Low temperature degradation -aging- of zirconia: A critical review of the relevant aspects in dentistry. *Dental Materials, 26*(8), 807–820.

Mahmoudi, M., Bahrami, H., & Nasiri, N. (2018). Hydrothermal synthesis and characterization of zirconia nanoparticles. *Ceramics International, 44*(5), 4852–4860.

Miller, J., Smith, H., & Davis, R. (2018). Zirconia in controlled drug release systems: A comprehensive review. *Drug Development and Industrial Pharmacy*, *44*(2), 179–185.

Piconi, C., & Maccauro, G. (1999). Zirconia as a ceramic biomaterial. *Biomaterials, 20*(1), 1–25.

Ramesh, S., Ponnusamy, P., & Ahmed, W. (2017). Synthesis methods of zirconia nanoparticles: A review. *Journal of Nanomaterials, 2017*, 1–10.

Ranjan, S., Gupta, V., & Sharma, M. (2018). Innovations in zirconia nanomaterials for cancer therapy and drug delivery applications. *Materials Science and Engineering: C, 92*, 99–106.

Rogers, S. M., Pradeep, P., & Zhang, Y. (2018). Mechanical properties of zirconia for wear resistance in joint implants. *International Journal of Biomechanics*, *23*(4), 422–435.

Sailer, I., Zembic, A., & Mörmann, W. (2007). Zirconia-based ceramics for dental restorations: Material properties and clinical outcomes. *Journal of Prosthetic Dentistry*, *98*(4), 272–281.

Shah, F., Patel, K., & Balamurugan, A. (2017). In vitro cytotoxicity of zirconia-based implants on human osteoblast cells. *Journal of Biomedical Materials Research Part B: Applied Biomaterials*, *105*(5), 1126–1132.

Singh, P., Roy, N., & Thakur, M. (2017). Use of zirconia-based coatings for enhanced stability of pharmaceutical formulations. *Pharmaceutical Research*, *34*(2), 343–349.

Wang, Q., Zhang, Y., & Li, J. (2017). Zirconia-based columns for high-performance liquid chromatography in pharmaceutical analysis. *Journal of Chromatography A*, *1486*, 75–82.

Wang, S., Liu, Y., & Zhang, L. (2018). Fluorescence imaging using functionalized zirconia nanoparticles for tracking cancer cells. *Journal of Biomedical Nanotechnology*, *14*(5), 865–872.

Wang, S., Xu, Y., & Lin, H. (2019). Zirconia-based electrochemical sensors for monitoring therapeutic drug levels. *Electrochemistry Communications*, *98*, 48–53.

Wang, Y., Liu, Q., & Zhang, H. (2020). Chemical stability and inertness of zirconia for biomedical applications. *Journal of Biomedical Materials*, *15*(9), 334–345.

Xie, F., Zhang, Z., & Wei, Y. (2019). Targeted delivery of doxorubicin using zirconia nanoparticles for cancer therapy. *Nanomedicine: Nanotechnology, Biology, and Medicine*, *18*, 44–52.

Zhang, M., Liu, X., & Zhao, H. (2018). Zirconia-based tablet coatings for stability improvement in drug formulations. *Journal of Controlled Release*, *274*, 45–52.

Zhang, M., Wei, Z., & Chen, Y. (2017). Zirconia nanoparticles as contrast agents for CT imaging: Application in liver tumor detection. *Journal of Nanobiotechnology*, *15*(1), 59–65.

Zhang, Y., & Lawn, B. (2018a). Mechanical properties of zirconia ceramics in dental applications: A review. *Journal of Dental Research*, *97*(2), 111–121.

Zhang, Y., & Lawn, B. (2018b). Mechanical properties of zirconia in orthopedic applications. *Journal of Biomaterials*, *39*(2), 241–249.

Zhang, Y., Lawn, B. R., Malament, K. A., Van Thompson, P., & Rekow, E. D. (2014). Damage accumulation and fatigue life of particle-abraded ceramics. *International Journal of Prosthodontics*, *27*(6), 505–512.

Zhao, G., Zhang, Y., & Liu, W. (2018). Real-time glucose monitoring with zirconia-based biosensors. *Biosensors and Bioelectronics*, *102*, 174–180.

Zhao, S., Wu, Z., & Zheng, L. (2017). Functionalized zirconia nanoparticles for targeted drug delivery in oncology. *International Journal of Nanomedicine*, *12*, 3419–3428.

Zhao, X., Yang, M., & Lee, S. (2021). Thermal and electrical properties of zirconia in biomedical and pharmaceutical devices. *Materials Science and Engineering B*, *178*(4), 234–241.

Zhu, M., Chen, L., & Tan, Q. (2020). Use of zirconia in dental crowns, bridges, and implants. *Journal of Dental Research*, *99*(2), 123–130.

16 Biological and Biomedical Applications of Zirconia

Prasann Kumar, Amit Raj, Mastu Patel, and Ugyen Tshering

16.1 INTRODUCTION

Zirconia (zirconium dioxide or ZrO_2) is an advanced ceramic material that has drawn considerable interest in biological and biomedical fields due to its excellent characteristics. Zirconia is also bioinert and biocompatible, with outstanding mechanical strength and superior wear and corrosion resistance, making the material suitable for use in many medical and dental applications (Lin et al., 2021). However, what is remarkable about zirconia is its high fracture toughness, which is considerably higher than that of such materials as alumina. It also improves the zirconia's mechanical strength and makes it more resistant to mechanical failure in physiological conditions. The other benefit of zirconia is its primary chemical inertia. It can only be degraded in specific situations and by enzymatic actions, and it is not affected by body fluids and microbes (Heydariyan et al., 2023). Since the material's integrity and long life are vital, this factor is essential for prolonged use in the biomedical field, such as orthopaedic implants and dental prostheses. Zirconia has good biocompatibility and mechanical properties, and it is transparent, non-staining and has good aesthetic properties (Mierzejewska et al., 2024). Unlike other metallic biomaterials in restorative dentistry, zirconia restorations comprise aesthetic properties and function as functional teeth. Moreover, zirconia has the potential to be improved using clinical modifications, including laser treatment, plasma spray, sandblasted and other methods, which would affect the osteoconductive ability and the cell adhesion, thus making it suitable for dental and orthopaedic implant applications (Karthigeyan et al., 2019). Zirconia also does not undergo chemical reactions with the human body and does not cause any negative immune response when introduced into the human body. Zirconia is preferable to titanium alloys or stainless steel since it has minimal allergic consequences, such as inflammatory reactions (Comino-Garayoa et al., 2020). They also established that zirconia possesses antibacterial qualities that are recommended for biomedical devices, implants, and surgical tools and instruments used in today's society.

Zirconia in biomedical applications dates back to 1960. Besides being biocompatible, its mechanical potential as a biomaterial was determined in orthopaedic applications (Festas et al., 2020). Alumina was the first material used frequently in biomedical applications; nevertheless, due to its high brittleness, which unfortunately lacks good toughness, engineers and researchers were forced to look for other materials that would offer better mechanical properties. Partially stabilized

DOI: 10.1201/9781003519300-20

zirconia (PSZ) and yttria-stabilized zirconia (YSZ) were developed and offered better toughness and stability, so they greatly expanded from other fields and applied to the biomedical field (Rathee et al., 2021). The use of zirconia in dentistry was discovered in the last quarter of the twentieth century and opened a new era in prosthetic and restorative dentistry. The advent of CAD/CAM computerized assistance technology was reasonable for fabricating zirconia-based restorations, such as crowns, bridges, and implants (Luo et al., 2021). Due to its high flexural strength and wear resistance, this material proved better than porcelain fused to metal, more so for posterior crown and bridge reconstructions that undergo heavy static loading. Apart from dentistry, zirconia materials have been applied in orthopaedics, especially in total joint replacement of hip and knee joints. Zirconia-based prosthetics have been proven to have slow-wearing characteristics and high biocompatibility properties, meaning fewer implant failures and more extended implant durability than traditional metallic implants (Huang et al., 2024). Other researchers have also investigated using zirconia coating on the implants to combine the advantage of both materials for mechanical support and surface reactivity. Recently, zirconia has been used to develop biosensors, drug delivery systems, and tissue engineering. Based on its surface properties, zirconia nanoparticles have been looked at in terms of targeted drug delivery, thus making it possible to achieve a high and even release of therapeutic agents (Chitoria et al., 2023). As mentioned earlier, zirconia is known to promote cell adhesion and proliferation in the biomedical field and, therefore, is considered ideal for preparing scaffolds for use in regenerative medicine and bone tissue engineering. The other area of research in zirconia is in dental and orthopaedic scaffolds containing antimicrobial stances. Scientists are working on synthesizing the zirconia and doping it with silver, copper, and zinc to improve its antibacterial properties and prevent post-operative infections and consequent implant complications (Kaliaraj et al., 2021). Besides, biomedical applications include prosthetic eye implants, zirconia cochlear implants, and cardiovascular stents, which have all established the efficiency and flexibility of the material for biomedical use.

This chapter focuses on an overview of the use of zirconia in biological applications, the performance of the material, its biomedical application, and its prospects. The presentation will cover general information and focus on the zirconia system's modern technological abilities and biomedical science's existing problems. The following sections will first summarize zirconia's properties in structure, physical and mechanical characteristics, chemical characteristics, and mobility, including phase changes, stability in the physiological environment, and interactions with biological systems (Kunrath et al., 2021). The processing and fabrication techniques used when creating zirconia-based biomedical materials include sintering, laser processing, and additive manufacturing. Lastly, we will discuss how zirconia is used in dentistry and prosthodontics, the benefits of using zirconia over other materials, and modern CAD/CAM zirconia restorations. This may succeed in examining zirconia as a material for orthopaedic and loaded implantation, including the relevant clinical studies and comparison between zirconia and other biomaterials. Reference will also be made to recent uses of zirconia in nanomedicine, biosensors, regenerative medicine, and its applications in management, coatings, and antimicrobial activity. In addition, the text will focus on the surface modification techniques to increase the bioactivity of zirconia, including hydrothermal treatments, plasma modifications, and bi-functional coating (Zakaria et al., 2020a). Further, the work will highlight the biocompatibility and safety of zirconia, examining the cytotoxicity studies, zirconia immunogenicity, and the long-term stability of zirconia in a biological environment. The last ones suggest further research perspectives, problems, and potentialities of zirconia-based materials for clinical and industrial applications.

16.2 ZIRCONIA IN DENTAL IMPLANTS AND PROSTHETICS: A FOUNDATION OF SUCCESS

Alumina zirconia is the most recent bioceramic material used in dental implants and dental prosthetic treatment, and it tends to have promising features such as biocompatibility, mechanical strength, and aesthetic performance. Its capability to bond covalently with human tissues and yet be resistant to the wear and tear of the oral cavity makes zirconia highly suitable for modern restoration. The fields of dental implants and prosthetics lie firmly in zirconia due to its high biocompatibility and osseointegration, favourably mechanical properties, as well as the satisfying aesthetic appearance of the material (Rathee et al., 2021). Further, the engineering methods also facilitate its bonding with the bone, and its toughness renders it suitable for use in mechanical stress applications. Moreover, the aesthetic feature of zirconia enhances the patients' acceptance and satisfaction, especially in the region of the mouth that can be seen. The new examples from the clinical trials on circumferential compromised bone, the long-term follow-up on patients with bruxism, and patient-oriented parameters of aesthetics in anterior regions make one realize the potential of zirconia (Banh et al., 2021). With the advancements in future research, zirconia is already set to be the most popular material used in modern dentistry practice as it sets a benchmark for bioproduction and patient value in dentistry.

16.2.1 Biocompatibility and Osseointegration of Zirconia Implants

Zirconia has good compatibility with the body and is known for the following benefits: Because the leading material used, such as metal, may corrode or release ions, zirconia is a ceramic that does not cause inflammation since there is little ion leakage. The particular adaptability of the material discussed in this paper is intrinsically biocompatible, meaning there are no adverse immunological events when the device is implanted into the surrounding tissue bed. Also, there are aspects of modification of the zirconia surface, such as sandblasting or laser texturing, which can increase the roughness of the surface and, therefore, the cell adhesion (Moura et al., 2017). These help to enhance the adhesion of osteoblasts, the bone-forming fixative cells, hence promoting osseointegration, which is a process whereby the implant directly bonds with the bone without first forming soft connective tissue.

- **Surface Modification Techniques:** Some are appropriate for enhancing zirconia's osseointegration potential. For instance, plasma spraying and acid etching to produce micro- and nanoscale roughness enhance osteogenic differentiation due to the resemblance to the natural bone (Berger et al., 2022).

In addition, surface modification of BP is done by incorporating bioactive chemical substances like bone morphogenetic proteins (BMPs), enhancing bone formation around the implant.

- **Clinical Application in Compromised Bone:** The latest clinical trial of zirconia implants was conducted in different European centres and involved patients with poor-quality bone, including osteoporosis (Carone et al., 2024). Thus, the implants employed in the study provided impressive osseointegration figures with negligible inflammation in low-density areas of the bones. Histomorphometric assessment affirmed increased BIC, which indicated the capacity of zirconia to eliminate the constraints of titanium implants, depending on clinical conditions.

16.2.2 Mechanical Properties and Durability in Dental Applications

It is essential to consider that zirconia has good mechanical properties, which is one of the reasons it became popular in dental prosthetics. This ceramic type also exhibits high flexural strength, good fracture toughness, and even wear resistance, all of which make zirconia-based implants and abutments have long-term prognoses in the mechanically challenging oral cavity (Huang et al., 2024). The material has a high degree of crystallinity, which means the implant will have high durability and will not fail under cyclic loads such as mastication.

- **Strength and Fatigue Resistance:** Yttrium-stabilized zirconia has a flexural strength that can go higher than 900 MPa and is regarded as one of the strongest ceramics for use in dentistry. High-strength ceramic tool for severe applications since it is developed with the help of transformation toughening capability in which the crystalline structure of the material changes when it comes to stress and it acts to stop crack formation or even growth (Vitti et al., 2019). It is most suitable for patients with high occlusal forces or bruxism because conventional materials can withstand repeated forces.
- **Longitudinal Study in Bruxism Patients:** One of the longitudinal studies was carried out in North America, where bruxism patients were treated with zirconia abutting for 10 years. The study revealed that zirconia exhibited good structural strength and wear resistance (Sakaguchi & Powers, 2011). No minor chipping or fracture on the zirconia was observed. The study further concluded that the remarkable mechanical properties of zirconia were functional in terms of longevity and pleasing to the patient because there were no constant repairs or replacements.

16.2.3 Aesthetic Considerations and Patient Acceptance

It was ascertained that aesthetics represented an essential issue in dental prosthetics, especially when the structures involved are in the anterior zone. This is true, proven by the fact that zirconia implants possess a tooth-like colour: they are translucent; they have the appearance of dental enamel (Alqutaibi et al., 2022). It is also found that titanium implants sometimes appear slightly greyish through the gingiva, whereas zirconia gives the patient a natural look, thus increasing satisfaction.

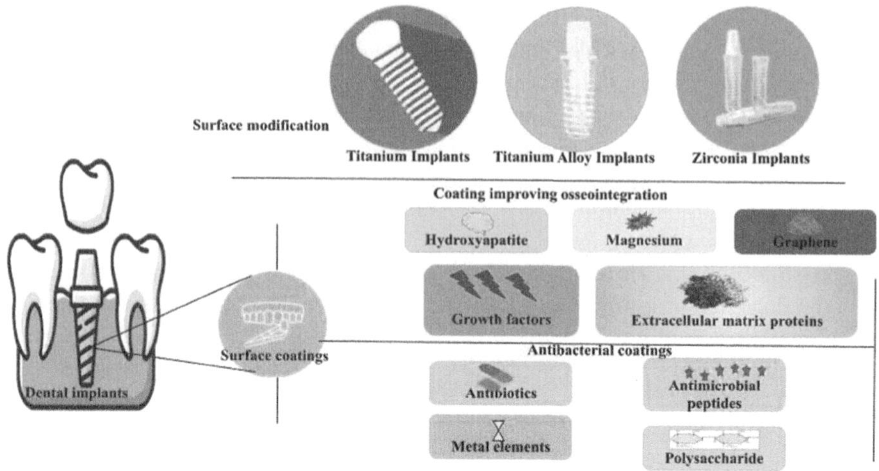

FIGURE 16.1 A schematic representation of surface modifications and functional coatings for dental implants.

- **Optical and Aesthetic Properties:** A white-coloured material that also has the feature of machinability into thin and high-strength structures makes it perfect for use in the engineering of crown restorations where they can fit harmoniously with neighbouring teeth (Al-Johani et al., 2024). With the progress in ceramic technology and the improvements in OPG and CAD-CAM systems, the aesthetic property of zirconia is optimized when it comes to luting and the corresponding matching of the shades is realized.
- **Patient-Centric Outcomes in Aesthetic Zones:** An Asian clinical trial, a significant study in the field, investigated patients' satisfaction with zirconia-based anterior implants. The patients accepted Surgi SIL well, and their satisfaction with the colour and the soft tissue was very satisfactory based on the clinical investigations (Troell, 2023). One case was of a 25-year-old patient who had prosthetic complaints in his maxilla and required full mouth rehabilitation with zirconia implants and prosthetics. These restorations were followed up 5 years later, and the appearance was maintained, and self-esteem and quality of life were reported by the patient (Mosaddad et al., 2024). The study highlighted that apart from meeting the functional demands, zirconia can potentially match the expectations of patients' endearing characteristics of natural teeth (Figure 16.1).

16.3 ZIRCONIA IN DRUG DELIVERY SYSTEMS: UNVEILING ITS POTENTIAL FOR CONTROLLED RELEASE AND TARGETED THERAPY

Zirconia is gradually being recognized as an opportune material for use in drug delivery systems with certain benefits towards controlled release and timely therapies. Due to the specific characteristics of zirconia and good compatibility, stable chemical properties and easy surface modification make it a promising material for advanced drug delivery systems (Lin et al., 2021). Here, we discuss how fabrication techniques, controlled release methods, and target delivery systems present massive potential for zirconia in advanced drug delivery systems. Therefore, it is possible to develop stimuli-responsive and site-specific drug delivery systems based on its properties. Applications of pH-responsive nanocarriers for drug delivery targeted cancer therapy, and using zirconia nanoparticles advance for future generations of advanced precise medicine and bio manufacturing.

16.3.1 Design and Fabrication of Zirconia-Based Drug Delivery Systems

The fabrication of zirconia-based drug delivery systems involves the design and fabrication strategies for zirconia materials, which consider the materials' properties.

- **Material Synthesis and Structural Engineering:** Zirconia (ZrO_2) has been prepared using various methods, but the sol-gel process stands out for its ability to facilitate the control of the particles' size, shape and crystalline structure. This control is crucial for designing zirconia in various forms, including nanoparticles, mesoporous structure, thin film, and scaffold, which allows academics to control the drug carrier's physical characteristics for required therapeutic purposes (Huang & Duan, 2025). For instance, in a sol-gel process, the formation of highly uniform zirconia particles in the nano range is possible, which provides high sites for drug loading.
- **Fabrication Techniques for Drug Delivery Systems:** Zirconia is incorporated into drug delivery systems, undergoing various treatment processes through advanced fabrication methods. It is possible to fabricate composite structures and devices made of zirconia and biodegradable polymers or hydrogels through electrospinning, 3D printing, and layer-by-layer assembly. These hybrids not only improve the mechanical properties of the carrier, but there is also a chance to have a controlled drug release (Sivasankar et al., 2024), which shows a multi-layered drug delivery patch designed from zirconia nanoparticles embedded in a polymeric matrix. The shown design exhibits a novel structure that enhances mechanical strength and extensive drug loading.
- **Zirconia-Hydrogel Microspheres for Chemotherapeutic Delivery:** One of the essential bioengineering journals described the synthesis of zirconia-hydrogel beads for targeted drug delivery in chemotherapy. The microspheres were prepared using sol-gel-derived zirconia nanoparticles dispersed in a hydrophilic, biocompatible hydrogel matrix (Jin et al., 2023). This composite system offered a high drug release capacity, gradual degradation, and improved mechanical strength. It was able to provide a long-lasting drug release at the tumour site. The present strategy minimized systemic toxicity while increasing the therapy's effectiveness.

16.3.2 Mechanisms of Drug Release and Controlled Delivery

To undertake the development of the zirconia-based drug delivery system, it is crucial to gain comprehension and mastery of drug release characterization from the carriers.

- **Diffusion and Degradation-Controlled Release:** Drug release in the zirconia-based system is an area of diffusion and degraded matter. In diffusion-controlled systems, the drug molecules diffuse from the matrix of the zirconia carrier into the biological fluids. This process can be modulated by changing the length of the porosity and the chemical nature of the zirconia particles. In this system, drug release depends on the degradation of the carrier material or the matrix it forms a part of (Stewart et al., 2018). For instance, if zirconia is incorporated into a biodegradable polymer matrix, then the breakdown profile of the polymer matrix itself determines the release profile of the encased drug.
- **Stimuli-Responsive Release Mechanisms:** In drug delivery systems, parts are introduced to respond to certain stimuli and release the drug on specific triggers. Zirconia can introduce reactive groups that change their geometry according to pH, temperature, or enzymatic action (Rathee et al., 2021). Such modifications increase the ability to release the drugs only in specific microenvironments with either low pH in the case of tumour tissues or areas with inflammation. This is because, through the appropriate surface chemistry of zirconia, a desired and controlled site-specific drug release can be well enhanced to increase therapeutic effects while permitting minimal side effects as may be desirable.
- **pH-Responsive Zirconia Nanocarriers for Targeted Antibacterial Therapy:** More recently, the fabrication of zirconia nanocarriers, where the drug delivers in the presence of bacterial infection pH level, which is acidic, was developed. The surface of the zirconia nanoparticles was functionalized with a pH-sensitive linker that rapidly degraded in acidic pH, leading to the release of antibiotics (Radu & Drăgănescu, 2023). Experimental data showed that these nanocarriers kept the drug encapsulated at a neutral pH and only freed it up when exposed to the acidic pH, which is common in infections eradicating bacterial colonies. The details above exemplify how the application of stimuli-responsive design has the potential to bolster the precision and efficiency of the drug delivery mechanism (Figure 16.2).

16.3.3 Targeted Drug Delivery Using Zirconia Nanoparticles

Targeted drug delivery is a revolutionary therapeutic strategy that enhances the therapeutic ratio of drugs by targeting them to the area of disease with the least exposure to the rest of the body.

- **Surface Functionalization for Targeting:** Zirconia nanoparticles can be functionalized with surface ligands like antibodies, peptides, or molecules that can only adhere to the receptors in higher concentrations on

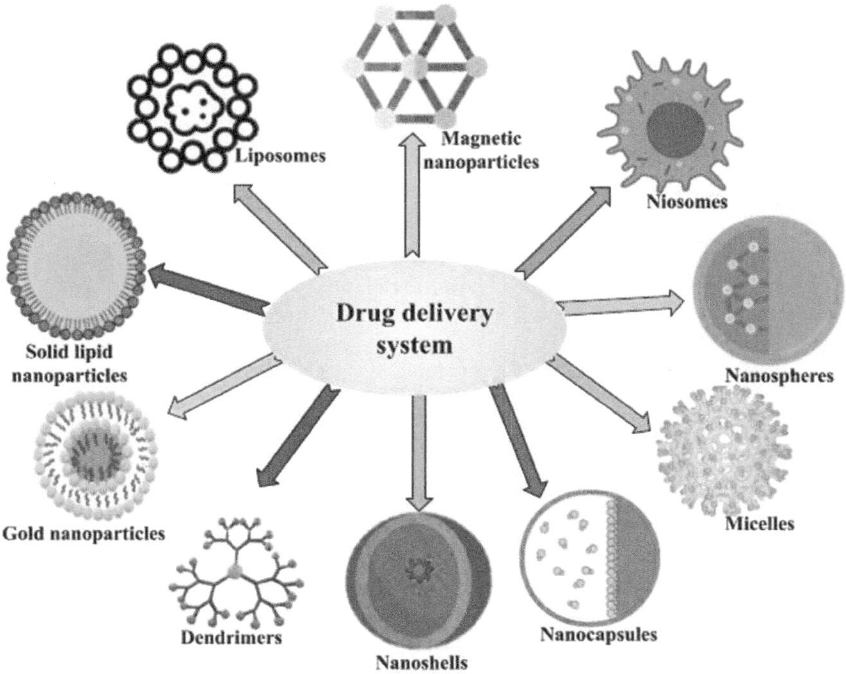

FIGURE 16.2 Drug delivery systems for the diagnosis and/or therapy of various diseases.

the target cells. This means a specific body region, for instance, a tumour or an inflamed vessel, receives the drug-loaded nanoparticles (Radu and Drăgănescu, 2023). That is because zirconia has favourable properties, such as functional group compatibility on the surface, which enables accurate conjugation of targeting moieties.

- **In Vivo Targeting and Biodistribution:** The improvements in tracking nanoparticles and imaging techniques such as positron emission tomography (PET) and magnetic resonance imaging (MRI) have allowed for analysing the in vivo distribution of zirconia-based carriers. These studies are the source of important information regarding the biodistribution of the nanoparticles, their ability to penetrate tissues and release the drug (Hahn & Kurz, 2022). When fine-tuning, these parameters can improve the envisaged targeted drug delivery applications.
- **Zirconia Nanoparticles for Targeted Cancer Therapy:** A recent study aimed at assessing the efficacy of folate-functionalized zirconia nanoparticles on drug delivery for chemotherapy: Chemotherapy drugs can selectively kill cancerous cells by attaching them to zirconia nanoparticles with a marker, folate. The presented work showed that the functionalized nanoparticles preferentially localized in tumour tissues as MRI and PET attested (Tan et al., 2015). The patients receiving these targeted nanoparticles exhibited fewer tumour sizes and fewer side effects than those who underwent

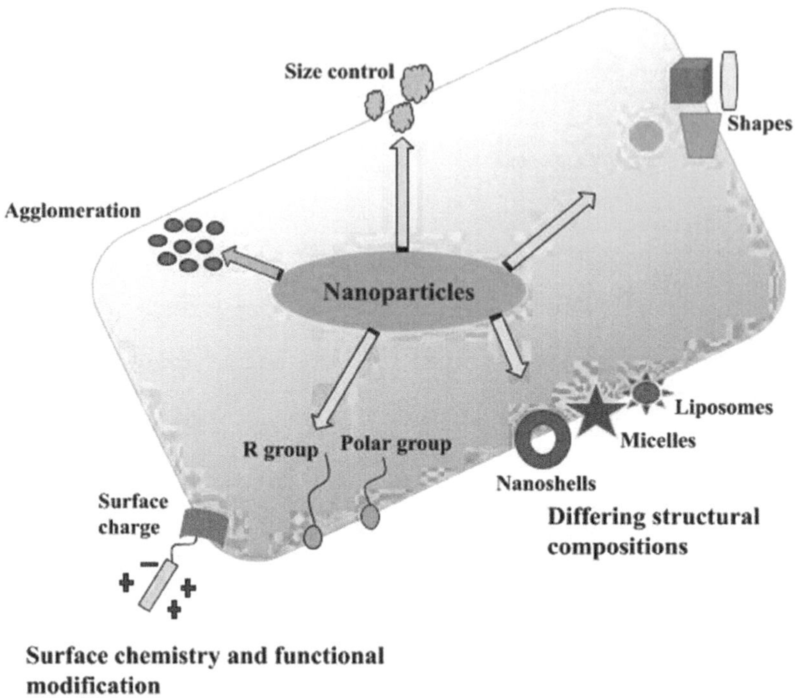

FIGURE 16.3 Physicochemical properties of nanoparticles.

standard chemotherapy. This clinical example shows how zirconia-based nanocarriers can change targeted therapy due to the release and targeted delivery of the affected tissues (Figure 16.3).

16.4 ZIRCONIA NANOPARTICLES IN BIOMEDICAL APPLICATIONS: IMAGING, DIAGNOSTICS, AND THERAPEUTICS

Zirconia nanoparticles have recently been found to possess several beneficial properties like chemical stability, biocompatibility, and a unique optical property, placing them in a unique position to break through biomedical applications. High mechanical strength, biocompatibility, and adjustable surface chemistry make zirconia nanoparticles potentially useful for imaging, diagnostics, and treatment. With the ability to modify the surface, imaging, and drug delivery properties, these nanomaterials shall revolutionize the current healthcare systems (Chitoria et al., 2023). The premise of having two imaging modes is that they can increase a disease's diagnosis and treatment capabilities. With further development in the sphere of research, the usage of zirconia nanoparticles in solving multidimensional patient-related problems will enhance the effectiveness of the technology in giving patients innovative therapy approaches. Three key areas of zirconia nanoparticle applications in biomedicine are: synthesis and characterization, applications in medical imaging and diagnostics, and therapeutic applications.

16.4.1 Synthesis and Characterization of Zirconia Nanoparticles

One of the most important parameters of zirconia (ZrO_2) nanoparticles preparation is to decide their medical application potential. These methods include sol-gel, hydrothermal synthesis, flame spray pyrosis, etc., to get nanoparticles with specific sizes, shapes, and crystalline structures (Figure 16.4).

- **Synthesis Techniques and Size Control:** The sol-gel method is preferred due to unique advantages like shooting well-dispersed nanoparticles under moderate conditions. Some factors that can be manipulated include the concentration of the precursor, pH of the solution, and temperature of the reaction; these play an important role in determining the size of the particle, which is essential in deciding its biodistribution and uptake in cells. Another way of producing zirconia nanoparticles with well-defined morphologies is through hydrothermal synthesis to improve the stability and functionality of the material for use in biological platforms (Bumajdad et al., 2018).
- **Surface Functionalization and Characterization:** Zirconia nanoparticles' surface modification is crucial to applying zirconia in biomedical fields. Targeting ligands, imaging agents, or regarded molecules can be conjugated to functional groups like carboxyl, amine, or hydroxyl grafted onto the nanoparticle surface. Analysing techniques such as TEM, XRD, and DLS help determine particle size, phase and crystallinity, and zeta potential. Such parameters are crucial in the reproducibility and operation of additional applications in the biomedical field (Shah & Rather, 2021).
- **Advanced Nanoparticle Platform for Multimodal Imaging:** A study conducted in the recent past also synthesized zirconia nanoparticles of

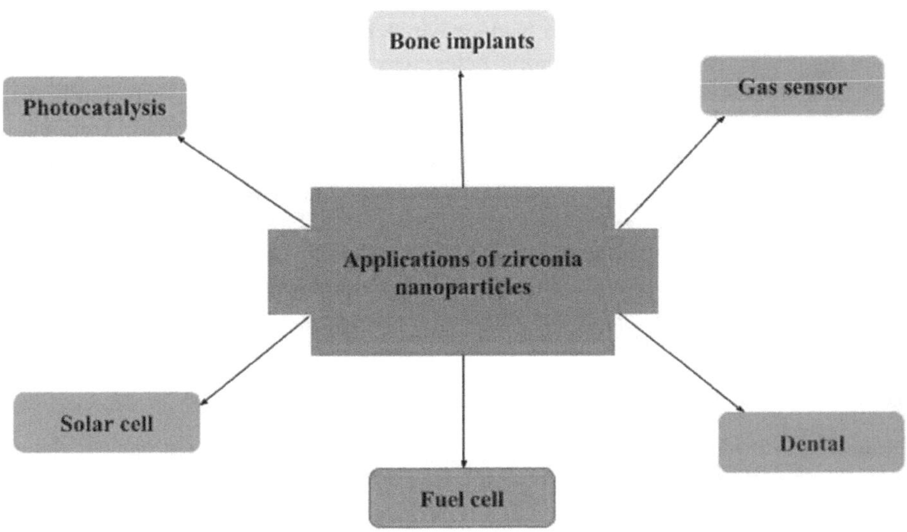

FIGURE 16.4 Applications of zirconia nanoparticles.

almost uniform size (~20 nm) by sol-gel method and later PEGylated and conjugated with a near-infrared fluorescent dye (Feng et al., 2017). It was also observed that the TEM and XRD analysis presented the required size and crystalline phase, and the DLS showed that the size distribution was narrow, which is essential for enhancing the performance in vivo. This platform, amenable to multimodal imaging applications, emphasizes the synthesis and characterization of clinically relevant nanomaterial.

16.4.2 Applications in Medical Imaging and Diagnostics

This paper focuses on zirconia nanoparticles, which are widely used as contrast agents and diagnostic tools in medical imaging in the UAE due to their optical and magnetic properties.

- **Contrast Enhancement in Imaging Modalities:** Its high value of X-ray attenuation coefficient makes zirconia very suitable for CT imaging. Nanoscale zirconia and an optimal surface modification offer a higher contrast than traditional agents when engineered into nanoparticle form (Kumari et al., 2022). Also, it can be functionally modified for other imaging purposes; for example, merging CT contrast with fluoroscopy imaging is helpful due to the added diagnostic information.
- **Biosensing and Diagnostic Platforms:** Zirconia is also beneficial for biosensing because of its chemical stability and electrical inertness. Using zirconia-based sensors, biomolecular, pathogenic, and metabolic marker detectors have been made (Rathee et al., 2021). These properties make them ideal for microfluidic diagnostic devices since they can withstand high chemical compatibility and have rough surfaces.
- **Zirconia Nanoparticles for Dual-Modality Imaging in Cancer Diagnostics:** In one clinical trial, researchers successfully created an imaging probe from zirconia nanoparticles, which possess two imaging modalities. These nanoparticles were fluorescently labelled by near-infrared dye and coupled with a CT contrast agent, allowing in vivo imaging of tumour tissues by CT and optical way. In the mouse model of breast cancer, the zirconia-based probe was taken up by tumour tissue according to CT images and in vivo fluorescence imaging experiments (Zhan et al., 2018). Such an integrated imaging approach helped achieve better anatomical imaging and molecular information regarding the corresponding efficacy of zirconia nanoparticles for diagnostics and interventions.

16.4.3 Therapeutic Applications of Zirconia Nanoparticles

In addition to diagnosis, zirconia nanoparticles are of even greater interest due to their therapeutic uses, primarily in target drug delivery and therapy.

- **Controlled Drug Release Systems:** This makes it a drug carrier since it has no chemical reactivity and is non-toxic to the human body. Using drugs in conjunction with zirconia-based matrix or adsorbing them onto the nanoparticle surface allows the formulation of systems for controlled release (Ifijen et al., 2023). Other changes include the pH-sensitive linker, which allows drug release at a given pH of the environment, for instance, a tumour or infected site.

- **Targeted Therapeutic Delivery:** Ligands used in designing zirconia nanoparticles may include antibodies, peptides, or small molecules that enhance the drug's delivery to diseased cells (Duvenhage et al., 2024). This approach reduces the whole system's toxicity and improves the loaded drug's effectiveness. They have been most effective in cancer treatment since the delivery of medicines is frequently a critical factor.
- **Zirconia Nanoparticles for Targeted Chemotherapy:** An actual study in a preclinical lung cancer model demonstrated the therapeutic application of zirconia nanoparticles. In the present work, zirconia nanoparticles were prepared and conjugated with folic acid to have a specific affinity for the folate receptors on the tumour cells. The chemotherapeutic compound was encapsulated in the nanoparticles and has been observed to release the drug when exposed to an acidic environment found in tumour tissues (Feng et al., 2017). The targeted formulation enhanced the concentration of the drug in the tumour tissue when tested in a lung cancer-bearing murine model without adverse effects on the rest of the body compared to the non-targeted formulations. Such a system was illustrated in the example of zirconia nanoparticles that can be used for controlled release to find a new direction in precision therapy.

16.5 ZIRCONIA-BASED COMPOSITES AND HYBRID MATERIALS: ENHANCED BIOACTIVITY AND FUNCTIONALITY

Zirconia ceramics and their combinations with different materials have become significant innovations in biomedical engineering, especially tissue engineering. Because PZT hinges possess excellent mechanical properties, biocompatibility, and biofunctionalizability, they are particularly suitable for applications involving structural and biological functions. Integrating zirconia with other materials and bioactive agents is thus an innovative approach towards developing multifaceted biomaterials for tissue engineering and regenerative medicine (Muzzio et al., 2021). With the design of enhancement in the composite materials' mechanical properties, attachment of bioactive molecules on the surface, and use of these hybrid systems in tissue engineering, researchers worldwide are in the process of providing a perfect solution for biomedical devices. Evidence from preclinical and clinical research supports these materials and reveals their potential application in regenerative medicine and overcoming many healthcare problems. This discussion is structured into three core sections: (i) the development of zirconia-based composites with improved properties, (ii) the integration of bioactive molecules and functional groups, and (iii) the applications of hybrid materials in regenerative medicine.

16.5.1 Development of Zirconia-Based Composites with Improved Properties

Zirconia (ZrO_2) is known for its best mechanical properties and chemical stability. Nevertheless, zirconia may not offer enough functionalities to meet the demands of various biomedical uses. Therefore, there has been research to carry out zirconia-based composites since adding or reinforcing other materials with zirconia could enhance the properties of the composites.

- **Material Synergy and Composite Fabrication:** Post-treatment phases like co-reinforcements, hydroxyapatite (HA), bioactive glass, or polymeric matrices are used to bring improvement in the mechanical properties of zirconia including fracture toughness, wear resistance as well as surface traits (Salari et al., 2019). These composites are prepared through complex sintering or sol-gel methods so that the second phase can be uniformly dispersed in the zirconia matrix. These methods do not cause a leachable to lose its mechanical stability, which is essential for stripping fragile materials, particularly zirconia. The roughness of the zirconia and the creation of porosity are two critical factors of cellular adhesion and proliferation.
- **Zirconia-Hydroxyapatite-Bioactive Glass Composite for Bone Regeneration:** The fabrication of the advanced composite scaffold is based on YSZ and it contains HA and body active glass (Ziaee et al., 2018). This nanocomposite material was synthesized from a sol-gel process and spark plasma sintering, which was accompanied by better compressive strength and controlled scaffold porosity. The composite scaffold replicated the form of natural bone, offering better osteoconductiveness and mechanical strength that suited load-bearing functions. These vitro findings and vitro evaluations demonstrated that the scaffold supported osteoblast adhesion and proliferation. In vivo evaluation using animal models also revealed the rapid formation of new bone tissues and the integration of the scaffold with new formation (Sartori et al., 2017). It also demonstrates how the combined application of zirconia and bioactive phases promotes the functional characteristics of the composites (Figure 16.5).

16.5.2 Integration of Bioactive Molecules and Functional Groups

It is, however, the second aspect where the potential of zirconia-based materials in biomedical applications seems to be even more exciting as it encompasses interaction with biological systems. Introducing bioactive molecules to the functional groups onto the surface of zirconia will enhance its biological properties greatly.

- **Surface Functionalization Strategies:** A functional group on the zirconia surface, such as carboxyl, amine, or phosphate groups. These changes allow for coupling to growth factors, peptides, or other bioactive agents, such as proteins, that can elicit specific cellular reactions. For instance, the

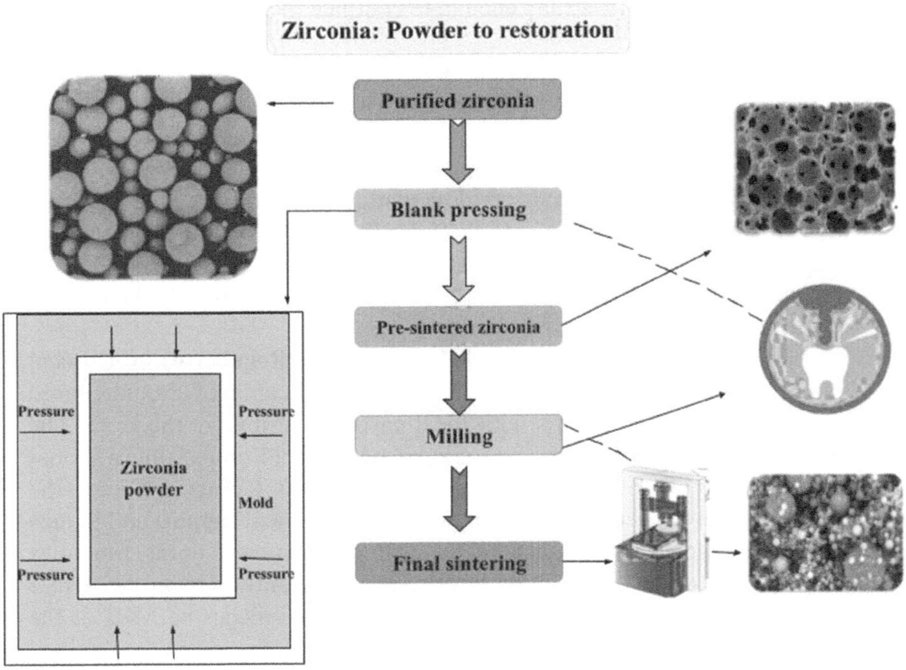

FIGURE 16.5 Development of zirconia-based composites with improved properties.

application of BMPs to the zirconia surface or the functionalization of RGD peptides (arginine-glycine-aspartate) enhances the osteogenic differentiation and the integration of the surface with the adjacent tissues (Zang et al., 2023). Moreover, anti-inflammatory agents could be incorporated into the implant to allow for decreased immune response in the area of implantation to prevent rejection and guarantee implant stability in the long term.

- **Functionalized Zirconia Nanoparticles for Enhanced Osseointegration:** Synthesized zirconia nanoparticles that included coating layers made of RGD peptides and BMP-2. The RGD motif improved cell adhesion; however, BMP-2 is a decisive osteoinductive factor because of its capability to stimulate osteogenesis. It is essential to note that nanoparticles of chitosan/gelatin have been incorporated into a composite scaffold, and the treatment was used in a critical-size bone defect model. Compared to non-functionalized nanoparticles, these functionalized nanoparticles enhanced the MSCs attachment and differentiation, new bone formation, and desired osseointegration. This approach reveals that surface modification can construct synthetic material/biological tissue interfaces (Zhao et al., 2020).

16.5.3 Applications of Hybrid Materials in Regenerative Medicine

Another aim of synthesizing zirconia-based composites and hybrids is to use them in regenerative medicine, as they should be capable of promoting healing tissue for growth and be mechanically strong enough to continue to serve their function for an extended period.

- **Tissue Engineering and Scaffold Design:** Zirconia and other biocompatible polymers or ceramics are incorporated in scaffolds used in tissue engineering. Some offer the required mechanical support and form the basis for growing cells. The product's degradation rate, porosity, and bioactivity can be controlled to generate scaffolds that facilitate tissue regeneration and gradually bear the loading on the tissue (Weng et al., 2021). This is especially the case in orthopaedic applications, mainly due to the need to engineer tissue like bone and cartilage.
- **Zirconia-Based Hybrid Scaffolds for Cartilage Regeneration:** Clinical pilot research was carried out to ascertain the application of zirconia-based hybrid scaffolds in patients with focal cartilage lesions in the knee. The used scaffolds, namely, the zirconia-based scaffold containing a biodegradable polymer and having chondrogenic growth factors grafted to the surface, were expected to offer immediate mechanical support and induce cartilage regeneration concurrently. All the patients' joint functions and pain over the lower limbs improved significantly during the 1-year follow-up; the changes were further validated based on the MRI of the affected joints and histological sections of the biopsy samples obtained from the affected tissues. The scaffold also formed hyaline-like cartilage and had a good interface adhesion with surrounding tissues, indicating that zirconia-based materials could promote future regenerative medicine (Pina et al., 2018).

16.6 SURFACE MODIFICATION OF ZIRCONIA FOR ENHANCED BIOCOMPATIBILITY

Zirconia is familiar as a biocompatibility and mechanical property that applies to dentistry and orthopaedics applications. However, it is possible to optimize it by making specific amendments on the outermost layer to increase compatibility with living tissues (Germaini et al., 2022). These are not only addendum cell adherence as well as proliferation but also the integration of tissue, and hence, zirconia becomes even more effective in implant and prosthetic applications. The method of surface modification of zirconia is a complicated process due to the ability to enhance its biocompatibility and function for biomedical applications. Most researchers have been able to alter cellular response and tissue integration through chemical modification, surface modification, and the incorporation of bioactive molecules. In addition, these surface properties must be tailored for specific applications in dental, orthopaedic, and/or regenerative medicine; several outcomes are then steered towards providing the best possible performance and timeline for the zirconia-based implant and/or devices. Another example of such thin film bioactive coating is the multifunctional dental implants that perfectly illustrate the countless possibilities of these approaches regarding enhancing patient's quality of life and better clinical outcomes (Borges et al., 2024).

16.6.1 Strategies for Modifying Zirconia Surfaces to Enhance Biocompatibility

Generally, improving the biocompatibility of zirconia is based on modifying its chemical and physical surface properties. To that end, several approaches have been created to realize these changes.

- **Chemical Functionalization:** Chemical modification, which involves attaching bioactive molecules or functional groups to the surface of zirconia, is one of its most efficient ways of modification. For instance, attaching carboxyl, hydroxyl, or amine groups onto zirconia surfaces could improve hydrophilicity and protein absorbance, an essential factor in cell attachment (Sharifi et al., 2024). Similarly, bioactive molecules such as peptides, for instance, RGD sequences or growth factors, or BMPs that enhance the differentiation of the osteogenic lineage and, therefore, osseointegration can be covalently grafted to the surface.
- **Physical Surface Treatments:** Some examples of physical modifications include sandblasting, etching, laser texturing, and plasma treatment. These methods form micro- and nanoscale relieves on the material surface resembling the dimensions of the natural extracellular matrix and increase the surface area. At the same time, features on the scale of several tens of nanometres guide cell adhesion and replication (Chen et al., 2018). Creating micro- and nano-topographies on substrates, for example, laser surface texturing can generate desired cell adhesion and orientation, bestowing differentiation on cells to improve tissue compatibility.

16.6.2 Impact of Surface Modifications on Cell Behaviour and Tissue Integration

The impact of the changes made to the surface of zirconia on factors such as the cells and the implantation of the implant in the host tissue.

- **Enhanced Cell Adhesion and Proliferation:** Modifying the zirconia surface will most often lead to increased wettability and better protein adsorption, enhancing cell attachment (Karthigeyan et al., 2019). For example, RGD peptides to surface successfully enhanced osteoblasts' adhesion and, thereby, the initial osseointegration stage. Increasing cell adhesion improves proliferation and differentiation, setting the necessary foundation for tissue regeneration around the implant.
- **Improved Osseointegration:** It should be noted that the modification of implant surfaces aims to realize the osseointegration process as effectively and as soon as possible. Scientific research has shown that micro-/nano-surface roughness in the form of zirconia has a larger bone-implant contact area than the smooth type. These are irregular in increasing the salvage of osteoprogenitor cells and promoting the synthesis of extracellular matrices, which are crucial for developing new bone tissue.

Furthermore, it incorporates various growth factors, including BMP, to boost osteoinduction and guarantee a firm implant connection into the bone (Georgeanu et al., 2023).

- **Orthopaedic Implant Success**: One study implemented the zirconia-based implant in orthopaedics, with a laser-fitted surface with BMP-2 bound to the surface. The modified implants were placed in a rabbit's femoral shaft model in the present study. Histologically, the new bachelor implant surface showed more sound contact area with the implant and more mature bone than the smooth zirconia implant (Zafar et al., 2020). The micro- and nano-topographical features created by laser with BMP-2 functionalization enhanced properties such as cell adhesion, cell proliferation, and differentiation, increasing the osseointegration rate. They are a good example of how specific surface modifications could influence the integration with the tissue and the implant's durability in the orthopaedic application.

16.6.3 Optimizing Surface Properties for Specific Applications

Surface attributes should be tailored for use in zirconia in biomedical applications, whether dental or orthopaedic implants or other regenerative medicine, to enhance its utilization.

- **Application-Specific Tailoring:** There are various situations where a certain kind of surface characteristic is necessary. For instance, practical and functional dental implants must have glamorous surfaces that encourage osseointegration. In detail, orthopaedic implants focus on the mechanical interlock and very fast osteointegration. Thus, the distinctions in roughness, porosity, and chemical modification of the zirconia surfaces can be tailored to meet these varied needs. For instance, incorporating nanoscale and microscale topographical features may help increase mechanical interlock and cellular interaction between the implant and the host tissue (Liu et al., 2021).
- **Integration with Emerging Technologies:** The optimization procedure also entails incorporating zirconia surfaces with other advanced technologies like nanotechnology and bioactive coatings. Incorporating zirconia with nanoparticles, such as silver or bioactive glass, can achieve additional properties, including antimicrobial or increased bioactivity. These hybrid materials create a path for the creation of new advanced implants, which can foster tissue integration while actively acting as barriers against the formation of infection and enhancing the overall rate of healing (Accioni et al., 2022).
- **Multifunctional Dental Implants:** An example of such a product is one of the relatively recent developments in the industry of dental implants. The researchers from Japan proposed a zirconia dental implant with a dual-modified surface; they conducted acid etching on the implant to make the surface micro-rough and provided a thin coating of the bio-glass

nanoparticles incorporated with antibiotic agents (Nikolova & Apostolova, 2022). Clinical trials also revealed that this multifunctional implant had improved osseointegration and minimized the complication, known as peri-implantitis, through its effectiveness against bacterial adhesion. This innovative concept illustrates that enhancing surface characteristics through combined treatments may improve zirconia implants' biological and clinical outcomes and solve various issues simultaneously.

16.7 TOXICITY AND SAFETY CONSIDERATIONS OF ZIRCONIA

16.7.1 Assessment of Zirconia's Biocompatibility and Potential Toxicity

Zirconia (zirconium dioxide or ZrO_2) is well known for its high biocompatibility. It is thus used in biomedical applications, including dental prosthetics, orthopaedic implantation, and biosensors (Tharmaraj et al., 2025). This is because its biocompatibility with living tissues is negligible and unlikely to provoke immunological allergic and inflammatory reactions in the human body. Some earlier investigations have proved that zirconia exhibits no cytotoxicity, genotoxicity, or mutagenicity if applied in the biomedical industry; hence, it has a non-toxic status. However, the toxicity of zirconia has given some concerns since the use of zirconia has been increased in different medical and industrial applications, mainly in the form of zirconia nanoparticles (ZrO_2 NPs). Generally, bulk zirconia is known to be bio-inert. However, the properties of nanoscale zirconia particles on a macroscale, especially their surface properties, are different partly due to the interfacial behaviour and chemical activity (Duraccio et al., 2015). The toxicity level of zirconia nanoparticles strongly depends on the size, charge, concentration, and time the above nanoparticles are exposed to the cells. At low concentrations, investigations suggest that the ZrO_2 NPs possess good biocompatibility attributes, including cell viability and cell proliferation, making them ideal for drug delivery and tissue engineering usage. However, in some cases, nanoparticles can provoke the reactive oxygen species, which may degenerate the cell membrane and lead to either apoptosis or necrosis due to disruption of mitochondria. These toxic effects are mainly attributed to ROS generation, which can lead to DNA damage and inflammation. Animals also contribute to understanding zirconia's toxic effects on the human body. The studies conducted on the implantation of zirconia-based biomaterials have shown no significant side effects on the tissues when the implantation is long-term (Lin et al., 2021). Inhalation or ingestion of zirconia nanoparticles has raised much concern regarding their accumulation in important organs such as the liver, lungs, and kidneys. It is possible to develop inflammatory reactions, fibrosis, or metabolic changes if exposed for a long time, depending on the dosage mode of the nanoparticle.

16.7.2 Factors Influencing the Toxicity of Zirconia Nanoparticles

It is crucial to consider that zirconia nanoparticles' toxicity is mainly determined by their physicochemical nature, particle concentration, and exposure mode.

- **Particle Size and Surface Area**: It is, therefore, imperative to indicate that the size of zirconia nanoparticles influences their biological interactions. Other factors also abound, including smaller nanoparticles having a larger surface-area-to-volume ratio than larger particles, making the smaller particles more reactive and quickly taken up by cells (Bera et al., 2022). ROS generation becomes more frequent and can reach the level of oxidative stress and toxicity. In this case, smaller nanoparticles have higher mobility and bioavailability and can cause systemic distribution and transport to various organs.
- **Surface Charge and Functionalization:** The state of charge on the surface of zirconia nanoparticles is essential to controlling their behaviour with the biological membrane. Studies have suggested that positively charged nanoparticles are internalized more effectively into the cells; however, they are cytotoxic due to their stronger interaction with the negative charge of the cell membrane (Bannunah, 2023). People have applied polymer coatings or biofunctionalization of nanomaterials to overcome toxicity issues, improve biocompatibility, and minimize side biological impacts.
- **Aggregation and Dispersion Stability:** It was also observed that the agglomeration or dispersion of zirconia nanoparticles in the biological fluids determines their toxicity. While it is well known that well-dispersed nanoparticles can have higher solubility and better pass-through barrier of cells, large aggregates may have the opposite effect by having less interaction with cells (Sun et al., 2021). Nevertheless, the formation of large structures in biological systems can change the properties of nanoparticles; thus, it is obligatory to consider measures to enhance the stability in biomedical use.
- **Dose and Exposure Duration:** Similar to many other nanomaterials, the toxicological potential of ZrO_2 NPs has been found to depend on the dose administered. It was established that low doses of ZrO_2 NPs are biocompatible and can stimulate cell division, while high doses yield cytotoxicity, inflammation, or apoptosis (Barbasz & Dyba, 2024). It may also build up to tissues, requiring long-term toxicity studies for the nanoparticles in biomedical applications.
- **Route of Exposure:** How one is exposed to zirconia nanoparticles affects its toxicity level. The workers employed in industries using zirconia nanoparticles, such as ceramics, coating, or biomedical, may expose themselves to respiratory toxicity. According to some experiments dealing with inhaled ZrO_2 NPs, there are consequences like pulmonary inflammation, oxidative stress, and lung fibrosis that have to be further investigated for chronic effects (Hichem et al., 2022). Some ZrO_2 NPs used to develop dental prosthetics and in food contact surfaces may migrate to the gastrointestinal tract. However, according to research studies, chronic consumption has been seen to cause slight effects on the toxic nature of the human body while affecting gut flora and metabolism. This means that zirconia nanoparticles are generally non-irritating to the skin; therefore, they present low dermal toxicity, which is ideal when applied on the skin for cosmetic and topical biomedical

uses. Nevertheless, adverse effects might be experienced with continued or prolonged usage, but this depends on the degree of exposure to the chemical. Ceramic zirconia-based implants are biocompatible and capable of integrating with the osseous tissue, causing almost no harm of systemic effects (Rathee et al., 2021). However, more concern is needed regarding the long-term effects of nanoparticle wear debris.

16.7.3 Regulatory Guidelines and Safety Standards

Several approximations and safety-chartered bodies have been advised and evaluated to control using zirconia ceramics for biomedical and industrial applications. These include:

- **Food and Drug Administration (FDA):** The FDA categorizes zirconia under the category of medical devices – class II; this is an appreciation of its safety and efficiency in the dental and orthopaedic industry. Cytotoxicity assessment of osteocompatible zirconia is done to ensure its biocompatibility; a genotoxicity test, as well as the rate of degradation of the implant, is also considered (Al-Zyoud et al., 2023).
- **International Organization for Standardization (ISO)**
  - **ISO 13356:** Specifies requirements for the chemical composition, mechanical properties, and biocompatibility of zirconia ceramics for surgical implants.
  - **ISO 10993:** Gives information on a biological evaluation of medical devices and the requirements for performing systemic toxicity evaluations on zirconia ceramics (Khaskhoussi et al., 2020).
- **Occupational Safety and Health Administration (OSHA):** OSHA regulates zirconia nanoparticle emission, especially for facilities where particulate matter, such as fine particles, can be inhaled (Zapór & Oberbek, 2020). To protect against health effects, permissible exposure limits (PELs) and protective measures need to be employed.
- **European Medicines Agency (EMA):** In the European Union, zirconia-containing biomaterials must conform to the CE regulation laws regarding the safety, quality, and performance of biomaterials in medical processes (Pezzotti, 2014).
- **National Institute for Occupational Safety and Health (NIOSH):** Currently, NIOSH has issued guidelines for working with nanomaterials, the use of zirconia nanoparticles and their exposure, protective equipment, and technical measures (Zapór & Oberbek, 2020) (Table 16.1).

16.8 CHALLENGES AND FUTURE DIRECTIONS IN ZIRCONIA RESEARCH

Zirconia (zirconium dioxide or ZrO_2), is increasingly becoming a remarkable material for different uses, especially in biomedical engineering, electronic devices, and structural ceramics. One of them is called α-Al2O3, the key features of which are

TABLE 16.1
Toxicity and Safety Considerations of Zirconia

Aspect	Description	Impact on Health	Examples	Future Directions	References
Biocompatibility assessment	Evaluating zirconia's compatibility with biological tissues.	Determines suitability for medical use	In vitro and in vivo studies	Standardized testing protocols	Tharmaraj et al. (2025)
Toxicity studies	Research on the toxic effects of zirconia on cells and tissues.	Identifies potential health risks	Cytotoxicity assays	Long-term toxicity studies	Duraccio et al. (2015)
Particle size influence	Effects of nanoparticle size on toxicity levels.	Smaller particles may exhibit higher toxicity	Size-dependent toxicity evaluations	Investigating size reduction techniques	Lin et al. (2021)
Surface chemistry	Role of surface modifications in influencing toxicity.	Affects cellular interactions	Coating with biocompatible materials	Development of safer surface treatments	Bera et al. (2022)
Dosage and exposure	Relationship between zirconia dosage and toxicity outcomes.	Higher doses may increase toxicity	Determining safe exposure levels	Establishing safe dosage guidelines	Bannunah, (2023)
Route of administration	Impact of different administration routes on toxicity (e.g., inhalation, injection).	Varies toxicity profile	Inhalation toxicity studies	Comparative studies on routes	Sun et al. (2021)
Chronic vs. acute toxicity	Differences in toxicity effects over short- and long-term exposure.	Important for risk assessment	Acute exposure vs. chronic exposure studies	Focus on chronic exposure studies	Hichem et al. (2022)
Immune response	Investigating how zirconia affects the immune system.	Potential for inflammatory responses	Cytokine release assays	Evaluating long-term immune effects	Rathee et al. (2021)
Environmental impact	Assessing the ecological effects of zirconia nanoparticles.	Potential risks to wildlife	Aquatic toxicity studies	Environmental monitoring strategies	Zapór and Oberbek (2020)
Regulatory guidelines	Overview of existing regulations on zirconia use.	Ensures public safety	FDA and EPA guidelines	Advocacy for updated regulations	Khaskhoussi et al. (2020)
Safety standards	Establishing safety standards for zirconia in various applications.	Protects consumers and workers	ISO standards for dental applications	Development of comprehensive safety protocols	Al-Zyoud et al. (2023)

(Continued)

TABLE 16.1 *(Continued)*
Toxicity and Safety Considerations of Zirconia

Aspect	Description	Impact on Health	Examples	Future Directions	References
Risk assessment frameworks	Developing frameworks to assess risks associated with zirconia exposure.	Informs safety measures	WHO risk assessment guidelines	Integrating new research findings	Zapór and Oberbek (2020)
Clinical applications	Evaluating safety in medical uses of zirconia (e.g., dental implants).	Direct implications for patients	Clinical trials on zirconia implants	Post-market surveillance studies	Zapór and Oberbek (2020)
Public awareness	Educating the public about the safety of zirconia products.	Increases informed consumer choices	Informational campaigns	Ongoing community education initiatives	Zapór and Oberbek (2020)
Research collaboration	Partnerships between industry, academia, and regulatory bodies.	Enhances safety evaluations	Collaborative research projects	Increased multi-disciplinary studies	Khaskhoussi et al. (2020)
Alternative materials	Exploring substitutes for zirconia with better safety profiles.	Reduces potential health risks	Biodegradable alternatives	Comparative safety studies	Al-Zyoud et al. (2023)
Analytical methods	Techniques for detecting and quantifying zirconia in biological systems.	Essential for toxicity assessments	Mass spectrometry, electron microscopy	Advancements in detection technologies	Zapór and Oberbek (2020)
Future research directions	Identifying gaps in current knowledge and future research needs.	Guides regulatory and safety efforts	Focus on long-term effects	Funding for comprehensive studies	Zapór and Oberbek (2020)

its mechanical strength, biocompatibility, corrosion resistance, and high thermal stability. Nonetheless, issues in the processing, fabrication, and performance of the material restrict the application of this material in new fields that are developing. This section discusses the existing issues in zirconia research and suggests possible ways to address them in order to develop its uses further. Zirconia remains an object of intensive investigations, which focus on the issues connected with the material processing, phase stability, cutting, and, in particular, its biomimetic properties (Yin et al., 2024). Subsequent generations will be on nanoengineering, functionalization, and new applications in the biomedical fields so that zirconia will remain a relevant material option in biomedical engineering, pharmaceutical and medical applications, and bi-electronics. New characterization techniques shall be developed and applied to improve the knowledge of the long-term behaviour of zirconia and create new markets and applications for this material.

16.8.1 Addressing Limitations in Zirconia Processing and Fabrication

Zirconia possesses the desired properties for applications requiring high strength, oxidation and corrosion resistance, and low thermal conductivity. However, it has several drawbacks from a technical fabrication and processing point of view. These include problems relating to phase transformation, densification, grain growth control, and mechanical failure (Zakaria et al., 2020b).

16.8.1.1 Phase Transformation and Stability Issues

One of the main problems in investigating the zirconia properties is its phase stability. There are three types of zirconia; all of them are well-established and include the following:

- **Monoclinic Phase (*m*-ZrO_2):** This phase is stable at room temperature but transitions to a different phase at higher temperatures (Garanin et al., 2024).
- **Tetragonal Phase (*t*-ZrO_2):** DIMM soluble at room temperature and turns to monoclinic phase at lower temperature freaking and plaque formation due to volume increase (Garanin et al., 2024).
- **Cubic Phase (*c*-ZrO_2):** It is stable at high temperatures but not at lower temperatures, so it requires doping to make it stable (Tian et al., 2024).

Such a phase transformation is damaging at the mechanical level, specifically in biomedical implants and structural ceramics. As pointed out before, other dopants like yttria (Y_2O_3), magnesia (MgO), ceria (CeO_2), or alumina (Al_2O_3) can be introduced so that the phase is stabilized in tetragonal and cubic. Still, maintaining the file stability regarding cyclic loading and physiologic conditions to lock the desired integrity for the long term remains challenging.

16.8.1.2 Sintering and Densification Difficulties

The dense, high-strength properties of zirconia satisfy the ceramic materials' characteristics, but they need to be sintered at temperatures above 1400°C. However, a significant downside to using the ordinary sintering process is the observed effect, which is undesirable grain growth as it weakens the material's toughness. Precursors of nano-sized zirconia powders bring specific difficulties connected with agglomeration behaviour, inhomogeneous densification, and porosity control.

- **Spark Plasma Sintering (SPS):** This process is a relatively modern method for carrying out this process at relatively high temperatures and pressures, but at lower temperatures and in a short period, thus minimizing grain growth and other properties that improve the mechanical properties of the material (Cavaliere et al., 2019).
- **Cold Sintering Process (CSP):** A promising method has recently been developed for sintering zirconia at relatively low temperatures (Guo et al., 2017).
- **Additive Manufacturing (3D Printing):** Stereolithography (SLA) and direct ink writing (DIW) are new technologies that can improve the mould-making of zirconia-based components (Sokola et al., 2024).

16.8.1.3 Machining and Surface Modification Challenges

Another challenge of working with zirconia is that it is abrasive, harsh, and brittle. This results in increased tool wear and high processing costs. Such post-treatment may include grinding and polishing, which cause residual stresses and phase transformation, affecting the material's durability in the long run.

- **Laser-Assisted Machining (LAM):** LAM is one potential means of opening new horizons in enhancing the possibility of machining zirconia components while ensuring that the ensuing residual stress is adequately controlled (Shin et al., 2020).
- **Plasma and Ion Beam Treatments:** These techniques can be used to treat samples to smoothen the surface and increase surface finish and mechanical reliability in cases such as biomedical implants (Thakur et al., 2022).

16.8.2 Exploring Novel Applications of Zirconia in Biomedical Engineering

The present study uses zirconia in dentistry as crowns, bridges, posy implants, orthopaedic implants, and even biosensors. However, studies are slowly venturing into other areas in the biomedical field, such as drug delivery systems, biosensors, and tissue engineering.

16.8.2.1 Zirconia-Based Drug Delivery Systems

Zirconia in nano forms is used in targeted drug delivery and release systems. The chemical stability and biocompatibility reveal that ZrO_2 NPs are suitable as drug delivery systems for anticancer drugs, antibiotics, and gene delivery applications.

- **Functionalized Zirconia Nanoparticles:** To enhance drug loading and targeting capabilities, functionalized zirconia nanoparticles may be achieved using polymeric coatings, peptides, or antibodies (Rathee et al., 2021).

- **pH-Responsive Zirconia Carriers**: Zirconia carriers were developed to deliver drugs to cancerous or inflamed tissue where the pH differs from that of healthy tissues (Arcos & Vallet-Regi, 2013).

16.8.2.2 Zirconia in Bioelectronics and Neural Interfaces

Due to implantable bioelectronics, zirconia is emerging as a candidate material for neural implants and biosensors (Ramesh et al., 2024). It is nonconductive, highly stable, and chemically unreactive and is compatible with implantation into the brain for long-term interfacing.

- **Flexible Base Electrodes made of Zirconia** may apply to BMI and neural prostheses.
- **Zirconia-Coated Electrodes:** Reduce the risk of corrosion and inflammatory responses in chronic implants.

16.8.2.3 Tissue Engineering and Regenerative Medicine

Zirconia's characteristics offer the composite the required properties suitable for constructing bone tissue engineering scaffolds (Sivasankar et al., 2024). Most of the work is concentrated on enhancing cell adhesion, proliferation, and differentiation for bone tissue engineering on zirconia-based scaffolds.

- **Porosities Zirconia Scaffolds:** Excellent for creating supporting tissue and bone growth.
- **Zirconia-Hydroxyapatite Composites:** Combining zirconia's mechanical strength with hydroxyapatite's bioactivity for advanced bone graft materials.

16.8.3 Developing Advanced Characterization Techniques for Zirconia-Based Materials

Several works have described the application of zirconia in various fields. As the number of practical applications increases, better characterization techniques are needed to study the material's phase stability, mechanical properties, and biocompatibility.

16.8.3.1 High-Resolution Electron Microscopy (HREM) and Spectroscopy

XRD is the most common method for identifying the phase of zirconia; however, techniques at the atomic level are required for further analysis (Mayoral et al., 2019).

- **Transmission Electron Microscopy (TEM)**: It can provide more information on the details of zirconia grain boundaries and defects.
- **X-ray Photoelectron Spectroscopy (XPS)**: It assists in deducing the chemical states of the outermost layers of the resultant zirconia layers and the interactions between dopants.

16.8.3.2 In Situ Mechanical Testing Techniques

For the biomedical application of zirconia, real-time variation of the mechanical properties, especially under dynamic loads, is essential (AbdelGawad et al., 2022).

- **Nanoindentation Investigations:** This technique investigates the local mechanical characteristics and phase transformation.
- **Synchrotron-Based XRD:** It allows for in situ phase transformation analysis during mechanical stress application.

16.8.3.3 Biodegradation and Long-Term Stability Studies

Zirconia is well known for its chemical inactivity, but prolonged exposure to physiological fluids degrades the material and causes phase transformation (Sharma et al., 2024).

- **Accelerated Aging Studies:** In this method, mechanical fatigue tests and simulated body fluids can predict the rate at which a given implant ages.
- **Wear Resistance Analysis:** This assesses the impacts of tribological wear and friction on zirconia-based joint replacements (Table 16.2).

16.9 CONCLUSION

It has been reported that zirconia as a material has found application in activities related to biological and biomedical uses, which has been considered a breakthrough in biomaterials. It has gone beyond the existing applicability in dental ceramics and is presently used in orthopaedic implants, tissue engineering scaffolds, and biosensors. Incorporating other factors, such as laser treatments and nanopatterning, has increased the surface treatment of zirconias' bioactivity and mechanical properties, enhancing its compatibility with biological tissues. The material's biocompatibility, corrosion, and wear properties have further boosted its usage in dentistry, whereby it displays better aesthetics and resilience with different restored teeth than the conventional materials. In addition, the capacity of zirconia in orthopaedic implants is being improved by improved integration of implants with bone tissue and deferred inflammation, which is of great importance for the inception of the implants. Upon further consideration in the biomedical fields for zirconia, there is much potential in the future. Current studies are focused on continuing the alterations of its microstructure and surface properties to provoke certain cell behaviours that are fundamental for tissue engineering approaches, such as increased stem cell attachment and growth. One can anticipate that integrating zirconia with bioactive molecules and growth factors will create innovative implants capable of participating in the healing process, the prospects of which will bring about revolutionary change in treating diseases involving joint replacement and bone regeneration. Apart from its structural function, zirconia is under investigation for application in diagnostic and therapeutic instruments. Its stability and unique optical properties make it ideal for synthesizing advanced biosensors to identify biomarkers at ultra-low levels. Such technologies would greatly help early disease identification and monitoring, enhancing customized therapy options. Furthermore, its extensive application in clinics relies mainly on programmes aiming at simplifying manufacturing techniques and reducing production costs. In biology and biomedical engineering, expressed, expanding boundaries of zirconia have a significant capacity to change medicine. Zirconia is already on the road to becoming the driving force for next-generation medical devices and

TABLE 16.2
Challenges and Future Directions in Zirconia Research

Aspect	Description	Challenges	Future Directions	References
Processing limitations	Challenges in the synthesis and processing of zirconia materials.	Variability in material properties	Development of standardized processing methods	Zakaria et al. (2020a)
Fabrication techniques	Issues with existing fabrication methods for zirconia components.	Difficulty in achieving complex geometries	Exploring additive manufacturing techniques	Garanin et al. (2024)
Mechanical properties	Enhancing strength and toughness of zirconia materials.	Brittleness in certain zirconia forms	Research into composite zirconia materials	Tian et al. (2024)
Surface modifications	Need for effective surface treatments to improve biocompatibility.	Inconsistency in surface properties	Development of new coating technologies	Cavaliere et al. (2019)
Thermal stability	Addressing thermal degradation of zirconia at high temperatures.	Loss of mechanical integrity	Investigating high-temperature zirconia formulations	Guo et al. (2017)
Cost of production	High costs associated with zirconia synthesis and processing.	Economic feasibility of large-scale production	Exploring cost-effective production methods	Sokola et al. (2024)
Novel applications	Identifying new uses for zirconia in various fields.	Limited knowledge of zirconia's full potential	Expanding research into different sectors, e.g., electronics	Rathee et al. (2021)
Biomedical engineering	Assessing zirconia's potential in medical implants and devices.	Long-term biocompatibility concerns	Clinical trials for new biomedical applications	Ramesh et al. (2024)
Characterization techniques	Advanced methods are needed to analyse zirconia materials at the nanoscale.	Current techniques may lack precision	Development of high-resolution imaging methods	Sivasankar et al. (2024)
Analytical methods	Improving techniques for evaluating zirconia's properties.	Lack of standardized testing protocols	Establishing comprehensive testing frameworks	Sharma et al. (2024)
Environmental impact	Assessing the ecological footprint of zirconia production.	Inadequate environmental assessments	Conducting lifecycle analyses of zirconia products	Zakaria et al. (2020b)
Regulatory challenges	Navigating complex regulations for zirconia use in medical applications.	Delays in approval processes	Engaging with regulatory bodies for more precise guidelines	Garanin et al. (2024)

(Continued)

TABLE 16.2
Challenges and Future Directions in Zirconia Research

Aspect	Description	Challenges	Future Directions	References
Interdisciplinary collaboration	Encouraging collaboration across various research fields.	Silos in research efforts	Building multidisciplinary research teams	Tian et al. (2024)
Market demand	Understanding the market needs for zirconia products.	Fluctuating demand for zirconia items	Market studies to inform research priorities	Cavaliere et al. (2019)
Public perception	Addressing misconceptions about zirconia safety and efficacy.	Misinformation in media	Educational outreach to improve public understanding	Guo et al. (2017)
Ageing and degradation studies	Investigating the long-term behaviour of zirconia in specific applications.	Lack of data on long-term performance	Longitudinal studies on zirconia products	Sokola et al. (2024)
Integration with other materials	Exploring hybrid materials incorporating zirconia.	Compatibility issues between materials	Research into composite structures	Rathee et al. (2021)
Tailored properties	Developing zirconia with customized properties for specific applications.	Complexity in achieving desired traits	Focused research on property optimization	Ramesh et al. (2024)
Scaling up production	Transitioning from lab-scale to industrial-scale zirconia production.	Maintaining quality at scale	Pilot projects to test scalability	Sivasankar et al. (2024)
Future research directions	Identifying key areas for future investigation in zirconia research.	Prioritizing research focus	Funding initiatives for emerging areas	Sharma et al. (2024)

regenerative medicine—which eventually translates into improved patient care and quality of life. This road is marked by growing multidisciplinary cross-talk between doctors, bioengineers, and material scientists.

REFERENCES

AbdelGawad, M., Usman, C. A., Shunmugasamy, V. C., Karaman, I., & Mansoor, B. (2022). Corrosion behaviour of Mg-Zn-Zr-RE alloys under physiological environment–impact on mechanical integrity and biocompatibility. *Journal of Magnesium and Alloys*, *10*(6), 1542–1572. https://doi.org/10.1016/j.jma.2021.11.025

Accioni, F., Vázquez, J., Merinero, M., Begines, B., & Alcudia, A. (2022). Latest trends in surface modification for dental implantology: Innovative developments and analytical applications. *Pharmaceutics*, *14*(2), 455. https://doi.org/10.3390/pharmaceutics14020455

Al-Johani, H., Haider, J., Satterthwaite, J., Borba, M., & Silikas, N. (2024). A comprehensive review of the multifaceted characterisation approaches of dental ceramics. *Prosthesis*, *6*(5), 1055–1090. https://doi.org/10.3390/prosthesis6050077

Alqutaibi, A. Y., Ghulam, O., Krsoum, M., Binmahmoud, S., Taher, H., Elmalky, W., & Zafar, M. S. (2022). Revolution of current dental zirconia: A comprehensive review. *Molecules*, *27*(5), 1699. https://doi.org/10.3390/molecules27051699

Al-Zyoud, W., Haddadin, D., Hasan, S. A., Jaradat, H., & Kanoun, O. (2023). Biocompatibility testing for implants: A novel tool for selection and characterization. *Materials*, *16*(21), 6881. https://doi.org/10.3390/ma16216881

Arcos, D., & Vallet-Regi, M. (2013).Bioceramics for drug delivery. *ActaMaterialia*, *61*(3), 890–911. https://doi.org/10.1016/j.actamat.2012.10.039

Banh, W., Hughes, J., Sia, A., Chien, D. C., Tadakamadla, S. K., Figueredo, C. M., & Ahmed, K. E. (2021). Longevity of polymer-infiltrated ceramic network and zirconia-reinforced lithium silicate restorations: A systematic review and meta-analysis. *Materials*, *14*(17), 5058. https://doi.org/10.3390/ma14175058

Bannunah, A. M. (2023). Biomedical applications of zirconia-based nanomaterials: Challenges and future perspectives. *Molecules*, *28*(14), 5428. https://doi.org/10.3390/molecules 28145428

Barbasz, A. M., & Dyba, B. (2024). Direct interaction of zirconia nanoparticles with human immune cells. *Biophysica*, *4*(1), 83–91. https://doi.org/10.3390/biophysica4010006

Bera, K., Ghosh, B., & Mukhopadhyay, M. (2022).Functional principal of nanotechnology in clinical research. In *Nanomaterials in Clinical Therapeutics: Synthesis and Applications*, 33–73. https://doi.org/10.1002/9781119857747.ch2

Berger, M. B., Slosar, P., Schwartz, Z., Cohen, D. J., Goodman, S. B., Anderson, P. A., & Boyan, B. D. (2022). A review of biomimetic topographies and their role in promoting bone formation and osseointegration: Implications for clinical use. *Biomimetics*, *7*(2), 46. https://doi.org/10.3390/biomimetics7020046

Borges, J., Zeng, J., Liu, X. Q., Chang, H., Monge, C., Garot, C., … & Picart, C. (2024). Recent developments in layer-by-layer assembly for drug delivery and tissue engineering applications. *Advanced Healthcare Materials*, *13*(8), 2302713. https://doi.org/10.1002/adhm.202302713

Bumajdad, A., Nazeer, A. A., Al Sagheer, F., Nahar, S., & Zaki, M. I. (2018).Controlled synthesis of ZrO_2 nanoparticles with tailored size, morphology and crystal phases via organic/inorganic hybrid films. *Scientific Reports*, *8*(1), 3695. https://doi.org/10.1038/s41598-018-22088-0

Carone, C., Lauria, P., Cardarelli, F., Del Vecchio, M., Memè, L., Bambini, F., … & Palermo, A. (2024). Advancing the clinical use of zirconia dental implants: A comprehensive review of outcomes, innovations and challenges. *Oral and Implantology: A Journal of Innovation and Techniques for Oral Health*, *16*(3 (S1) Special Issue), 190–203.

Cavaliere, P., Sadeghi, B., & Shabani, A. (2019). Spark plasma sintering: Process fundamentals. In *Spark Plasma Sintering of Materials: Advances in Processing and Applications* (pp. 3–20). https://doi.org/10.1007/978-3-030-05327-7_1

Chen, L., Yan, C., & Zheng, Z. (2018). Functional polymer surfaces for controlling cell behaviors. *Materials Today*, *21*(1), 38–59. https://doi.org/10.1016/j.mattod.2017.07.002

Chitoria, A. K., Mir, A., & Shah, M. A. (2023). A review of ZrO_2 nanoparticles applications and recent advancements. *Ceramics International*, *49*(20), 32343–32358. https://doi.org/10.1016/j.ceramint.2023.06.296

Comino-Garayoa, R., Cortés-BretónBrinkmann, J., Peláez, J., López-Suárez, C., Martínez-González, J. M., & Suárez, M. J. (2020). Allergies to titanium dental implants: What do we really know about them? A scoping review. *Biology*, *9*(11), 404. https://doi.org/10.3390/biology9110404

Duraccio, D., Mussano, F., & Faga, M. G. (2015). Biomaterials for dental implants: Current and future trends. *Journal of Materials Science*, *50*, 4779–4812. https://doi.org/10.1007/s10853-015-9056-3

Duvenhage, J., Kahts, M., Summers, B., Zeevaart, J. R., & Ebenhan, T. (2024, October). Highlighting new research trends on zirconium-89 radiopharmaceuticals beyond antibodies. In *Seminars in Nuclear Medicine*. WB Saunders. https://doi.org/10.1053/j.semnuclmed.2024.10.003

Feng, L., Gai, S., He, F., Dai, Y., Zhong, C., Yang, P., & Lin, J. (2017). Multifunctional mesoporous ZrO_2 encapsulated upconversion nanoparticles for mild NIR light activated synergistic cancer therapy. *Biomaterials*, *147*, 39–52. https://doi.org/10.1016/j.biomaterials.2017.09.011

Festas, A. J., Ramos, A., & Davim, J. P. (2020). Medical devices biomaterials–A review. *Proceedings of the Institution of Mechanical Engineers, Part L: Journal of Materials: Design and Applications*, *234*(1), 218–228. https://doi.org/10.1177/1464420719882458

Garanin, Y., Shakirzyanov, R., Borgekov, D., Kozlovskiy, A., Volodina, N., Shlimas, D., & Zdorovets, M. (2024). Study of morphology, phase composition, optical properties, and thermal stability of hydrothermal zirconium dioxide synthesized at low temperatures. *Scientific Reports*, *14*(1), 29398.

Georgeanu, V. A., Gingu, O., Antoniac, I. V., & Manolea, H. O. (2023).Current options and future perspectives on bone graft and biomaterials substitutes for bone repair, from clinical needs to advanced biomaterials research. *Applied Sciences*, *13*(14), 8471. https://doi.org/10.3390/app13148471

Germaini, M. M., Belhabib, S., Guessasma, S., Deterre, R., Corre, P., & Weiss, P. (2022). Additive manufacturing of biomaterials for bone tissue engineering–A critical review of the state of the art and new concepts. *Progress in Materials Science*, *130*, 100963. https://doi.org/10.1016/j.pmatsci.2022.100963

Guo, H., Bayer, T. J., Guo, J., Baker, A., & Randall, C. A. (2017). Current progress and perspectives of applying cold sintering process to ZrO_2-based ceramics. *ScriptaMaterialia*, *136*, 141–148. https://doi.org/10.1016/j.scriptamat.2017.02.004

Hahn, A., & Kurz, F. T. (2022). Advanced imaging/MRI for tissue engineering. In *Tissue Engineering Using Ceramics and Polymers* (pp. 281–343). Woodhead Publishing. https://doi.org/10.1016/B978-0-12-820508-2.00016-7

Heydariyan, Z., Soofivand, F., Dawi, E. A., Abd Al-Kahdum, S. A., Hameed, N. M., & Salavati-Niasari, M. (2023). A comprehensive review: Different approaches for encountering bacterial infection of dental implants and improving their properties. *Journal of Drug Delivery Science and Technology*, *84*, 104401. https://doi.org/10.1016/j.jddst.2023.104401

Hichem, N., Hadjer, Z., Fateh, S., Feriel, L., & Wang, Z. (2022). The potential exposure and hazards of zirconia nanoparticles: A review. *Ecotoxicology and Environmental Contamination*, *17*(1), 1–21. https://doi.org/10.5132/eec.2022.01.01

Huang, B., Chen, M., Wang, J., & Zhang, X. (2024). Advances in zirconia-based dental materials: Properties, classification, applications, and future prospects. *Journal of Dentistry*, 105111. https://doi.org/10.1016/j.jdent.2024.105111

Huang, Y., & Duan, Z. (2025). Research progress on Cu-ZrO_2 catalysts: A systematic review of their preparations, doping, and applications. *Journal of Materials Chemistry A*, *13*, 4043–4089.

Ifijen, I. H., Uzoekwe, N. M., Ohiocheoya, E. B., Osarobo, J. A., & Omonmhenle, S. I. (2023). Nano-zirconia synthesis methods and their pioneering applications in dentistry. *Momona Ethiopian Journal of Science*, *15*(2), 122–152. https://doi.org/10.4314/mejs.v15i2.1

Jin, M., Sun, N., Weng, W., Sang, Z., Liu, T., Xia, W., … & Yang, H. (2023). The effect of GelMA/alginate interpenetrating polymeric network hydrogel on the performance of porous zirconia matrix for bone regeneration applications. *International Journal of Biological Macromolecules*, *242*, 124820. https://doi.org/10.1016/j.ijbiomac.2023.124820

Kaliaraj, G. S., Siva, T., & Ramadoss, A. (2021). Surface functionalized bioceramics coated on metallic implants for biomedical and anticorrosion performance–A review. *Journal of Materials Chemistry B*, *9*(46), 9433–9460.

Karthigeyan, S., Ravindran, A. J., Bhat, R. T., Nageshwarao, M. N., Murugesan, S. V., & Angamuthu, V. (2019). Surface modification techniques for zirconia-based bioceramics: A review. *Journal of Pharmacy and Bioallied Sciences*, *11*(Suppl 2), S131–S134.

Khaskhoussi, A., Calabrese, L., Bouaziz, J., & Proverbio, E. (2020). Performances and aging stability of new Al_2O_3-ZrO_2-TiO_2 ternary ceramic composites. *Materials Chemistry and Physics*, *243*, 122586.

Kumari, N., Sareen, S., Verma, M., Sharma, S., Sharma, A., Sohal, H. S., … & Mutreja, V. (2022). Zirconia-based nanomaterials: Recent developments in synthesis and applications. *Nanoscale Advances*, *4*(20), 4210–4236. https://doi.org/10.1039/D2NA00367H

Kunrath, M. F., Gupta, S., Lorusso, F., Scarano, A., & Noumbissi, S. (2021). Oral tissue interactions and cellular response to zirconia implant-prosthetic components: A critical review. *Materials*, *14*(11), 2825. https://doi.org/10.3390/ma14112825

Lin, H., Yin, C., & Mo, A. (2021). Zirconia based dental biomaterials: Structure, mechanical properties, biocompatibility, surface modification, and applications as implant. *Frontiers in Dental Medicine*, *2*, 689198. https://doi.org/10.3389/fdmed.2021.689198

Liu, Z., Liu, X., & Ramakrishna, S. (2021). Surface engineering of biomaterials in orthopedic and dental implants: Strategies to improve osteointegration, bacteriostatic and bactericidal activities. *Biotechnology journal*, *16*(7), 2000116. https://doi.org/10.1002/biot.202000116

Luo, F., Hong, G., & Wan, Q. (2021). Artificial intelligence in biomedical applications of zirconia. *Frontiers in Dental Medicine*, *2*, 689288. https://doi.org/10.3389/fdmed.2021.689288

Mayoral, Á., del Angel, P., & Ramos, M. (2019). Electron microscopy techniques to study structure/function relationships in catalytic materials. In *Advanced Catalytic Materials: Current Status and Future Progress* (pp. 97–128). https://doi.org/10.1007/978-3-030-25993-8_5

Mierzejewska, Ż. A., Rusztyn, B., Łukaszuk, K., Borys, J., Borowska, M., & Antonowicz, B. (2024). The latest advances in the use of nanoparticles in endodontics. *Applied Sciences*, *14*(17), 7912. https://doi.org/10.3390/app14177912

Mosaddad, S. A., Talebi, S., Keyhan, S. O., Fallahi, H. R., Darvishi, M., Aghili, S. S., … & Fathi, A. (2024). Dental implant considerations in patients with systemic diseases: An updated comprehensive review. *Journal of Oral Rehabilitation*, *51*(7), 1250–1302. https://doi.org/10.1111/joor.13683

Moura, C. G., Pereira, R., Buciumeanu, M., Carvalho, O., Bartolomeu, F., Nascimento, R., & Silva, F. S. (2017). Effect of laser surface texturing on primary stability and surface properties of zirconia implants. *Ceramics International*, *43*(17), 15227–15236. https://doi.org/10.1016/j.ceramint.2017.08.058

Muzzio, N., Moya, S., & Romero, G. (2021). Multifunctional scaffolds and synergistic strategies in tissue engineering and regenerative medicine. *Pharmaceutics*, *13*(6), 792. https://doi.org/10.3390/pharmaceutics13060792

Nikolova, M. P., & Apostolova, M. D. (2022). Advances in multifunctional bioactive coatings for metallic bone implants. *Materials*, *16*(1), 183. https://doi.org/10.3390/ma16010183

Pezzotti, G. (2014). Bioceramics for hip joints: The physical chemistry viewpoint. *Materials*, *7*(6), 4367–4410. https://doi.org/10.3390/ma7064367

Pina, S., Rebelo, R., Correlo, V. M., Oliveira, J. M., & Reis, R. L. (2018).Bioceramics for osteochondral tissue engineering and regeneration. In *Osteochondral Tissue Engineering: Nanotechnology, Scaffolding-Related Developments and Translation* (pp. 53–75). https://doi.org/10.1007/978-3-319-76711-6_3

Radu, R. D., & Drăgănescu, D. (2023).Present and future of ZrO_2 nanostructure as reservoir for Drug loading and release. *Coatings*, *13*(7), 1273. https://doi.org/10.3390/coatings13071273

Ramesh, B., Anandhi, R. J., Arun, V., Singla, A., Chandra, P. K., Sethi, V. A., & Abood, A. S. (2024). A review on biomaterials for neural interfaces: Enhancing brain-machine interfaces. In *E3S Web of Conferences* (Vol. 505, p. 01005). EDP Sciences.

Rathee, G., Bartwal, G., Rathee, J., Mishra, Y. K., Kaushik, A., & Solanki, P. R. (2021). Emerging multimodel zirconia nanosystems for high-performance biomedical applications. *Advanced NanoBiomed Research*, *1*(9), 2100039. https://doi.org/10.1002/anbr.202100039

Sakaguchi, R. L., & Powers, J. M. (2011). *Craig's Restorative Dental Materials-E-Book: Craig's Restorative Dental Materials-E-Book*. Elsevier Health Sciences.

Salari, M., MohseniTaromsari, S., Bagheri, R., & FaghihiSani, M. A. (2019). Improved wear, mechanical, and biological behavior of UHMWPE-HAp-zirconia hybrid nanocomposites with a prospective application in total hip joint replacement. *Journal of Materials Science*, *54*(5), 4259–4276. https://doi.org/10.1007/s10853-018-3146-y

Sartori, M., Pagani, S., Ferrari, A., Costa, V., Carina, V., Figallo, E., … & Giavaresi, G. (2017). A new bi-layered scaffold for osteochondral tissue regeneration: In vitro and in vivo preclinical investigations. *Materials Science and Engineering: C*, *70*, 101–111. https://doi.org/10.1016/j.msec.2016.08.027

Shah, A. H., & Rather, M. A. (2021). Effect of calcination temperature on the crystallite size, particle size and zeta potential of TiO_2 nanoparticles synthesized via polyol-mediated method. *Materials Today: Proceedings*, *44*, 482–488. https://doi.org/10.1016/j.matpr.2020.10.199

Sharifi, S., Javadpour, J., Rezaie, H., Jamshidi, M., & Ghamarpoor, R. (2024). Developing a deposited calcium-phosphate layer on zirconia surface by chemical grafting of L-Serine molecules. *Journal of Materials Research and Technology*, *28*, 66–77. https://doi.org/10.1016/j.jmrt.2023.11.285

Sharma, S. K., Gajević, S., Sharma, L. K., Pradhan, R., Miladinović, S., Ašonja, A., & Stojanović, B. (2024). Magnesium-titanium alloys: A promising solution for biodegradable biomedical implants. *Materials*, *17*(21), 5157. https://doi.org/10.3390/ma17215157

Shin, Y. C., Wu, B., Lei, S., Cheng, G. J., & Lawrence Yao, Y. (2020).Overview of laser applications in manufacturing and materials processing in recent years. *Journal of Manufacturing Science and Engineering*, *142*(11), 110818. https://doi.org/10.1115/1.4048397

Sivasankar, M. V., Chinta, M. L., & Rao, P. S. (2024). Zirconia based composite scaffolds and their application in bone tissue engineering. *International Journal of Biological Macromolecules*, 130558. https://doi.org/10.1016/j.ijbiomac.2024.130558

Sokola, P., Ptáček, P., Bafti, A., Panžić, I., Mandić, V., Blahut, J., & Kalina, M. (2024). Comprehensive study of stereolithography and digital light processing printing of zirconia photosensitive suspensions. *Ceramics*, *7*(4), 1616–1638. https://doi.org/10.3390/ceramics7040104

Stewart, S. A., Domínguez-Robles, J., Donnelly, R. F., & Larrañeta, E. (2018). Implantable polymeric drug delivery devices: Classification, manufacture, materials, and clinical applications. *Polymers*, *10*(12), 1379. https://doi.org/10.3390/polym10121379

Sun, T., Liu, X., Zhan, X., Ou, L., & Lai, R. (2021). Hepatic distribution and toxicity of zirconia nanoparticles in vivo and in vitro. *Process Safety and Environmental Protection*, *147*, 134–145. https://doi.org/10.1016/j.psep.2020.09.021

Tan, S., Wu, T., Zhang, D., & Zhang, Z. (2015). Cell or cell membrane-based drug delivery systems. *Theranostics*, *5*(8), 863. https://doi.org/10.7150/thno.11852

Thakur, A., Kumar, A., Kaya, S., Marzouki, R., Zhang, F., & Guo, L. (2022).Recent advancements in surface modification, characterization and functionalization for enhancing the biocompatibility and corrosion resistance of biomedical implants. *Coatings*, *12*(10), 1459. https://doi.org/10.3390/coatings12101459

Tharmaraj, M., Radhakrishnan, A., Ramani, A., & Srinivasan, N. (2025). Investigation and in vitro studies of a ZrO_2/g-C_3N_4 composite coated on 316L stainless steel for biomedical applications. *Materials and Corrosion*, *76*(2), 284–294. https://doi.org/10.1002/maco.202414525

Tian, G., Xu, L., Fang, H., Geng, F., Zhou, Y., & Wei, S. (2024). Synergistic enhancement of strength-ductility balance of TZM alloy through c-ZrO_2 combined with large deformation and annealing. *Journal of Alloys and Compounds*, *996*, 174877. https://doi.org/10.1016/j.jallcom.2024.174877

Troell, R. J. (2023). Optimizing the lip aesthetic ratio using surgically placed silastic implants for lip augmentation. *The American Journal of Cosmetic Surgery*, *40*(4), 252–262. https://doi.org/10.1177/07488068221104573

Vitti, R. P., Catelan, A., Amaral, M., & Pacheco, R. R. (2019). Zirconium in dentistry. In *Advanced Dental Biomaterials* (pp. 317–345).Woodhead Publishing. https://doi.org/10.1016/B978-0-08-102476-8.00014-1

Weng, W., Wu, W., Hou, M., Liu, T., Wang, T., & Yang, H. (2021). Review of zirconia-based biomimetic scaffolds for bone tissue engineering. *Journal of Materials Science*, *56*, 8309–8333. https://doi.org/10.1007/s10853-021-05824-2

Yin, Y., Xu, J., & Chen, M. (2024). A review on surface texturing of zirconia ceramics for dental applications. *The International Journal of Advanced Manufacturing Technology*, *130*(11), 5109–5135. https://doi.org/10.1007/s00170-024-12948-7

Zafar, M. S., Fareed, M. A., Riaz, S., Latif, M., Habib, S. R., & Khurshid, Z. (2020). Customized therapeutic surface coatings for dental implants. *Coatings*, *10*(6), 568. https://doi.org/10.3390/coatings10060568

Zakaria, A., Shukor, H., Todoh, M., & Jusoff, K. (2020a). Bio-functional coating on Ti6Al4V surface produced by using plasma electrolytic oxidation. *Metals*, *10*(9), 1124. https://doi.org/10.3390/met10091124

Zakaria, Z., Abu Hassan, S. H., Shaari, N., Yahaya, A. Z., & Boon Kar, Y. (2020b). A review on recent status and challenges of yttria stabilized zirconia modification to lowering the temperature of solid oxide fuel cells operation. *International Journal of Energy Research*, *44*(2), 631–650. https://doi.org/10.1002/er.4944

Zang, S., Xiao, C., He, M., Chen, B., Liu, B., Yi, S., … & Jin, L. (2023). RGD and rhBMP-7 immobilized on zirconia scaffold with interweaved human dental pulp stem cells for promoting bone regeneration. *Materials & Design*, *232*, 112052. https://doi.org/10.1016/j.matdes.2023.112052

Zapór, L., & Oberbek, P. (2020). Nanomaterials in the work environment. In *Emerging Chemical Risks in the Work Environment* (pp. 5–72). CRC Press.

Zhan, Y., Ehlerding, E. B., Shi, S., Graves, S. A., Goel, S., Engle, J. W., … & Cai, W. (2018). Intrinsically zirconium-89-labeled manganese oxide nanoparticles for in vivo dual-modality positron emission tomography and magnetic resonance imaging. *Journal of biomedical nanotechnology*, *14*(5), 900–909. https://doi.org/10.1166/jbn.2018.2498

Zhao, J., Santino, F., Giacomini, D., & Gentilucci, L. (2020). Integrin-targeting peptides for the design of functional cell-responsive biomaterials. *Biomedicines*, *8*(9), 307. https://doi.org/10.3390/biomedicines8090307

Ziaee, F., Zebarjad, S. M., & Javadpour, S. (2018). Compressive and flexural properties of novel polylactic acid/hydroxyapatite/yttria-stabilized zirconia hybrid nanocomposite scaffold. *International Journal of Polymeric Materials and Polymeric Biomaterials*, *67*(4), 229–238. https://doi.org/10.1080/00914037.2017.1320659

17 Influence of Thickness on the Mechanical Strength of Monolithic Zirconia Dental Crowns

Praveen Kumar Verma,
Hitesh Vasudev, and Manpreet Singh

17.1 INTRODUCTION

Monolithic zirconia has transformed contemporary dentistry by offering a highly durable and aesthetic substitute for the conventional metal-ceramic restorations. Its widespread popularity is based on its exceptional mechanical properties such as high flexural strength, fracture toughness, and wear resistance. In contrast to layered zirconia restorations that tend to develop veneer chipping, monolithic zirconia does not have this limitation because it consists of a single material. The development of high-translucency zirconia further supports the aesthetic attraction, rendering it a good option for anterior and posterior restorations. This development overcomes the aesthetic shortcomings hitherto attributed to zirconia, allowing for its wider use in contemporary prosthodontics. The thickness of monolithic zirconia crowns is one of the most important factors determining their performance. Restoration thickness determines the mechanical stability, fracture resistance, and load-carrying capacity. Practitioners should thoughtfully address thickness needs depending on the placement of the restoration and the functional stress to which it will be subjected. While thin restorations enhance translucency and leave more tooth structure intact, they are perhaps at greater risk of mechanical fracture. Conversely, increasing thickness does improve strength but is often necessitated by the need for greater tooth reduction and potential compromise in long-term vitality. The relationship between the thickness of zirconia and its mechanical response has been widely researched. Studies have shown that restorations with a thickness that is too low are prone to chipping, delamination, and bulk fractures, while those that are too thick can cause occlusal interferences and patient discomfort. Finite element analysis (FEA) and in vitro tests have proven that optimal thickness improves stress distribution, minimizing crack initiation and propagation. Figure 17.1 shows the 3D representation of monolithic and layered restorations as designed with CAD software. The digital image facilitates accurate visualization and personalized adaptation of dental restorations prior to production, providing a perfect fit and functioning [1–5].

 DOI: 10.1201/9781003519300-21

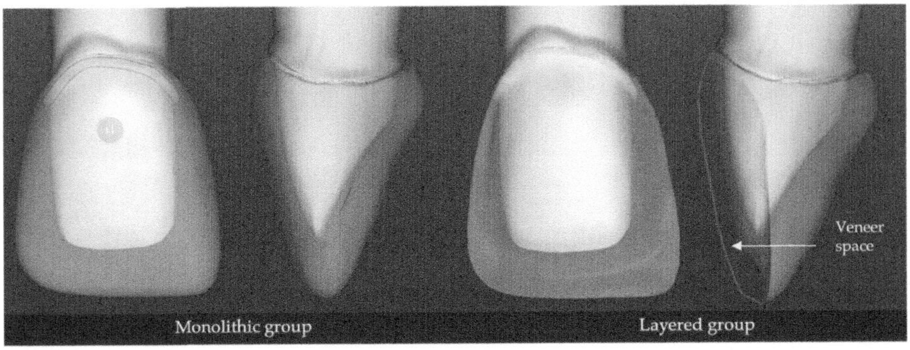

FIGURE 17.1 3D model of monolithic and layered restorations in CAD software [6]. (CC By 4.0 attribution.)

The fracture resistance of zirconia restorations depends significantly on the interaction between the material thickness and functional loads imposed. Research indicates that thickness increases directly relate to increased fracture resistance due to enhanced occlusal force dissipation. Excessive thickness, though, can bring about stress concentration points because of non-uniform force transmission and subsequent premature failure. Also, the internal adaptation of zirconia crowns to prepared tooth structure is important in overall restoration longevity. Inadequate fit can create areas of localized stress that can foster crack formation over time. Long-term longevity of monolithic zirconia restorations is not just dependent on material choice but also accurate optimization of thickness. Clinicians ought to follow established guidelines for minimal thickness according to the location-specific demands. One of the critical factors in deciding thickness is the functional load-bearing demand, especially for posterior teeth, where higher occlusal forces are involved. To achieve clinical longevity, practitioners also need to take into account factors like cementation strategies, preparation design, and occlusal adjustment. Computer-aided design and manufacturing (CAD/CAM) developments have further enabled the precision of restorations with controlled thickness. Digital processes enable the design of customized zirconia crowns that are tailored to patient-specific requirements while ensuring the optimal mechanical properties. Surface treatments like glazing and polishing can also affect the wear resistance and stress distribution, eliminating possible disadvantages of thin restorations [7–9]. Figure 17.2 illustrates the major factors that can affect color variation in monolithic zirconia ceramics. They encompass sintering temperature, dopant content, effects of aging, surface treatments, and transmission properties in the visible spectrum, all of which affect the final aesthetic and functional behavior of the restoration.

Recent research is still developing new avenues to maximize the mechanical performance of monolithic zirconia restorations. Research is underway to determine the influence of microstructural alterations, dopants, and hybrid-layering processes to increase strength and translucency without compromising on material thickness. Progress in gradient composition-based multi-layered zirconia also provides promising avenues by providing high strength in occlusal areas coupled with better

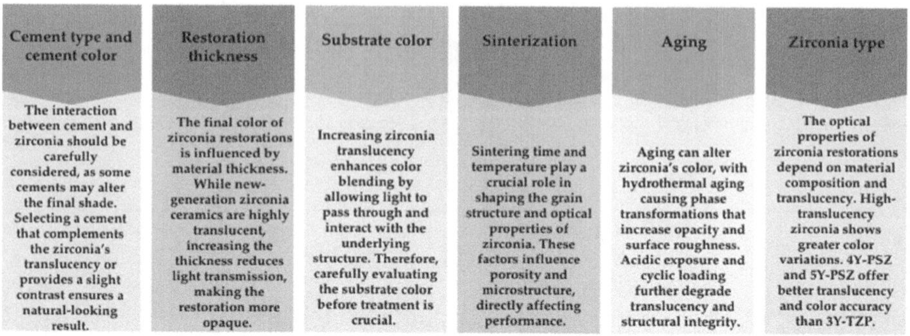

FIGURE 17.2 A brief summary of the factors affecting the color change in monolithic zirconia ceramics [10]. (CC By 4.0 attribution.)

translucency in aesthetic areas. Monolithic zirconia restoration thickness is a defining factor in restorative performance, durability, and aesthetic harmony. A carefully considered strategy to choose the proper thickness will lead to optimal mechanical properties with minimized tooth structure compromise. With future advances in the development of new zirconia compositions and forming methods, restorative dentistry will continue to see increased utilization of versatile monolithic zirconia [11].

17.2 MATERIAL PROPERTIES OF MONOLITHIC ZIRCONIA

17.2.1 Composition and Structure

Monolithic zirconia, with a high proportion of yttria-stabilized tetragonal zirconia polycrystals (Y-TZP), is well known for its high mechanical properties and adaptability in dental and biomedical uses. The stabilizing process incorporates yttrium oxide (Y_2O_3) into the zirconium dioxide (ZrO_2) to provide the maintenance of the metastable tetragonal phase at ambient temperature. The phase is fundamental to the toughness and strength of the material as it is characterized by a stress-induced transformation toughening mechanism. With an applied external stress, the tetragonal structure transforms into monoclinic, resulting in volumetric expansion within localized regions. This expansion develops compressive stresses that serve to stop crack advancement, greatly increasing the fracture toughness of the material. Zirconia's microstructure is responsible for its overall performance, as it governs mechanical properties, optical response, and long-term stability. Grain size, density, and phase distribution are the most important factors that control its functionality. Reducing grain size enhances strength by limiting defect propagation, while material density controls wear resistance and structural integrity. The structure of well-sintered zirconia with high density has low porosity, resulting in improved mechanical reliability and durability [1–3,9]. The yttria concentration has a dual function in altering the zirconia properties. While the higher content of yttria enhances the proportion of the cubic phase, enhancing translucency, it also decreases the mechanical strength. This strength-translucency trade-off is especially significant in dental applications, where both durability and aesthetics are key factors. In order to maximize these properties,

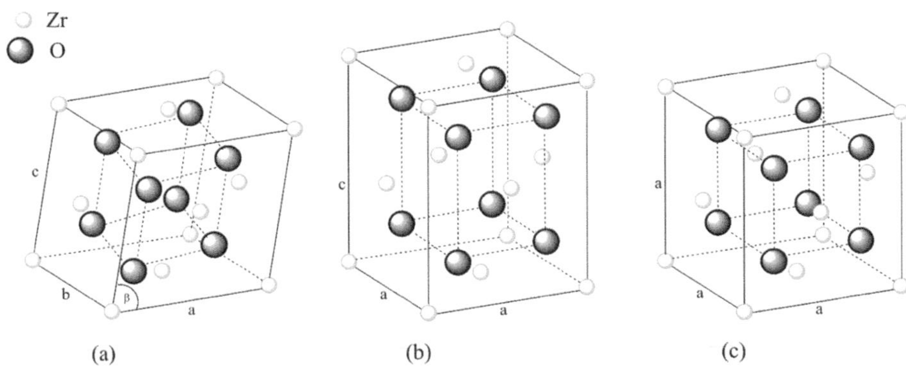

FIGURE 17.3 Schematic representation of polymorphisms of zirconia: (a) monoclinic, (b) tetragonal, and (c) cubic phase [12]. (CC By 4.0 attribution.)

sophisticated processing methods like hot isostatic pressing (HIP), two-step sintering, and microwave sintering have been established. These processes improve the microstructural properties by enhancing strength with or without improving translucency. Hot isostatic pressing (HIP) is a subsequent sintering process involving high pressure and temperature to minimize residual porosity and microstructural refinement. The process leads to better density, increased fracture toughness, and enhanced long-term performance. In the same way, two-step sintering provides greater control over grain growth through the use of an initial high-temperature step for densification followed by a low-temperature step to avoid too much grain enlargement. This technique provides a fine-grained microstructure, which is desirable both for mechanical properties and optical performance [4,5,7]. Figure 17.3 shows the schematic illustration of the three polymorphic phases of zirconia: (a) monoclinic, (b) tetragonal, and (c) cubic. The phase transformations are very important for the mechanical properties, stability, and performance of zirconia-based dental restorations.

Aging behavior is also a key consideration in determining the durability of monolithic zirconia. Low-temperature degradation (LTD) phenomenon, also called hydrothermal aging, is related to slow metastable tetragonal phase transformation to the monoclinic phase in humid environments. It causes microcracking, surface roughness, and mechanical degradation. In order to prevent LTD, scientists have studied dopant modifications, improved sintering methods, and surface treatments like alumina coating and laser surface modification. These methods increase the tetragonal phase's stability, thus enhancing the long-term success of zirconia restorations. Besides its optical and mechanical benefits, monolithic zirconia has excellent biocompatibility and is a candidate of choice for dental and orthopedic implants. The fact that there is no veneering layer, unlike in conventional zirconia restorations, removes the chipping risk, thus adding to its clinical reliability. Current research continues to improve the performance of zirconia with nanostructured adjustments, new combinations of dopants, and hybrid fabrication methods. Through the improvement of its microstructure and solutions to issues of aging and translucency trade-offs, monolithic zirconia remains an advancing material in biomedical applications [8].

17.2.2 Mechanical Properties

The superior mechanical properties of monolithic zirconia render it the most sought-after material for dental restorations. Its great flexural strength, fracture toughness, and hardness directly increase its life and durability in applications in the clinical setting. Unlike traditional ceramic-based dental restorative materials, monolithic zirconia does not need veneering, thus removing the possibility of chipping or delamination that is usually experienced in bilayered zirconia restorations. This strength and integrity make it especially well-suited for posterior restorations, where masticatory forces are much greater. Another main reason for the extensive use of monolithic zirconia is that its flexural strength is exceptionally high, typically in excess of 1000 MPa. This high strength enables the production of thin restorations without compromising durability, allowing it to be used in minimally invasive dental procedures. Moreover, its higher fracture toughness, generally between 5 and 10 $MPa\sqrt{m}$, offers greater resistance to crack growth, guaranteeing long-term stability under cyclic loading conditions. Such properties are especially advantageous for bruxing or parafunctional habit patients, as monolithic zirconia is more resistant to occlusal forces compared to conventional porcelain-based restorations [9,11].

Another important benefit of monolithic zirconia is its high hardness, which makes it wear-resistant and durable. Its potential to induce wear on opposing dentition, particularly when the occlusal surfaces are highly polished or poorly adjusted, has been a concern. Research has indicated that careful surface finishing, such as polishing instead of glazing, can reduce the abrasive nature of zirconia, making it safer for opposing natural teeth. In addition to its mechanical characteristics, monolithic zirconia also has superior biocompatibility, which makes it a suitable material for intraoral use. It is non-adherent to bacteria, minimizing the risk of plaque formation and secondary caries development around restorations. Additionally, its low thermal conductivity makes it comfortable for patients by preventing sensitivity to temperature changes.

More recent developments in translucent zirconia formulations have even widened its use clinically, correcting the aesthetic constraints normally linked to traditional opaque zirconia. With improved microstructural composition and lowering of alumina content, the optical quality of zirconia has been increased, so that it has emerged as a viable substitute for lithium disilicate and other glass ceramics in anterior restorations. In total, monolithic zirconia's high strength, fracture toughness, wear resistance, and biocompatibility make it among the most trustworthy materials for dental restorations. With ongoing material advancements and better manufacturing methods, it is likely to continue being a foundation for restorative and prosthetic dentistry, delivering both functional and aesthetic advantages to patients globally [1–3,9].

17.2.2.1 Flexural Strength

Monolithic zirconia is best known for its high flexural strength, normally between 900 and 1200 MPa, which is among the highest for dental ceramics. This characteristic is of essential importance in determining the longevity and durability of dental restorations, especially in high-load-bearing structures like posterior crowns and bridges.

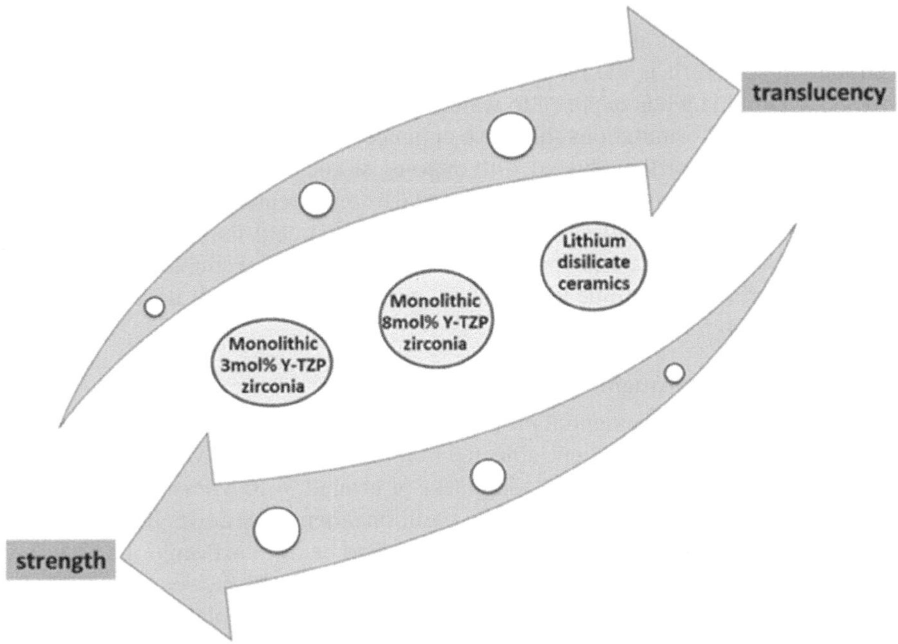

FIGURE 17.4 Correlation of translucency and strength of contemporary zirconia and lithium disilicate ceramics [13]. (CC By 4.0 attribution.)

In comparison to lithium disilicate, whose flexural strength is around 360–400 MPa, zirconia shows greater resistance to mechanical stress, lowering the chances of catastrophic failure when subjected to occlusal forces. The flexural strength of zirconia is affected by several important factors, such as grain size, sintering temperature, and phase stability. Smaller grain sizes also lead to better mechanical properties through the limitation of crack paths, while optimum sintering temperatures (in the range of 1450°C–1600°C) provide an adequately densified microstructure with low porosity. Phase stability is also essential; zirconia exists predominantly in the tetragonal form at room temperature, stabilized by the presence of yttria (Y_2O_3). The phase-controlled transformation maintains zirconia strength and toughness upon mechanical loading. Figure 17.4 illustrates the relationship between translucency and strength in modern zirconia and lithium disilicate ceramics. Although zirconia provides greater strength with reduced translucency, lithium disilicate offers greater translucency but reduced mechanical strength, which affects material choice for aesthetic and functional restorations [4,5].

The surface treatments play a major role in the flexural strength of the zirconia restorations. Polishing and glazing are used largely to improve surface texture, lower surface defects, and increase mechanical performance. The polished surface of zirconia contains fewer microcracks and is also less prone to crack initiation and thus maintains material strength over long periods. It is the opposite when the zirconia is left unpolished or poorly finished, as that can form microcracks rendering its structural support weak. The other significant factor influencing the flexural strength of zirconia is degradation due to aging, also known as low-temperature degradation (LTD). It is the result of phase transformation and roughening of the surface of zirconia when exposed to water over some time. Even with the advancements in zirconia formulations that have enhanced resistance to LTD, its long-term clinical influence is still evaluated with ongoing studies. The union of high flexural strength, phase stability, and optimized surface finish renders monolithic zirconia suitable for dental restorations requiring high mechanical durability. With ongoing advancements in material processing and fabrication technologies, zirconia's flexural strength will continue to be a characteristic that marks its universal clinical acceptance [7,14].

17.2.2.2 Fracture Toughness

Fracture toughness is an essential property that establishes the capacity of a material to resist crack propagation when subjected to mechanical stress. Monolithic zirconia possesses an outstanding fracture toughness of around 5–10 $MPa{\cdot}m^{1/2}$, rendering it considerably more resistant compared to traditional dental ceramics like feldspathic porcelain and lithium disilicate, which range around below 3 $MPa{\cdot}m^{1/2}$. Its high fracture toughness is among the main reasons zirconia finds extensive applications for high-stress restorations like posterior crowns, bridges, and implant-supported prostheses [15]. One of the primary mechanisms behind zirconia's high fracture toughness is stress-induced phase transformation toughening. Zirconia differs from most ceramics, in that it transforms from the tetragonal to monoclinic phase upon application of tensile stress. Upon transformation, the process is accompanied by a localized volume expansion (approximately 3%–5%). This volume expansion acts against crack propagation by providing compressive stresses around the crack tip. This self-healing process makes the material more resistant to mechanical loading and inhibits early failure. Some variables affect the fracture toughness of zirconia, such as yttria composition, grain size, and thermally processed variables. Yttria-stabilized tetragonal polycrystals of zirconia (Y-TZP) are favored for dental applications because they contain an optimal mixture of strength and toughness. Increased yttria content can promote excessive cubic phase development, and this diminishes transformation toughening and decreases general toughness. Another consideration in maintaining fracture toughness is aging-related degradation, where exposure to moisture over time can lead to progressive phase transformation, surface roughening, and crack initiation. Research has shown that surface modifications, such as polishing rather than glazing, can enhance resistance to aging while preserving mechanical integrity. Further, restoration thickness and connector design also have a very important role to play in averting mechanical failure of zirconia restorations. Designing the proper restoration, as well as the appropriate connector sizes in multi-unit prostheses, can improve stress distribution and the overall long-term survival of zirconia restorations. By utilizing its inherent fracture toughness, monolithic zirconia continues to be a first choice for applications where high durability and long-term reliability are necessary. With continued improvement in microstructural refinement and aging-resistant compositions, zirconia is likely to continue to be a leading material for dental restorations [16–18].

17.2.2.3 Hardness and Wear Resistance

Monolithic zirconia has a Vickers hardness of around 12–14 GPa, rendering it among the hardest materials to be found in dentistry. Its high hardness is responsible for its good wear resistance, guaranteeing that restorations of zirconia retain their surface quality and structure over a very long time. In comparison to other dental ceramics like lithium disilicate (5–6 GPa) and feldspathic porcelain (4–6 GPa), zirconia is more resistant to mechanical abrasion and wear, and is therefore best suited for high-load-bearing restorations like molars and bridges. Although zirconia has excellent durability, its hardness has been a source of concern regarding its possible influence on opposing dentition. Overly hard and abrasive zirconia surfaces may result in excessive wear of natural enamel, particularly in bruxing or high occlusal force patients. Yet, surface finishing procedures, such as polishing rather than glazing, have been found to decrease enamel wear dramatically. Research shows that the well-polished zirconia surface is not as abrasive against opposing teeth compared to glazed zirconia, since glazing layers may disintegrate eventually, leaving the rougher underlying surface exposed [19].

Microstructural factors, including grain size and phase content, affect the wear resistance of zirconia. A finely controlled grain size distribution guarantees a dense and homogeneous microstructure that supports mechanical stability. The presence of the tetragonal phase also makes zirconia more resistant to wear-induced degradation under occlusal forces. One benefit of zirconia's high hardness is its resistance to surface degradation and discoloration. In contrast to lower-hardness ceramics, which tend to accumulate microfractures or surface flaws with time, zirconia retains its smoothness and aesthetic qualities even after extended clinical service. This renders it highly appropriate for anterior restorations where durability and aesthetics are of equal concern. Clinically, an ideal compromise between hardness and biocompatibility is crucial for long-term success. Appropriate occlusal adjustments and the incorporation of sophisticated polishing techniques can reduce the negative effects on opposing dentition while optimizing the durability of zirconia restorations. With continued research aimed at enhancing translucency and lowering abrasiveness, zirconia remains one of the most dependable materials in restorative and prosthetic dentistry [20,21].

17.3 INFLUENCE OF THICKNESS ON MECHANICAL PERFORMANCE

17.3.1 Effect on Flexural Strength

Flexural strength is an important mechanical characteristic of zirconia dental restorations, as it reflects the material's resistance to tensile and compressive forces without fracturing. It has been established that the increase in thickness of zirconia increases flexural strength by decreasing the chances of critical flaw propagation within the material structure. This enhancement is largely due to the tetragonal-to-monoclinic phase transformation toughening mechanism inherent in zirconia, which is responsible for resisting crack propagation. The minimum thickness for anterior zirconia crowns has been recommended to be typically 0.5 mm. This thickness is still sufficient to provide good mechanical properties of the material while providing enough translucency to achieve proper aesthetics. In areas where there are high stresses like posterior restorations, it is typically recommended to have a minimum thickness of 1.0 mm to support occlusal forces without early failure. Research has shown that restorations with thinner zirconia thicknesses (<0.5 mm) are more susceptible to flexural failure caused by stress concentration sites that enable crack initiation and propagation [22].

In addition, experimental studies employing three-point and four-point bending tests have shown that flexural strength is progressively higher with increasing material thickness. Once the thickness of zirconia is more than 1.0 mm, its mechanical deformation resistance is significantly enhanced. But too thick restorations (>1.5 mm) can cause increased stiffness, which will reduce the material's capacity to dissipate occlusal forces effectively and result in stress transmission to the tooth structure [23].

17.3.2 Fracture Toughness and Failure Resistance

Fracture toughness is a principal factor in zirconia's resistance to crack propagation and catastrophic failure under mechanical loading. It is affected by restoration thickness, as thickness variations affect the material's ability to dissipate stresses. Thin zirconia restorations (<0.5 mm) are more prone to crack initiation because of localized stress concentrations. Thin restorations do not have enough bulk to resist tensile forces, and thus there is a greater chance of catastrophic failure under functional loads. The poor resistance to crack propagation in thinner zirconia layers is a significant issue, especially in posterior restorations where high masticatory forces are present [24].

Moderate-thickness zirconia restorations (0.5–1.0 mm) represent a compromise between strength and translucency. The material is sufficient in terms of load resistance at this thickness, while also having acceptable aesthetic qualities, especially for anterior crowns. It has been reported that moderate-thickness zirconia can resist physiological occlusal forces without major risk of fracture, as long as the tooth structure beneath the restoration is properly prepared to accommodate the restoration. For thick restorations of zirconia (>1.0 mm), greater fracture toughness is noted as a result of greater resistance to stress-induced failure. The increased bulk of the material absorbs and redistributes mechanical loading, reducing the possibility of crack propagation. There is a trade-off, however, in terms of aesthetics and increased tooth reduction required to place the thicker material. Too thick zirconia restorations have the potential to produce an opaque appearance and are therefore less ideal for anterior indications where translucency is important. Generally, fracture toughness and resistance against failure increase with thicker zirconia, yet there has to be a balance. The thickness needs to be between 0.5 and 1.0 mm for anterior as well as posterior restorations to provide adequate mechanical performance while maintaining aesthetics and tooth structure [25].

17.3.3 Load-Bearing Capacity and Occlusal Stress

The resistance to load of zirconia restorations is directly related to their thickness, with more thick restorations showing greater masticatory force resistance. This is especially critical for posterior crowns, which are subjected to high occlusal loads when chewing. FEA research has helped elucidate the effect of zirconia thickness on load transfer and resistance to stress. Thicker zirconia crowns (>1.0 mm) have better resistance to occlusal forces without fracturing. The added bulk provides better stress distribution throughout the restoration, minimizing the risk of stress concentrations that may cause failure. FEA models indicate that an optimal thickness of 1.0–1.5 mm is best for posterior restorations since this thickness provides sufficient strength without excessive transmission of stress to the underlying tooth structure and cement layer. Thin zirconia restorations (<0.5 mm) are very much at increased risk of failure because they possess a severely constrained load-bearing capability. Subject to heavy masticatory loads, these restorations will experience premature crack generation, leading to material deterioration and ultimate failure. Stress concentration around the point contacts can further promote mechanical deficiencies and render thin restorations not appropriate for applications involving high stress levels. Moderate-thickness zirconia restorations (0.5–1.0 mm) represent an acceptable compromise between strength and aesthetics. They provide adequate load resistance for anterior and posterior crowns with desirable translucency. Nevertheless, to maximize mechanical performance, attention to preparation technique and cementation protocols is necessary to maximize the restoration's lifespan [26,27].

17.4 CLINICAL IMPLICATIONS AND RECOMMENDATIONS

17.4.1 Optimal Thickness Selection

Choosing the ideal thickness for zirconia-based restorations is essential to strike a balance between aesthetics, mechanical strength, and longevity. The thickness needed will depend on the site of the restoration and the functional loads placed on it. Anterior crowns need to find a balance between strength and translucency to look natural. A thickness of 0.5–0.7 mm is usually preferred since it maximizes the translucency required for aesthetic acceptability without impeding mechanical performance. Current high-translucency zirconia materials support such lower thicknesses without excessively affecting strength. Additionally, a thinner restoration conserves more tooth structure, minimizing the danger of excessive preparation-related complications including dentin exposure and postoperative sensitivity [28].

Posterior teeth are subjected to much greater masticatory forces than anterior teeth, requiring more material thickness for strength. The 1.0–1.5 mm thickness is recommended for posterior crowns to provide resistance to occlusal loads and avoid fractures. The additional thickness offers greater mechanical integrity with room for required adjustments during occlusal equilibration. Monolithic zirconia crowns with a smooth, polished surface also reduce excessive wear on opposing teeth, which adds to long-term clinical success [29]. Bridges and multi-unit restorations need even more thickness to support occlusal loads and avoid structural failure. At least 1.5 mm of thickness is needed to maintain mechanical stability and longevity. The connectors used between pontics and abutments should also be properly designed to resist masticatory forces. Maintaining optimal thickness in these restorations reduces the risk of catastrophic failure and aids in their long-term clinical success [30].

17.4.2 Preparation Guidelines

Successful tooth preparation is necessary to maximize the fit, function, and longevity of zirconia restorations. The design of the tooth preparation must balance material thickness requirements against the removal of as much natural tooth structure as possible. Minimally invasive preparation is preferred to preserve the integrity of the underlying pulp and dentin. Over-preparation may result in greater dentin exposure, sensitivity, and decreased bond strength. By preserving only the required amount of reduction, clinicians can maximize retention and support for zirconia restorations without compromising tooth vitality [31].

A chamfer or rounded shoulder margin is strongly advised for zirconia restorations. These margin types have better stress distribution, minimizing marginal chipping and fractures. A chamfer margin, in fact, presents the best compromise between structural support and ease of fabrication, providing a well-sealed and stable restoration. Sufficient occlusal and axial reduction is essential in order to create space for the zirconia material without losing too much tooth structure. An even reduction pattern ensures that the restoration has even thickness, which improves its durability and aesthetic qualities. Preventing sharp internal line angles in preparation reduces points of stress concentration, lowering the risk of initiation of fracture in the zirconia structure. Adhesive or conventional cementation is utilized based on the type of zirconia restoration. Conventional cementation with resin-modified glass ionomer is acceptable for monolithic zirconia, whereas adhesive bonding is preferred for high-translucency zirconia to ensure maximum retention and marginal seal [32].

17.4.3 Long-Term Performance and Wear Considerations

Although zirconia restorations are very strong and durable, long-term function is determined by wear dynamics, surface quality, and interactions with opposing dentition. Zirconia restorations are highly wear-resistant and, thus, are the restorations of choice in high-load-bearing regions. Nonetheless, excessive material thickness may cause excessive wear to opposing enamel, especially when the occlusal surface is not polished or has a rough finish. To counteract this danger, it is essential to ensure a well-polished surface and employ antagonist-friendly zirconia materials with optimally structured microstructures [33].

A well-polished zirconia surface minimizes frictional forces and decreases wear on the opposing teeth. Research suggests that polished zirconia is more conservative compared to glazed zirconia because glaze tends to wear away after a period, revealing a coarser underlying surface. Thus, a careful polishing regimen should be employed to increase longevity and ensure occlusal harmony. With time, zirconia restorations can degrade low temperatures, which creates microstructural modifications that impact the mechanical properties. Moisture contact, temperature, and occlusal forces have the potential to cause phase transformation from tetragonal to monoclinic zirconia and compromise aesthetics and strength. New yttria-stabilized zirconia compositions demonstrate enhanced resistance against LTD, ensuring that they can be a safer choice for extended use. Impeccable oral maintenance procedures, including regular check-ups by the dentist and elimination of parafunctional behaviors like bruxism, enhance the longevity of zirconia restorations. Bruxism patients can be helped with occlusal splints to minimize excessive loading on their restorations. Regular polishing at recall sessions can also be used to keep the surfaces of zirconia smooth and avoid early wear of opposing teeth [34,35].

17.5 CONCLUSION

The thickness of monolithic zirconia crowns plays a significant role in determining their mechanical properties such as strength, fracture resistance, and durability. Thicker zirconia restorations have greater resistance to occlusal forces and chipping and are therefore best suited for posterior teeth that bear heavy masticatory loads. Nevertheless, too much thickness can affect the aesthetic value by minimizing translucency and can also require excessive tooth preparation, risking the integrity of the underlying dental structure. Hence, a perfect equilibrium between strength, thickness, and aesthetics is essential for success in the clinical field. Today's clinical practice guidelines suggest variable thickness depending upon functional needs and aesthetic needs. Anterior monolithic zirconia crowns may be as thin as 0.5 mm, utilizing sophisticated high-translucency zirconia to achieve the appearance of natural teeth. For posterior restorations, where mechanical strength is the priority, a thickness of 1.0–1.5 mm is recommended to resist occlusal forces without sacrificing translucency. Zirconia-based bridges, which are subjected to higher stress concentrations, usually need at least 1.5 mm thickness for structural stability.

Continuous investigation and material innovations are sought to improve the optical and mechanical qualities of zirconia for making ultra-thin restorations with stronger quality. Innovative compositions with added dopants in the form of alumina, yttria, and ceria are in pursuit to augment translucency-strength with the requirement for insufficient thickness. Also, processing improvements like hot isostatic pressing and graded microstructures are exciting prospects for developing greater durability and wear resistance without sacrificing material volume. Future studies will need to examine bioinspired zirconia composites mimicking the natural enamel's hierarchical structure in optimizing both aesthetic and mechanical toughness. The evolution of digital workflows and precision manufacturing techniques, including CAD/CAM milling and 3D printing, will further refine zirconia restoration designs, allowing for minimally invasive yet robust prosthetic solutions.

REFERENCES

1. Zhang Y, Lawn BR. Novel zirconia materials in dentistry. *Journal of Dental Research.* 2018;97(2):140–7.
2. Denry I, Kelly JR. State of the art of zirconia for dental applications. *Dental Materials.* 2008;24(3):299–307.
3. Sulaiman TA. Materials in digital dentistry—A review. *Journal of Esthetic and Restorative Dentistry.* 2020;32(2):171–81.
4. Miyazaki T, Nakamura T, Matsumura H, Ban S, Kobayashi T. Current status of zirconia restoration. *Journal of Prosthodontic Research.* 2013;57(4):236–61.
5. Al-Amleh B, Lyons K, Swain MV. Clinical trials in zirconia: A systematic review. *Journal of Oral Rehabilitation.* 2010;37(8):641–52.

6. Gasparik C, Manziuc MM, Burde AV, Ruiz-López J, Buduru S, Dudea D. Masking ability of monolithic and layered zirconia crowns on discolored substrates. *Materials*. 2022 Mar 17;15(6):2233.
7. Zhang F, Van Meerbeek B, Vleugels J. Importance of tetragonal phase in high-translucent zirconia for dental restorations. *Dental Materials*. 2020;36(4):491–500.
8. Sulaiman TA, Abdulmajeed AA, Donovan TE, Vallittu PK, Närhi TO, Lassila LV. The effect of thickness and background color on the optical properties of monolithic zirconia. *Journal of Dentistry*. 2017;58:67–72.
9. Alghazzawi TF. Advancements in CAD/CAM technology: Options for practical implementation. *Journal of Prosthodontic Research*. 2016;60(2):72–84.
10. BiniciAygün E, KaynakÖztürk E, Tülü AB, TurhanBal B, KarakocaNemli S, BankoğluGüngör M. Factors affecting the color change of monolithic zirconia ceramics: A narrative review. *Journal of Functional Biomaterials*. 2025 Feb 11;16(2):58.
11. Zhang Y, Lee JJW, Srikanth R, Lawn BR. Edge chipping and flexural resistance of monolithic ceramics. *Dental Materials*. 2013;29(12):1201–8.
12. Sorrentino R, Navarra CO, Di Lenarda R, Breschi L, Zarone F, Cadenaro M, Spagnuolo G. Effects of finish line design and fatigue cyclic loading on phase transformation of zirconia dental ceramics: A qualitative micro-Raman spectroscopic analysis. *Materials*. 2019 Mar 14;12(6):863.
13. Kontonasaki E, Rigos AE, Ilia C, Istantsos T. Monolithic zirconia: An update to current knowledge. Optical properties, wear, and clinical performance. *Dentistry Journal*. 2019 Sep 2;7(3):90.
14. Sulaiman TA, Abdulmajeed AA, Shahramian K, Lassila L. Effect of different treatments on the flexural strength of fully versus partially stabilized monolithic zirconia. *The Journal of Prosthetic Dentistry*. 2017 Aug 1;118(2):216–20.
15. Schatz C, Strickstrock M, Roos M, Edelhoff D, Eichberger M, Zylla IM, Stawarczyk B. Influence of specimen preparation and test methods on the flexural strength results of monolithic zirconia materials. *Materials*. 2016 Mar 9;9(3):180.
16. Candido LM, Miotto LN, Fais LM, Cesar PF, Pinelli LA. Mechanical and surface properties of monolithic zirconia. *Operative Dentistry*. 2018 May 1;43(3):E119–28.
17. Carrabba M, Keeling AJ, Aziz A, Vichi A, Fonzar RF, Wood D, Ferrari M. Translucent zirconia in the ceramic scenario for monolithic restorations: A flexural strength and translucency comparison test. *Journal of Dentistry*. 2017 May 1;60:70–6.
18. Asli HN, Rahimabadi S, Falahchai M. Flexural strength of monolithic zirconia after different surface treatments. *Space*. 2019;3:8–11.
19. Malkondu Ö, Tinastepe N, Akan E, Kazazoğlu E. An overview of monolithic zirconia in dentistry. *Biotechnology & Biotechnological Equipment*. 2016 Jul 3;30(4):644–52.
20. Kontonasaki E, Giasimakopoulos P, Rigos AE. Strength and aging resistance of monolithic zirconia: An update to current knowledge. *Japanese Dental Science Review*. 2020 Nov 1;56(1):1–23.
21. Basso GR, Moraes RR, Borba M, Griggs JA, Della Bona A. Flexural strength and reliability of monolithic and trilayer ceramic structures obtained by the CAD-on technique. *Dental Materials*. 2015 Dec 1;31(12):1453–9.
22. Juntavee N, Attashu S. Effect of different sintering process on flexural strength of translucency monolithic zirconia. *Journal of Clinical and Experimental Dentistry*. 2018 Aug 1;10(8):e821.
23. Özkurt-Kayahan Z. Monolithic zirconia: A review of the literature. *Biomedical Research*. 2016 Jan 1;27(4):1427–36.
24. Muñoz EM, Longhini D, Antonio SG, Adabo GL. The effects of mechanical and hydrothermal aging on microstructure and biaxial flexural strength of an anterior and a posterior monolithic zirconia. *Journal of Dentistry*. 2017 Aug 1;63:94–102.

25. Zandinejad A, Das O, Barmak AB, Kuttolamadom M, Revilla-León M. The flexural strength and flexural modulus of stereolithography additively manufactured zirconia with different porosities. *Journal of Prosthodontics*. 2022 Jun;31(5):434–40.
26. Cardoso KV, Adabo GL, Mariscal-Muñoz E, Antonio SG, ArioliFilho JN. Effect of sintering temperature on microstructure, flexural strength, and optical properties of a fully stabilized monolithic zirconia. *The Journal of Prosthetic Dentistry*. 2020 Nov 1;124(5):594–8.
27. Nossair SA, Salah T, Ebeid KK. Biaxial flexural strength of different types of monolithic zirconia. *Brazilian Dental Science*. 2019 Jan 31;22(1):118–23.
28. Karbasian N, Fathi A, Givehchian P, Nosouhian S, Jamshidian M, Almassi F, Fazeli A. Comparative assessment of flexural strength of monolithic zirconia with different thicknesses and two sintering techniques. *Clinical and Experimental Dental Research*. 2024 Jun;10(3):e856.
29. Weigl P, Sander A, Wu Y, Felber R, Lauer HC, Rosentritt M. In-vitro performance and fracture strength of thin monolithic zirconia crowns. *The Journal of Advanced Prosthodontics*. 2018 Apr 1;10(2):79–84.
30. Alghazzawi TF, Janowski GM, Eberhardt AW. An experimental study of flexural strength and hardness of zirconia and their relation to crown failure loads. *The Journal of Prosthetic Dentistry*. 2024 Feb 1;131(2):320–8.
31. Fischer J, Stawarczyk B, Hämmerle CH. Flexural strength of veneering ceramics for zirconia. *Journal of Dentistry*. 2008 May 1;36(5):316–21.
32. Strasser T, Wertz M, Koenig A, Koetzsch T, Rosentritt M. Microstructure, composition, and flexural strength of different layers within zirconia materials with strength gradient. *Dental Materials*. 2023 May 1;39(5):463–8.
33. Zadeh PN, Lümkemann N, Sener B, Eichberger M, Stawarczyk B. Flexural strength, fracture toughness, and translucency of cubic/tetragonal zirconia materials. *The Journal of Prosthetic Dentistry*. 2018 Dec 1;120(6):948–54.
34. Church TD, Jessup JP, Guillory VL, Vandewalle KS. Translucency and strength of high-translucency monolithic zirconium oxide materials. *General Dentistry*. 2017 Jan 1;65(1):48–52.
35. Öztürk C, Can G. Effect of sintering parameters on the mechanical properties of monolithic zirconia. *Journal of Dental Research, Dental Clinics, Dental Prospects*. 2019;13(4):247.

18 Performance of Zirconia Coatings in Dental Applications

Praveen Kumar Verma,
Hitesh Vasudev, and Bhupinder Singh

18.1 INTRODUCTION

Zirconia coatings have been a notable breakthrough in dental applications, providing the best balance between mechanical toughness, biocompatibility, and aesthetic appearance. The durability and longevity of dental materials have improved significantly over the past few decades, thus enhancing oral health. Zirconia (ZrO_2) has attracted great attention because of its excellent properties relative to traditional ceramics and metal-based films. Zirconia is a polymorphic ceramic that has three crystalline phases: monoclinic, tetragonal, and cubic. The transformation toughening potential of tetragonal zirconia polycrystal (TZP) offers superior fracture resistance, thus making it the best material to be used in dental applications. Zirconia coatings are also renowned for having superior wear resistance, low bacterial adhesion, and good corrosion resistance, which are essential in the aggressive oral environment. The application of zirconia-based coatings in dentistry is necessitated by the requirement of enhanced performance of dental implants, crowns, bridges, and other prosthetic devices. Traditional metal-based dental restorations are usually plagued by aesthetic restrictions and biocompatibility issues, while zirconia provides a metal-free solution with a natural tooth-like appearance. In addition, zirconia coatings assist in improving the mechanical stability and durability of dental restorations by minimizing surface degradation and wear [1–4]. Figure 18.1 shows the surface modifications and functional coating of dental implants.

A range of deposition processes are used in the production of zirconia coatings, including physical vapour deposition (PVD), plasma spraying, sol-gel processing, and laser cladding. These deposition methods enable controlled coating thickness, surface topology, and strength of adhesion, all of which are important in assessing the overall function of coatings in clinical procedures. The selection of deposition technique has a major impact on the microstructure and properties of coatings and hence their functionality in dental restorations. In addition to their mechanical benefits, zirconia coatings have been widely researched for their biological performance. Biocompatibility and osseointegration are critical for the success of dental implants and prosthetic restorations. The results of research have proved that zirconia surfaces encourage positive interaction with oral tissues, with increased cell adhesion

DOI: 10.1201/9781003519300-22

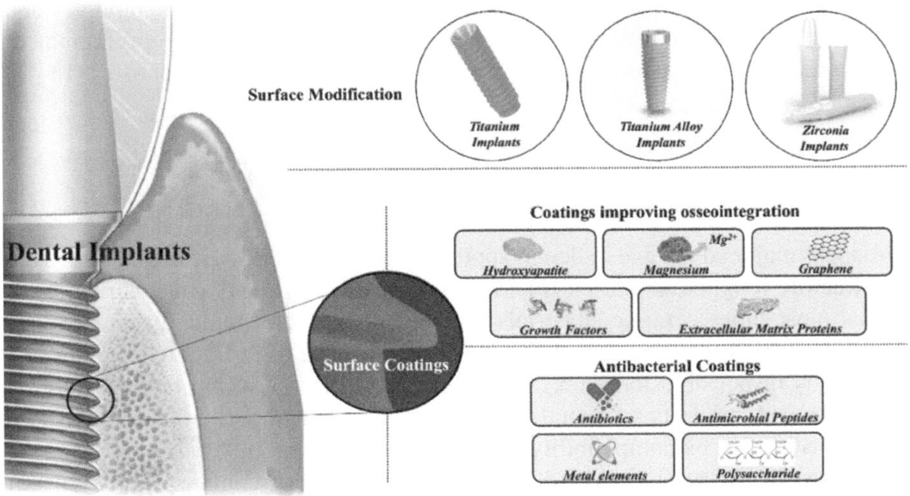

FIGURE 18.1 Schematic illustration of surface modifications and functional coating of dental implants [5]. (CC BY 4.0 license.)

and proliferation as well as reduced bacterial adhesion and biofilm maturation. These characteristics are responsible for enhanced healing and long-term survival of zirconia-coated implants [6–8].

Though they have numerous benefits, there are also certain drawbacks to zirconia coatings. The first of these concerns is the likelihood of coating failure or delamination due to harsh masticatory forces. Providing good adhesion between the substrate and zirconia coating is an active area of research. Second, processing limitations in the form of thermal degradation, phase transformation, and costliness of fabrication need to be overcome for effective clinical applications. With advancing research and technology, the prospects of zirconia coatings in dentistry are encouraging. New ideas such as nanostructured zirconia coatings, hybrid bioactive coatings, and intelligent coatings with antibacterial properties are being researched for further improvement in the performance of coatings. New surface treatment methods such as laser texturing and ion implantation are also under investigation to make the mechanical and biological properties of zirconia coatings more effective. Technologies such as nanostructured zirconia coatings, hybrid zirconia bioactive coatings, and sophisticated surface modification processes, including laser texturing and ion implantation, have the potential to improve the performance and biological compatibility of zirconia coatings. These technologies will help improve the mechanical properties, adhesion strength, and wear resistance of zirconia coatings, while decreasing the chance of delamination. In addition, the incorporation of bioactive substances and drug-releasing properties into zirconia coatings is also expected to enhance osseointegration and minimize the risk of implant infection [9–12].

In summary, zirconia coatings are a revolutionary advancement in dental materials that offer greater durability, aesthetics, and biocompatibility. Their capacity to resist the extreme conditions of the oral environment while providing superior functional and biological performance makes them a worthwhile option for contemporary dental restorations. Nevertheless, research and development is needed to overcome current limitations and realize the full potential of zirconia coatings in dentistry [13–15].

18.2 ZIRCONIA AS A DENTAL COATING MATERIAL

18.2.1 Properties of Zirconia

Zirconia's major advantage in dental applications is its excellent mechanical strength and fracture toughness. With a fracture toughness of 8–10 MPa·m$^{1/2}$, zirconia surpasses most other ceramic materials that have been conventionally employed in dentistry, including alumina and lithium disilicate. The high fracture toughness of zirconia is responsible for its resistance to crack growth, which helps ensure the durability and longevity of dental restorations. Another important characteristic of zirconia is its high biocompatibility. In contrast to some metal-based restorations that can cause allergic reactions or adverse tissue responses, zirconia is highly biocompatible with the oral environment. It does not elicit inflammatory reactions or induce cytotoxic effects and is therefore safe for long-term use in patients with sensitivity to metal-based dental materials [16,17].

Zirconia has low bacterial adhesion, which is critical in maintaining oral health. Research has proved that zirconia surfaces have lower bacterial biofilm accumulation than conventional metal- or polymer-based restorations, which decreases the incidence of peri-implantitis and secondary caries. Reduced bacterial colonization is a factor that overall extends the lifespan of zirconia-based dental applications by reducing infection risks and improving oral cleanliness. Aesthetically, zirconia has good translucency, which closely resembles a natural tooth in appearance. In contrast to metal restorations that can lead to an unnatural metallic colour or gum discoloration, zirconia restorations are a more aesthetic option. The development of processing techniques has further improved the translucency of zirconia, and it is now a great option for anterior restorations where aesthetics is the primary concern. Another important advantage of zirconia is its chemical stability within the oral environment. It is highly resistant to acid and alkaline conditions found within the mouth and will not degrade or discolour over time. This chemical inertness is partly responsible for the long-term stability and dependability of zirconia restorations, even in patients who have fluctuating oral pH levels due to food habits or disease states [18–21].

18.2.2 Types of Zirconia Used in Dentistry

Zirconia has become a central material in dentistry because of its superior mechanical strength, biocompatibility, and aesthetic appeal. Of the multiple types of zirconia that have been used, TZP, cubic zirconia (c-ZrO_2), and zirconia composites have been well received. Each type possesses characteristics that determine its applications in dentistry [22].

TZP is the most common form utilized in dentistry, largely because of its transformation toughening capacity. This is a significant benefit in dental restorations, as it provides enhanced fracture resistance. Under stress, TZP experiences a phase transformation from the metastable tetragonal to the monoclinic phase, coupled with a local volume expansion that creates compressive stresses, preventing crack growth. This characteristic makes TZP suitable for applications that involve high load-bearing capacity, such as crowns, bridges, and implant abutments. Furthermore, TZP has a good strength–aesthetic balance, making it possible to fabricate restorations closely resembling natural dentition. Nevertheless, despite its high toughness, TZP can have restricted translucency, necessitating the use of additional veneering materials to fulfil its optical requirements [23].

Figure 18.2 shows a schematic representation of microstructure and light scattering of dental zirconia.

Cubic zirconia is another important category for dental applications, which is especially appreciated for its increased translucency. It does not undergo phase transformation under stress, and thus, it has reduced toughness. It is stable at high temperatures and does not play a role in the transformation toughening mechanism. Its optical characteristics, however, make it extremely popular for anterior restorations where aesthetics is of the highest priority. The higher translucency of c-ZrO_2 results in a more aesthetic appearance, reducing the demand for veneering materials. Despite this benefit, the reduced fracture resistance of cubic zirconia restricts its application in posterior areas that are subject to heavy occlusal forces. To overcome this limitation, researchers are investigating a number of methodologies, including dopant modifications, to improve the mechanical properties of cubic zirconia without compromising its excellent aesthetics [25,26].

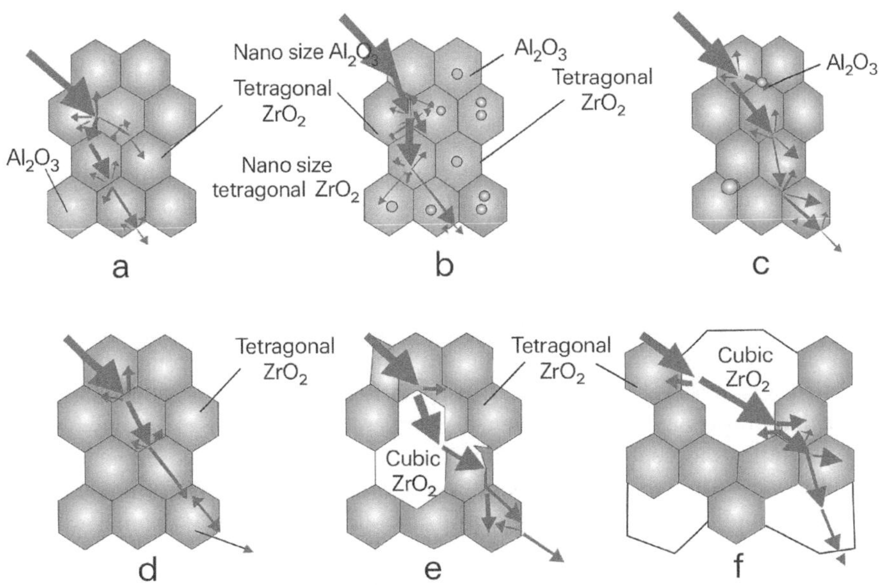

FIGURE 18.2 Schematic illustration of microstructure and light scattering of dental zirconia. (a) alumina-toughened zirconia (ATZ); (b) Ce-TZP/alumina nanocomposite (Nano-ZR); (c) conventional TZP (3Y-HA); (d) high-translucent TZP (3Y); (e) high-strength PSZ (4Y); and (f) high-translucent PSZ (5Y) [24]. (CC BY 4.0 license.)

Zirconia composites are a promising development in dental materials in which secondary phases are added to improve the overall properties of zirconia. These composites are engineered to counter the intrinsic drawbacks of monolithic zirconia by adding components such as alumina, bioactive glass, and other reinforcing phases. Alumina-reinforced zirconia composites, for example, have better mechanical properties by suppressing grain growth and improving fracture toughness. The incorporation of alumina stabilizes the tetragonal phase and minimizes the risk of low-temperature degradation, a process that can compromise the long-term durability of zirconia restorations. Likewise, bioactive glass-reinforced zirconia composites are increasingly popular for their ability to enhance osseointegration and improve the biocompatibility of dental implants. The addition of bioactive glass not only enhances bonding with the surrounding tissues but also provides antibacterial properties, minimizing the risk of peri-implant infections [27,28].

The choice between TZP, c-ZrO_2, and zirconia composites is based on the individual clinical needs and the balance between mechanical durability and aesthetic demands. Although TZP is still considered the gold standard for load-bearing restorations because of its remarkable toughness, c-ZrO_2 is used in situations where higher translucency is needed. Zirconia composites, however, provide a customized way of tailoring both strength and bioactivity, opening the door to the next generation of dental materials. Advances in processing techniques, surface treatments, and material composition refinements all enhance the performance of zirconia-based restorations, warranting their long-term success and functionality in today's dentistry [29].

18.3 DEPOSITION TECHNIQUES FOR ZIRCONIA COATINGS

18.3.1 Physical Vapour Deposition

PVD is a popular method for the deposition of thin, adherent films with well-defined microstructure. It involves several techniques, such as sputtering and electron beam evaporation, to deposit material on substrates, which enhance their surface characteristics for particular applications [30].

Sputtering is one of the most popular PVD methods. Here, a high-energy ion beam hits a target material, which leads to the ejection of atoms their deposition onto a substrate. This process results in uniform coatings with strong adhesion and controlled thickness. This technique is especially useful in applications that require high-density coatings with improved mechanical and tribological behaviour. The ability to tailor the coating composition by cosputtering multiple target materials further enhances its versatility, making it suitable for a wide range of industrial and biomedical applications. Electron beam evaporation is another crucial PVD technique that utilizes high-energy electron beams to vaporize the coating material. The vaporized atoms then condense onto the substrate, forming a thin and uniform film. This process is extremely effective in creating high-purity coatings with little or no contamination. Electron beam evaporation finds widespread applications in optical coating manufacturing, protective coatings, and biomedical field, where purity and

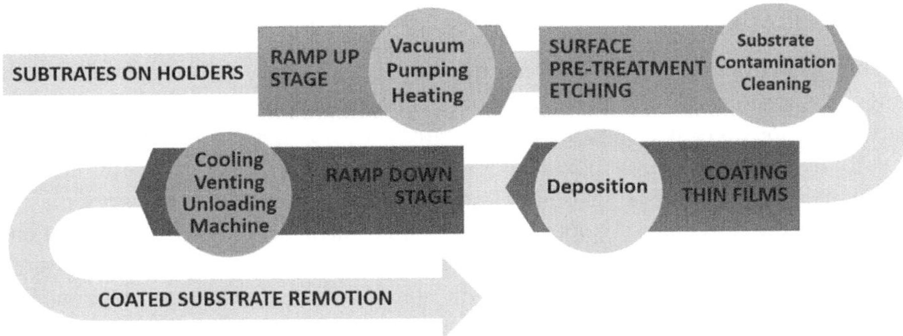

FIGURE 18.3 Processing flowchart for a classic PVD sputtering process [33]. (CC BY 4.0 license).

adhesion are important. Control of deposition parameters such as temperature and deposition rate makes it possible to produce coatings with optimized properties for targeted functional purposes [31,32]. Figure 18.3 presents a processing flowchart for a classic PVD sputtering process.

One of the most significant benefits of PVD coatings is that they can improve mechanical, chemical, and biological surface properties. By controlling microstructure with great accuracy, PVD coatings can achieve better hardness, wear resistance, and corrosion protection. In the biomedical field, PVD coatings are important to increase the performance and lifespan of implants, dental restorations, and surgical tools. The use of zirconia coatings with PVD methods, for instance, has been demonstrated to enhance the biocompatibility and mechanical stability of implant surfaces, reducing the risk of implant failure and enhancing patient outcomes. The microstructure of PVD coatings is highly sensitive to deposition parameters, including substrate temperature, deposition rate, and vacuum. Optimization of these parameters can help researchers obtain coatings with desired properties that are suitable for various applications. The possibility of depositing multilayered or functionally graded coatings extends the versatility of PVD methods to more advanced engineering and biomedical applications. PVD coatings also have environmental advantages over traditional coating methods. In contrast to electroplating, which uses toxic chemicals, PVD processes are cleaner and generate little waste, making them a more environmentally friendly method for surface modification. This benefit has contributed to the growing use of PVD coatings across industries such as aerospace, automotive, and medical device production [34–36].

PVD processes such as sputtering and electron beam evaporation are vital to creating high-performance coatings with precision-controlled microstructure. The versatility of PVD coatings, environmental benefits, and capacity to augment surface properties position them as the primary choice in different industrial and biomedical applications. While future research enhances deposition methods and widens the scope of available materials, PVD coatings will be a technology central to pushing material performance and endurance to greater levels [37].

18.3.2 Plasma Spraying

Plasma spraying is a commonly used thermal spraying method for coating a wide range of substrates. It is especially popular in biomedical applications because of its high deposition rate, capability of forming thick coatings, and excellent adhesion to the substrate. It is critical for increasing the durability, biocompatibility, and functionality of biomedical prosthetics and implants. Plasma spraying uses a plasma torch to produce an intensely hot plasma jet. This plasma jet is employed to heat and accelerate powder coating materials, melt them, and drive them onto the substrate's surface. On impact, molten particles rapidly solidify and create a dense and bondable coating. The flexibility in plasma spraying means that a diverse array of materials such as metals, ceramics, and composite coatings can be deposited, making plasma spraying highly applicable in many medical and industrial practices [38]. Figure 18.4 shows a schematic diagram of the solution precursor plasma spraying process, illustrating numerous significant operational spray parameters that may affect the physical and chemical properties of cermet coatings.

One of the most significant strengths of plasma spraying is that it has a high deposition rate, which can allow for rapid coating of thick coatings over large surfaces. This aspect is specifically valuable in biomedical uses, where implants need thick and uniform coatings in order to optimize their function. For instance, hydroxyapatite (HA) coatings that replicate the mineral structure of bone are often used in orthopaedic and dental implants with plasma spraying. These coatings enhance osseointegration, increasing the implant–bone interface bond, thereby improving the longevity and stability of the implant. Besides HA, other biocompatible substances such as zirconia and titanium-based coatings are also coated using plasma spraying.

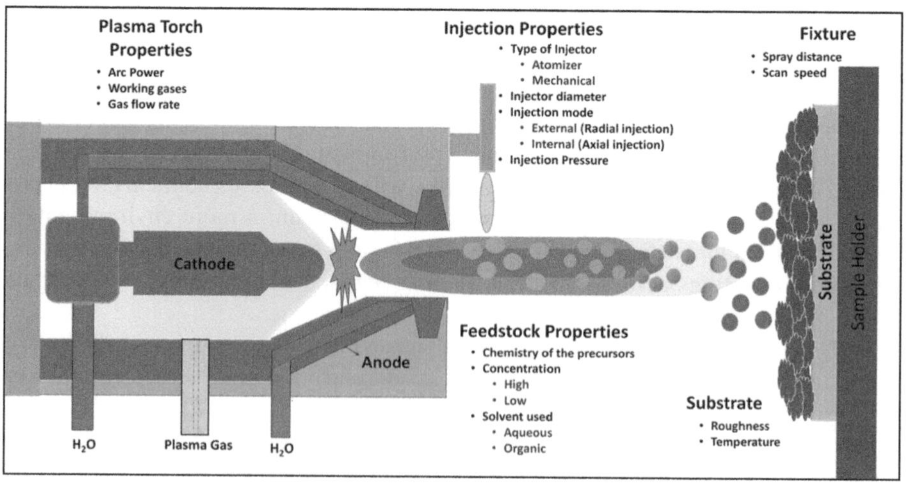

FIGURE 18.4 Schematic diagram of the solution precursor plasma spraying process showing the numerous important operational spray parameters that could influence the physical and chemical characteristics of cermet coatings [39]. (CC BY 4.0 license.)

Zirconia coatings have been found to possess high strength, corrosion resistance, and bioinertness and are therefore applicable in load-bearing implants. Titanium and titanium alloys, however, have good mechanical compatibility with human bone and demonstrate high adhesion when deposited using plasma spraying. Customization of coating compositions enables the creation of multifunctional surfaces that enhance the mechanical and biological performance of implants [40].

Plasma-sprayed coatings have improved substrate adhesion as a result of the rapid solidification of molten particles upon impact. This process forms a robust mechanical bond between the coating and the substrate, which is essential for maintaining the long-term stability of biomedical implants. By optimizing spraying parameters such as particle velocity, plasma temperature, and substrate preparation, researchers can produce coatings with enhanced adhesion strength and durability. Plasma spraying is also characterized by an interesting capability of coating porosity control. Porosity coatings are also normally required in biomedicine, where tissue ingrowth and incorporation of implants within the host bone structure are favoured. It is possible, through a process parameter manipulation, to control plasma spraying to obtain coatings with porosities in which biological properties can be controlled as desired. Besides its myriad benefits, there are some challenges in plasma spraying. One of the most important issues is the possibility of material property modification under high-temperature processing. Some materials, such as HA, can experience phase transformations during spraying, which can influence their bioactivity and function. To overcome these effects, researchers are continuously optimizing process parameters and investigating different feedstock materials that are less susceptible to high-temperature degradation. In addition, the adhesion strength of plasma-sprayed coatings can differ depending on the surface preparation method. Appropriate surface roughening and cleaning before coating deposition are necessary to ensure maximum adhesion between the coating and substrate. Improvements in surface treatment techniques, including grit blasting and chemical etching, have contributed to better adhesion strength, and thus, plasma-sprayed coatings are increasingly being used in biomedical applications with confidence [41–43].

18.3.3 Sol-Gel Processing

Sol-gel processing is a common chemical synthesis technique used to produce homogeneous coatings with fine control over crystallinity and porosity. It is especially beneficial in many domains, such as biomedical applications, because it can generate high-purity, homogenous coatings with specific properties. In the sol-gel process, a system changes from a liquid "sol" state to a solid "gel" state via hydrolysis and condensation reactions of precursor materials. Most often, precursors such as metal alkoxides or inorganic salts are utilized, which undergo reactions under controlled conditions to yield a stable colloidal suspension. The solution is subsequently coated on a substrate using dip coating, spin coating, or spraying and dried, followed by heat treatment to produce the final coating. Since reaction parameters, e.g. pH, solvent mixture, and drying conditions, can be controlled, it is possible to precisely control the microstructure of the coating [44].

One of the key benefits of sol-gel processing is that it can produce coatings with controlled porosity. Porosity in the resulting coating can be controlled by altering precursor concentration, ageing time, and drying conditions. In the field of biomedicine, porous coatings find special applications in cell attachment, tissue integration, and drug delivery systems. For instance, HA coatings derived from sol-gel processing have been widely studied for improving the bioactivity of dental and orthopaedic implants. These coatings are similar in composition to natural bone and promote enhanced osseointegration, resulting in higher implant stability and lifespan. Crystallinity control is another important advantage of sol-gel processing. Through precise choice of heat treatment temperatures and processing conditions, researchers can create coatings with a wide range of crystallinity, which has a direct influence on their mechanical strength and bioactivity. Crystalline coatings have higher wear resistance and chemical stability and are thus best suited for long-term implant applications. Amorphous sol-gel coatings, on the other hand, find applications in bioactive glass, where controlled degradation and ion release are favourable characteristics. Besides biomedical applications, sol-gel coatings are common in the electronics, optical, and energy sectors. Their capability of creating nanostructured coatings with enhanced mechanical and thermal properties makes them excellent for applications such as antireflective coatings, corrosion-resistant coatings, and protective barriers against severe conditions. The addition of dopants or functional nanoparticles further increases the variability of sol-gel coatings, allowing for tailored properties for defined purposes [45,46]. Figure 18.5 presents the general scheme of a sol-gel synthesis method.

Despite its benefits, sol-gel processing also has some drawbacks. It is sensitive to environmental conditions such as temperature and humidity, which can affect the reproducibility of coatings. Drying and sintering processes should also be optimally controlled to avoid cracking and ensure uniformity. Researchers are working on novel methods, including hybrid sol-gel techniques and templating routes, to overcome these limitations and improve the performance of sol-gel coatings. New developments in sol-gel technology have contributed to the emergence of multifunctional coatings that show improved biological, mechanical, and chemical properties. For example, antibacterial sol-gel coatings containing silver or copper nanoparticles have been introduced to inhibit infections associated with implantation. Moreover, bioactive glass sol-gel coatings that have ion release profile control are being explored as promising candidates in bone regeneration and tissue engineering studies [48–50].

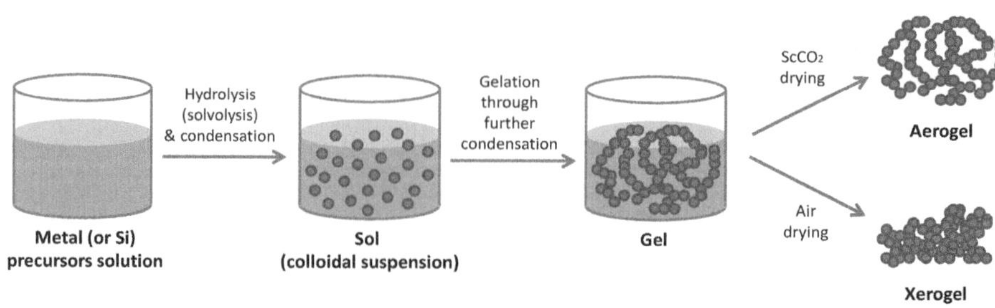

FIGURE 18.5 General scheme of a sol-gel synthesis method [47]. (CC BY 4.0 license.)

18.3.4 Laser Cladding

Laser cladding is a sophisticated surface treatment method that achieves a dense, strongly bonded coating with low porosity. It is extensively used to improve wear and corrosion resistance and is especially beneficial in industrial, biomedical, and aerospace applications. In laser cladding, a high-power laser beam is used to melt a coating material in powder or wire form, while at the same time being supplied into the laser beam and onto a substrate. The high-concentration heat of the laser promotes solid-state bonding between the coating and substrate, with the formation of a dense and defect-free layer. As opposed to traditional thermal spraying methods, laser cladding provides improved adhesion and control of the microstructure, eliminating the porosity and enhancing the durability of coatings [51,52]. Figure 18.6 presents a schematic of the laser cladding process.

One of the greatest benefits of laser cladding is that it can improve wear resistance. Parts that are exposed to high friction and mechanical loads, such as turbine blades, engine parts, and cutting tools, are highly benefited by laser-clad coatings. This technique enables the deposition of hard and wear-resistant materials such as carbides, nitrides, and metal matrix composites, which can greatly increase the service life of key components. Furthermore, laser cladding allows for the refurbishment and repair of exhausted components, reducing replacement costs and the demand for total replacements. Apart from wear resistance, laser cladding dramatically enhances corrosion resistance, and as such, applications under extreme conditions are perfect. Corrosion-resistant coatings in the form of stainless steel, nickel-based alloys, and titanium can be applied to safeguard structural parts from degradation due to exposure to chemicals, high temperatures, and moisture. This is especially useful in the oil and gas sector, where pipes and drilling gear need to be protected from aggressive corrosive environments. Likewise, biomedical implants can be enhanced with laser-clad coatings that make them longer lasting and more biocompatible [54,55].

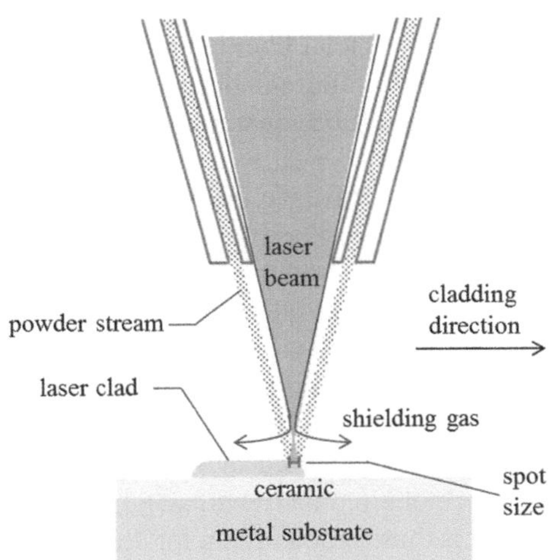

FIGURE 18.6 Schematic of the laser cladding process [53]. (CC BY 4.0 license.)

Laser cladding is a highly efficient process due to the precision and control it delivers. The provision of thin or thick coatings with minimal heat-affected zones means that distortion or damage to the substrate is not desired. Such a benefit is important for fragile components that demand minimal thermal stress, such as medical devices and aerospace parts. Furthermore, the process is highly automated, leading to consistency and repeatability in coating quality. Although it has many advantages, laser cladding also has some drawbacks. The initial cost of high-powered laser systems and equipment is substantial. Moreover, it is a challenging task to achieve optimal process parameters such as laser power, scanning speed, and powder feed rate. However, with improvements in laser technology, such as fibre and diode lasers, the process has become more cost-effective and easily available over the years. Some of the recent advances in laser cladding involve the incorporation of artificial intelligence and real-time monitoring, making it more efficient by facilitating adaptive control and detection of defects while depositing coatings. Research is also ongoing to develop new materials for coatings and hybrid processes involving a combination of laser cladding with other surface engineering processes to enhance performance further [56].

18.4 MECHANICAL AND BIOLOGICAL PERFORMANCE OF ZIRCONIA COATINGS

Zirconia coatings have outstanding mechanical properties, hence their suitability for dental and biomedical use. Among the most distinguishing characteristics of zirconia coatings is high hardness, which ranges from 12 to 14 GPa. Hardness confers good resistance to mechanical wear and thus suitability of zirconia coatings for occlusal surfaces that are subjected to high masticatory loads. Furthermore, zirconia exhibits high wear resistance, which can prevent material loss due to abrasion during long-term exposure. This characteristic is particularly useful in dental restorations, where surface integrity is essential for sustained performance. Another essential mechanical property of zirconia coatings is their high adhesion to both metal and ceramic substrates. The high interfacial bonding guarantees coating stability, minimizing the risk of delamination or failure under functional loading [57]. Figure 18.7 presents composite materials and coating materials for bioengineered zirconia-based scaffolds.

Besides their mechanical properties, zirconia coatings also display excellent corrosion and chemical stability, which is critical for prolonged exposure in the oral environment. Zirconia is chemically stable when subjected to oral fluids and does not leach ions that will damage material integrity or induce unwanted biological response. This stability proves helpful in avoiding the release of potentially toxic ions that will interfere with surrounding tissues. In addition, zirconia coatings are resistant to degradation even under extremely acidic conditions, providing durability in adverse conditions like those caused by dietary acids and bacterial metabolism. The chemical resistance of zirconia to withstand challenges without degrading increases its lifespan, and it is thus a good option for dental implants, prosthetics, and other biomedical devices. Apart from its chemical and mechanical durability, zirconia is extremely biocompatible, supporting good interaction with biological tissues.

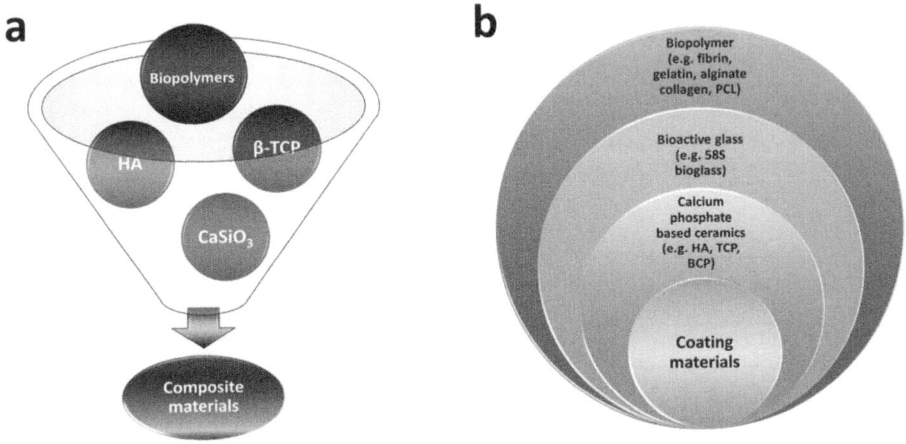

FIGURE 18.7 (a) Composite materials and (b) coating materials for bioengineered zirconia-based scaffolds [58]. (CC BY 4.0 license.)

The surface characteristics of zirconia coatings support excellent cell adhesion and growth, essential for effective integration in implantation applications. This biocompatibility supports the healing process, such that zirconia-coated implants have a positive interaction with adjacent bone and soft tissues. Another significant advantage of zirconia is its low bacterial adhesion, which helps reduce plaque formation and the risk of infections. The reduced bacterial accumulation on zirconia surfaces contributes to improved oral hygiene and reduces the likelihood of peri-implantitis, a common complication associated with dental implants. Moreover, zirconia coatings support osseointegration, the direct structural and functional connection between the implant surface and bone. It is critical for the stability of implants and long-term success in dental and orthopaedic applications. In general, zirconia coatings provide an integrated balance of mechanical strength, chemical stability, and biological compatibility. Their chemical stability, hardness, and wear resistance make them suitable for load-bearing dental surfaces, while corrosion resistance underlines the durable performance of zirconia coatings in the oral environment. Moreover, their osseointegration and biocompatibility properties also encourage their applications in implants, further leading to improved patient outcomes. With continued research that improves zirconia-based coatings, their usage in biomedical applications will continue to grow further, cementing them as a prime material for long-lasting and trustworthy implants and restorations [59–61].

18.5 CLINICAL APPLICATIONS AND PERFORMANCE IN DENTISTRY

Zirconia coatings have also gained considerable interest in dental and orthopaedic implants because of their superior properties that improve function and longevity. Zirconia is a biocompatible ceramic that is highly recognized for its capability to

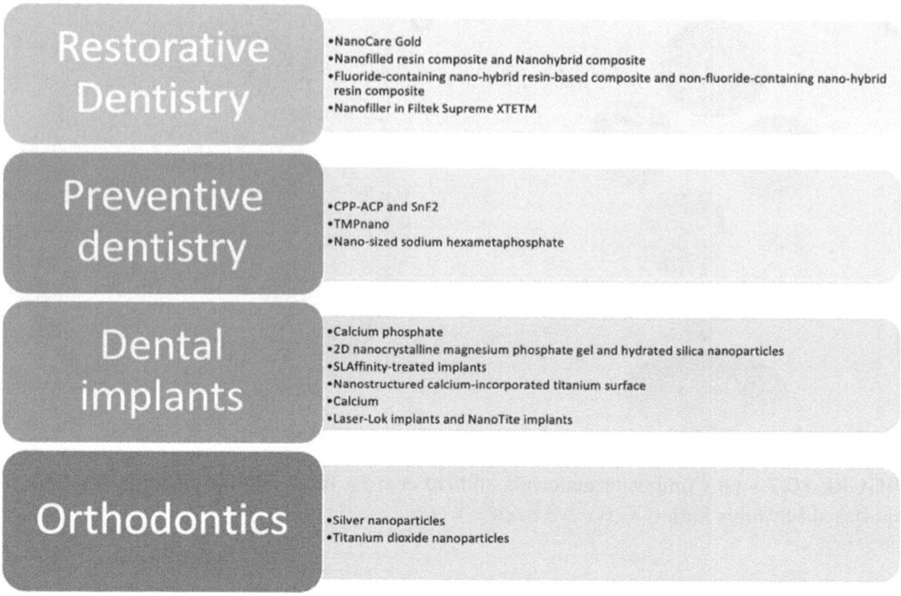

FIGURE 18.8 Applications of nanotechnologies and biomaterials in dentistry [63]. (CC BY 4.0 license.)

enhance osseointegration, being the direct structural and functional contact between the implant and the surrounding bone tissue. With regard to implants, zirconia coatings help provide better corrosion and wear resistance, important aspects that increase the longevity of implants in aggressive body environments. The biocompatibility of zirconia also minimizes the risk of peri-implantitis, a frequent complication that may arise due to bacterial infection around the implant. The use of zirconia coatings on implants results in improved bone integration and reduced implant failure, and thus, zirconia is a preferred material for dental and orthopaedic use [62]. Figure 18.8 illustrates the applications of nanotechnologies and biomaterials in dentistry.

Zirconia coatings are also commonly applied in dental prostheses, such as crowns, bridges, and veneers. They are greatly appreciated for their aesthetic appeal since zirconia is translucent by nature, which makes it resemble the look of natural tooth enamel. This feature makes zirconia coatings a great option for tooth restoration since they integrate well with natural teeth, making the overall appearance and feel of the prosthesis better. In addition, the mechanical properties of zirconia, including its high flexural strength and fracture toughness, are responsible for the durability and longevity of dental restorations, which experience constant stress due to masticatory forces. Zirconia coatings prevent wear and tear and ensure that dental prostheses remain functional and aesthetically pleasing over a period of time. In addition, the resistance of zirconia to staining and discoloration contributes to its attractiveness as it maintains its aesthetic properties for a long duration. Despite these advantages, the use of zirconia coatings is associated with some challenges and drawbacks. One of the major issues is the risk of coating delamination, particularly when prostheses are subjected to extreme masticatory forces. This problem is especially common in areas of the mouth that are subject to high stress, like the molars. In such situations, the adhesion between the zirconia coating and substrate can degrade, resulting in delamination or even total failure of the coating. This issue can undermine the functionality and durability of zirconia-coated prostheses, making material properties and load distribution critical considerations during the design process [63,64].

Another major challenge in the use of zirconia coatings is producing defect-free coatings in the manufacturing process. Zirconia coatings need to be deposited under controlled and precise conditions in order to obtain uniformity, adhesion, and the intended microstructure. Surface preparation, coating thickness, and deposition process can, however, introduce defects such as cracks, voids, and non-uniformity that can compromise the mechanical properties of the coating as well as its overall performance. In addition, maintaining the best possible adhesion between the zirconia coating and substrate material, which can be either metal or ceramic, is still an issue. Loss of bonding will result in premature failure of the coating due to mechanical load or environmental exposure. Lastly, the cost factors associated with zirconia coatings are a major obstacle, especially when compared with uncoated zirconia prostheses. The processing and deposition methods needed to deposit zirconia coatings are frequently more intricate and expensive, which contributes to the general cost of dental and implant restorations. Furthermore, the time and technical expertise involved in producing high-quality coatings can increase both manufacturing and labour costs. All these factors can render zirconia-coated prostheses more costly for patients, especially in areas where cost is a key component in treatment planning. Although the advantages of zirconia coatings, including enhanced performance and durability, might be worth the increased expense in some scenarios, patients can choose uncoated zirconia prostheses, which are cheaper but also provide satisfactory function [65].

18.6 FUTURE PERSPECTIVES

Zirconia coatings have become a likely option to enhance the mechanical and biological functionality of implants and dental prostheses, and the prospect will get brighter in the years ahead with continuing progress in research and development. Next-generation zirconia coatings promise to adopt futuristic technologies that also enhance their biological interaction qualities in addition to improving their mechanics. Among these advances are nanostructured zirconia coatings, hybrid zirconia bioactive coatings, surface modification, and creating smart coatings that possess added functions such as antibacterial behaviour and drug release. These advances have the potential to extend the limits of zirconia coatings, making them more effective, long-lasting, and biologically compatible, and they have the potential to improve clinical outcomes in both orthopaedic and dental implantology considerably. The future of biomedical applications of zirconia coatings is very bright, with increased research and advances in technology paving the way for coatings that are stronger, longer-lasting, and more biologically compatible than they have ever been. Nanostructured zirconia coatings, hybrid zirconia bioactive coatings, novel surface modifications, and smart coatings with antibacterial and drug release functionalities are destined to take zirconia coating performance to a new level in dental and orthopaedic applications. These breakthroughs will fill existing gaps, enhance clinical efficacy, and pave the way for new applications for zirconia coatings across medical and dental specializations. Zirconia coatings are set to become an increasingly central component of contemporary implantology as these technologies continue to progress, providing patients with enhanced quality of life and superior long-term outcomes [66,67].

18.7 CONCLUSION

Zirconia coatings are being increasingly valued for their substantial benefits in dental applications, combining superior mechanical strength, biocompatibility, and aesthetics. Used primarily in dental implants, crowns, bridges, and veneers, these coatings feature outstanding durability and wear resistance, which is necessary in materials that undergo the daily strain of chewing and biting. The mechanical properties of zirconia, such as its high flexural strength and fracture toughness, render it a perfect candidate for dental restorations as these prostheses can withstand long-term functional loading without failure. In addition, the natural translucency of zirconia gives it an aesthetic benefit as restorations can be made to closely resemble the form of natural teeth, which is particularly important in patients who require both functionally and aesthetically successful dental rehabilitation. One of the primary issues with the application of zirconia coatings in dental practice is that coatings have a tendency to delaminate when subjected to excessive masticatory forces, particularly when subjected to stresses associated with biting or grinding. The adhesion between the zirconia coating and substrate has to be strong enough to withstand such failures, but existing coating technologies continue to have difficulties in providing uniform adhesion, especially in areas subject to high forces. The expense of processing zirconia coatings is also high, which continues to be a limitation to their widespread application, particularly for budget-constrained patients. The complex deposition techniques required for high-quality zirconia coatings, such as chemical vapor deposition or plasma spraying, increase the overall expense of manufacturing, making it more difficult for some patients to afford these advanced dental solutions.

Despite these challenges, the future of zirconia coatings in dentistry is bright. Ongoing developments in deposition methods and material science are likely to overcome the existing limitations, enhancing the adhesion strength, wear resistance, and overall performance of zirconia coatings. For example, nanostructured zirconia coatings can be developed to enhance mechanical properties, and hybrid coatings with bioactive materials can enhance osseointegration and minimize the risk of peri-implantitis. In addition, new surface modification methods, such as laser texturing and ion implantation, promise to enhance the bonding of zirconia coatings with the substrate, as well as biological responses. With further advances in these technologies, the expense of zirconia coatings will reduce, bringing them within the reach of a wider population of patients. In summary, despite the limitations, continuous research and development in zirconia coating technologies holds great promise for greatly improving their performance, thereby making zirconia coatings an increasingly viable choice for dental restorations in future.

REFERENCES

1. O'Brien WJ. Dental materials and their properties. *J Dent Res*. 2005;84(2):156–163.
2. Zhang Y, Yang X, Huo K, Zhang X, Wu Z. Zirconia-based ceramics in dental applications: A review of their physical properties and performance. *Dent Mater.* 2021;37(7):1167–1177.
3. Sailer I, Zembic A, Jung RE, et al. Randomized controlled trial of zirconia and titanium implants in the maxillary posterior region: 5-year results. *Clin Oral Implants Res.* 2016;27(1):103–109.
4. Guazzato M, Albakry M, Ringer SP, et al. Strength, fracture toughness and microstructure of a selection of all-ceramic materials. Part II. Zirconia-based dental ceramics. *Dent Mater.* 2004;20(5):450–456.
5. Dong H, Liu H, Zhou N, Li Q, Yang G, Chen L, Mou Y. Surface modified techniques and emerging functional coating of dental implants. *Coatings*. 2020 Oct 22;10(11):1012.
6. Denry I, Kelly JR. Emerging ceramic-based materials for dentistry. *J Dent Res.* 2008;87(7):155–165.
7. Swain MV, Xie D. The mechanical properties of zirconia for dental applications. *Aust Dent J.* 2012;57(4):1–10.
8. Kaga M, Kurashige H, Kuroda S, et al. Biocompatibility of zirconia ceramics: Interactions with oral tissues and bacteria. *J Mater Sci: Mater Med.* 2017;28(5):1–8.
9. Habelitz S, Marshall SJ, Marshall GW. Microhardness and fracture toughness of zirconia ceramics. *J Dent Res.* 2002;81(8):478–482.
10. Turner CW, Papadopulos NA, Mantell J, et al. Clinical outcomes of zirconia implants. *J Clin Periodontol.* 2015;42(1):49–54.
11. Kamijo R, Kawai K, Furuya Y, et al. Development of zirconia-based dental materials: Processing and performance. *J Prosthodont Res.* 2015;59(2):63–70.
12. Boening KW, Arola DD, Kruzic JJ. Wear and fracture resistance of zirconia ceramics. *J Dent Res.* 2007;86(7):665–669.
13. You X, Xu J, Zhang J, et al. Application of zirconia coatings in the medical field: Progress and challenges. *J Biomater Appl.* 2021;35(5):630–648.
14. Buser D, Sennerby L, De Bruyn H. Modern implant dentistry and the role of zirconia-based materials. *Int J Oral Maxillofac Implants.* 2007;22(6):101–105.
15. Ribeiro AR, Ribeiro C, Ramos AP, et al. The state of the art of zirconia ceramics in dental restorations. *J Prosthodont.* 2020;29(2):156–163.
16. Lynch CD, McConnell RJ. Clinical considerations for the use of zirconia crowns in dental practice. *J Dent.* 2014;42(2):112–118.
17. Yamada Y, Sato Y, Hirano S, et al. Evaluation of biocompatibility and mechanical properties of zirconia ceramics for dental applications. *J Mater Sci: Mater Med.* 2007;18(2):249–256.
18. Della Bona A, Moser JB. Zirconia as a dental biomaterial. *J Appl Biomater Funct Mater.* 2009;7(4):168–175.
19. Wang L, Liu Y, Yu X, et al. Evaluation of the antimicrobial properties and mechanical behavior of zirconia coatings. *J Biomed Mater Res B.* 2017;105(4):837–844.
20. Flury S, Luthy H, Helbling C, et al. In vitro evaluation of bacterial adhesion to zirconia ceramics in comparison with titanium. *Clin Oral Implants Res.* 2008;19(5):527–533.
21. Honma S, Harada Y, Okada Y. Esthetic and chemical properties of zirconia ceramics for dental use. *J Prosthodont.* 2013;22(7):577–583.
22. Tinschert J, Natt G, Möller S, et al. Chemical stability of zirconia and its application in dental restorations. *J Prosthet Dent.* 2000;83(5):548–552.
23. Sousa F, Lima E, Cunha M, et al. Zirconia in dental applications: Properties, types, and clinical considerations. *J Prosthet Dent.* 2020;124(1):12–16.

24. Ban S. Classification and properties of dental zirconia as implant fixtures and superstructures. *Materials*. 2021;14:4879.
25. Yilmaz B, Sabiha E. Transformation toughening in zirconia ceramics. *J Prosthodont*. 2009;18(6): 424–428.
26. Krell A, Suttor H, Borrmann L, et al. Flexural strength and optical properties of zirconia and its veneering materials. *J Prosthet Dent*. 2011;105(4):279–285.
27. Silveira A, McDonnell D, Corazza P. The use of cubic zirconia for dental restoration. *Int J Oral Maxillofac Implants*. 2014;29(2):432–439.
28. Denry I, Kato H, Sasaki T. The influence of cubic zirconia on dental restoration. *Dent Mater*. 2011;27(7):713–717.
29. Jang J, Lee I, Kim S. Zirconia composites for dental applications: Strengthening strategies and clinical applications. *J Biomed Mater Res B*. 2018;106(4):1331–1342.
30. Anga R, Rahman N, Silva J, et al. Zirconia composites: Enhancing the mechanical properties and bioactivity of dental materials. *J Dent Res*. 2019;98(6):1275–1284.
31. Zhang Y, Yang X, Wang M, et al. Advances in zirconia ceramics for dental applications. *J Dent Res*. 2016;95(6):629–635.
32. Li D, Ouyang F, Liu Y, et al. Physical vapor deposition of zirconia coatings for dental applications. *J Mater Sci*. 2015;50(2):205–213.
33. Baptista A, Silva F, Porteiro J, Míguez J, Pinto G. Sputtering physical vapour deposition (PVD) coatings: A critical review on process improvement and market trend demands. *Coatings*. 2018 Nov 14;8(11):402.
34. Lin L, Guo X, Shi W, et al. Electron beam evaporation for thin film deposition: Application to zirconia coatings. *Appl Surf Sci*. 2018;432:1–7.
35. Choi Y, Cho C, Lee C. Sputtering and electron beam deposition of zirconia coatings for biomedical applications. *J Biomed Mater Res B*. 2019;107(7):1970–1976.
36. Teixeira G, Mello L, Silva A, et al. Application of physical vapor deposition (PVD) in dental materials: The case of zirconia coatings. *J Dent Mater*. 2021;38(9):1234–1242.
37. Ferracane JL. PVD coatings for dental applications: A review of physical properties. *Dent Mater*. 2017;33(6):714–722.
38. Liu Z, Xiao Y, Li Q, et al. Plasma spraying of zirconia coatings for medical applications: A review of process and materials. *J Appl Biomater Funct Mater*. 2020;18(4):48–55.
39. Unabia R, CandidatoJr R, Pawłowski L. Current progress in solution precursor plasma spraying of cermets: A review. *Metals*. 2018 Jun 4;8(6):420.
40. Taguchi M, Watanabe I, Imai Y, et al. Properties and applications of plasma-sprayed zirconia coatings for dental implants. *Biomaterials*. 2012;33(33):7851–7858.
41. Zhang L, Guo C, Liu J. New advances in zirconia-based coatings for implant applications: Sputtering, laser, and plasma spray. *J Mater Sci Mater Med*. 2022;33(5):49.
42. Wong Y, Lee S, Yoon S, et al. Plasma spraying of zirconia coatings for enhanced implant functionality. *Dent Mater*. 2015;31(8):882–890.
43. Figueiredo M, Guimaraes L, Cabral J. Plasma spraying of zirconia coatings: Performance in dental applications. *Int J Biol Sci*. 2019;15(9):1781–1789.
44. Li H, Zhang S, Fu Z. Plasma spraying of zirconia coatings for dental implants: A review. *J Mater Sci Mater Med*. 2018;29(11):174–182.
45. Xie H, Zhang J, Tang C, et al. Plasma-sprayed hydroxyapatite coatings for biomedical applications: An overview. *J Biomed Mater Res Part B Appl Biomater*. 2021;109(2):1033–1042.
46. Liu S, Chen Z, Yu L, et al. Influence of plasma-spraying parameters on the properties of hydroxyapatite coatings. *Mater SciEng C*. 2020;116:111188.
47. Tao Y, Pescarmona PP. Nanostructured oxides synthesised via $scCO_2$-assisted sol-gel methods and their application in catalysis. *Catalysts*. 2018 May 17;8(5):212.
48. Zheng X, Yang X, Han X, et al. Effect of substrate roughness on adhesion strength of plasma-sprayed hydroxyapatite coatings. *J Mater Sci Mater Med*. 2021;32(3):43.

49. Zhang W, Li Y, Chen J, et al. Sol-gel processing for biomedical coatings: Synthesis and applications. *J Sol-Gel Sci Technol.* 2020;95(3):637–645.
50. Tan J, Shen Y, Xu D, et al. Sol-gel derived hydroxyapatite coatings for dental and orthopedic applications: A review. *J Biomed Mater Res Part A.* 2022;110(2):218–227.
51. Wang J, Li H, Sun B, et al. The role of sol-gel derived coatings in enhancing bioactivity and mechanical performance in biomedical implants. *Biomater Sci.* 2021;9(11):3685–3697.
52. Chen Y, Zhang Y, Sun H, et al. Sol-gel derived porous coatings for bone tissue engineering. *J Mater Sci.* 2020;55(14):5487–5497.
53. Zhang Y, Mack DE, Mauer G, Vaßen R. Laser cladding of embedded sensors for thermal barrier coating applications. *Coatings.* 2018 May 4;8(5):176.
54. Liu Y, Wang X, Lin Q, et al. Functional sol-gel coatings for improved bacterial resistance in biomedical implants. *J Biomed Mater Res Part B Appl Biomater.* 2023;111(6):1203–1211.
55. Yang Z, Li J, Li X, et al. Sol-gel derived bioactive glass coatings for bone regeneration: A review. *Biomaterials.* 2022;46:175–185.
56. Zhang H, Wei X, Zhang L, et al. Bioactive sol-gel coatings with controlled ion release for bone regeneration. *Biomater Sci.* 2024;12(5):1081–1093.
57. Wang S, Huang M, Lu F, et al. Laser cladding for surface engineering of biomedical implants: A review. *J Laser Appl.* 2020;32(4):042003.
58. Sakthiabirami K, Soundharrajan V, Kang JH, Yang YP, Park SW. Three-dimensional zirconia-based scaffolds for load-bearing bone-regeneration applications: Prospects and challenges. *Materials.* 2021 Jun 10;14(12):3207.
59. Liu F, Zhang H, Zhao Y, et al. Laser cladding with corrosion-resistant coatings for biomedical applications. *Mater SciEng C.* 2021;118:111448.
60. Lin Z, Xu T, Deng X, et al. Laser cladding of wear-resistant coatings on medical implant materials. *J Mater Sci.* 2022;58(12):4565–4574.
61. Li Y, Zhou S, Zhang J, et al. Laser cladding of corrosion-resistant coatings for biomedical implants: A review. *Biomaterials.* 2021;25(7):1569–1578.
62. Wang J, Zhang X, Li F, et al. Laser cladding for improving the biocompatibility and wear resistance of biomedical implants. *J Biomed Mater Res Part B Appl Biomater.* 2022;110(9):2438–2448.
63. Dipalma G, Inchingolo AD, Guglielmo M, et al. Nanotechnology and its application in dentistry: A systematic review of recent advances and Innovations. *J Clin Med.* 2024 Sep 5;13(17):5268.
64. Zhang W, Li X, Huang B, et al. Recent advances in laser cladding for biomedical applications: Hybrid processes and real-time monitoring. *J Mater Sci.* 2024;59(4):2305–2317.
65. Verma PK, Singh S, Kapoor M, Singh S. A review on the surface topography and corrosion behavior of Mg-alloy coatings for biomedical implants. *Results Surf Interfaces.* 2024 May 9:100227.
66. Zhang J, Yu Y, Gao X, et al. Wear resistance and mechanical properties of zirconia coatings for biomedical applications. *J Mater Sci Mater Med.* 2022;33(1):56–67.
67. Chen F, Wang R, Sun J, et al. Biomechanical and biological performance of zirconia coatings for dental implants. *J Biomed Mater Res Part A.* 2023;111(5):1402–1412.

Section V

Emerging Technologies

19 Advancements in Zr and ZrO_2 Technologies
Additive Manufacturing and Emerging Fields

Amrinder Mehta, Hitesh Vasudev, and Rajwinder Singh

19.1 INTRODUCTION

Zirconium and zirconium dioxide (ZrO_2) are among the materials that are apparent in the application of additive manufacturing (AM) and supplement the properties of 3D printed parts. Zirconium is a lustrous, grayish-white metal that is very resistant to corrosion and has excellent mechanical characteristics, and therefore, is widely used in manufacture. On the other hand, ZrO_2 or zirconia is a ceramic material that has high strength and toughness as well as excellent thermal stability. Regarding the AM, ZrO_2 has recently received increased interest as a reinforcement in polymer-based materials used for 3D printing [1].

Figure 19.1 presented in the timeline below depicts the impressive advancement in medical and bioprinting 3D printing technology from the year 1984 to 2020. The creation of stereolithography (SLA) in 1984 and, subsequently, selective laser sintering

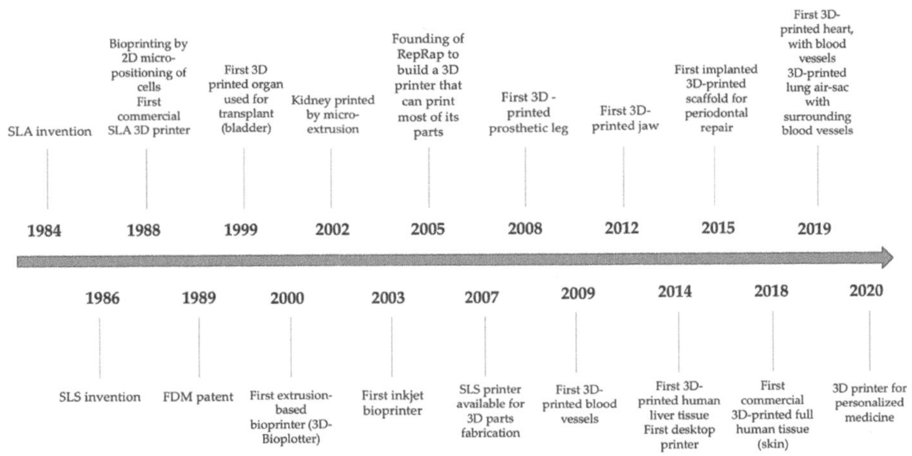

FIGURE 19.1 Evolution of 3D printing technology [2]. (Under CC BY 4.0.)

DOI: 10.1201/9781003519300-24

(SLS) in 1986, has been constant. The first commercial SLA 3D printer and basic tissue engineering bioprinting using 2D micro-positioning of cells occurred in 1988. The advancement of technology continued to this progress through 1990s and 2000s and reached some important achievements including the first ever 3D printed bladder transplant in 1999 and the first kidney transplant through 3D printing in 2002. Extraordinary progress was obvious in the medical area in the 2010s; for instance, prosthetic legs, jaws, and blood vessels were developed. Such a decade ended up in triumphs like the creation of the first 3D printed heart with blood vessels and lung air sac in 2019. The evolution of printing was summed up by marking 3D printers for use in the personalized medicine in 2020 and this is because the advancement of technology shows that basic print has metamorphosed into a medical practice [2].

The mechanical properties of such nanocomposites composed of PLA and PA12, in particular, are enhanced when ZrO_2 nanoparticles are added. For instance, the introduction of ZrO_2 nanoparticles, when used at the correct dosage, enhances the strength of PA12 by 47.7% and PLA by 20.1%. These enhancements in mechanical behavior led to the application of ZrO_2-reinforced polymers in the engineering structures which have better performances. The incorporation of ZrO_2 nanoparticles into 3D printing materials is normally done by the means of melt mixing compounding techniques [3]. This process ensures homogeneity of the nanoparticles within the polymer matrix, thus ensuring homogeneity of material properties of the printed object. The type of base polymer affects the loading of ZrO_2 nanoparticles; according to the research, concentrations of between 1.0 and 3.0 wt.% are most effective in terms of the improvement of mechanical properties. Besides the mechanical reinforcement, the ZrO_2 nanoparticles have excellent other uses in the nanocomposite designs of the micro-optics systems used in 3D-printed micro-optical elements [4]. Recently, by adding zirconia (ZrO_2) nanoparticles to photopolymers that can be used in femtosecond two-photon photolithography, researchers have been able to obtain increased spread of the optical characteristics of material based on polymer matrixes [5]. This has facilitated the possibility to tailor opthalmic materials to meet standard or desired refractive index and dispersion to produce quality optical parts such as achromatic lenses. Specifically, nanoparticles of zirconium oxide have been incorporated in dental restoration application along with functional halloysite nanotubes containing a layer of silver nanoparticles for applying improved 3D printing resins [6]. Figure 19.2a reveals the yttria-stabilized zirconia (YSZ) powder. It also shows that the particles look like smooth and spherical equally and has a field of view of 50 μm. This powder finds wide application in certain processes and technologies because of its thermal stability and its ionic conduction ability. Figure 19.2b shows the graphite particles incorporated as the absorbing additive in the formulation. These particles are irregular in form, flaky, and the external surfaces are heterogeneous; the field of view is 10 μm. Graphite is used widely due to its electrical conductivity nature, and also for its lubricating nature. These images aid in the analysis of the microstructure and the morphology of the surface of the materials, which influence their performances in various applications [7].

Nano-scaled plasticity characteristics of 10% and 30% YSZ embedded in the alumina ceramics produced by additive manufacturing based on the SLA innovation have been investigated using techniques of nanoindentation and scanning electron

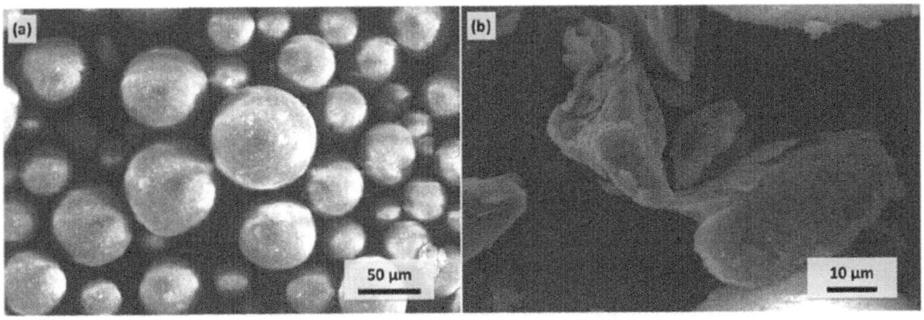

FIGURE 19.2 (a) Scanning electron microscopy images of the yttria-stabilized zirconia powder utilized in the study and (b) the graphite particles employed as an absorptive additive [7]. (Reproduction permissions by Publisher.)

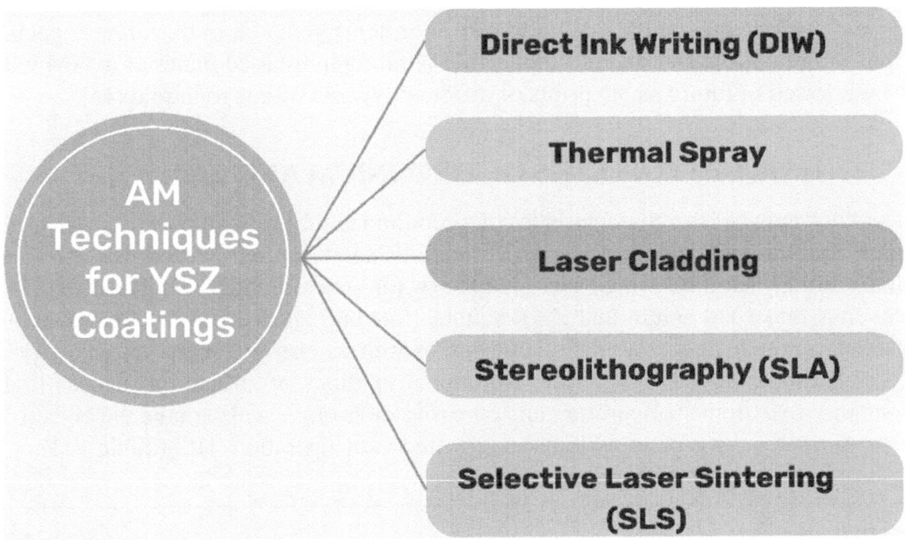

FIGURE 19.3 AM methods used for YSZ coatings.

microscopy (SEM) [8]. Figure 19.3 shows different AM methods applied to YSZ coatings itself. Contained in the circle of the mind map, it is proceeded with five specifically different methods as follows: AM Techniques for YSZ Coatings, Direct Ink Writing (DIW), Thermal Spray, Laser Cladding, SLA, and SLS [9–11]. Each method is documented in the different colored rounded rectangle so that they look fairly redundant and organized. It also presents the configuration and manufacturing choices on the different ways of depositing YSZ coatings with different types of additive manufacturing. Since the design is simple and uncluttered, one is able to quickly grasp what various technological strategies are about [12].

The flexural strength, the flexural modulus, the fracture toughness, and the microhardness of these modified resins are higher than that of the unmodified resins. According to the literature, the ideal loading levels generally range between 4 wt.% for the ZrO_2 particles and approximately 5 wt.% for the silver-halloysite nanotubes that have been incorporated into epoxy matrices, significantly improving the mechanical properties [13]. Of the pure metals, zirconium has also been investigated as a material to be directly deposited in additive manufacturing applications. Biosensor has employed laser additive manufacturing with injected powder and has successfully applied it in fabrication of zirconium components where injection of powder enables more geometrical variety with the advantages of the zirconium metal [14]. This approach allows new opportunities for fabricating components from zirconium in custom geometries and inner cavities that are extremely challenging to create using conventional techniques. Zirconium and ZrO_2 as materials used in 3D printing are regarded as one of the new innovations in material science and additive manufacturing. From such an understanding of its properties, researchers and engineers can develop components that are stronger, more durable, and functionally improved for applications in structures, automotive, aircraft, biomedical applications, and for people's health and comfort. With the advancements going on in this area, expectations are that more and diverse application of zirconium-based material in AM will be witnessed in future as 3D prints of different types continue to unfold [15].

19.2 UNIQUE PROPERTIES OF ZIRCONIUM AND ZrO_2

The knowledge of the characteristics of zirconium and ZrO_2 is important in order to apply them in practice, because these materials have very high mechanical strength and thermal stability, which are mandatory for severe conditions. These properties thus make zirconium and ZrO_2 suitable for the required use in facets such as aerospace, medical device manufacturing, as well as energy generation where performance is accepted as a virtue. Additional qualities of zirconium such as high tolerance to extreme temperature and corrosive environment also make the objective more attractive by promoting long-lasting use in key operations [16] (Table 19.1).

TABLE 19.1
Information on the Biocompatibility of 3D Printed Zirconia Scaffolds

Raw Materials	Properties	Outcomes
Zirconia, Hydroxyapatite (HA), Tricalcium Phosphate (TCP)	High mechanical strength, biocompatibility, compressive strength of 7–8 MPa	Effective in promoting bone regeneration, significant new bone formation observed in vivo [17]
Hydroxyapatite (HA), Tricalcium Phosphate (TCP), Yttria-stabilized Tetragonal Zirconia Polycrystals (Y-TZP)	High mechanical strength (flexural strength >1000 MPa), biocompatibility, interconnected porous structure	Scaffolds support bone regeneration, enhance cell adhesion, proliferation, and osteogenic differentiation, especially with BMP-2 incorporation [18]

(Continued)

TABLE 19.1 (*Continued*)
Information on the Biocompatibility of 3D Printed Zirconia Scaffolds

Raw Materials	Properties	Outcomes
BCP powder (60 wt% hydroxyapatite and 40 wt% β-tricalcium phosphate) and 10 wt% ZrO_2	Enhanced mechanical strength and biocompatibility, demonstrated by cell proliferation tests with MG63 cells and hMSCs	Improved mechanical properties and biocompatibility, with increased expression of BMP-2 in hMSCs under dynamic culture conditions [19]
Dental resin reinforced with zirconia and glass silica nanoparticles	Improved flexural strength and biocompatibility due to the addition of zirconia and glass silica nanoparticles	Biocompatibility testing showed cell viabilities greater than 80% in all tested groups, indicating clinical potential for restorative dentistry [20]
3-mol% yttria-stabilized tetragonal zirconia		Printing orientation affects physio mechanical characteristics; milled group had highest BFS, tilted group had highest surface roughness; confirms biocompatibility for dental applications [21]
Hydroxyapatite, Tricalcium Phosphate, Zirconium Oxide	High mechanical strength, biocompatibility, mimics mechanical properties of natural bone	Promotes bone regeneration, effective for clinical applications [22]

19.3 APPLICATIONS OF ZIRCONIUM

Zirconium, for individuals with knowledge in chemistry, is a transition metal that's famous for its versatility-rich properties; this metal is used widely in many sectors. Zirconium's application is most crucial in the nuclear power industry where it is used in the production of fuel rod cladding and reactor core structures. It being poor absorber of neutron, its high thermal coefficient of expansion, and its capability to resist expressive mechanical destroy due to neutron impact make it suitable for use in these vital parts [23]. Figure 19.4 reveals two separate uses for zirconia which are in highly technical areas of engineering and bioengineering. In the first application as shown in Figure 19.4a, an operating gas turbine engine demonstrates the application of thermal barrier coatings on gas turbine components. The cross-sectional view shows that the coating has a complex layered distribution of the top coat, bond coat, and the nickel-based sub/super alloy substrate. This coating system is essential for protecting the features of engine, especially the first-stage vanes and rotating blades, from erosion because of high temperatures in the combustor section. Details regarding the application of the powder thermal barrier coating are provided, including references to thermal gradients, zirconia-based layers, and the lifespan of components within the turbine's four-stage operational environment [24].

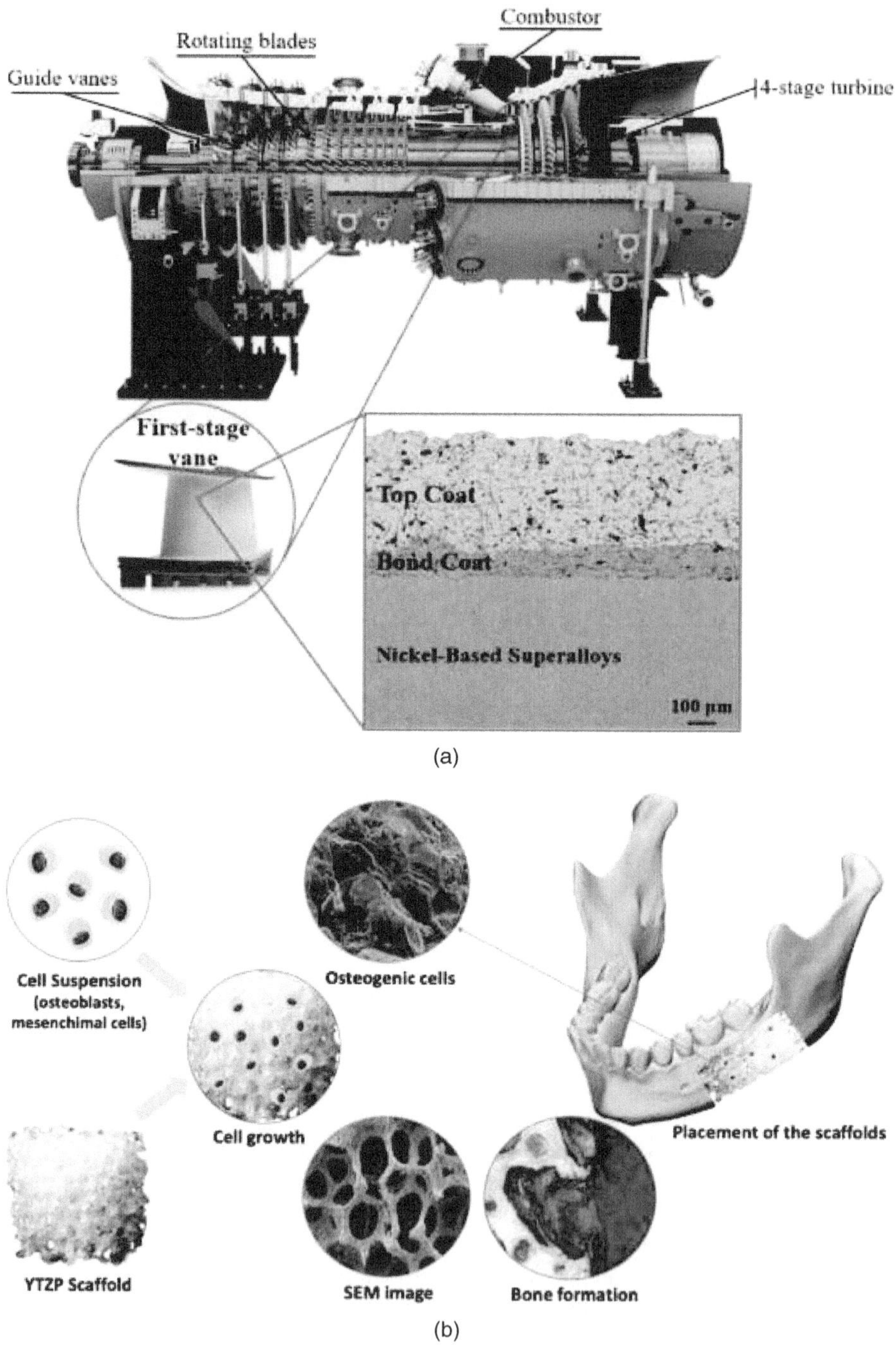

FIGURE 19.4 The application of zirconia in (a) high-temperature thermal barrier coatings [24] and (b) cell growth in the porous zirconia structures [25]. (Reproduction permissions by Publisher.)

In the second application, Figure 19.4b describes the biomedical application of zirconia for tissue engineering through an yttria-stabilized tetragonal zirconia polycrystal (Y-TZP) scaffold system. The process starts with the cultivation of a cell suspension containing osteoblasts and mesenchymal cells, which are then put into the porous zirconia scaffold. The SEM image also shows conventional interconnected porous network of the scaffold which is crucial for cell adhesion, viability, and proliferation. The sequence displays the exact manner that cells attach on the scaffold structure and result to the evolution of osteogenic cells to produce new bones. The last application is demonstrated on the theatrically represented jaw model with the scaffold placed for the regeneration of the bones. These SEM pictures illustrate that the cells penetrated the interconnected pores properly and generate new bone tissue based on the histological section of the color contrasted and the highlighted section which shows the integration of new bone tissue in the region [25].

This applies to the chemical processing industry as zirconium has proven to have some of the best corrosion resistance properties. It's employed for building equipment, including reboilers, evaporators, tanks, reactor vessels, pumps, valves, and piping that have to perform in severe chemical environments. Zirconium also has outstanding corrosion resistance to most organic and inorganic acids, salt solutions, strong alkalis, and several molten salts, and it is used in petrochemical, pharmaceutical, and water treatment industries. The oxide ZrO_2 (zirconia) holds an important position in the field of technical ceramics. Because of its high hardness, wear resistance, and thermal stability, this metal is utilized in cutting tools, grinding media, and other operating components in elevated-temperature equipment [26]. Figure 19.5 uses a clean, simple design with a light background and white boxes connected to the central oval by solid lines, effectively showing how zirconia nanoparticles play a crucial role across multiple technological and medical domains [27].

The biomedical industry employs zirconia in dental applications, such as implants and crowns, and orthopedic uses, such as hip and knee replacements. It may be noted that zirconium finds use in super alloys and even in newer forms of material and technology. They are ingredients in high-temperature superconductors and amorphous

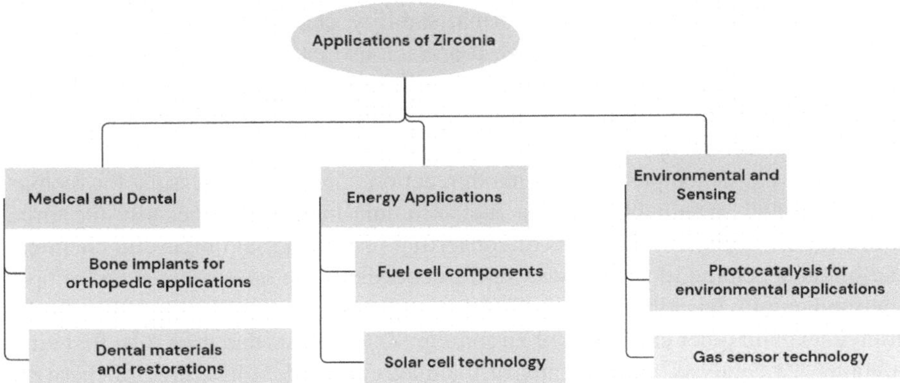

FIGURE 19.5 Applications of zirconia nanoparticles.

metal (metallic glass) and materials which possess high wear resistance and hardness [28]. Zirconium alloys are used in aerospace engineering for the manufacturing of pieces of space engines and rockets due to their high temperature and environment resistance. Due to the optical properties of zirconium compounds, especially CZ, zirconium is used in jewelry as a diamond imitation. In the electronics industry, it is used in the manufacturing of capacitors and getters utilized for vacuum tubes [29]. Because it is highly capable for the binding of urea, it has been used in the treatment of chronic kidney disease. Zirconium is widely used in the chemical industry since the equipment intended for use in the chemical industry always comes into contact with corrosive agents [30]. It is used in the manufacture of acetic acid, chlorine, caustic soda, and a number of other chemicals. In the paper and packaging industry, zirconium compounds are employed with surface coatings because of water resistance and strength. Due to its durability and resistance to corrosion, the metal is used for manufacturing of surgical instruments, cases for watches, and filaments for lamps [31]. Within the optical industry employing zirconium, materials assist in manufacturing superior performing lenses as well as coating that need to be both clear and durable. It also has uses in the preparation of pigments, ceramics, and refractories. Its compounds are used as an opacifier in the glazes used in ceramics and as high-grade refractory material. In the textile industries, zirconium compound is used in treating fabrics for waterproofing and for fire resistance. The applications of zirconium also include the use in solid oxide fuel cells, oxygen sensors, as well as in several catalytic processes either as a catalyst or as a catalyst supporter [32]. Due to its resistance to high temperatures and pressures, physical and chemical non-degradation can easily be used in industries such as space technology and electronics production. Due to its ability to resist corrosion, having high melting point and mechanical strength plus biocompatibility, zirconium is used in variety of complex applications across various sectors such as nuclear, chemical, aerospace, and biomedical products [33].

19.4 ZIRCONIUM DIOXIDE IN CATALYSIS

Ceramic material, ZrO_2 also called zirconia, is fast becoming the most efficient catalyst and catalyst support for a variety of chemical processes in many industries. Its main advantages such as high thermal stability, changeable phase composition, and qualified surface properties make it a precious material in catalysis. As one of the key uses of zirconia as a catalytic material, the synthesis of sustainable fuels and chemicals is one of the best-known applications of zirconia for catalysis [34]. Nonetheless, due to its inherent stability and its resistance to chemical degradation, it will not decompose during harsh reaction conditions necessary for biomass conversion and sustainable aviation fuel. Zirconia materials, especially the zirconia catalysts, improve the process of converting plant material into useful chemicals and biofuels helping to shrink the world's dependence on oil resources and support environmentally friendly economy [35]. Figure 19.6 presents a detailed circular diagram to explain other categories of zirconium (Zr) catalysts and uses. The first circle contains "Zr catalysts" at the center of the diagram, while it has three further rings enclosing it and divided into sections in different colors. Different forms of zirconium compounds are within the innermost rim, which include inorganic zirconium,

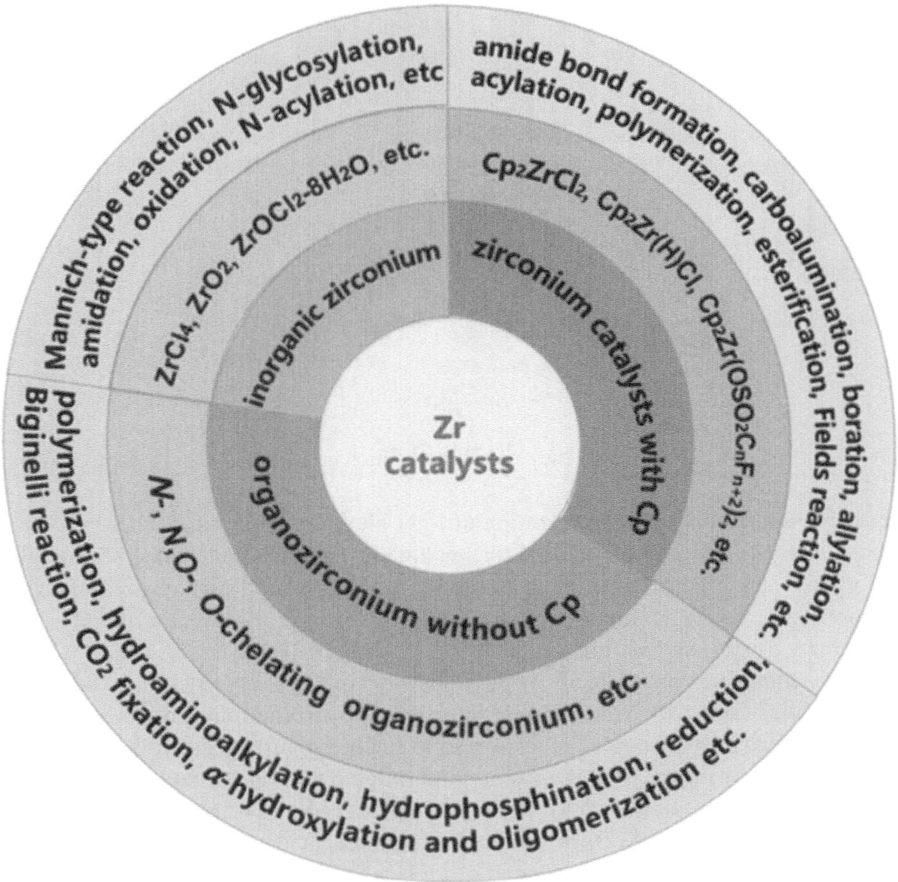

FIGURE 19.6 Zirconium catalysts in organic reaction [36]. (Reproduction permissions by Publisher.)

organozirconium without CP, and a zirconium catalyst that achieved CP. The second ring depicts several zirconium complexes such as $ZrCI_4$, ZrO_2, $ZrOCI_2{\cdot}8H_2O$, and Cp_2ZrCI_2. The outermost ring overviews the chemistry and uses of each kind of zirconium catalyst. Some of the reactions that occur in the upper left section are Mannich-type reactions, N-glycosylation, oxidation, and N-acylation. The upper right area elaborates on amide bond formation, carboalumination, boration, and polymerization reactions. Polymerization, hydroaminoalkylation, hydrophosphination and CO_2 fixation, α hydroxylation, and oligomerization processes are grouped under the lower section. Usually, when dealing with such a broad field, it is difficult to put all the applications of zirconium catalysts in the same diagram, but this one presents various applications of the catalyst in the synthesis of organic compounds as well as industrial uses with clear presentation [36].

In addition, zirconia has the capability of being in many forms and various compositions. Tungstated and sulfated zirconia-doped systems have displayed tremendous

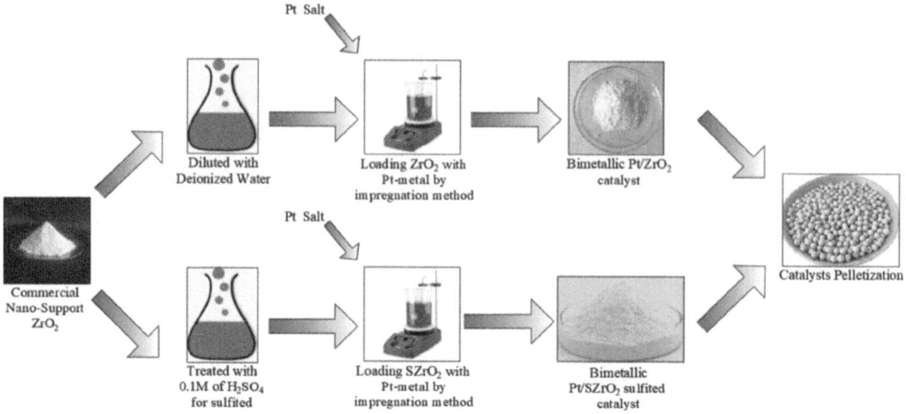

FIGURE 19.7 Steps followed in the catalyst preparation procedures [39]. (Under CC BY 4.0.)

activity essential for the isomerization of light alkanes and similar other processes. These catalysts extend the potential for sustainable fuel and chemical synthesis, paving way to new green chemistry processes [37]. The specific surface area is high in zirconia which offers many active sites through which molecules can interact freely and enhance catalytic reactions. When coupled with the fact that it is immune to fouling from impurities in the feeds, it can thus sustain high catalytic performance for long periods of time. To synthesis, methods such as hydrothermal and mechanochemical treatments have also been investigated to further improve catalytic applicability of zirconia [38]. Figure 19.7 represents a parallel synthesis method for developing bimetallic catalysts supported by ZrO_2. The process commences with commercially available nano-support ZrO_2 powder, which is subjected to two distinct treatment pathways. In the upper path, deionized water is added to the ZrO_2, while in the lower path 0.1 M H_2SO_4 is added to the ZrO_2 for sulfidation. Both paths then have platinum salt added through an impregnation, similar to the process of the first step; the materials are mixed on magnetic stirrers. This results in the generation of two chemically distinct bimetallic catalysts Pt/ZrO_2 in the upper cycle and sulfided $Pt/S/ZrO_2$ in the lower cycle. The last step demonstrates that the both catalysts pass through a step of pelletization and form of pellets of constant size. It is quite educative as it illustrates the various steps of catalyst synthesis systematically depending on a specific synthesis method, while other synthesized products are illustrated in shapes of beakers, mixing tools, etc. The parallel synthesis method described here enables investigators to compare the normal and sulfided forms of the catalyst having been prepared under the same conditions [39].

These treatments can bring in defect states into the structure, which enhance the photocatalytic activity of the structure under visible light illumination and broaden the application prospects. Catalytic properties of zirconia can also be modified by doping and phase stabilization of the material, for the purpose that best serves the application. This versatility has placed them in a wide variety of catalytic applications including pollution control of automobile and industrial exhausts, refining and chemical reactions including hydrogenation, oxidation, amination, and isomerization among others [40]. There has also been action to develop zirconia-based catalysts for use in CO_2 application area as well. These conceptual frameworks have been applied, tested in the quest to catalyze the conversion of CO_2 to useful chemical molecules and fuels, thus playing a role in greenhouse emission reductions. At high temperatures, the oxygen ion conductivity makes it a suitable material for expensive applications, such as solid oxide fuel cells and oxygen sensors for environmental control, also backing clean energy. Reactor development case of fine chemical synthesis is that zirconia catalysts have been shown to be outstanding in selective oxidation reactions and in specialty chemicals manufacture [41]. This further enhances their role in catalyzing reaction under mild condition and in green solvents as embraced by green chemistry. Indeed, the application of zirconia in catalysis is expected to expand in the future due to the current studies' efforts on the preparation of new types of zirconia materials with better catalytic functionality [42]. These include nanostructured zirconia, zirconia supported metal catalysts, and zirconia containing mixed oxides. With the continuous rise of the importance of efficiency and environmental friendliness in chemical processes, ZrO_2 serves as a significant enabler of this chemical change [43].

19.5 CHALLENGES AND LIMITATIONS OF USING ZIRCONIUM IN ADDITIVE MANUFACTURING

As with most stimuli, there are constraints associated with the use of additive manufacturing with zirconium and its compounds, especially ZrO_2, limiting the potential of these materials in 3D printing applications as advances in future research and manufacturing processes. However, one of the most significant difficulties arises from the inability to form fully dense materials with porosities. A characteristic feature of the AM process is the high porosity, typical for the printed parts with dangerous influence on its mechanical characteristics and general application [44]. Figure 19.8 shows a mind map of different challenges observed in what seems to be a manufacturing or production setting. In the middle, there is an oval inscribed with the word: "Challenges"; from this oval stem eight separated problems. Among them are "Distortion and shrinkage" where product dimensions can be affected and "Geometrical overgrowth" where structural precision may be impacted. Other challenges labeled in the diagram include "Particle size and distribution," which indicates the size and dispersion of the particles being used in manufacturing; hence, the quality of materials used in creating the layers of the product. The next major challenge is "Internal defects" and "Surface finish" points at issues to do with appearance or quality of the product. New to the 2012 version is the "Staircase effect" which probably relates to layering in additive manufacturing. "Technical shortcoming" probably broadens the operational restraint. New challenges are outlined in ovals of different colors and hence you get to see a neatly arranged set of challenges that need to be met on a project. The diagram in this case is useful to show the interrelatedness of these manufacturing challenges as well as offer an appreciation of the possible areas of concern in the manufacturing process [45].

The porosity issue stems from two main sources: the powder production technique and the 3D printing process itself. Gas atomization during powder production can lead to pore formation in the raw material. During printing, insufficient laser energy

FIGURE 19.8 Major challenges of AM zirconia-based ceramics [45]. (Under CC BY 4.0.)

may fail to properly melt the metal, while excessive energy can cause melted material to splatter, both resulting in porous structures. These tiny cavities and holes not only reduce the density of the part but also make it prone to cracks and other damages, particularly under high loads [46]. Another significant challenge is the control of residual stress in AM zirconium parts. The extreme thermal changes during the printing process, involving rapid heating and subsequent cooling, can lead to residual stress within the material. This stress can cause various forms of deformation, with the highest concentration typically found at the contact area between the printed part and the print bed. Residual stress can result in cracking and warping, especially during the cooling phase when the melted metal contracts [47]. The reactive nature of zirconium poses additional difficulties in AM processes. Zirconium has a high affinity for common gases like oxygen and hydrogen at elevated temperatures, making annealing and other heat treatment practices challenging. This reactivity necessitates stringent control of the printing environment to prevent contamination and maintain the desired material properties. Dimensional accuracy and surface finish are also areas of concern in AM of zirconium. Achieving precise dimensions and smooth surfaces often requires extensive post-processing, which can be time-consuming and costly [48]. The layer-by-layer nature of AM can result in a rough surface finish, requiring additional machining, grinding, or polishing to meet the required specifications. The mechanical properties of AM zirconium parts, particularly their strength, often fall short of those produced by conventional manufacturing methods [49]. This limitation is a significant hurdle in adopting AM for critical applications where high strength and reliability are essential. The insufficient mechanical strength, combined with long carbon-debinding times in some AM processes, currently limits the widespread adoption of 3D printed zirconia in certain industries. Raw material preparation and process control present further challenges [50]. The characteristics of the ceramic powders and resins used, including their concentrations, particle size distribution, and shape, significantly influence the rheological properties of the printing suspension. Optimizing these parameters during maintaining consistent quality across batches can be complex. Lastly, the design and optimization of printing parameters such as velocity, layer height, orientation, and light source characteristics require careful consideration [51]. These factors, along with post-printing treatments like debinding and sintering, play crucial roles in determining the final properties of the AM zirconium parts. While additive manufacturing of zirconium and zirconia offers exciting possibilities for creating complex geometries and customized parts, several challenges must be addressed to improve the overall quality, reliability, and performance of the printed components [52]. Ongoing research and development efforts are focused on enhancing material properties, refining printing processes, and optimizing post-processing techniques to overcome these limitations and expand the application range of AM zirconium in various industries.

19.6 FUTURE TRENDS

The application of zirconium and ZrO_2 is expected to be in the vanguard of new technologies in a wide spectrum of industries due to its advantageous characteristics. As industries advance and develop new products, ignoring the laws of material science, these materials are being applied in new and increasingly versatile ways throughout advanced technologies. In the electronics sector, ZrO_2 is becoming a new platform solution with great potential. The features such as high thermal stability and good dielectric properties make the material a suitable improvement for the next generation of semiconductor devices and high-performance electronics [53]. Figure 19.9 presents multiple chemical bonds oriented in relation to the surface of ZrO_2 marked

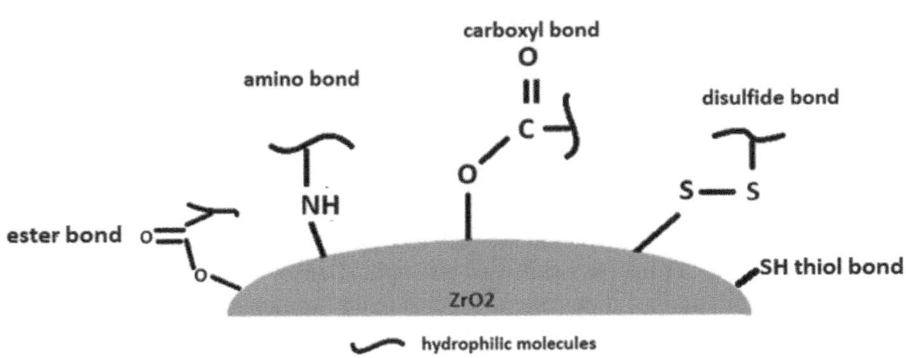

FIGURE 19.9 Functionalization of the surface of ZrO_2 nanostructures with functional groups [54]. (Under CC BY 4.0.)

by the semicircle. The diagram shows five distinct types of molecular bonds: an ester bond, an amino bond, a carboxyl bond, a disulfide bond, and an SH thiol bond. These bonds are depicted in rather simple forms: ester bond is depicted as O-O, the amino bond is depicted as NH, the carboxyl bond as C=O with an indication of an oxygen bond, the disulfide bond is depicted as S-S, and the thiol bond as SH. They also give information about the existence of hydrophilic molecules, represented by curved lines, that are attached to the ZrO_2 surface. This representation shows that various organic functional groups may form chemical bonds or interactions with the ZrO_2 surface, which is useful to study in surface molecular attachment in materials science as well as chemical uses [54].

Miniaturization and improved performance with electronic equipment are the key factors boosting the use of ZrO_2 in sophisticated industries such as semiconductor manufacturing. So far nanostructured electronics from the titanium dioxide and zirconia ceramics has been offering improved performance characteristics which will extend the possibilities for developing new and smaller efficient electronic devices [55]. Another industry where zirconium and ZrO_2 are enjoying a great deal of progress is the energy sector. In the area of clean energy, one of the most important properties of zirconia that has a high oxygen ion transmission rate at high temperatures is effectively used in solid oxide fuel cells (SOFCs). Zirconia for use in fuel cells: With the increasing trend toward environmentally friendly energies, zirconia is placed ideally to become an enabling technology for this market [56]. In addition, current research in the advancement and production of new and improved zirconium-based materials for use in the next-generation batteries and energy storable systems may redefine the manner in which we store and deliver energy [57]. In the specialty of AM, ZrO_2 could be considered as one of the rising stars of high-performance materials for 3D printing. Due to high mechanical properties, biocompatibility, and nice appearance, PEEK is especially suitable for dental and medical applications. Due to the improvement in the 3D printing technology as well as material science, we get to see more complex and bespoke zirconia-based implants and prosthetic devices with better performance and a favorable impact on the patient. Automotive industry is also using the properties of zirconium and ZrO_2 for the improvement of their automobile's performance and fuel efficiency [58]. Zirconia has been used in oxygen sensors and thermal coatings. Now, it is also used in high-performance catalytic converters and engine systems. As more vehicles become electric, zirconium-based alloys are being considered to improve electric car batteries. In the field of advanced ceramics, ZrO_2 is still being tested for new uses. New zirconia-based materials are being developed to work better in aerospace vehicles and deep-sea mining [59]. Hypersonic vehicle and advanced propulsion systems are the most suitable areas of application of ultra-high temperature ceramics (UHTCs) containing zirconia. Biomedical field is also another area where zirconium and ZrO_2 are finding their application in a very big way. Apart from their well-known application in the production of dental implants, titanium and its alloys are now considered for numerous other biomedical purposes such as drug delivery vehicles, biosensors, and functional imaging systems [60]. The characteristics of zirconia offer the potential of the material for long-term applications such as implantation of device and tissue engineering scaffolds.

19.7 CONCLUSION

In the present day, zirconium and ZrO_2 have come out as materials of flexibility with remarkable performance that is very crucial in modern uses and the new product developments. That along with good mechanical strength, high thermal stability, and biocompatibility have put it at par with leading technological trends in various fields.

In the area of AM, it has been found that ZrO_2 nanoparticles have promised in improving the mechanical characteristics of polymer-based 3D printing materials. Reflecting the desirable mechanical characteristics of ZrO_2, its addition to polymers, such as PLA and PA12, leads to an increase in strength and performance for its application in engineering structures and mini/micro-optical systems. Zirconium fuel rod cladding and core structures in the nuclear power industry, as well as equipment used in chemical processing are resistant to the harsh environments.

In biomedical applications, zirconia finds its use in the fabrication of dental implants, artificial joints replacement, and tissue engineering scaffolds. In aerospace industry, zirconium alloys are used in space engines and rockets because of their high-temperate strength. In catalysis, zirconia is also an active catalyst and supports clean and efficient production of fuel and chemicals. Nevertheless, there is still much to be achieved in the area, with the continued problems associated with the AM of zirconium and ZrO_2 density, residual stress, reactivity, and dimensional stability. However, ongoing studies aim to fix these problems and expand the use of AM zirconium in areas like aerospace, automotive, and medical fields. medical. Analyzing the forecast, it is possible to state that both zirconium and zirconia will perform significant functions in perspective developments. In electronics, ZrO_2 is viewed as the material of choice for the development of the next-generation semiconductors. In the energy perspective, zirconia is being viewed as solid oxide fuel cell and advanced battery application. The automotive industry is using zirconium materials to enhance efficiency of its vehicles and reduce fuel consumption. New findings and developments are expected to enhance the possibilities and the quality of the zirconium materials to be firmly established in novel technologies in the electronics, clean energy, additive manufacturing, automotive, ceramics, as well as the biomedical industries.

REFERENCES

1. Zhu Y, Liu K, Deng J, Ye J, Ai F, Ouyang H, Wu T, Jia J, Cheng X, Wang X. 3D printed zirconia ceramic hip joint with precise structure and broad-spectrum antibacterial properties. *International Journal of Nanomedicine*. 2019:5977–87.
2. Sufaru I-G, Macovei G, Stoleriu S, Martu M-A, Luchian I, Kappenberg-Nitescu D-C, Solomon SM. 3D printed and bioprinted membranes and scaffolds for the periodontal tissue regeneration: A narrative review. *Membranes*. 2022;12:902.
3. Petousis M, Moutsopoulou A, Korlos A, Papadakis V, Mountakis N, Tsikritzis D, Ntintakis I, Vidakis N. The effect of nano zirconium dioxide (ZrO_2)-optimized content in polyamide 12 (PA12) and polylactic acid (PLA) matrices on their thermomechanical response in 3D printing. *Nanomaterials*. 2023;13:1906.
4. Hu C, Sun J, Long C, Wu L, Zhou C, Zhang X. Synthesis of nano zirconium oxide and its application in dentistry. *Nanotechnology Reviews*. 2019;8:396–404.

5. Nasiri S, Khosravani MR. Machine learning in predicting mechanical behavior of additively manufactured parts. *Journal of Materials Research Technology*. 2021;14:1137–53.
6. Aati S, Shrestha B, Fawzy A. Cytotoxicity and antimicrobial efficiency of ZrO_2 nanoparticles reinforced 3D printed resins. *Dental Materials Journal*. 2022;38:1432–42.
7. Ferrage L, Bertrand G, Lenormand P. Dense yttria-stabilized zirconia obtained by direct selective laser sintering. *Additive Manufacturing*. 2018;21:472–8.
8. Afrasiabi M, Bambach M. Modelling and simulation of metal additive manufacturing processes with particle methods: A review. *Virtual Physical Prototyping*. 2023;18:e2274494.
9. Ali SRA- S, Hussein AHA, Nofal A, Elnaby SIH, Elgazzar H. A contribution to laser cladding of Ti-6Al-4V titanium alloy. *Metallurgical Research Technology*. 2019;116:634.
10. Ajdary R, Kretzschmar N, Baniasadi H, Trifol J, Seppälä JV, Partanen J, Rojas OJ. Selective laser sintering of lignin-based composites. *ACS Sustainable Chemistry Engineering*. 2021;9:2727–35.
11. Huang J, Qin Q, Wang J. A review of stereolithography: Processes and systems. *Processes*. 2020;8:1138.
12. Gao M, Li S, Guan W, Xie H, Wang X, Liu J, Wang H. Excellent thermal shock resistance of NiCrAlY coatings on copper substrate via laser cladding. *Journal of Materials Science Technology*. 2022;130:93–102.
13. Aboushelib MN, Kleverlaan CJ, Feilzer AJ. Selective infiltration-etching technique for a strong and durable bond of resin cements to zirconia-based materials. *The Journal of Prosthetic Dentistry*. 2007;98:379–88.
14. AbuShanab WS, Moustafa EB, Taha MA, Youness RA. Synthesis and structural properties characterization of titania/zirconia/calcium silicate nanocomposites for biomedical applications. *Applied Physics A*. 2020;126:1–12.
15. Ahmadi-Pidani R, Shoja-Razavi R, Mozafarinia R, Jamali H. Evaluation of hot corrosion behavior of plasma sprayed ceria and yttria stabilized zirconia thermal barrier coatings in the presence of Na_2SO_4+ V_2O_5 molten salt. *Ceramics International*. 2012;38:6613–20.
16. Ahmed S, Atia NN. Fast and efficient zirconia-based reversed phase chromatography for selective determination of triptans in rat plasma. *Journal of Pharmaceutical Biomedical Analysis*. 2017;143:241–51.
17. Alahnoori A, Badrossamay M, Foroozmehr E. Characterization of hydroxyapatite powders and selective laser sintering of its composite with polyamide. *Materials Chemistry Physics*. 2023:127316.
18. Aruna S, Kulkarni S, Chakraborty M, Kumar SS, Balaji N, Mandal C. A comparative study on the synthesis and properties of suspension and solution precursor plasma sprayed hydroxyapatite coatings. *Ceramics International*. 2017;43:9715–22.
19. Almeida V, Balzaretti N, Costa T, Machado G, Gallas M. Surfactants for CNTs dispersion in zirconia-based ceramic matrix by sol–gel method. *Journal of Sol-Gel Science Technology*. 2013;65:143–9.
20. Arantes V, Coutinho R, Martins S, Huang S, Vleugels J. Solid state sintering behavior of zirconia-nickel composites. *Ceramics International*. 2019;45:22120–30.
21. Bacchelli B, Giavaresi G, Franchi M, Martini D, De Pasquale V, Trirè A, Fini M, Giardino R, Ruggeri A. Influence of a zirconia sandblasting treated surface on peri-implant bone healing: An experimental study in sheep. *Acta Biomaterialia*. 2009;5:2246–57.
22. Balla VK, Das M, Bose S, Ram GJ, Manna I. Laser surface modification of 316L stainless steel with bioactive hydroxyapatite. *Materials Science Engineering: C*. 2013;33:4594–8.
23. Baig M, Khalid F, Khan F, Rehman K. Properties and residual stress distribution of plasma sprayed magnesia stabilized zirconia thermal barrier coatings. *Ceramics International*. 2014;40:4853–68.

24. Song J-B, Wang L-S, Dong H, Yao J-T. Long lifespan thermal barrier coatings overview: Materials, manufacturing, failure mechanisms, and multiscale structural design. *Ceramics International*. 2023;49:1–23.
25. Resende-Gonçalves CI, Sampaio N, Moreira J, Carvalho O, Caramês J, Manzanares-Céspedes MC, Silva F, Henriques B, Souza J. Porous zirconia blocks for bone repair: An integrative review on biological and mechanical outcomes. *Ceramics*. 2022;5:161–72.
26. Baino F, Minguella-Canela J, Korkusuz F, Korkusuz P, Kankılıç B, Montealegre MÁ, De los Santos-López MA, Vitale-Brovarone C. In vitro assessment of bioactive glass coatings on alumina/zirconia composite implants for potential use in prosthetic applications. *International Journal of Molecular Sciences*. 2019;20:722.
27. Ban S. Classification and properties of dental zirconia as implant fixtures and superstructures. *Materials*. 2021;14:4879.
28. Bhowmick A, Pramanik N, Jana P, Mitra T, Gnanamani A, Das M, Kundu PP. Development of bone-like zirconium oxide nanoceramic modified chitosan based porous nanocomposites for biomedical application. *International Journal of Biological Macromolecules*. 2017;95:348–56.
29. Biswas S, Chen Y, Xie Y, Sun X, Wang Y. Polypyrrole merged zirconium-based metal-organic framework NU-1000 for detection of levodopa. *Microchimica Acta*. 2020;187:1–7.
30. Prashar G, Vasudev H, Mehta A. Application of Bimodal stuructured thermal sprayed coatings in power plant components. *Advanced Ceramic Coatings for Energy Applications*. 2024:273–87.
31. Bosetti M, Vernè E, Ferraris M, Ravaglioli A, Cannas M. In vitro characterisation of zirconia coated by bioactive glass. *Biomaterials*. 2001;22:987–94.
32. Brandon JR, Taylor R. Thermal properties of ceria and yttria partially stabilized zirconia thermal barrier coatings. *Surface and Coatings Technology*. 1989;39:143–51.
33. Bu L, Jiang D, Song Q, Shan X, Wang W, Chen Z. A novel three-dimensional molecularly imprinted polypyrrole electrochemical sensor based on MOF derived porous carbon and nitrogen doped graphene for ultrasensitive determination of dopamine. *Analyst*. 2022;147:5194–202.
34. Caravaca C, Shi L, Balvay S, Rivory P, Laurenceau E, Chevolot Y, Hartmann D, Gremillard L, Chevalier J. Direct silanization of zirconia for increased biointegration. *Acta Biomaterialia*. 2016;46:323–35.
35. Carinci F, Pezzetti F, Volinia S, Francioso F, Arcelli D, Farina E, Piattelli A. Zirconium oxide: Analysis of MG63 osteoblast-like cell response by means of a microarray technology. *Biomaterials*. 2004;25:215–28.
36. Peng L, Zhao Y, Yang T, Tong Z, Tang Z, Orita A, Qiu R. Zirconium-based catalysts in organic synthesis. *Topics in Current Chemistry*. 2022;380:41.
37. Chattopadhyay K, Mandal M, Maiti DK. A review on zirconium-based metal–organic frameworks: Synthetic approaches and biomedical applications. *Materials Advances*. 2024;5:51–67.
38. Chavez JJG, Naraparaju R, Mikulla C, Mechnich P, Kelm K, Ramana C, Schulz U. Comparative study of EB-PVD gadolinium-zirconate and yttria-rich zirconia coatings performance against Fe-containing calcium-magnesium-aluminosilicate (CMAS) infiltration. *Corrosion Science*. 2021;190:109660.
39. Khalil M, Al-Zaidi BY, Shakor ZM, Hussein SJ, Al-Shathr A. Experimental and kinetic study of the catalytic behavior of sulfate-treated nanostructured bifunctional zirconium oxide catalysts in n-heptane hydroisomerization reactions. *ChemEngineering*. 2023;7:115.
40. Chen H, Fu Y, Feng K, Zhou Y, Wang X, Huang H, Chen Y, Wang W, Xu Y, Tian H. Polydopamine-coated UiO-66 nanoparticles loaded with perfluorotributylamine/tirapazamine for hypoxia-activated osteosarcoma therapy. *Journal of Nanobiotechnology*. 2021;19:1–18.

41. Chen X, Li Y, Hodgson PD, Wen C. In vitro behavior of human osteoblast-like cells (SaOS2) cultured on surface modified titanium and titanium–zirconium alloy. *Materials Science Engineering:* C. 2011;31:1545–52.
42. Verma PK, Mehta A, Vasudev H, Kumar VJSR. Performance of thermal spray coated metallic materials for bio-implant applications. *Surface Review Letters*. 2023:2250017.
43. Choudhary S, Islam A, Mukherjee B, Richter J, Arold T, Niendorf T, Keshri AK. Plasma sprayed Lanthanum zirconate coating over additively manufactured carbon nanotube reinforced Ni-based composite: Unique performance of thermal barrier coating system without bondcoat. *Applied Surface Science*. 2021;550:149397.
44. Chuang CH, Kung CW. Metal–organic frameworks toward electrochemical sensors: Challenges and opportunities. *Electroanalysis*. 2020;32:1885–95.
45. Kumaresan S, Vaiyapuri S, Kang J-H, Dubey N, Manivasagam G, Yun K-D, Park S-W. Perspective chapter: Additive manufactured zirconia-based bio-ceramics for biomedical applications. IntechOpen, 2022. doi: 10.5772/intechopen.101979
46. Cura ME, Kim S-H, Muukkonen T, Varjus S, Vaajoki A, Söderberg O, Suhonen T, Kanerva U, Lee SW, Hannula S-P. Microstructure and tribological properties of pulsed electric current sintered alumina–zirconia nanocomposites with different solid lubricants. *Ceramics International*. 2013;39:2093–105.
47. Das I, De G. Zirconia based superhydrophobic coatings on cotton fabrics exhibiting excellent durability for versatile use. *Scientific Reports*. 2015;5:18503.
48. Demir H. The effects on thermal efficiency of yttria-stabilized zirconia and lanthanum zirconate-based thermal barrier coatings on aluminum heating block for 3D printer. *Coatings*. 2021;11:792.
49. Saxena A, Mehta A, Vasudev H, Prashar G, Jasim NY. A short review on machine learning for the purpose of optimizing and predicting the properties of polymeric nanocomposites. *Materials Today: Proceedings*. 2023. doi: 10.1016/j.matpr.2023.07.347
50. Di Girolamo G, Blasi C, Brentari A, Schioppa M. Microstructural, mechanical and thermal characteristics of zirconia-based thermal barrier coatings deposited by plasma spraying. *Ceramics International*. 2015;41:11776–85.
51. Di Girolamo G, Blasi C, Pagnotta L, Schioppa M. Phase evolution and thermophysical properties of plasma sprayed thick zirconia coatings after annealing. *Ceramics International*. 2010;36:2273–80.
52. Di Girolamo G, Blasi C, Schioppa M, Tapfer L. Structure and thermal properties of heat treated plasma sprayed ceria–yttria co-stabilized zirconia coatings. *Ceramics International*. 2010;36:961–8.
53. Ding S-J, Chu Y-H, Chen P-T. Mechanical biocompatibility, osteogenic activity, and antibacterial efficacy of calcium silicate–zirconia biocomposites. *ACS Omega*. 2021;6:7106–18.
54. Radu R-D, Drăgănescu D. Present and future of ZrO_2 nanostructure as reservoir for drug loading and release. *Coatings*. 2023;13:1273.
55. Du Q, Rao R, Bi F, Yang Y, Zhang W, Yang Y, Liu N, Zhang X. Preparation of modified zirconium-based metal-organic frameworks (Zr-MOFs) supported metals and recent application in environment: A review and perspectives. *Surfaces Interfaces*. 2022;28:101647.
56. Elraggal A, Silikas N, Aboushelib M. Influence of surface treatments and cyclic fatigue on subsurface defects and mechanical properties of zirconia frameworks. *Dental Materials Journal*. 2021;37:905–13.
57. Singh N, Mehta A, Vasudev H, Samra PS. A review on the design and analysis for the application of wear and corrosion resistance coatings. *International Journal on Interactive Design Manufacturing*. 2023;18:1–25. doi: 10.1007/s12008-023-01411-8

58. Esmaeilkhanian AH, Sharifianjazi F, Ahmadi E, Ijadi A, Meskher H, Zarei R, Nili-Ahmadabadi M, Irandoost M, Karimi N, Ghiasvand A. Thermal barrier coating with improved durability: An overview of doped, nanostructured, multilayered, and gradient-structured zirconia-based thermal barrier coatings. *Materials Today Communications*. 2023;37:107514.
59. Fang H, Wang W, Yang Z, Yang T, Wang Y, Huang J, Ye D. Phase stability, thermal shock behavior and CMAS corrosion resistance of Yb_2O_3-Y_2O_3 co-stabilized zirconia thermal barrier coatings prepared by atmospheric plasma spraying. *Surface Coatings Technology*. 2021;427:127864.
60. Fei C, Lei T, Qian Z, Shu Z. Piston thermal analysis of heavy commercial vehicle diesel engine using lanthanum zirconate thermal-barrier coating. *Energies*. 2022;15:4225.

Index

W

X

Y

Z

For Product Safety Concerns and Information please contact our
EU representative GPSR@taylorandfrancis.com Taylor & Francis
Verlag GmbH, Kaufingerstraße 24, 80331 München, Germany